41st UIT International Heat Transfer Conference 2024

Journal of Physics: Conference Series Volume 2940

Napoli, Italy
19-21 June 2024

Editors:

Nicola Bianco
Vincenzo Naso
Sara Rainieri

ISBN: 979-8-3313-1427-9
ISSN: 1742-6588

Printed from e-media with permission by:

Curran Associates, Inc.
57 Morehouse Lane
Red Hook, NY 12571

Some format issues inherent in the e-media version may also appear in this print version.

Additional copies of this publication are available from:

Curran Associates, Inc.
57 Morehouse Lane
Red Hook, NY 12571 USA
Phone: 845-758-0400
Fax: 845-758-2633
Email: curran@proceedings.com
Web: www.proceedings.com

TABLE OF CONTENTS

Journal of Physics: Conference Series **2940** (2025) 011001 doi:10.1088/1742-6596/2940/1/011001

41st UIT International Heat Transfer Conference

June 19 – 21, 2024 - Napoli, Italy
Congress Centre of the Università degli Studi di Napoli Federico II

Preface

It is our great pleasure to present the Proceedings of the **41st UIT International Heat Transfer Conference**, held from 19 to 21 June 2024 at the Congress Centre of the Università degli Studi di Napoli Federico II, organized by the Dipartimento di Ingegneria Industriale. This year, the conference held special significance as it took place within the events celebrating the eight hundredth anniversary of the foundation of the Università degli Studi di Napoli Federico II, a milestone underscoring the university's long-standing tradition of excellence in education and research.

The conference was attended by approximately 180 participants, including 85 members of the UIT community, 40 non-members, and 55 students. The sessions were dedicated to key topics such as Computational Fluid Dynamics, Conduction Radiation and Thermophysical Properties, Heat and Mass Transfer in Energy Systems and Nuclear Plants, Natural, Forced, and Mixed Convection, Micro and Nano-Scale Heat Transfer, and Multiphase Heat Transfer.

The program featured three keynote lectures, during which distinguished speakers shared their insights on pressing issues and emerging challenges in the field. Additionally, the Program Manager of the European Innovation Council presented valuable opportunities for transferring research into innovation, bridging the gap between academia and industry.

A total of 102 papers were presented during the oral sessions, complemented by 28 contributions in the poster session, reflecting the vibrant research activity within the heat transfer community.

We extend our sincere gratitude to the Scientific and Organizing Committees for their tireless efforts in ensuring the success of the conference.

We also wish to acknowledge the generous support of our sponsors, whose contributions were instrumental in making this event possible.

Finally, we thank all authors, presenters, and participants for their valuable contributions and active engagement, as well as the reviewers for their dedication to maintaining the high standards of this publication.

Sincerely, the Editors
Nicola Bianco
Vincenzo Naso
Sara Rainieri

Journal of Physics: Conference Series **2940** (2025) 011001 doi:10.1088/1742-6596/2940/1/011001

Conference Chairs

Nicola Bianco (Università degli Studi di Napoli Federico II)
Vincenzo Naso (Università degli Studi di Napoli Federico II)

Scientific Committee

Ciro Aprea (Università degli Studi di Salerno)
Dario Ambrosini (Università degli Studi dell'Aquila)
Antonio Barletta (Alma Mater Studiorum Università di Bologna)
Pietro Asinari (Politecnico di Torino)
Giovanni Sebastiano Barozzi (Università degli Studi di Modena e Reggio Emilia)
Nicola Bianco (Università degli Studi di Napoli Federico II)
Gian Piero Celata (ENEA)
Wilson K. S. Chiu (University of Connecticut)
Michele Ciofalo (Università degli Studi di Palermo)
Franco Cotana (Università degli Sudi di Perugia)
Davide Del Col (Università degli Studi di Padova)
Marco Dell'Isola (Università degli Studi di Cassino e del Lazio Meridionale)
Filippo de' Rossi (Università degli Sudi di Napoli Federico II)
Livio De Santoli (Università degli Studi di Roma La Sapienza)
Paolo Di Marco (Università degli Studi di Pisa)
Alberto Fichera (Università degli Studi di Catania)
Mariarosa Giardina (Università degli Studi di Palermo)
Oronzio Manca (Università della Campania Luigi Vanvitelli)
Nicola Massarotti (Università degli Studi di Napoli Parthenope)
Rita Maria Antonietta Mastrullo (Università degli Studi di Napoli Federico II)
Gian Luca Morini (Alma Mater Studiorum Università di Bologna)
Vincenzo Naso (Università degli Studi di Napoli Federico II)
Perumal Nithiarasu (Swansea University)
Alfonso Niro (Politecnico di Milano)
Enrico Nobile (Università degli Studi di Udine)
Carlo Nonino (Università degli Studi di Udine)
Alberto Ortona (Scuola Universitaria Professionale della Svizzera Italiana)
Fabio Polonara (Università Politecnica delle Marche)
Sara Rainieri (Università degli Studi di Parma)
Luisa Rossetto (Università degli Studi di Padova)
Maurizio Sasso (Università degli Studi del Sannio)
Luca Tagliafico (Università degli Studi di Genova)
Giovanni Tanda (Università degli Studi di Genova)
Paolo Tartarini (Università degli Studi di Modena e Reggio Emilia)
Giuseppe Peter Vanoli (Università degli studi del Molise)

Organizing Committee

Marilena Musto (Università degli Studi di Napoli Federico II) - Coordinator
Assunta Andreozzi (Università degli Studi di Napoli Federico II)
Fabrizio Ascione (Università degli Studi di Napoli Federico II)
Marcello Iasiello (Università degli Studi di Napoli Federico II)
Teresa Iovane (Università degli Studi di Napoli Federico II)
Margherita Mastellone (Università degli Studi di Napoli Federico II)
Alfonso William Mauro (Università degli Studi di Napoli Federico II)
Francesco Pelella (Università degli Studi di Napoli Federico II)
Luca Viscito (Università degli Studi di Napoli Federico II)

Journal of Physics: Conference Series **2940** (2025) 011001 doi:10.1088/1742-6596/2940/1/011001

Sponsors

Institutions

UNIVERSITÀ DEGLI STUDI DI NAPOLI FEDERICO II
www.unina.it

neaPòliS
Università degli Studi di Napoli Federico II
Scuola Politecnica e delle Scienze di Base
www.scuolapsb.unina.it

DIPARTIMENTO DI INGEGNERIA INDUSTRIALE
www.dii.unina.it

Companies

Gold Sponsor

S.I.I.P
Servizi Ingegneria Innovativa Personalizzati
www.siip.one

Silver Sponsor

cogepi S.r.l.
www.cogepisrl.com

TVP SOLAR
www.tvpsolar.com

Bronze Sponsor

EURISCO CONSULTING
www.eurisconsulting.com

Journal of Physics: Conference Series **2940** (2025) 011001 doi:10.1088/1742-6596/2940/1/011001

Editors contact details

- ✓ Nicola Bianco, Università degli Studi di Napoli Federico II, Napoli, Italy, nicola.bianco@unina.it
- ✓ Vincenzo Naso, Università degli Studi di Napoli Federico II, Napoli, Italy, vincenzo.naso@unina.it
- ✓ Sara Rainieri, Università degli Studi di Parma, Parma, Italy, sara.rainieri@unipr.it

Journal of Physics: Conference Series **2940** (2025) 011002 doi:10.1088/1742-6596/2940/1/011002

Peer Review Statement

All papers published in this volume have been reviewed through processes administered by the Editors. Reviews were conducted by expert referees to the professional and scientific standards expected of a proceedings journal published by IOP Publishing.

- **Type of peer review:** Single Anonymous
- **Conference submission management system:** Morressier
- **Number of submissions received:** 36
- **Number of submissions sent for review:** 34
- **Number of submissions accepted:** 34
- **Acceptance Rate (Submissions Accepted / Submissions Received × 100):** 94.4
- **Average number of reviews per paper:** 2
- **Total number of reviewers involved:** 31
- **Contact person for queries:**
  Name: Vincenzo Naso
  Email: vincenzo.naso@unina.it
  Affiliation: Università degli Studi Federico II - Dipartimento di Ingegneria industriale

Heat Transfer Enhancement: Exploring Nanofluids and Devices

Antonio C. M. Sousa [1,2,*]

[1]Professor Emeritus, Department of Mechanical Engineering, University of New Brunswick, Fredericton N.B., Canada E3B 5A3

[2]Principal Investigator, TEMA, Department of Mechanical Engineering, University of Aveiro, 3810-193 Aveiro, Portugal

[*] e-mail: asousa@unb.ca

Abstract. Effective heat transfer is crucial for energy conservation, efficiency of industrial processes, and the development of advanced heating and cooling systems. Heat transfer enhancement techniques aim to facilitate faster and more efficient thermal energy transport. The first part of the lecture focuses on the emerging field of nanofluids, which, as a novel class of heat transfer fluids, has gained significant attention in recent years due to their unique properties and potential for improving heat transfer effectiveness. It is presented a brief overview of the experimental methods involved in the preparation of the nanofluids along with their combination with other heat transfer enhancers such as turbulators and vortex generators. Their application to small-scale solar thermal systems is also reported. Current challenges, such as nanoparticle agglomeration and long-term stability, are addressed, while exploring new nanoparticle materials for multi-purpose use. The second part of the lecture stresses the urgent need for further heat transfer enhancement by device development in two specific and very distinct areas: thermal management of battery assemblies in electric vehicles and geothermal energy extraction using super-long gravity heat pipes. The thermal management of battery assemblies in electric vehicles is vital to maintain safe operating temperatures under different weather conditions and prevent thermal runaway. Effective heat transfer within the battery pack is critical for optimizing battery performance, increasing its operational lifespan, and ensuring safety. A particular CFD study is reported along its main conclusions. In what concerns geothermal energy extraction, the super-long gravity heat pipe, with lengths over 3,000 meters, offer unique promise. Data from field tests for a prototype are reported along with preliminary conclusions. They are also noted several challenges, which include, among others, design of the heat pipe internals, reduction of the thermal resistance between pipe and surrounding ground, insulation of the adiabatic region, and reservoir recovery.

1. Introduction

The pervasive impact of climate change, predominantly driven by the emission of greenhouse gases (GHG), poses an imminent threat to the well-being of our planet and its inhabitants. Carbon dioxide, a significant contributor to global GHG emissions, has been identified as a primary culprit in the escalating environmental crisis. Addressing this challenge necessitates a transformative shift from fossil fuels to sustainable energy alternatives, with renewable sources, notably solar, wind and geothermal energy, emerging as pivotal players in this transition.

While the adoption of renewable energy is crucial, the journey toward a sustainable future is a prolonged one. Even as societies endeavor to distance themselves from reliance on fossil fuels, a continuous demand for high-efficiency heat-exchange equipment persists, particularly in sectors which still largely rely on fossil fuels like industries engaged in processing and manufacturing. They are responsible for a staggering 24.2% of the 50 billion tonnes of annual greenhouse gas emissions.

By recognizing the dual challenge which the global energy system faces - the need for more energy and less carbon emission, endeavors to enhance heat transfer mechanisms become paramount, not only for economic considerations but as a pivotal strategy in mitigating climate change. Improving heat transfer efficiency not only translates to energy savings and reduced operational costs but also extends the operational lifespan of heat transfer systems, contributing significantly to the all-encompassing goal of diminishing greenhouse gas emissions. In this context, the paper reports on the personal experience of the author in different interplaying areas of heat exchange technology and alternative forms of energy.

Research on heat transfer enhancement has been pursued for centuries; however, it started being systematized with the pioneering work back in 1965 by Bergles and Morton [1], who conducted a survey on the techniques to enhance single phase forced convection heat transfer. In their work, the enhancement techniques were classified into two categories: passive and active. In this work, only passive techniques in conjunction with single phase flow will be analysed.

The paper is organized in three distinct sections: Nanofluids and their applications, Li-ion battery cooling – a case study, and Innovative geothermal energy exploration. These three sections cover a wide range of issues, hopefully not too disparate, which have the commonality of the need for heat transfer enhancement to achieve a future with greener energy.

2. Nanofluids and their utilization

2.1 Nanofluids

2.1.1 Introductory remarks. Working fluids, such as water, ethylene glycol, propylene glycol, engine oil, mineral oil, kerosene oil and silicon oil, are commonly used in heat transfer equipment in a vast range of engineering applications. The thermal effectiveness of this equipment can be greatly enhanced by improving the thermal conductivity of the working fluids. This can be achieved by dispersing a small quantity of solid particles into the base fluids as discovered by Maxwell [2].The augmentation of thermal conductivity of conventional fluids through the suspension of solid particles, such as millimetre or micrometre-sized particles has not been of interest for practical applications, due to problems associated with sedimentation, erosion, fouling, and increased pressure drop through the flow passages. However, dispersion of nanometer sized solid particles in base fluids, named as 'nanofluids' by Choi and Eastman [3], has attracted much interest in the past two decades. In a seminal study by Choi *et al* [4], they investigated the thermal conductivity of multi-walled carbon nanotubes dispersed in synthetic poly (α-olefin) oil nanofluids, and they found an 160% enhancement as compared to the value of the base fluid with no particles; this is a clear indication that well dispersed nanometer sized solid particles in base fluids have the potential of dramatically altering the thermal properties of the base fluid.

The literature dealing with nanofluids is vast and it has been expanding at an increasingly higher rate; therefore, the present work is restricted to reporting only the direct experience of the author's research group at University of Aveiro.

2.1.2 Characterization of the nanoparticles. Typically, the structure and morphology of the single and synthesized binary and ternary nanoparticles were identified with X-ray diffraction (XRD) and scanning electron microscope (SEM). For completeness, the formation of functional groups was evaluated by Fourier transform infrared spectroscopy (FTIR).

2.1.3 Preparation of the nanofluids. The preparation is exemplified for the nanofluid resulting from the suspension of reduced graphene oxide (rGO) in water.

The 2-dimensional reduced graphene oxide nanoparticles were prepared using the technique proposed by Hummers and Offeman [5]. After mixing 5 g of graphite flakes with 100 mL of H2SO4 and 0.025 mg of NaNO3, the mixture was cooled using ice water. Potassium permanganate (0.039 M) was added to the solution, which was thoroughly stirred for 20 min at 0°C. The presence of an oxidizing agent (MnO3+) caused the solution hue to change. With the addition of water, the temperature returned to normal. The solution was further mixed with water and cleaned with 70 mL of H_2O_2 for half an hour. Water was added to the solution, which was then repeatedly centrifuged and washed. For eight hours, the rGO was heated to 70°C. A greater volume of rGO was prepared using the same process. With the application of 1/10th of the weight of rGO of the SDBS (sodium dodecylbenzene sulfonate) surfactant, stable rGO/water nanofluids were produced.

2.1.4 Experimental facility. Different test facility configurations have been used by our research group; however, when investigating a new nanofluid, in general, it is experimentally analyzed its flow through a horizontal tube. The schematic of a typical test facility used is depicted in Fig. 1. In particular, this test facility was used in the study by Sundar *et al* [6], and it consists of a test tube, pump, cooler, heating element, flow control valve and u-tube manometer. The inner and the outer diameter of the test tube are 10 mm and 12 mm, respectively, and the length of the tube is 1500 mm. The J-type thermocouples are placed at the inlet, outlet, and at tube locations of 250, 500, 750, 1000, and 1250 mm from the left end of the tube to measure the temperatures, which were used for heat transfer and exergy calculations. The heating element has maximum capacity of 2,000 W, and it uses 20 mm gauge (nichrome). The thermocouples are brazed on the tube at the locations mentioned above, and at inlet and outlet; the needles of the thermocouples are connected to a data logger. The heating element is wrapped on to the test tube and connected to the control panel. The whole test tube is insulated with asbestos rope insulation. The receiving tank has 30 liters capacity, and the galvanized-iron pipe is connected to the tank and test tube through a pump, flow control valve and flow meter.

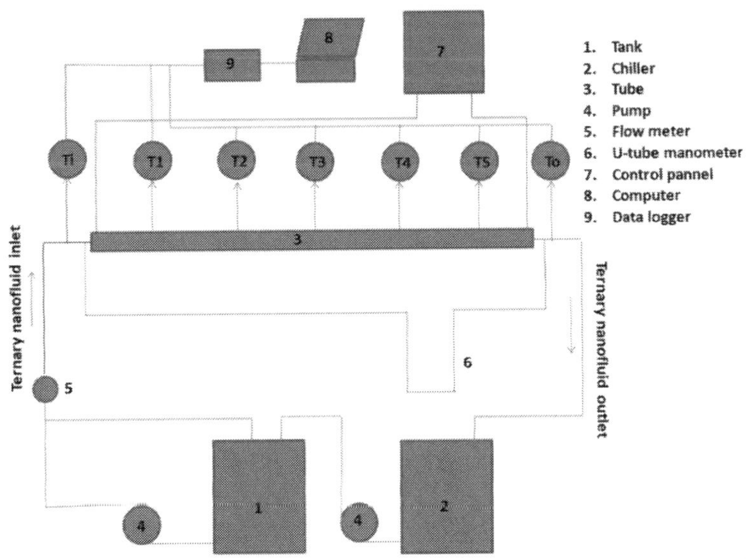

Figure 1 Schematic representation of a typical experimental test facility [6].

Journal of Physics: Conference Series **2940** (2025) 012001 doi:10.1088/1742-6596/2940/1/012001

Since a constant head AC pump is used, only the required amount of fluid is allowed to enter the test tube, while the excess flow is sent back to the receiving tank. The flow runs in a closed-loop and the outflow, before returning to the receiving tank, goes through the cooler to bring its temperature to that of the inflow. The pressure difference between outlet and intlet, which is required for the evaluation of friction and eventually pumping power, is determined by using the mercury U-tube manometer. Temperatures and fluid flow rates are recorded when the system reaches steady state.

2.1.5 Thermophysical properties. The thermophysical properties, in particular thermal conductivity and viscosity, were measured for a specified temperature range of the base fluid and of the nanofluid with different particle loadings. In general, density and specific heat were determined by using analytical estimates. The thermal conductivity and viscosity measurements were conducted using the following instruments: KD-2 Pro Thermal Conductivity Analyzer and AR-100 Rheometer, respectively.

2.1.6 Results.

*2.1.6.1 Introduction.*This section reports specific examples for mono, binary and tertiary nanofluids. This classification is based on the number of types of nanoparticles used, namely, mono, binary and tertiary, which refers to one type of nanoparticle, and two and three types of nanoparticles, respectively.

2.1.6.2 Mono nanofluids. The case selected to exemplify the results for the mono nanofluid is the magnetic nanofluid prepared by dispersing magnetic Ni nanoparticles in distilled water (Sundar *et al* [7]). The nanoparticles were synthesized by chemical co-precipitation method and characterised by x-ray diffraction and atomic force microscopy. The average particle size was measured by the dynamic light scattering method. Thermal conductivity and absolute viscosity of the nanofluid were experimentally determined as a function of particle concentration and temperature. In addition, the Nusselt number and friction factor were experimentally estimated as a function of particle concentration and Reynolds number for constant heat flux conditions in the forced convection apparatus already described. The experiments were conducted for a Reynolds number range of 3000 to 22000, and for a particle concentration range from 0% to 0.6%. The results indicate that heat transfer coefficient (Fig. 2a), Nusselt number (Fig. 2b) and friction factor (Fig. 3) of the nanofluid

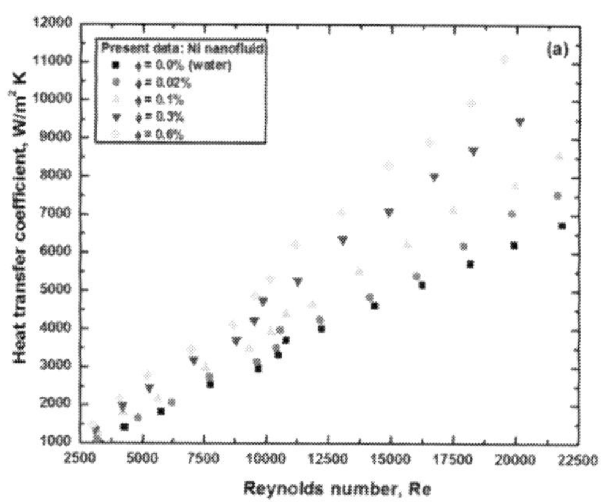

Figure 2 (a) Experimental heat transfer coefficient (h_{Exp}) of Ni nanofluid as a function of particle concentration and Reynolds number at constant heat flux [7].

$[Q = \dot{m} \times C_p \times (T_o - T_i)\ (Heat\ absorbed),\ h_{Exp} = \frac{Q}{A\,(T_{wall} - T_b)}$ (Internal area of the test tube: $A = \pi\ D\ L$, $T_{wall} = \frac{T_1+T_2+T_3+T_4+T_5}{5}\ and\ T_b = \frac{T_o+T_i}{2}; D = 0.014\ m, L = 1.5\ m]$

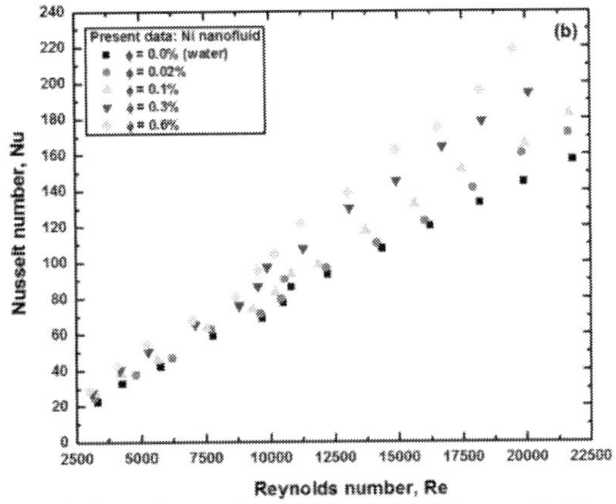

Figure 2 (b) Experimental Nusselt number (*Nu*) of the Ni nanofluid as a function of particle concentration and Reynolds number at constant heat flux [7].

increase with increasing particle volume concentration and Reynolds number. For 0.6% volume concentration at Reynolds number of 22000, the increase of heat transfer coefficient, Nusselt number and friction factor is 62.9%, 39.18% and 19.12%, respectively, as compared to distilled water under the same flow conditions.

The effectiveness of the heat transfer enhancement of the Ni nanofluid can be evaluated by using the Thermal Performance Factor (*TPF*), which was proposed by Bergles *et al* [8], and it is the ratio of the relative effect of change in the heat transfer rate to change in friction factor. The Thermal Performance Factor is defined by the following relation:

$$TPF = \frac{Nu/Nu_0}{(f/f_0)^{1/3}}$$

where the subscript "0" refers to the base fluid, which in the present case is water.

It should be noted this expression is only valid for the turbulent flow regime. For the highest values of concentration and Reynolds number tested in this study, the value of *TPF* is approximately 1.32, which is comparable, if not superior, to the *TPF* values for the majority of inserts, and in particular the butterfly, jagged and classic inserts, reviewed extensively by Maradiya *et al* [9].

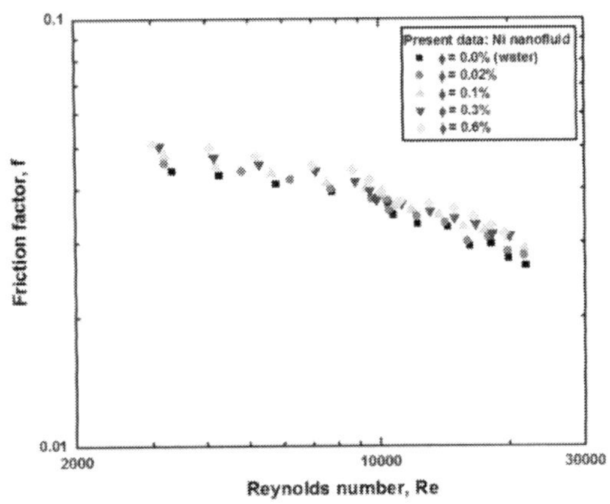

Figure 3 Experimental friction factor (*f*) of the Ni nanofluid as a function of particle concentration and Reynolds number at constant heat flux [7].

Journal of Physics: Conference Series **2940** (2025) 012001 doi:10.1088/1742-6596/2940/1/012001

2.1.6.3 Hybrid nanofluids. As briefly analyzed in the previous section, the heat transfer enhancement by the nanofluids depends on particle concentration, thermal conductivity of nanoparticles and mass flow rate. The thermal conductivity of the nanoparticles can be changed by synthesizing two or more nanoparticles leading to hybrid (nanocomposite) nanoparticles and the aim is expected that fluids prepared with hybrid nanoparticles yield further heat transfer enhancement. Hybrid nanoparticles represent an area of nanotechnology, which is experiencing a marked research interest due to its potential impact in materials science and engineering. Their use in nanofluids was the subject of an earlier review by Sundar *et al* [10]. A recent book edited by Zafar Said [11] offers a comprehensive state-of-the-art in what concerns the preparation, characterization, and applications of hybrid nanofluids. As stated in Sundar *et al* [12], hybrid nanofluids, as compared to mono nanofluids, offer the advantage of wide absorption range, lower extinction, remarkable thermal conductivity, and lower frictional losses leading to minimal pressure drop and pumping power.

2.1.6.3.1 Binary hybrid nanofluids. Early work by Sundar *et al* [13] on hybrid nanofluids used as nanoparticles multiwalled carbon nanotubes (MWCNT) (purity ~95%, outer diameter 10-30 nm and length 0.5-500 μm) and Fe_3O_4. The selection of these nanoparticles was motivated by the high thermal conductivity of MWCNT and the magnetic properties of Fe_3O_4. The MWCNT-Fe_3O_4 nanocomposite was prepared by the *in-situ* method, which includes dispersion of caboxylated carbon nanotubes in distilled water and followed by mixing of ferrous chloride and ferric chloride. Sodium chloride was used as a reducing agent to form the nanocomposite. Detailed surface characterization and magnetic properties were performed by x-ray diffraction and scanning electron microscopy and vibrating sample magnetometer, respectively. Furthermore, the stable hybrid nanofluids were prepared in distilled water; the heat transfer and friction factor experiments were conducted in the particle loadings of 0.1% and 0.3% in turbulent flow under constant heat flux conditions. The results showed that a maximum of 31.10% enhancement in Nusselt number with a penalty of 1.18 times pumping power under particle loading of 0.3% at a Reynolds number of 22000, as compared to water data.

Just for the sake of completeness, Figs. 4 and 5 are included to report the thermal conductivity and viscosity, respectively, of the nanocomposite nanofluid as a function of concentration and temperature.

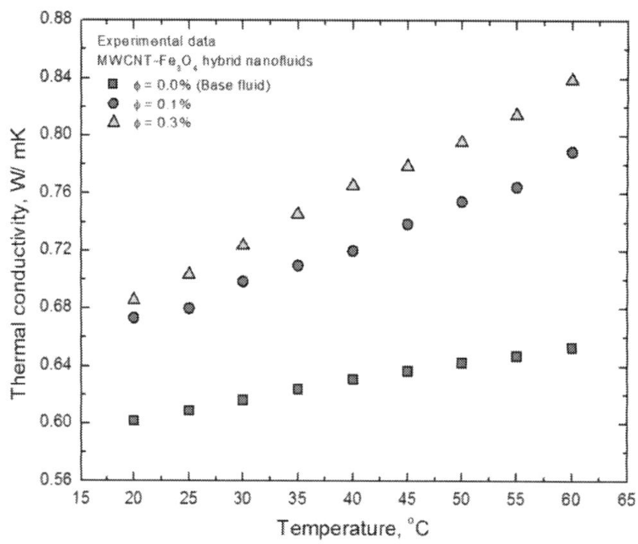

Figure 4 Thermal conductivity of MWCNT-Fe_3O_4/water hybrid nanofluid as a function of volume concentration and temperature [13].

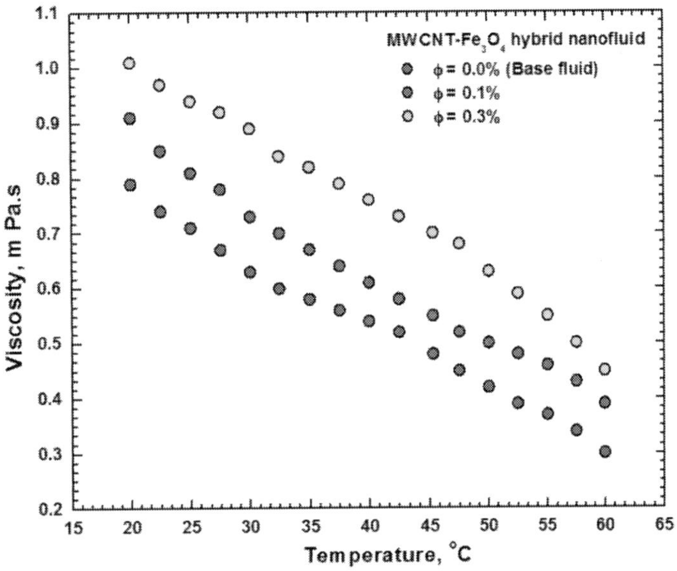

Figure 5 Experimental viscosity of MWCNT-Fe_3O_4/water hybrid nanofluid as a function of volume concentration and temperature [13].

It can be observed in Fig. 4 that at 60°C, a volume concentration of 0.3%, which is a fairly small amount, leads to a thermal conductivity enhancement of approximately 31% as compared to that of the base fluid (water) at the same temperature. Interestingly, this value is similar to the enhancement experienced by the Nusselt number at a particle loading of 0.3% and at a Reynolds number of 22000.

The increase in pumping power by using the composite nanofluid is relatively low (18%), despite the substantial viscosity rise (~45% at 60°C) due to the use of the nanofluid instead of the base fluid (distilled water).

2.1.6.3.2 Ternary hybrid nanofluids. Just to illustrate the properties of a ternary hybrid nanofluid, it is presented a brief overview of the experimental work of Sundar *et al* [14]. The main aim of the experiments was the evaluation of thermal entropy generation, frictional entropy generation and exergy efficiency of an rGO-Fe_3O_4-TiO_2 hybrid nanofluid based on ethylene glycol (*EG*) flowing in a circular tube under laminar flow conditions. The synthesized ternary nanoparticles were characterized by XRD, SEM, and FTIR. The ternary hybrid nanofluid was studied in terms of its thermophysical properties, heat transfer, friction, and pumping power for various particle weight concentration (ψ) ranging from 0.05% to 0.2%, and with values of Reynolds number varying from 211 to 2200. By increasing the particle loading and Reynolds number (*Re*), the Nusselt number, heat transfer coefficient, friction factor and exergy efficiency also increase. The values of thermal conductivity and viscosity rise by 10.6% and 108.3%, respectively, at ψ equal to 0.2% and at 60°C over the base fluid. At ψ equal to 0.2% and *Re* equal to 1548S, Nusselt number and heat transfer coefficient are enhanced by 17.78% and 24.76%, thermal entropy generation is reduced by 19.85%, and exergy efficiency increases by 6.23%, over the base fluid data. At the lowest Reynolds number evaluated, 221.1, and for ψ equal to 0.2% occurs the maximum rise in pressure drop, pumping power, and friction factor, which takes values of 13.65%, 11.33%, and 16%, respectively, as compared to the base fluid data. The overall thermal performance of the system is enhanced by 14.32%.

The pumping power is calculated based on the pressure drop (ΔP), mass flow rate ($\dot{m}$), and density of the fluid (ρ_f), as follows:

$$\text{Pumping power} = \left(\frac{\dot{m}}{\rho_f}\right) \times \Delta P$$

Figure 6 reports the pumping power required to circulate the base fluid, and the ternary nanofluid with 0.05%, 0.1% and 0.2% particle loadings as a function of the Reynolds number. For each experimental batch, the flow rate was maintained constant, while testing the base fluid and the nanofluid with different weight loadings the mass flow; therefore, the different values of kinematic viscosity of the fluids lead to different values of Reynolds number. As expected, the pumping power increases with increasing Reynolds number (*Re*) and particle concentration. The power required to pump the base fluid is 0.07 W at *Re* equal to 321, while is 0.08 W for 0.2 wt. % of nanofluid at *Re* equal to 221. For higher values of *Re*, the pumping power required to pump the base fluid at *Re* equal to 2250 is 3.6 W, while it is 3.77 W for the 0.2 wt. % nanofluid at *Re* of 1548. The use of the nanofluid leads to minor increase of the pumping power requirement, which is practically negligible in view of the potential benefits.

To obtain further understanding about the ternary nanofluid flowing in a tube heated at constant heat flux, the thermal ($\dot{S}_{g,T}$) and frictional ($\dot{S}_{g,F}$) entropy generation are computed. The relations are proposed by Huminic and Huminic [15] as follows:

$$\dot{S}_{g,T} = \frac{\dot{Q}^2}{Nu.\pi.k.T_i.T_o.L}$$

$$\dot{S}_{g,F} = \frac{8.f.\dot{m}^3.L}{\rho^2.\pi^2.D^5.(T_o - T_i)} ln\left(\frac{T_o}{T_i}\right)$$

The symbols have the same meaning as previously used in this text. *D* and *L* refer to the internal diameter and length of the tube, respectively. The heat input, $\dot{Q}$, is equal to 700 W.

The total entropy generation ($\dot{S}_g$) is given as:

$$\dot{S}_g = \dot{S}_{g,T} + \dot{S}_{g,F}$$

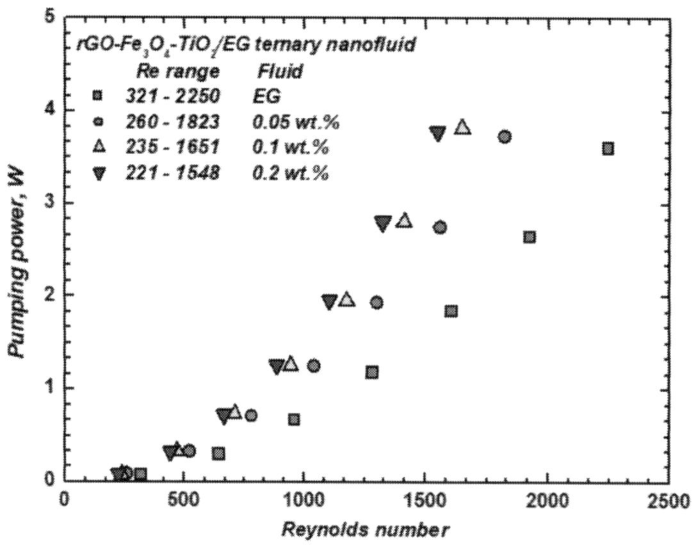

Figure 6 Pumping power as a function of Reynolds number and particle loading of the ternary hybrid nanofluid [14].

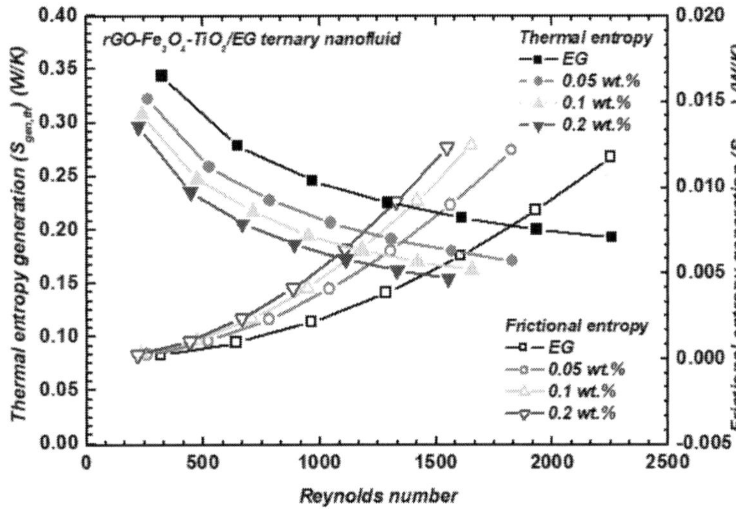

Figure 7 Thermal and frictional entropy generation of the rGO-Fe3O4-TiO2 hybrid nanofluid as a function of Reynolds number and concentration [14].

Figure 7 reports the thermal and frictional entropy generation as a function of the Reynolds number for the base fluid and for the nanofluid with different concentrations of the nanocomposite. It can be observed the thermal entropy generation decreases with increasing values of the concentration and Reynolds number. This observation can be explained by the consequent reduction of the difference between the temperature of the tube wall and the mean fluid temperature. For the present study, the thermal entropy generation of the rGO-Fe3O4-TiO2 hybrid nanofluid with concentration of 0.2% at Reynolds number of 1548 presents a decrease of 19.85%, when compared to that of the base fluid; the frictional entropy generation, as it can be noted in Fig. 7, follows a different trend, and it increases with increasing values of the concentration and Reynolds number. The frictional entropy generation at a Reynolds number of 1548 and a concentration of 0.2% nearly doubles the value of that for the base fluid.

Just to complement the data for the thermal and frictional entropy generation, the thermal $\left(\dot{E}x_{dest,T}\right)$ and frictional exergy destruction $\left(\dot{E}x_{dest,F}\right)$ are also reported; they are defined as follows:

$$\dot{E}x_{dest,T} = T_a \dot{S}_{g,T}$$

$$\dot{E}x_{dest,F} = T_a \dot{S}_{g,F}$$

In Fig. 8, the thermal and the frictional exergy destruction are presented as a function of the Reynolds number for the base fluid and the nanofluid with different loadings. As expected, the trends of the curves are similar to those of Fig. 7, increasing values of the Reynolds number and concentration lead to a rise of frictional exergy destruction and a reduction of the thermal exergy destruction.

This effect can be further noted by two indicators, namely the Bejan number (*Be*) and the entropy generation number (N_S).

The Bejan number is defined by the ratio of the thermal entropry generation to the total entropy generation, as follows:

$$Be = \frac{\dot{S}_{g,T}}{\dot{S}_{g,T} + \dot{S}_{g,F}}$$

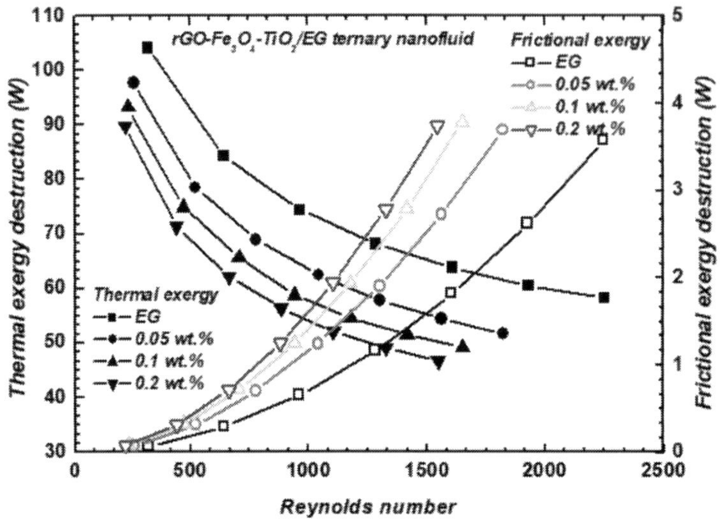

Figure 8 Thermal and frictional exergy destruction as a function of the Reynolds number and concentration [14].

A value of the Bejan number close to one indicates the influence of frictional entropy generation is negligible. In Fig. 9 (a), values of *Be* are presented for the base fluid and the nanofluid with three concentrations at different values of Reynolds number. The trend of *Be* clearly shows that for the range of *Re* studied, the thermal entropy generation largely dominates the frictional entropy generation, and the decrease of *Be* by increasing the concentration of the nanofluid denotes the rising importance of the contribution of the frictional entropy generation.

The entropy generation number(N_S)is calculated from the following equation [14]:

$$N_s = \frac{T_a \cdot S_{g,T}}{Q_h}$$

where T_a and Q_h are the ambient temperature and heat supplied to the fluid, respectively.

The entropy generation number, which is the reciprocal of exergy efficiency, can be a useful tool in the development and design of effective thermal systems. The entropy generation number of the base fluid and the hybrid nanofluid is reported in Fig. 9 (b). The limiting (ideal) value of the entropy generation number is zero, when thermal entropy generation does not occur, which is obviously impossible. Even so, the aim is the minimization of N_S. It can be noted in the figure that the lowest entropy generation number was achieved with the highest values of the particle concentration and Reynolds number of the present study.

2.1.6.4 Concluding remarks. There is extensive research literature on the nanofluids, but existing theories do not fully explain the role of the nanoparticles in enhancing the thermal conductivity of the base fluid. It is quite common in the literature to attribute to Brownian motion of the particles this enhancement. In a recent review by Apmann *et al* [16], the emphasis is on the Brownian motion to explain the thermal conductivity enhancement. It is not recognized that Brownian motion is too slow to transport significant amounts of heat through a nanofluid, as discussed by Keblinski *et al* [17]. These authors, based on results of MD (Molecular Dynamics) simulations, advanced the possibility that Brownian motion can have the role of leading to particle clustering, which may significantly enhance thermal conductivity. However, beyond the fact that particle clustering is undesirable for the nanofluid stability, for small particle concentrations, the addition of surfactant increases the thermal conductivity of the nanofluid. The presence of an interfacial layer between the particle and the base

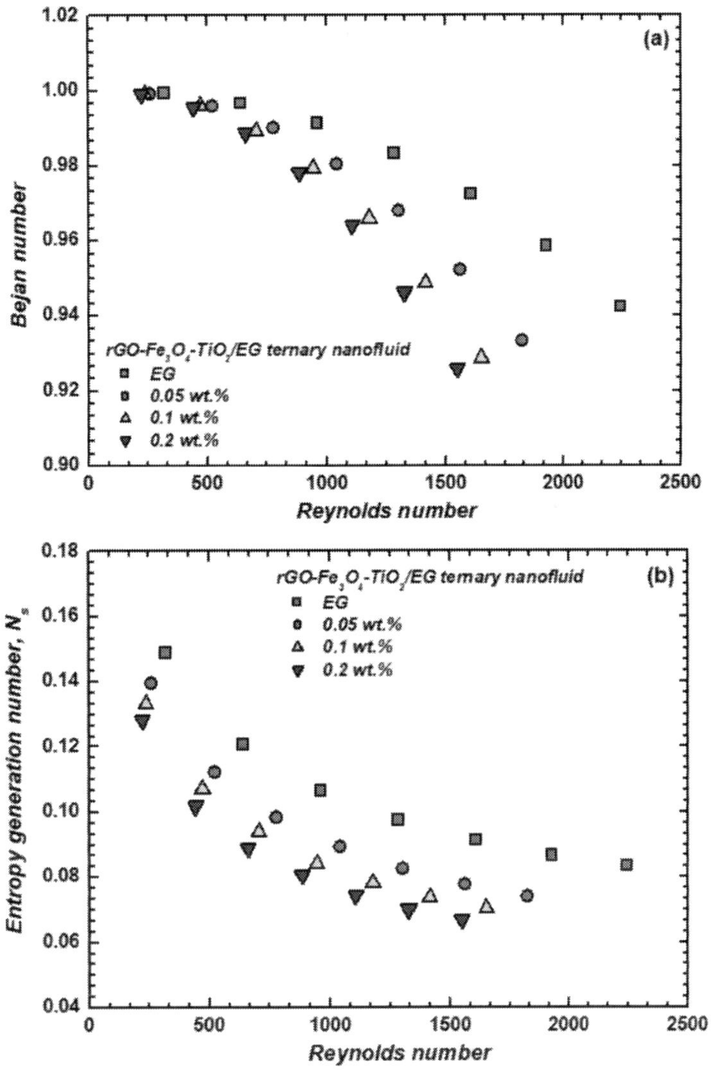

Figure 9 Bejan number (**a**) and entropy generation number (**b**) as a function of Reynolds number for the base fluid and the nanofluid with different concentrations [14].

fluid can play a role in this heat transport enhancement, although a relatively minor one. Moreover, it is not able to explain the anomalous increase of the thermal conductivity of the nanofluid with the decrease of particle size. A considerable amount of the theoretical work on the effective thermal conductivity is based on the approach of Hamilton and Crosser [18], which assumes the heat transport in each component by diffusion. The shortcoming of this approach is ignoring the contribution of the size of the nanoparticle to the thermal conductivity enhancement. An interesting possibility is that the heat transport in the nanofluid being accomplished by propagating lattice vibrations (phonons), as proposed by Joshi and Majumdar [19] for thin films. In fact, a recent paper by Oresta *et al* [20] proposes an interesting theoretical model, which was named undulatory theory of phonons. It includes the volume of fluid wrapping the particle, the thermal expansion coefficient, the velocity of sound, for both the liquid and solid phases. The model still relies on the macroscopic thermal conductivity of the base fluid and the particles. Remarkably, the agreement between the experiments and predictions is within 4%.

Journal of Physics: Conference Series **2940** (2025) 012001 doi:10.1088/1742-6596/2940/1/012001

2.2 Utilization

2.2.1 Introductory remarks. The body of work dealing with the utilization of nanofluids in heat transfer equipment has been increasing at a staggering rate. The reviews are also numerous, just for the year of 2019, Okonkwo *et al* [21] identified 29 reviews. These authors took a retrospective look at the year 2019 by reviewing the progress achieved in the application of nanofluids to various heat transfer devices such as solar collectors, heat exchangers, refrigeration systems, radiators, thermal storage systems and electronic cooling. For obvious reasons, most reviews are limited in scope, and they deal with specific applications. For instance, the review by Gupta and Sharma [22] is focussed on the adequacy of different nanofluids for use in three distinct types of applications, namely: electronic cooling, automobile radiators and refrigeration systems. Yang *et al* [23] reviewed specific applications of nanofluids in a tube with trapezoidal ribs, several types of heat exchangers and an electric heater. Their work is somewhat limited, but it is interesting that reports research on free convective flow of nanofluids in cavities with different configurations; in addition, the effect of magnetic fields on magnetic nanofluids contained in a square cavity and flowing through a straight tube and a corrugated structure. The review of Sundar *et al* l [24] aims to bring together the use of hybrid nanofluids in two solar applications: flat plate collector and parabolic trough collector. The main challenges of the use of hybrid nanofluids in these devices are their stability and cost; selection of the nanoparticles is a key element in making them viable.

In the present work, the two applications exemplifying the use of nanofluids are the thermosyphon solar flat plate collector and the flat plate exchanger.

2.2.2 Thermosyphon solar flat plate collector.

2.2.2.1 Experimental facility. The study of Sundar *et al* [25] is used to exemplify the overall performance enhancement of a thermosyphon solar flat plate collector by resorting to a nanofluid and inserts. Here, the preparation of the nanoparticles and of the nanofluid and the techno-economical aspects of the study will not be reported, as here, the aim is the performance of the device.

The schematic representation of the experimental facility is presented in Fig. 10.

The nanofluid was prepared with the suspension of nanoparticles of Cu using water as the base fluid for two different particle volume concentrations (ϕ) of 0.1% and 0.3%, respectively.

The wire coils (WC), which were inserted in the tubes of the collector, were tested for three different values of the pitch ratio (ratio of pitch, p, to diameter, d), namely: 1.79, 2.54, and 3.14, respectively. The three types of the wire coils are shown in Fig. 11.

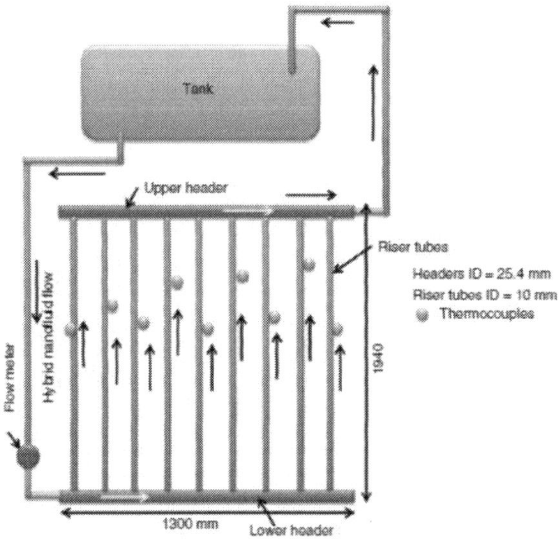

Figure 10 Schematic diagram of the experimental thermosyphon solar flat plate collector.

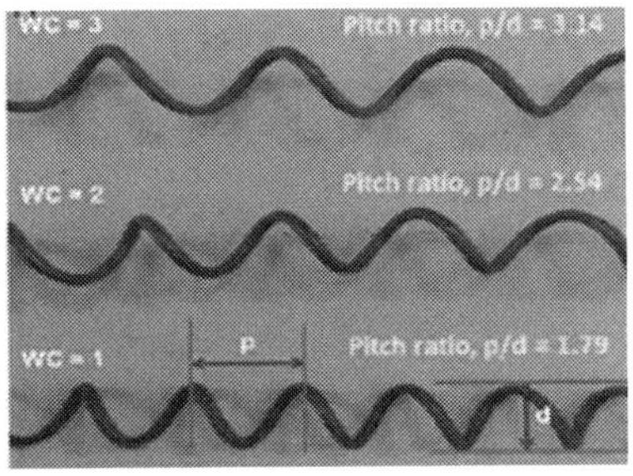

Figure 11 The wire coil inserts with the three different pitch ratios tested in the study.

2.2.2.2 Experimental conditions. The experiments were conducted during the day from 9h00 to 16h00. The solar radiation and mass flow rate increased from 09h00 to 13h00, and it started to decrease from 13h00 to 16h00. The recordings from 09h00 to 13h00 are grouped as Phase-1, while those from 13h00 to 16h00 as Phase-2. The solar radiation and mass flow rate are presented in Fig. 12 (a).

The inlet (T_i), outlet (T_o), surface and ambient (T_a) temperatures, global solar intensity (G_T), and mass flow rate ($\dot{m}$) were recorded.

2.2.2.3 Results.

Efficiency

The heat gained (Q) by the water or nanofluid with or without WC inserts was calculated by the following relation:

$$Q = \dot{m}\, C_p(T_o - T_i)$$

where, C_p is the specific heat of the water or of the nanofluid.

Based on the heat removal factor, absorptance–transmittance product ($\tau\alpha$), global solar irradiation (G_T), the heat gained by water or nanofluids can also be calculated as follows:

$$Q = A_c F_R[G_T \tau\alpha - U_L(T_i - T_a)]$$

where Q is in W, A_c is the surface area of the solar collector in m^2, F_R is the heat removal factor, G_T is in W/m^2, U_L is the overall loss coefficient of the solar collector, and T_i and T_a are in degrees Kelvin. The surface temperature is used to determine the absorptance and transmittance of the solar collector.

The instantaneous solar-flat plate collector efficiency (η_i) is commonly described in terms of the Hottel–Whillier–Bliss [26–28] equation, which can be formulated as:

$$\eta_i = \frac{Q}{A_c G_T} = \frac{\dot{m} C_p(T_o - T_i)}{G_T}$$

or

$$\eta_i = F_R \tau\alpha - F_R U_L \left(\frac{T_i - T_a}{G_T}\right)$$

The collector efficiency for water (base fluid) and the nanofluid with the concentrations of 0.1% and 0.3% is reported in Fig. 12 (b) over the testing period. The efficiency follows the gradual increase of the solar irradiation from 09h00 to 13h00, and then its decrease from 13h00 to16h00. The highest values of the efficiency for water and nanofluid with volume concentrations of 0.1% and 0.3% are 56.68%, 59.47% and 61.47%, respectively; this last efficiency value corresponds to an increase of about 8.5% over that for water.

Figure 12 (c) reports the collector efficiency with the WC inserts for water and the nanofluid with the concentrations of 0.1% and 0.3%. The efficiency increases with increasing particle loading and pitch ratio. For water as the working fluid, the collector efficiency with WC-3, WC-2, and WC-1 inserts takes the values of 58.59%, 60.59%, and 62.48%, respectively; for the nanofluid with concentration of 0.1, the values are 62.15%, 64.78%, and 66.49%, respectively, while for the nanofluid with concentration of 0.3%, they are 63.48%, 65.89%, and 68.47%, respectively. This last value represents an increase of about 21% over the efficiency with water and no inserts.

The WC inserts further promote mixing, which leads to higher outlet temperatures.

Friction factor

The use of the nanofluid and the wire coil inserts leads to a penalty in terms of increased friction factor. For Phase-1, in comparison to water data, the penalty for the 0.1% nanofluid is 4.8% and 8% at *Re* of 380 and 260, respectively, while, for the 0.3% nanofluid is 8.8% and 13% at *Re* of 380 and 260, respectively, as shown in Fig. 13 (a). For water, the friction factor penalty with WC-3, WC-2, and WC-1 inserts is 6.21%, 8.03% and 9.85%, respectively, at *Re* of 170; at *Re* of 600, the corresponding penalty is 10.46%, 14.12%, and 16.31%, respectively. For the 0.1% nanofluid at *Re* of 600, the WC-3, WC-2, and WC-1 inserts lead to friction factor increases of 15.54%, 16.89%, and

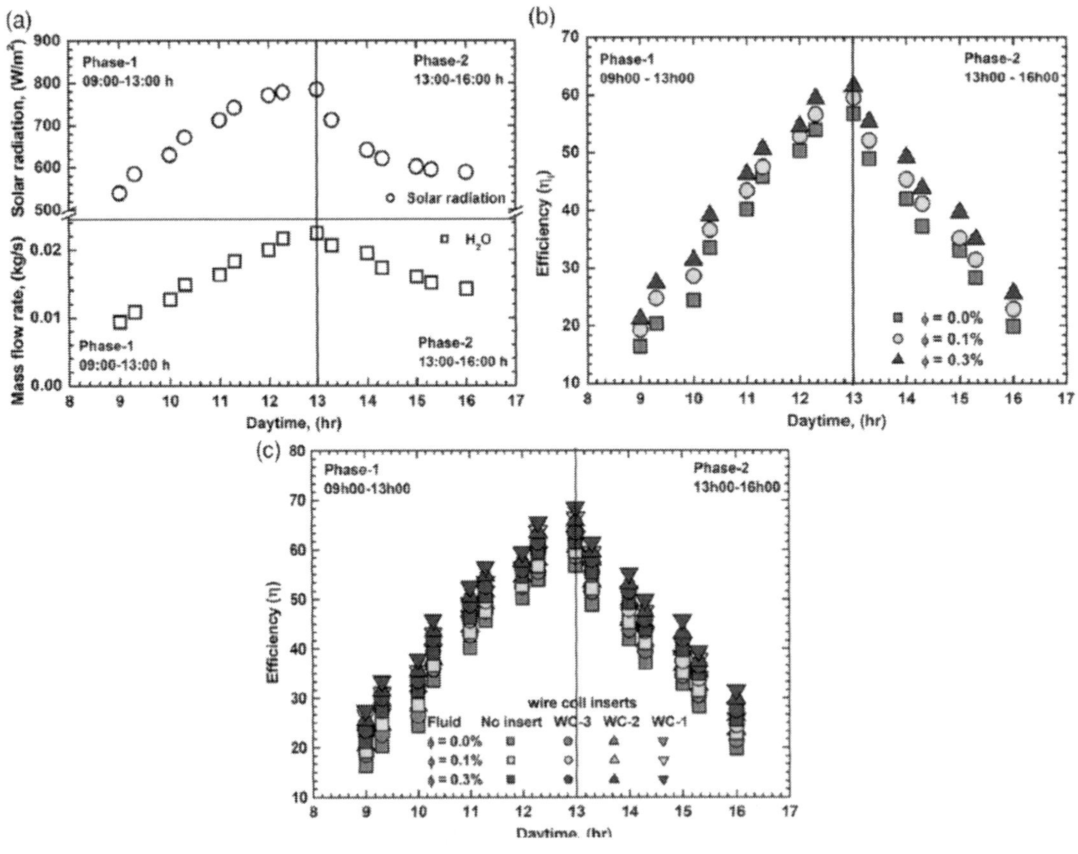

Figure 12 (a) Variation over the testing period of the solar irradiation and mass flow rate, **(b)** efficiency of the collector with no wire coil inserts, and **(c)** efficiency of the collector with wire coil inserts.

18.92%, respectively, as compared to 0.1% nanofluid data with no inserts. Similarly, for the 0.3% nanofluid at *Re* of 600, the use of the WC-3, WC-2, and WC-1 inserts results in friction factor penalty values of 21.15%, 23.08% and 27.56%, respectively, as compared to 0.3% nanofluid data with no inserts. The use of the 0.3% nanofluid with, as compared to the water data with no WC inserts, leads to friction factor penalty values of 23.24% and 45.24% at *Re* of 170 and 800, respectively.

For Phase-2, Fig. 13 (b), the friction factor penalty of using the WC-3, WC-2, and WC-1 inserts with water is 8.87%, 10.89% and 12.90%, respectively, at *Re* of 650; however, it decreases to 8.03%, 9.88%, and 11.27% at *Re* of 200. At *Re* of 650, the friction factor penalty values by using the WC-3, WC-2, and WC-1 inserts are: for the 0.1% nanofluid, 11.54%, 14.10%, and 16.67%, respectively, and for the 0.3 % nanofluid, 19.75%, 22.84%, and 29.63%, respectively; in both cases the comparison is with the respective nanofluid with no inserts. Compared to the water data with no WC inserts, the 0.3% nanofluid leads to friction factor penalty values of 41.32% and 34.65% at *Re* of 650 and 200.

2.2.2.4 Concluding remarks. The enhancement results appear to be modest; however, by taking into consideration the very minor modifications required and relatively low operating temperatures, it is remarkable a 21% peak increase of the collector efficiency. The results reported in this study are

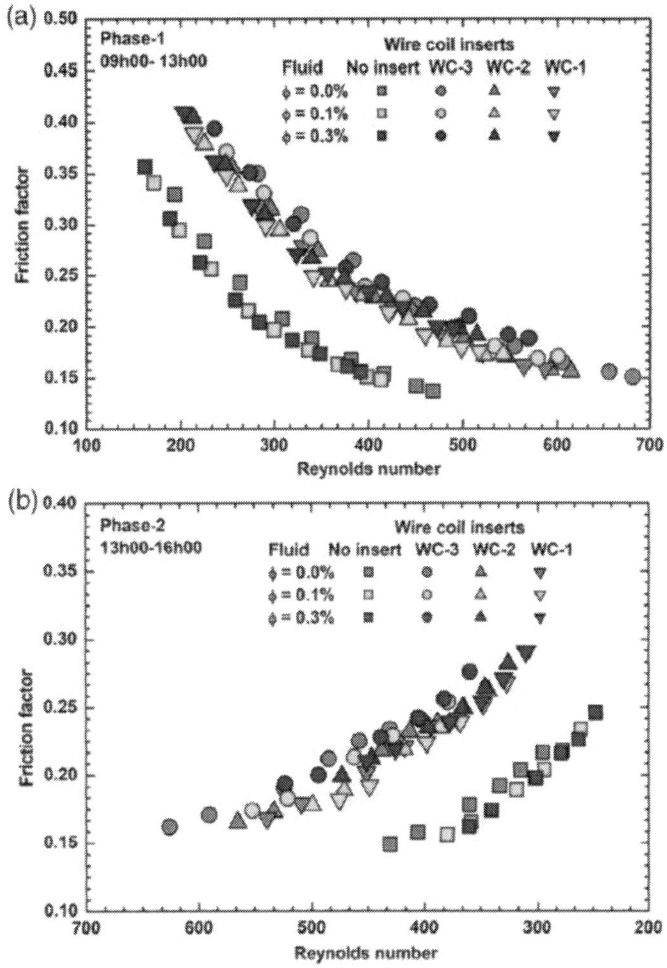

Figure 13 Friction factor of nanofluids with and without wire coil inserts as a function of the Reynolds number: **(a)** Phase-1 and **(b)** Phase-2

based on averaged values taken over one month. Obviously, there is considerable uncertainty associated with the data, considering that it was obtained outside laboratory-controlled conditions.

The study also reports that by using the 0.3% nanofluid and wire coil insert with a pitch ratio of 1for the same thermal output the collector area, which is 3 m^2, can be reduced by approximately 17%.

2.2.3 Plate heat exchanger

2.2.3.1 Experimental facility. The study conducted by Sundar *et al* [29] aimed to understand the performance of a plate heat exchanger by using water-based nanodiamond (ND) nanofluid. The experiments were performed for the volume concentration (ϕ), Reynolds number (Re) and mass flow rate ($\dot{m}$) range of $0 \leq \phi \leq 1.0\%$, $140 \leq Re \leq 610$ and 0.05 kg/s $\leq \dot{m} \leq 0.183$ kg/s, respectively.

The schematic representation of the testing facility of the plate heat exchanger (*PHE*) is presented in Fig. 14 (a), and the constructive details of the *PHE* are reported in Fig. 14 (b). A sample of the ND/ water nanofluid with a concentration of 1.0% is shown in Fig. 14 (c). The constructive specifications of the plate heat exchanger are given by Sundar *et al* [29]. The experimental setup consists of a hot water tank, cold nanofluid tank, two rotameters, plate heat exchanger, two pressure sensors, heater, and cooling systems. Hot water and cold nanofluid were used as working fluids in the *PHE* and both circulate in a counter flow arrangement. The water was heated in the tank with 2 kW immersion heaters, and the fluid temperature was controlled by a proportional–integral–derivative controller. The inlet temperature of the water (T_h) throughout the experiments was maintained at 80°C. The inlet temperature of the ND nanofluid (T_{nf}) for each concentration was maintained at 30°C. An AC pump was used for the circulation of the water and the nanofluid. The volume flow rate of the water was fixed at 3 L/min, while volume flow rate of the nanofluid varied from 3 L/min to 11 L/min. A pressure gauge was used at inlet and outlet of the *PHE* cold and hot streams to measure the pressure drop across each circuit.

Extensive testing was conducted to validate the methodology and the accuracy of the measurements.

2.2.3.2 Results

In what follows the results considered to be most relevant are succinctly discussed. The indicators are favourable to the nanofluid with exception those that are directly affected by the increase viscosity.

<u>Overall heat transfer</u>

For each nanofluid concentration, taking into consideration the change in kinematic viscosity, the range of Reynolds number also varies. The nanofluid Re ranges are: ϕ = 0.2%, 159.64 to 585.35; ϕ = 0.4%, 155.61 to 570.59; ϕ = 0.6%, 150.55 to 552.04; ϕ = 0.8%, 146.97 to 538.89; and ϕ = 1.0%, 143.55 to 526.36, respectively. The overall heat transfer coefficient of water and the nanofluid at different concentrations is presented in Fig. 15 as a function of Re. It can be noted the coefficient increases with increasing values of Reynolds number and particle loading. The coefficient enhancement values, as compared to water, are 12.22% (Re = 585.36), 17.42% (Re = 570.60), 22.41% (Re = 552.04), 25.59% (Re = 538.90) and 32.50% (Re = 526.37) for the nanofluid particle loadings of 0.2%, 0.4%, 0.6%, 0.8% and 1.0%, respectively.

<u>Nusselt number</u>

The Nusselt number (Nu) based on the experimental data of water and the nanofluid at different concentrations is presented in Fig. 16 for different values of Re. As expected, the trend of the Nusselt number is similar to that of the overall heat transfer coefficient, i.e., Nu increases with increasing values of Reynolds number and particle loading. The enhancement values of Nu, as compared to water, are 11.42% (Re = 585.36), 16.57% (Re = 570.60), 21.88% (Re = 552.04), 25.11% (Re = 538.90) and 35.11% (Re = 526.37) for the nanofluid particle loadings of 0.2%, 0.4%, 0.6%, 0.8% and 1.0%, respectively.

<u>Thermal entropy generation</u>

The thermal entropy generation is a key factor in the analysis of the potential benefits of a particular nanofluid. In Fig. 17, it is reported the thermal entropy generation for different values of Re and ϕ

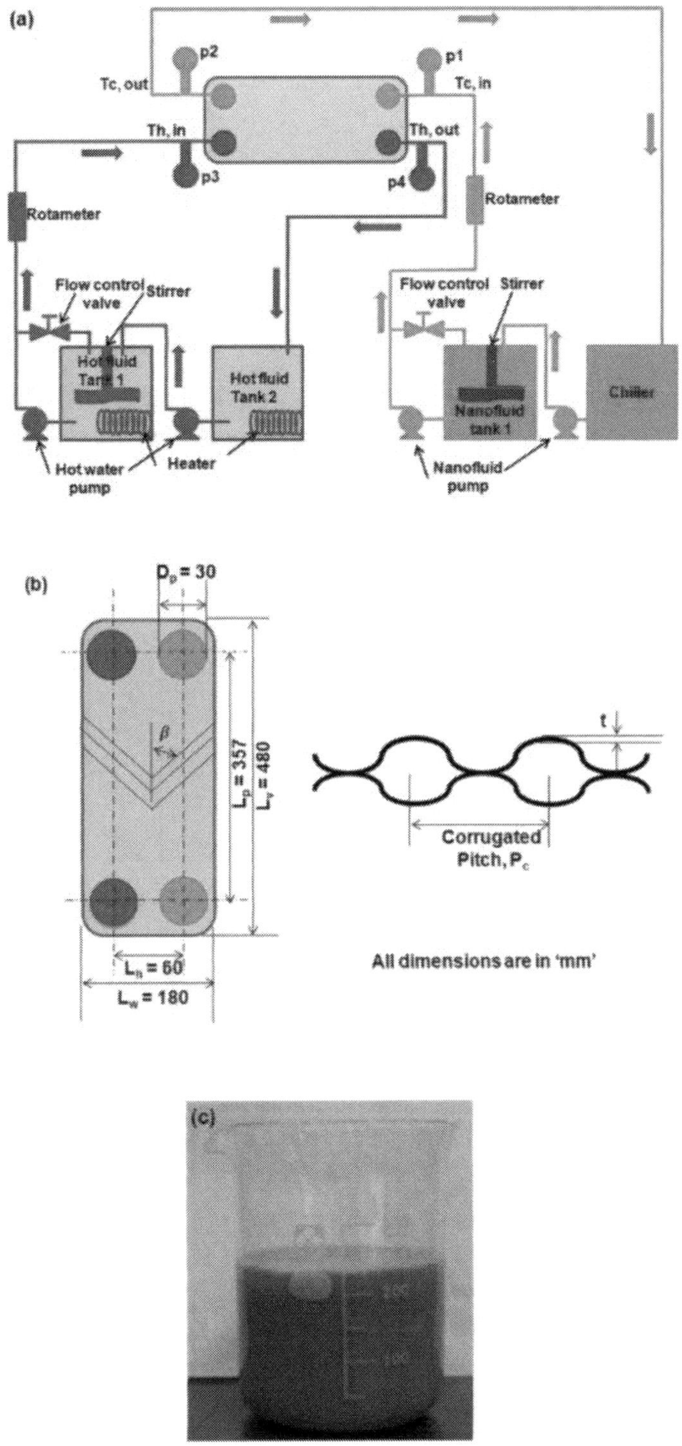

Figure 14 (**a**) Schematic arrangement of the *PHE* for heat transfer study; (**b**) Details of the *PHE* plates; and (**c**) Sample of the ND/water nanofluid with a concentration of 1%.

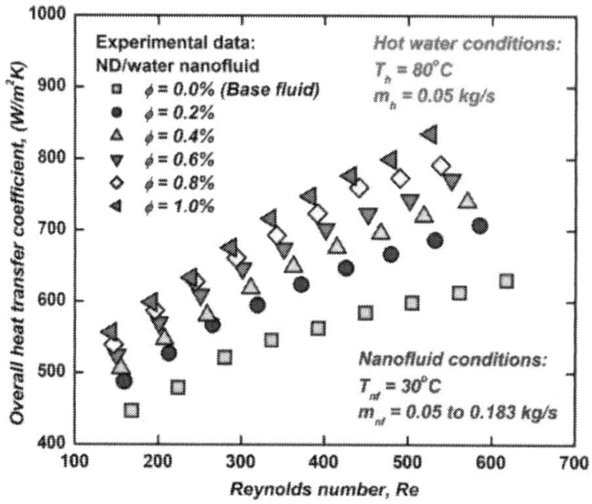

Figure 15 The overall heat transfer coefficient of ND/water nanofluid as a function of particle volume concentration and Reynolds number (the concentration of 0% corresponds to the base fluid).

and it can be noted that decreases with increasing values Re and ϕ. This effect is primarily due to the mitigation of the temperature difference between the hot stream (water) and the cold stream (nanofluid). The reduction values of the thermal entropy generation, as compared to water, are 6.38% (Re = 585.36), 7.64% (Re = 570.60), 9.51% (Re = 552.04), 15.63% (Re =538.90) and 20.66% (Re = 526.37) for the nanofluid particle loadings of 0.2%, 0.4%, 0.6%, 0.8% and 1.0%, respectively.

Bejan number

The Bejan number (Be), which is the ratio of the thermal entropy generation to the total entropy generation, is reported in Fig. 18 as a function of Re and ϕ. The trend of Be clearly reveals the increasing importance of the frictional entropy generation over the thermal entropy generation as the values of Re and ϕ increase. For water at Re of 617.3, Be is 0.652, while for the 1% nanofluid at 526.4, is 0.556, which represents a significant reduction of approximately 15%.

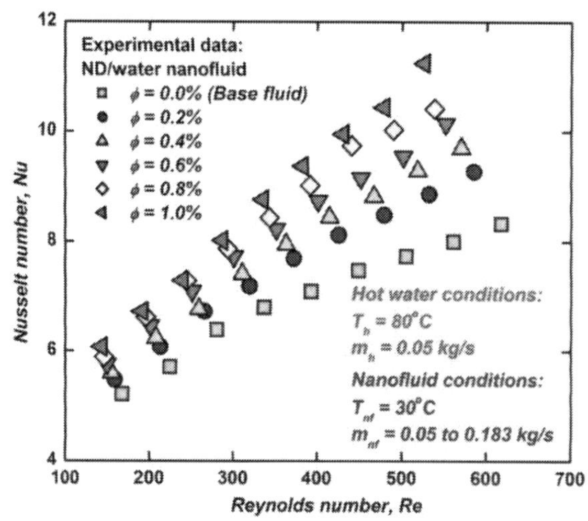

Figure 16 Nusselt number of the ND/water nanofluid as a function of particle volume concentration and Reynolds number.

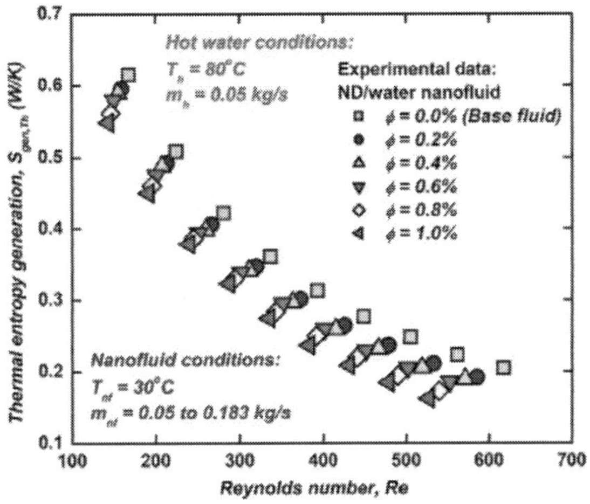

Figure 17 Thermal entropy generation of the ND/water nanofluid as a function of particle volume concentration and Reynolds number.

Performance index ratio

The experimental results of this study, while presenting very favourable trends in terms of the heat trasfer with the increase of ϕ, simultaneously, they show marked pressure drop rise. In view of these antagonistic effects, to elucidate when the utilization of the nanofluid is preferential to that of the base fluid, it is used the performance index ratio (*PIR*). The performance index was proposed by Tiwari *et al* [30], and *PIR* is the ratio between the performance indexes of the nanofluid and the base fluid, namely:

$$PIR = \frac{\frac{Q_{a,nf}}{P_{p,nf}}}{\frac{Q_{a,bf}}{P_{p,bf}}}$$

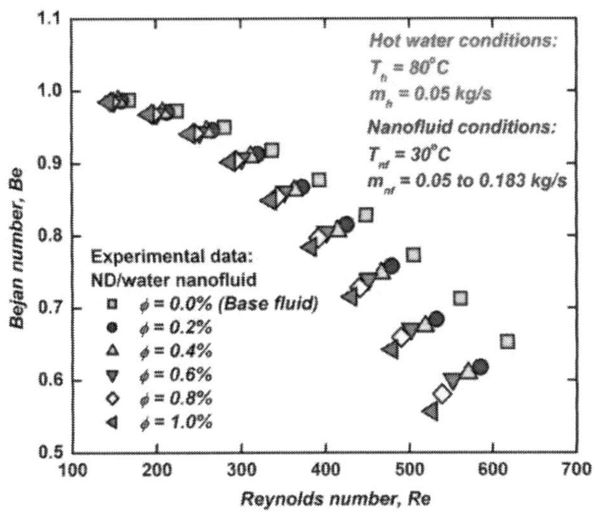

Figure 18 Bejan number of ND/water nanofluid as a function of particle volume concentration and Reynolds number.

Journal of Physics: Conference Series **2940** (2025) 012001 doi:10.1088/1742-6596/2940/1/012001

The symbols Q_a and P_p refer to the *PHE* absorbed heat and required pumping power, respectively; while the subscripts *nf* and *bf* denote the nanofluid and the base fluid, respectively.

A performance index higher than one indicates the heat transfer absorbed by the fluid surpasses the pumping power required to circulate the same fluid. The purpose of *PIR* is to compare the performance index of the nanofluid with that of the base fluid. Therefore, for a value of *PIR* equal or less than one, the use of the nanofluid does not bring any advantage over the base fluid. In Fig. 19, it is reported the variation of the *PIR* with different values of *Re* and ϕ. For the data obtained in the study, the highest *PIR* has the value of 1.086 at ϕ equal to 1.0%, at *Re* of 143.5. The threshold for the beneficial use of the nanofluid is at a value of ϕ equal to 0.6% at *Re* of ~500. The utilization of the nanofluid for values beyond this threshold offer no benefit as compared to the base fluid.

2.2.3.3 Concluding remarks. The use of the ND/water nanofluid presents interesting results in terms of thermal performance. In particular, the overall heat transfer coefficient and the Nusselt number are augmented by 32.5%, and 35.11% at ϕ equal to 1.0% and *Re* of 526.37, respectively, over the base fluid data. Also, the thermal entropy generation is lower by 20.66% than that of water at ϕ equal to 1.0% and *Re* of 526.37. In the opposite direction, the ND/water nanofluid at ϕ equal to 1.0% and *Re* of 526.37 leads to an increase, as compared to water, of 18.66% and 18.93% of the frictional entropy generation and pumping power, respectively. For this particular study, taking into consideration the specific configuration of the plate heat exchanger and assuming the performance index ratio is a reliable indicator, the threshold for the beneficial utilization of the nanofluid is ϕ equal to 0.6% at *Re* of ~500.

3. Cooling of lithium-ion battery packs

3.1 Introductory remarks

Global warming and the immediate pressing issues related to air pollution, primarily caused by vehicles powered by internal combustion engines, in the large cities will bring a ban and an eventual production end of these vehicles. As an obvious and front-runner alternative is electric vehicles (EVs), including plug-in hybrid vehicles. Lithium-ion batteries (LIBs) are generally considered as the best candidate for powering EVs taking into consideration their relatively high specific energy and energy density, long lifecycles, good lifespan, and reasonable cost. A battery pack consists of a large number of LIBs and, due to their electrochemical reactions and consequent heat generation, during its operation particular attention is required to deal with different thermal issues, among others,

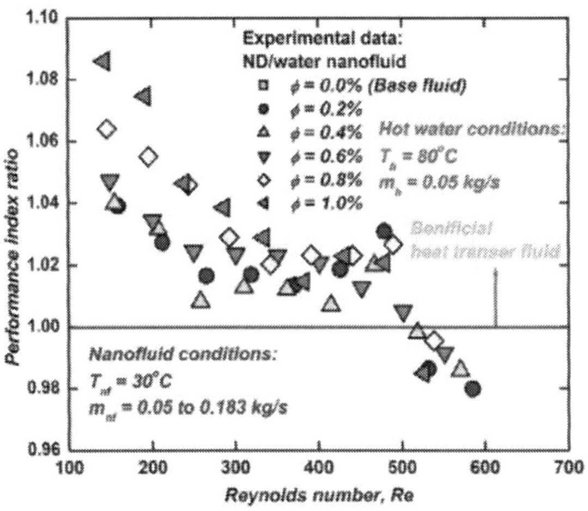

Figure 19 Performance index ratio for the ND/water nanofluid as a function of particle volume concentration and Reynolds number. The performance index ratio of the base fluid (ϕ = 0%) is equal to 1.

IOP Publishing
doi:10.1088/1742-6596/2940/1/012001

overheat and temperature non-uniformity. These issues are of critical importance in what concerns safety and, battery pack lifetime and performance; therefore, appropriate thermal management measures are required to resolve or, at least, alleviate them. Charge and discharge operations of LIBs will inevitably lead to heat generation while the optimum working temperature of LIBs is generally within the range of 25°C to 40°C. To avoid battery overheat, a common strategy involves liquid cooling.

An effective battery thermal management system should be capable of ensuring that every single LIB performs within the appropriate temperature range. To facilitate the evaluation of the BTMS (battery thermal management system), two simple indicators are introduced, and they are: maximum temperature and thermal uniformity, which is expressed in terms of the temperature difference. In the present work the primary focus is thermal uniformity, which in its absence will cause differences in the ohmic resistance of the LIBs, resulting in current unevenness leading ultimately to inhomogeneous heat generations of the LIBs. Therefore, to maintain thermal uniformity of the battery pack is of critical importance to its performance.

In this work, a simple but comprehensive analysis, which is reported by Zhao *et al* [31], is presented with the aim of further understanding the underlying mechanisms required for battery pack thermal uniformity improvement. Two cases involving two short channels and orderly magnified contact areas, respectively, were studied numerically. However, here, only the two-channel strategy will be discussed.

3.2 Physical model of the battery module

The structure of the battery pack, which consists of 71 18650-type LIBs, and the serpentine cooling channel are shown in Fig. 20.

The neighboring batteries just have a contact line with each other, and it is assumed that no heat transfer occurs between the batteries; therefore, under steady conditions, the heat generated by the module will be transferred to the coolant circulating through the serpentine channel. The cooling channel is made up of aluminum and its dimensions are: thickness of the channel top/bottom wall (a) is 1mm, the thickness of the channel side wall (b) is 0.5 mm, and height (c) and width (d) of the channel flow passage are 63 mm and 3 mm, respectively.

3.3 Numerical model

Battery lumped capacitance approximation

The commercial 18650-type LIB with NMC (nickel-manganese-cobalt) cathode used in this study has a capacity of 1.5 Ah. Its components are the can itself, electrolyte, anode copper foil, anode electrode, separator, cathode, and cathode aluminum foil. The thermo-physical properties of the

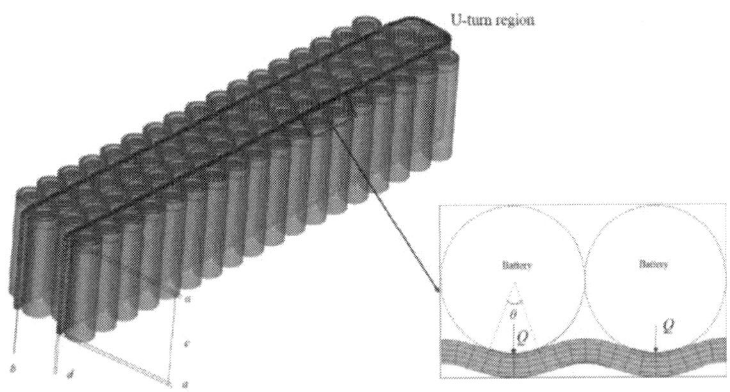

Figure 20 Physical lithium-ion battery module and the numerical meshing for the serpentine cooling channel [31].

battery components and its geometric parameters are well established; however, it is unrealistic to model individually each battery component for the full pack. Therefore, the LIB is regarded as a homogenous body with equivalent thermo-physical properties with no spatial variance and it is treated as thermally lumped, where the internal temperature gradient for each battery is neglected.

LIB heat generation rate

The 5C galvanostatic discharge was implemented as the protocol of discharging at constant current (7.5A) to the cut-off voltage of 2.5V. The single battery was placed into a thermostatic chamber at a temperature of 25°C, and the voltage, current and surface temperature of the 18650-type LIB were recorded during the 5C discharge operations. These values were used as input into the formula proposed by Bernardi *et al* [32] to determine the heat generation rate. Figure 21 presents the heat generation rate as function of DOD (depth-of-discharge).

Governing equations

The time-dependent governing equations describe the three subdomains - LIBs, aluminum channel and coolant. The flow in the channel is assumed to be laminar.

Boundary conditions and numerical strategy

The coolant inlet and outlet boundary conditions were set by specifying a fixed value for velocity and pressure, respectively, with the relative static pressure set to zero at the outlet; on the inner channel walls, the non-slip boundary condition was specified. Conduction along the channel walls is neglected; therefore, the heat flux at the interface between battery wall and outer wall of the channel is taken equal to that transferred to the coolant. All other boundary conditions are taken as adiabatic.

The initial temperature of the battery module, ambient temperature, and temperature of the fluid at inlet were all set at 25°C; all batteries are fully charged at 100% state of charge (SOC).

Upon completion of setting up the boundary and initial conditions, the governing equations to determine the 5 unknowns: u (three velocity components), P (pressure) and T (temperature) by using the commercial computational fluid dynamics (CFD) solver, FLUENT®, in which the discretization is based on the finite volume (FV) approximation. For the cases presented, it was used a mesh with 22240 hexahedral elements and a time step of 1.0 s.

3.4 Results

The results are presented only for the two short channels strategy. The simulations were conducted for 5C galvanostatic discharge and an inlet velocity of 0.1 m/s, which yields a Reynolds number value of 427.7, consequently satisfying the laminar flow assumption.

Double short channel strategy

The strategy to reduce temperature non-uniformity is to replace the single long channel (Fig. 22a) by two short channels (Fig. 22b). In this way, along the streamwise direction, the coolant temperature will be reduced, consequently, enhancing the cooling rate.

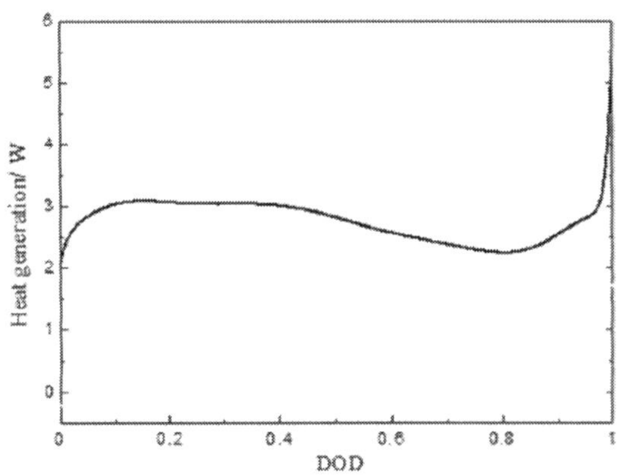

Figure 21 Heat generation rate as a function of DOD for a single LIB during 5C-rate discharge operation at a temperature ambient of 25°C [31].

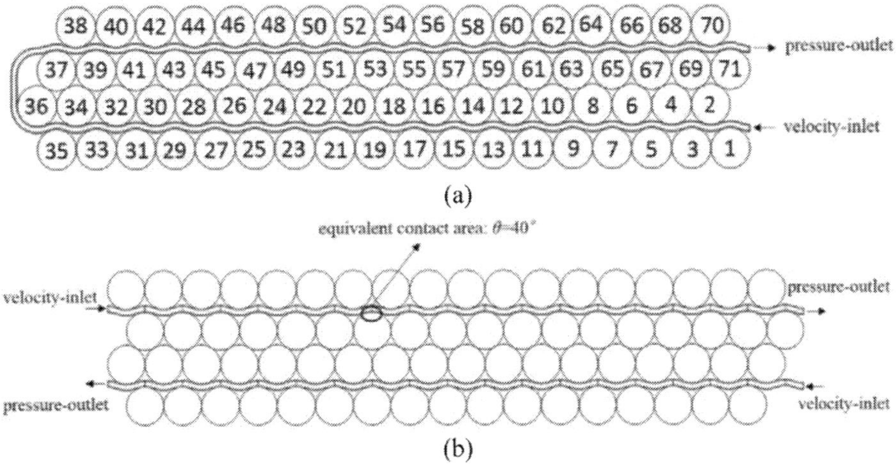

Figure 22 Schematic of single long channel (**a**) and two short channels (**b**) [31].

The two cases (single channel and two short channels) are both set with the same contact angle, θ, which is 40°. Evolution of maximum temperature and temperature difference of the battery module are depicted in Fig. 23 as a function of DOD. Figure 23 (a) presents the evolution of maximum temperature for both cases - two short channels and single long channel. It can be noted that both cases have similar trend with extremely high rate of maximum temperature increase from the beginning of discharge up to 0.4 DOD, then the rate decreases slowly in the range of 0.4 to ~0.9 DOD, and finally it increases from 0.9 DOD to the end of discharge. The evolution, as expected, is significantly influenced by the initial temperature condition (25°C; 298.15 K), particularly prior to 0.4 DOD; from this point to the end of discharge, the trend is primarily dictated by the heat generation rate. It can be noted that the two short channels arrangement yields lower values for the maximum temperature than those for the single long channel. In what concerns the temperature difference (Fig. 23b), the two-channel arrangement reduces it between 0 and ~1.88 K as compared to the single channel.

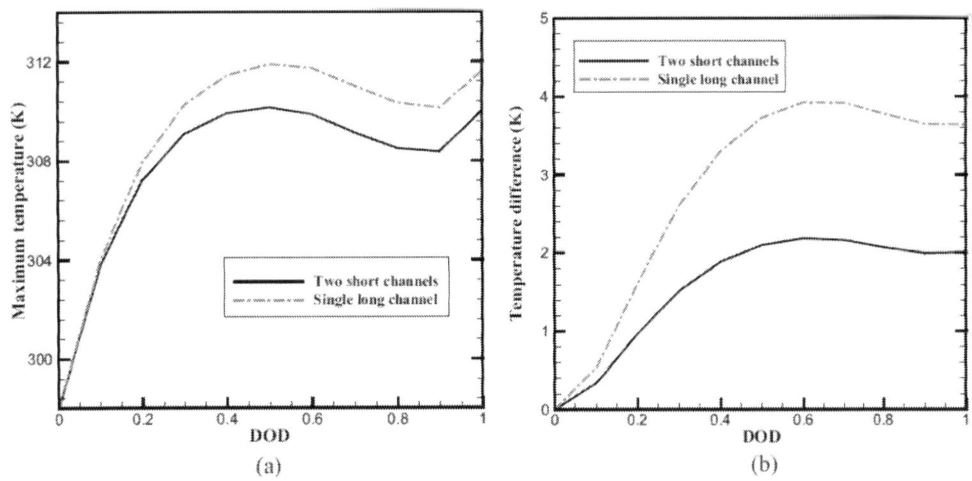

Figure 23 Evolution of thermal performance of the battery module for single long channel case and two short channels case: (**a**) maximum temperature; (**b**) temperature difference [31].

The two-channel arrangement provides better cooling performance than the single channel configuration, not only in what concerns the maximum temperature but also the thermal uniformity; the maximum temperature and the temperature difference are kept below 310.2 K and 2.2 K, respectively.

3.5 Concluding remarks

All simulated cases only consider 5C galvanostatic discharge operation with coolant inflow velocity at 0.1 m/s.

Further work will be required to further assess an appropriate BTMS taking into consideration capital costs associated with additional materials and manufacturing, and operating costs related to pumping for optimized coolant flow rates. In addition, the thermal effects should be considered for the full range of C-rates.

4. Super-long gravity heat pipe for geothermal energy exploitation

4.1 Introductory remarks

Geothermal energy can play a pivotal role in divesting the economies of the carbon-based energy production. As estimated by Rogner [33], the global annual potential of geothermal energy is on the order of 140,000,000 EJ. This value is limited to the accessible resource base, in which the heat could be theoretically tapped within a depth of 5 km. This amount would be sufficient to cover by several orders of magnitude the world annual energy consumption, which in 2019 was close to 583 EJ. The potential of geothermal resources also lies not only on their accessibility over large regions of the earth surface, but also on its long-term stable supply.

The Laboratory of Advanced Energy Systems (Guangzhou Institute of Energy Conversion, Chinese Academy of Sciences) has been leading in the development of a remarkable technology aiming at the hot dry rock (HDR) heat extraction: the super-long geothermal heat pipe (*SLGHP*).

In what follows it will be reported the characterization of the very first *SLGHP* test site and the key features of the *SLGHP* (Huang *et al* [34]); also, it will be given a brief overview of the ongoing research effort to improve the *SLGHP* performance.

4.2 Characterization of the test site

The geothermal well site where the tests were conducted is located in Matouying Uplift, Tangshan City, China. The geothermal well has a two-stage diameter, the first one up to a depth of 1400 m is 320 mm in diameter (cemented well), and the other one from a depth of 1400 m on is 224 mm in diameter (open well).

Two weeks after the drilling of the 4,000-meter well, the log data of the well was monitored in terms of temperature and thermal conductivity, as presented in Fig. 24.

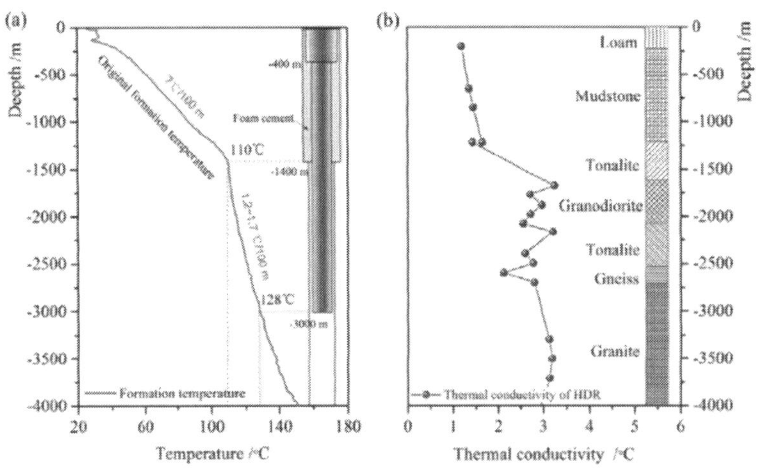

Figure 24 Temperature (**a**) and thermal conductivity (**b**) as a function of depth [34].

The different values of the temperature gradient along the geological strata, namely: ~7°C/100 m in the Cenozoic Erathem (0 to 1,371 m in depth), 1.2~1.7°C/100 m in the Archaeozoic (from the bottom of the Cenozoic Erathem to 3,000 m in depth) are reported in Fig. 24 (a). In addition, 29 rock core samples with 0.05 m in length were collected at different depths and their thermal conductivity was measured; the thermal conductivity data are given in Fig. 24 (b).

4.3 Description of the SLGHP system

The *SLGHP* used in the field test reported by Huang *et al* [34] is 3,000 metres long and it was designed and manufactured at the Laboratory of Advanced Energy Systems (LAES).

Prior to the actual field test there were several theoretical studies. In 2017, Jiang *et al* [35] proposed to extract the HDR geothermal energy by a super-long gravity heat pipe in a geothermal single-well. Huang *et al* [36] evaluated the heating and power generation potential of a 4,000-metre long *SLGHP* system through a numerical model combining the heat and mass transfer process in the gravity heat pipe with the heat flow model inside the HDR heat reservoir. Huang *et al* [37] further enhanced the numerical model; in fact, the field test aimed not only to be proof of concept, but also to provide test data for the benchmarking of the numerical model.

Figure 25 depicts the structure of the heat pipe, which was developed at LAES. Taking into consideration the high-pressure underground, the heat pipe is made up of hundreds of P110 oil casing tube; up to a depth of 400m, the tube is 244.5 mm in diameter and 11.99 mm in thickness and beyond 400 m is 177.8 mm in diameter and 9.19 mm in thickness. The heat pipe design specification is that it has a 30 MPa pressure-bearing ability. All casing tubes are connected by gas-tight premium connection to guarantee their tightness. The inset shows the interior of the casing, which consists of a patented stepped ladder structure (Jiang *et al* [38]). Each unit module of the stepped ladder consists of one inner pipe step and a small condensate guiding pipe. One step of the inner pipe and the outer casing pipe form an annular container, which can hold the *SLGHP* working fluid. The small guiding pipe directs the working fluid reflux and controls the depth of liquid accumulating in the annular container. Each ladder module is about 10 meters in height and the assembled small guiding pipe is about 3-4 meters long. Many such ladder modules are arranged inside the *SLGHP* evaporation section.

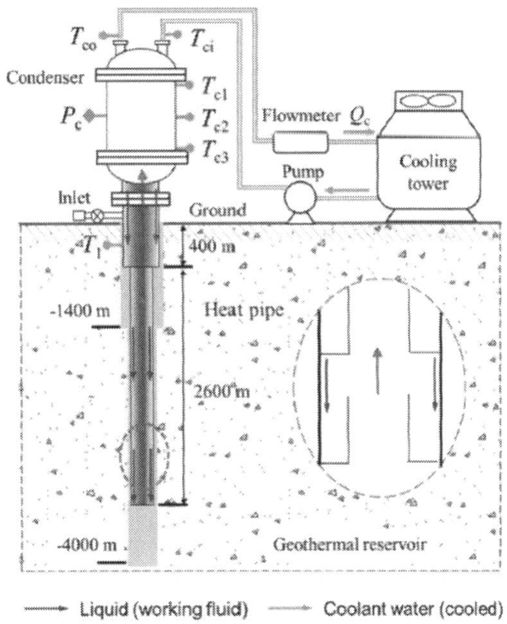

Figure 25 The *SLGHP* and the ground heat transfer system [34].

Along the *SLGHP* longitudinal direction there is a small gap left between neighboring modules, which is specially set to guide the vapour to flow into the central passage.

After running the heat pipe into the well, foam cement grouting with thermal conductivity close to 1 W/m.K was installed over the section running from 0 to 1400 m of the geothermal well depth. The heat transfer to this section is practically negligible, considering the low thermal conductivity of the foam cement; therefore, for practical purpose, this section can be assumed to be adiabatic. Consequently, the evaporation section runs from the depth of 1,400 m to 3,000m. In what concerns, the HDR formation temperature, its average value over the full heat pipe length is 92.6°C, and its maximum value is 128°C. The condensation section is composed by the condenser (Fig. 25), which is placed at ground level.

Heat transfer system at ground level

Upon the heat absorption from the HDR underground into the heat pipe evaporation section, the vapour flows upward and transfers the heat to the vertical shell-and-tube condenser at ground level. The condenser is 562 mm in diameter and 2120 mm in height. The condenser main body and the tube bundle are manufactured with SAE 304 stainless steel. The vapour inlet at the bottom of the condenser is 200 mm in diameter, allowing the upward flow of the vapour into the condenser and downward flow of the condensate into the heat pipe. The vapour flows in the shell side of the condenser, while the cooling water flows in the tube side. The inlet and outlet of the cooling water are both set on the top of the condenser. To determine the heat extraction, the inlet (T_{ci}) and outlet (T_{co}) temperature, and the flow rate of the cooling water are measured, as shown in Figure 2. The coolant flow is driven by a pump, and the flow rate is controlled by a valve. The outlet temperature of the cooling water is brought to its inlet temperature with the help of a countercurrent high temperature open cooling tower.

Parameters measurement

A temperature measurement system was used to monitor the temperature of the underground heat pipe. Four temperature sensors were fixed on the outer surface of the heat pipe and connected by cable to the data logger before the casing into the well bore was installed. However, due to the narrow gap between the well borehole and the heat pipe, three sensors were damaged, and only the one at the depth of 275 m remained operational. The temperature measured at 275 m is denoted as the adiabatic temperature T_1. For monitoring the operation, three thermocouples (Fig. 25) were installed at the upper (T_{c1}), middle (T_{c2}), and lower (T_{c3}) locations in the interior of the condenser, respectively. The average of these three temperatures, $(T_{c1}+T_{c2}+T_{c3})/3$, is denoted as the condensation temperature T_c. A pressure sensor was installed in the interior of the condenser to monitor the pressure in the condenser, and it is denoted as P_c in Fig. 25. A pipe with 6 mm inner diameter was used to connect the pressure sensor and the interior of the condenser.

The volumetric flow rate of the cooling water, V_s, was measured by a turbine flowmeter located at the outlet of the condenser. Two thermocouples were installed at the inlet and outlet of the condenser, respectively, to measure the inlet temperature T_{ci} and outlet temperature T_{co} of the cooling water. Therefore, the instantaneous heat extraction, Q, is determined as:

$$Q = V_s \rho c_{p,OH_2} (T_{co} - T_{ci})$$

where the symbols ρ and c_{p,OH_2} denote density and constant pressure specific heat of the cooling water, respectively.

4.4 Tests

Testing process

After running the pipe into the well and installing the ground equipment, vacuum was made in the heat pipe system; the system was allowed to stand for one week, after this period, tightness testing indicated that it was leak-free. Deionized water was used as the working fluid and filling into the heat pipe, and tap water was used as coolant. The test lasted for 4 months from December 2019 to April 2020, and the ambient temperature during this period had a wide variation ranging from -8°C to 8°C.

Journal of Physics: Conference Series **2940** (2025) 012001 doi:10.1088/1742-6596/2940/1/012001

Heat extraction pre-test

The heat extraction pre-test was conducted to verify the functional integrity of the system, before conducting the continuous heat extraction operation. This pre-test lasted nine days. The heat extraction rate for the first 12 hours is reported in Figure), and it can be observed that the heat extraction rate decreases from 230 kW to 170 kW, while the pressure decreases from 70 kPa to 64.5 kPa (Figure). Figure 26 (b) reports the measured condenser pressure P_c and the corresponding saturated pressure P_{H2O} calculated based on the condenser temperature T_c. It is apparent that the values P_c and P_{H2O} are initially close, but they diverge with time. Their trend presents three distinct stages, namely: in the first stage, both P_c and P_{H2O} decrease for about 3 hours; in the second stage, P_c keeps a stable value for about 4 hours, while P_{H2O} still decreases; finally, in the third stage, P_c gradually increases, while P_{H2O} keeps decreasing. The insert of Figure (b) illustrates the difference between P_c and P_{H2O}, which increase almost linearly with time indicating that the working fluid is departing from the equilibrium saturation state.

The three stages, I, II, and III, (Fig. 26b) of the temperature and pressure evolution can be explained as follows: I) the temperature and pressure drop is caused by the start of the condenser cooling water circulation, which yields fast removal of heat from the vapour phase of the working fluid and its consequent condensation; at this stage, the heat extraction rate is maintained at a relatively high level. II) Further decrease of the condensation temperature implies that the heat extraction rate becomes dependent on the new dynamic balance between the heat carried by the heat pipe working fluid and the heat transferred from the geothermal reservoir surrounding the heat pipe. III) The increasing difference between P_c and the P_{H2O} indicates that the thermodynamic state of the fluid in the condenser is gradually departing from the equilibrium state; this finding was unexpected, and it was verified later on that it was due to the accumulation of non-condensable gases at the top end of the condenser.

Continuous heat extraction over 30 days

In the 30-day continuous heat extraction, the condenser cooling water circulation was always turned on with a flow rate of 17m^3/h. To remove the accumulated non-condensable gases, a vacuum pump connected to the condenser top vent was operated every 12 hours. During the 30-day test, the condensation temperature varied in the range of 58.8 to 64.2°C. In Fig. 27 (a), it is reported the daily heat extraction rate and average ambient temperature over the 30-day heat pipe heat extraction testing. The heat output has no significant downward trend with time; however, it presents a somewhat irregular fluctuation. This observation can be explained on the basis that lower ambient temperature yields improved operating effectiveness of the cooling tower. Therefore, for the same coolant flow rate, higher heat dissipation can be obtained, with the consequent drop in condensation temperature at the condenser. Thus, the pressure of the condenser is reduced, resulting in increased steam

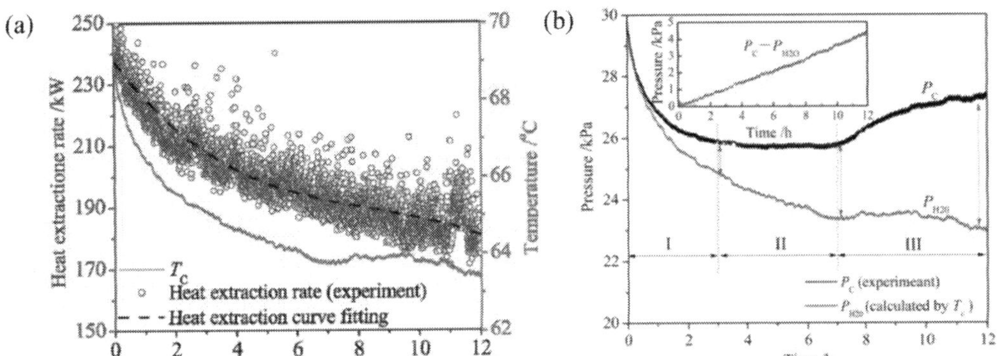

Figure 26 Condenser temperature and heat transfer (**a**), and condenser pressure (**b**) as a function of time during the 12-hour heat recovery cycle [34].

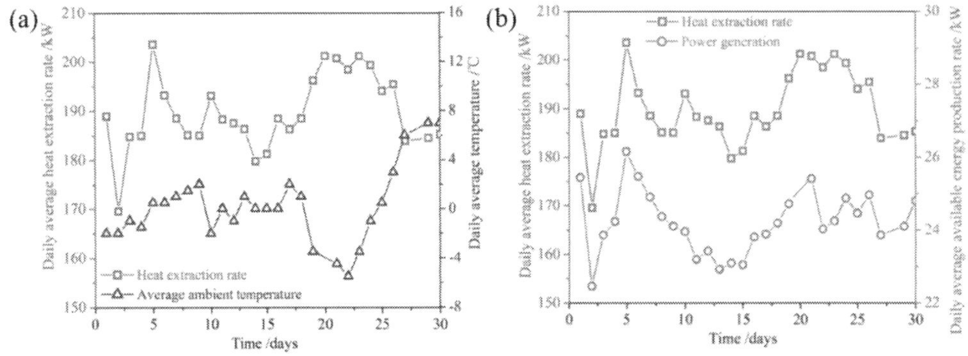

Figure 27 Heat extraction rate and average ambient temperature (**a**), heat extraction rate and available exergy production (**b**) during the 30-day continuous heat extraction test [34].

production by the heat pipe leading to enhanced heat extraction by the system. In Fig. 27 (a), it can be noted that the heat extraction and ambient temperature have opposite trends, which is consistent with the analysis, and supports the observation that the change in heat extraction is primarily caused by the variation of the ambient temperature. Figure 27 (b) reports heat extraction rate and the power generation as a function of time; as expected, they present a similar trend. The 30-day average heat extraction and available exergy output were 189.4 kW and 24.2 kW, respectively, and the total available exergy output was 17,424 kWh.

Variable working conditions test

The variable working condition test has as primary aim studying the relation between the condensation temperature and the heat extraction, to determine the condensation temperature leading to the best thermal performance. To this purpose, the cooling water temperature at the condenser inlet was adjusted by varying the flow rate of the cooling water pump and the rotation speed of the fans of the cooling tower. During the test, the vacuum pump connected to the condenser top vent was actuated every 12 hours for the removal of the accumulated non-condensable gases. The daily averaged experimental heat extraction data are presented in Fig. 28 (a) along with the simulation results, which will be discussed in the section 4.4 Numerical simulation. Figure 28 (b) reports the best fitting for the heat extraction rate and the available exergy rate based on the experimental data.

The available exergy, W, represents the maximum amount of thermal energy that can be converted to mechanical energy, and it is defined as follows:

$$W = Q \cdot \left(1 - \frac{T_{am}}{T_c}\right)$$

where Q is the heat; T_{am} and T_c are the ambient and condensation absolute temperatures, respectively; $(1-T_{am}/T_c)$ is the Carnot efficiency.

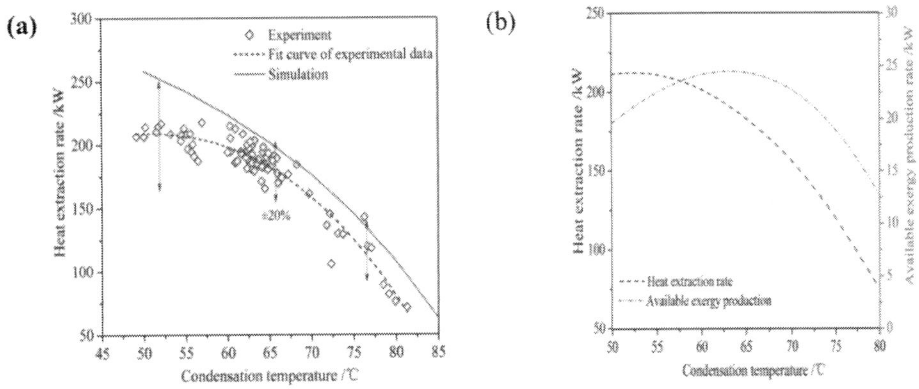

Figure 28 (a) Comparison of the averaged experimental condensation temperature and the simulation results, **(b)** best fitting of the experimental data for heat extraction rate and available exergy rate [34].

In Fig. 28 (b), it can be observed the available exergy rate of the system increases with the increase of the condensation temperature up to a value of approximately 63°C, when it reaches a maximum of 24.8 kW. Further increase of the condensation temperature leads to a decrease of the available exergy rate of the system. Also, in Fig. 28 (a), it can be noted that an increase of the condensation temperature from ~55°C, the heat extraction rate gradually decreases. It is apparent that the heat extraction rate is relatively insensitive to values of the condensation temperature below 55°C. In the temperature range of 55°C to 80°C, by decreasing the condensation temperature will reduce the temperature and pressure of the working fluid throughout the heat pipe, including its evaporation section. Consequently, the temperature difference between the heat pipe and the HDR increases, and it leads to enhanced heat extraction by the heat pipe.

4.5 Numerical simulation

4.5.1 Simulation model. The model has been upgraded and the progress has been reported in several publications of which the most relevant are Huang *et al* [39], Chen *et al* [40], Li *et al* [41] and Ma *et al* [42].

The model is based on the coupled heat transfer between the HDR and the in-pipe vapour-liquid two-phase flow. Also, taking into consideration that the gradient of the physical variables is far less in the radial direction of the heat pipe than that in the axial direction, the analysis of the heat pipe is conducted by using a simplified one-dimensional model. Therefore, change of the physical variables of the gravity heat pipe itself is only considered in the axial-direction. However, for the HDR, conduction is in the radial and axial direction. The interface between the heat pipe wall and the underground rock mass is a variable temperature boundary condition, which satisfies the heat transfer continuity condition to allow the coupling of the processes occurring inside and outside the heat pipe.

The computational domain is two-dimensional and axisymmetric, consequently no circumferential gradients are considered, and the central axis of the wellbore is the axis of symmetry; this domain is bounded by a 100 m radius from the heat pipe wall and a height measured from the ground to 250 m below the bottom of the heat pipe. This 250 m extension of the computational domain is taken to avoid that the temperature field within the reservoir be influenced by the far-away rock boundary condition. The top and bottom boundaries of the impermeable rock are assumed to be at constant temperature, while the lateral boundary condition of the impermeable rock is assumed to be adiabatic.

Journal of Physics: Conference Series **2940** (2025) 012001 doi:10.1088/1742-6596/2940/1/012001

The model of the heat transfer in the HDR outside the heat pipe only considers heat conduction and the HDR is assumed to be a porous medium with specified porosity (ε) and effective thermal conductivity (λ_{eff}).

In what concerns the *SLGHP* modelling are made the following assumptions:

No stagnant liquid pool occurs at the bottom of the heat pipe;

No liquid-vapour entrainment takes place along the heat pipe, so that the flow resistance in the *SLGHP* is mainly determined by the flow resistance of the vapour and follows an empirical of in-pipe turbulent flow resistance correlation;

The fluid inside the heat pipe is in thermodynamic phase equilibrium state;

The fluid condenses at constant temperature T_c at the top of the *SLGHP*;

The interphase momentum exchange of the gas-liquid phase is neglected.

Here the transient one-dimensional equations governing the conservation of mass, momentum and energy of the fluid in-pipe are omitted, as they are readily available in the references cited in this section.

The original formation temperature and the thermal conductivity of rock are set according to the logging data.

The model was implemented in the FLUENT software by resorting to the user-defined function (UDF) package. Grid refinement was set at the region in the vicinity of the wellbore to ensure sufficient resolution to describe the intervening physical processes. The size of the grids in the radial direction increases by a factor of 1.012 with the first value equal to 0.0125 m, while in the axial direction is maintained uniform at 20 m. The computational mesh selected on the basis of mesh independence tests encompasses 14125 structural grid elements.

4.5.2 Results. The simulation prediction for the heat extraction rate is presented in Fig. 28 (a) along with the heat pipe test data. The experimental results are close to the prediction when the condensation temperature is higher than 60°C. For a condensation temperature lower than 60°C, the prediction tends to overestimate the experimental results. This deviation between experiments and simulation results, as ongoing research is pointing out, is due to two main reasons, namely: (1) the simulation does not consider the presence of non-condensables; even by actuating the vacuum pump every 12 h, there is no guarantee of the complete removal of the non-condensable gases, and (2) in the actual site, foam cement was used for insulation in the insulated section and, although its thermal conductivity is low, the eventual heat loss can still occur.

Figure 29 shows the heat pipe temperature simulation results for the condensation temperature T_c equal to 40, 60, and 80°C, respectively. The temperature gradient along the depth for the three temperatures is less than that for the original formation temperature.

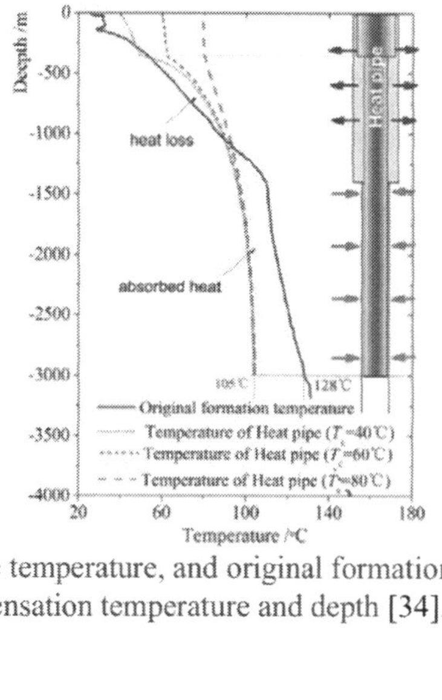

Figure 29 Predicted heat pipe temperature, and original formation temperature as a function of condensation temperature and depth [34].

The temperature gradient along the heat pipe dT/dy increases as the depth increases up to y equal to – 400 m, where the heat pipe diameter increases from 177.8 mm to 244.5 mm. The temperature gradient is less pronounced for values of the depth less than -1500 m, which agrees with the findings reported by Chen *et al* [39]. The increasing temperature gradient is attributed to the increase of the frictional pressure drop caused by the increased flow rate and the increase of dT_{equ}/dP_{equ} with the temperature decrease.

4.6 Concluding remarks

The development of the super long heat pipe for geothermal energy extraction is an ongoing project at LAES and the new results are very promising. However, this review only reports information, which is already in the public domain at the time of its preparation (March of 2024).

The testing results can be summarized as follows:

The highest condensation temperature achieved was ~90°C, while the average temperature of the underground formation along the full heat pipe length is 92.6°C, which is an indication of the good temperature uniformity of the heat pipe working fluid.

The heat pipe consistently produced approximately 190 kW of heat during the 30-day continuous heat extraction, although it presented some fluctuations due to the change of ambient temperature. After continuous extraction, and during one week of heat recovery, the geothermal reservoir returned to the original state prior to heat extraction.

The decrease of condenser temperature increases the heat pipe heat extraction; however, when the condenser temperature drops below 55°C, further heat extraction is practically negligible. For the condenser temperature of 63°C, it yiels the maximum available energy rate with a value of 24.8 kW.

The experimental results are in reasonable agreement with the predictions obtained using the numerical simulation model, which is based on quasi-ideal operating conditions.

The sustained heat extraction rate obtained during the 30-day continuous operation was 190 kW. Analyses based on the experimental results and simulation predictions demonstrate that the main factors restricting the heat extraction rate are related to the increase of the frictional pressure drop caused by the increased flow rate and the increase of dT_{equ}/dP_{equ} with the decrease of temperature.

The field tests confirmed the technical feasibility of using *SLGHP* to extract HDR geothermal energy. *SLGHP* is a novel technology with considerable potential in making geothermal energy more affordable with consequent wider utilization.

5. Conclusion

The lecture highlights the author's experience with two research groups dealing with different forms of heat transfer enhancement over the past decade.

There is considerable emphasis on the nanofluids and two low-temperature thermal applications are used as examples, namely the thermosyphon solar collector flat plate collector and the plate heat exchanger (*PHE*). As mixing promotors, coil wire inserts are used in the *PHE*. The nanofluids are presented with the single focus of their potential heat transfer enhancement. However, hybrid nanoparticles used in the preparation of hybrid nanofluids open new paths to other applications in quite different fields, in particular solid-state sensors, electrochromic devices, heterogeneous catalysts, imaging and medical devices. In what concerns nanofluids, their data base is extensive and although, in general, the scientific community recognizes their capability in terms of heat transfer enhancement, the theoretical work lags far behind the experimental work. Only recently other theories, such as ballistic transport, started being explored beyond diffusive transport and Brownian motion to explain the heat transfer enhancement.

A simple strategy for cooling of lithium-ion battery packs is described using a CFD approach. The challenge is to achieve good temperature uniformity for safe operation and increased lifespan of the battery pack. As part of the challenge, there is the need to maximize the heat removal from the pack.

Finally, it is succinctly discussed the super-long gravity heat pipe (*SLGHP*), which is a novel technology to extract geothermal energy at depths of several kilometers. The results discussed, although preliminary, already anticipate considerable promise. *SLGHP* requires careful design of its internals to avoid stagnant liquid, while at the interface, along the evaporation section the thermal resistance should be minimized, but on the upper section of the *SLGHP* heat losses should not occur and the section should be made adiabatic by applying appropriate thermal insulation.

As stated before, this lecture just reports the personal experience and vision of the author in the quest of heat transfer enhancement. Undoubtedly and fortunately, many other paths to this goal do exist.

Acknowledgment
The author acknowledges the scientific collaboration of Professor Fangming Jiang and Dr L. Syam Sundar.

The work was partially supported by a Visiting Scientist Fellowship awarded to the author under the Chinese Academy of Sciences President's International Fellowship Initiative, and by the FCT (Foundation for Science and Technology of Portugal) grant UID/EMS/00481/2013FCT and the infrastructures support CENTRO-01-0145-FEDER-022083.

References
[1] AE Bergles and HL Morton, Survey and evaluation of techniques to augment convection heat transfer, Engineering Projects Laboratory Report HTL 5382-34, Massachusetts Institute of Technology, Cambridge, MA.1965.
https://dspace.mit.edu/handle/1721.1/61479
[2] JC Maxwell, Treatise on Electricity and Magnetism, 3rd Edition, Volume I, Clarendon Press (Oxford), 435–441, 1904.
https://www.aproged.pt/biblioteca/MaxwellII.pdf
[3] SUS Choi, and JA Eastman, Enhancing thermal conductivity of fluids with nanoparticles. *International Mechanical Engineering Congress and Exhibition*, San Francisco CA, USA, Rep. No: ANL/MSD/CP-84938; Conf. 951135–29, 1995.
https://www.osti.gov/servlets/purl/196525
[4] SUS Choi, ZG Zhang, FE Lockwood and EA Grulke, Anomalous thermal conductivity enhancement in nanotube suspensions. *Appl. Phys. Lett.*, 79, 2252–2254, 2001.
https://doi.org/10.1063/1.1408272
[5] WS Hummers and RE Offeman, Preparation of graphitic oxide, *J. Am. Chem. Soc.*, 80, 6, 1339, 1958.
https://doi.org/10.1021/ja01539a017
[6] LS Sundar, VV Kotturu, C Mouli, Z Said and ACM Sousa, Heat transfer and second law analysis of ethylene glycol based ternary hybrid nanofluid under laminar flow. *J. Thermal Sci. Eng. Appl.*,13(5): 051021 (16 pages), Paper No: TSEA-20-1568, 2021.
https://doi.org/10.1115/1.4050228
[7] LS Sundar, MK Singh, I Bidkin and ACM Sousa, Experimental investigations in heat transfer and friction factor of magnetic Ni nanofluid flowing in a tube, *Int. J. Heat Mass Transf.*, 70, 224-234, 2014.
https://doi.org/10.1016/j.ijheatmasstransfer.2013.11.004
[8] AE Bergles, RL Bunn and GH Junkhan, Extended Performance Evaluation Criteria for Enhanced Heat Transfer Surfaces, *Letters in Heat and Mass Transfer*, 1, 113–120, 1974.
https://www.sciencedirect.com/science/article/pii/0094454874901477
[9] C Maradiya, J Vadher and R Agarwal, The heat transfer enhancement techniques and their Thermal Performance Factor, *Beni-Suef Univ J Basic Appl Sci*, 7, 1-21, 2018.
https://doi.org/10.1016/j.bjbas.2017.10.001

[10] LS Sundar, KV Sharma, MK Singh and ACM Sousa, Hybrid nanofluids preparation, thermal properties, heat transfer and friction factor - A review, *Renew. Sustain. Energy Rev.*, 68, 185-198, 2017.
http://dx.doi.org/10.1016/j.rser.2016.09.108

[11] Z Said (Ed), Hybrid Nanofluids - Preparation, Characterization and Applications, **ISBN**: 978-0-323-85836-6, Elsevier Inc., 2022.
https://shop.elsevier.com/books/hybrid-nanofluids/said/978-0-323-85836-6

[12] LS Sundar, EV Ramana, Z Said and ACM Sousa, Chapter 4 Hydrothermal properties of hybrid nanofluids. *Hybrid Nanofluids - Preparation, Characterization and Applications*, Ed. Zafar Said, **ISBN**: 978-0-323-85836-6, Elsevier Inc., 93-110, 2022.

[13] LS Sundar, MK Singh and ACM Sousa, Enhanced heat transfer and friction factor of MWCNT–Fe_3O_4/water hybrid nanofluids. *Int Comm in Heat and Mass Transfer*, 52:73–83, 2014. 10.1016/j.icheatmasstransfer.2014.01.012

[14] LS Sundar, VV Kotturu, C Mouli, Z and ACM Sousa, Heat transfer and second law analysis of ethylene glycol based ternary hybrid nanofluid under laminar flow. *J. Thermal Sci. Eng. Appl.*, 13(5): 051021 (16 pages), Paper No: TSEA-20-1568, 2021.
https://doi.org/10.1115/1.4050228

[15] G Huminic and A Huminic, "The Heat Transfer Performances and Entropy Generation Analysis of Hybrid Nanofluids in a Flattened Tube", *Int. J. Heat Mass Transf.*, 119, 813–827, 2018.
https://doi.org/10.1016/j.ijheatmasstransfer.2017.11.155

[16] K Apmann, R Fulmer, A Soto and S Vafaei. Thermal Conductivity and Viscosity: Review and Optimization of Effects of Nanoparticles. *Materials* (Basel), 14(5):1291 (71 pages), 2021.
10.3390/ma14051291

[17] P Keblinski, RS Phillpot, SUS Choi, and JA Eastman, Mechanism of heat flow in suspensions of nano-sized particles (nanofluids), *Int. J. Heat Mass Transf.*, 45, 855–863, 2002.

[18] RL Hamilton and OK Crosser, Thermal conductivity of heterogeneous two-component systems, *I & EC Fundamentals*, 1, 187-191, 1962.
10.1021/i160003a005

[19] AA Joshi and A Majumdar, Transient ballistic and diffusive phonon heat transport in thin films, *J. Appl. Phys.*, 74, 31-39, 1993.
https://doi.org/10.1063/1.354111

[20] P Oresta, F Micaliand and A De Risi, Undulatory theory of phonons on the nanofluid thermal conduction, *International Journal of Thermal Sciences*, 183, 107853, 2023.
https://doi.org/10.1016/j.ijthermalsci.2022.107853

[21] EC Okonkwo, I Wole-Osho, IW Almanassra, YM Abdullatif1 and T Al-Ansari1, An updated review of nanofluids in various heat transfer devices. *J Therm Anal Calorim*, 145, 2817–2872, 2021.
https://doi-org.proxy.hil.unb.ca/10.1007/s10973-020-09760-2

[22] SK Gupta and A Sharma, A brief review of nanofluids utilization in heat transfer devices for energy saving, *Materials Today: Proceedings* (in press, 31.01.2024).
https://doi.org/10.1016/j.matpr.2023.03.364

[23] J Yang, X Yang, J Wang, HH Chin and B Sundén Review on Thermal Performance of Nanofluids With and Without Magnetic Fields in Heat Exchange Devices. *Front. Energy Res.* 10:822776, 2022.
10.3389/fenrg.2022.822776

[24] LS Sundar, S Solomon, S Mesfin, YT Sintie, V Punnaiah, AJ Chamkha and ACM Sousa, A Review on the Use of Hybrid Nanofluid in a Solar Flat Plate and Parabolic Trough Collectors and Its Enhanced Collector Thermal Efficiency, *Journal of Nanofluids*, 10, 147–171, 2021.
10.1166/jon.2021.1783

[25] LS Sundar, TT Akanaw, MK Singh and ACM Sousa, Thermosyphon solar water heating system with Cu/water nanofluid and wire coil configurations: efficiency, energy, economic, environmental and heat transfer study, *Environ Prog Sustainable Energy,* 40:e13648, 1-15, 2021.
https://doi.org/10.1002/ep.13648
[26] A Whillier, Design factors influencing collector performance, RC Jordan (Ed.), *Low Temperature Engineering Application of Solar Energy*, ASHRAE, New York, N.Y. 10017, 1967.
[27] HC Hottel and A Whillier, Chapter 6. Evaluation of flat plate collector performance, *Transactions of the Conference on the Use of Solar Energy,* Volume II, Part 1, Section A, University of Arizona Press, 74-104, 1958.
Library of Congress Catalogue Card Number: 58-62504
[28] RW Bliss, The derivation of several plate efficiency factors useful in the design of flat plate solar heat collectors, *Solar Energy,* 3, 55–64, 1959.
https://doi.org/10.1016/0038-092X(59)90006-4
[29] LS Sundar, V Punnaiah, KV Sharma, AJ Chamkha and ACM Sousa, Thermal entropy and exergy efficiency analyses of nanodiamond/water nanofluid flow in a plate heat exchange*r, Diamond & Related Materials*, 120, 108648, 1-16, 2021.
https://doi.org/10.1016/j.diamond.2021.108648
[30] AK Tiwari, P Ghosh and J Sarkar, Performance comparison of the plate heat exchanger using different nanofluids, *Exp. Thermal Fluid Sci.*, 49 , pp. 141-151, 2013.
10.1016/j.expthermflusci.2013.04.012
[31] C. Zhao, ACM Sousa and F Jiang, Minimization of Thermal Non-uniformity in Lithium-ion Battery Pack Cooled by Channeled Liquid Flow, *Int. J. Heat Mass Transf.,* 129, 660-670, 2019.
https://doi.org/10.1016/j.ijheatmasstransfer.2018.10.017
[32] D Bernardi, E Pawlikowski and J Newman, A General Energy Balance for Battery Systems, *Journal of The Electrochemical Society*, 132, 5-12, 1985.
10.1149/1.2113792
[33] H-H Rogner, Chapter 5, Energy resources, World Energy Assessment report: energy and the challenge of sustainability, United Nations Development Programme, 165, 2000.
https://www.undp.org/sites/g/files/zskgke326/files/publications/World%20Energy%20Assessment-2000.pdf
[34] W Huang, J Cen, J Chen, W Cao, Z Li, F Li and F Jiang, Heat extraction from hot dry rock by super-long gravity heat pipe: A field test, Energy 247, 123492, 2022.
https://doi.org/10.1016/j.energy.2022.123492
[35] F Jiang, W Huang and W Cao, "Mining hot dry rock geothermal energy by heat pipe: conceptual design and technical Feasibility Study," *Advances in new and renewable energy,* 5, 426-434, 2017. (in Chinese).
[36] W Huang, W Cao, T Li and F Jiang, "Numerical study and economic analysis of heat pipe hot dry rock geothermal system " *CIESC Journal,* 72, 1-24, 2021. (in Chinese).
[37] W Huang, W Cao and F Jiang, "A Single-Well Geothermal System for Hot Dry Rock Geothermal Energy Exploitation," *Energy,* 162, 630-644, 2018.
https://doi.org/10.1016/j.energy.2018.08.05
[38] F Jiang, W Huang, W Cao and Y Wang, Ladder-structural gravity-assisted-heat-pipe geothermal energy recovery system without liquid-accumulation effect, EP 3 961 122 B1.
https://data.epo.org/publication-server/document?iDocId=7190372&iFormat=0
[39] W Huang, J Chen, J Cen, W Cao, Z Li, F Li and F Jiang, Heat extraction from hot dry rock by super-long gravity heat pipe: Effect of key parameters, *Energy*, 248,123527, 1-13, 2022.
https://doi.org/10.1016/j.energy.2022.123527

[40] J Chen, W Huang, J Cen, W Cao, Z Li, F Li and F Jiang, Heat extraction from hot dry rock by super-long gravity heat pipe: Selection of working fluid, *Energy*, 255,124531, 2022. https://doi.org/10.1016/j.energy.2022.124531

[41] Z. Li, W. Huang, J. Chen, J Cen, W Cao, F Li and F Jiang, An enhanced super-long gravity heat pipe geothermal system: conceptual design and numerical study, *Energy*, 267, 126524, 1-15, 2023.
https://doi.org/10.1016/j.energy.2022.126524

[42] Q Ma, J Chen, W Huang, Z Li, A Li and F Jiang, Power generation analysis of super-long gravity heat pipe geothermal systems, *Applied Thermal Energy,* 242, 122533, 1-12, 2024.

Journal of Physics: Conference Series **2940** (2025) 012002 doi:10.1088/1742-6596/2940/1/012002

Accuracy in evaluating convective heat transfer coefficient by RANS CFD simulations in a rectangular channel with high aspect ratio and V-shaped ribs

M Corti[1], P Gramazio[1], D Fustinoni[1] and A Niro[1,*]
[1] Politecnico di Milano, Department of Energy via Lambruschini 4, 20156 Milano, Italy
E-mail: *alfonso.niro@polimi.it

Abstract. In the framework of heat transfer enhancement inside channels with ribbed surfaces, V-shaped ribs are placed as the most promising among the standard-shaped ribs. This works aims to present a numerical analysis of heat transfer characteristics for forced convection inside a rectangular channel with high aspect ratio 1:10 equipped with V-shaped ribs. CFD simulations are carried out by means of Reynolds Averaged Navier-Stokes (RANS) equations, working with an air flow inside a narrow rectangular channel 120 mm wide and 840 mm long. V-shaped ribs have square cross section and they have been tested both with the tip of the V pointing downwards both upwards with respect to the channel inlet. For each configuration three different values of dimensionless pitch (10, 20 and 40) have been analysed. Comparisons between numerical and experimental data on convective global heat transfer coefficient and pressure drops are presented, showing very good agreement. Finding a properly working CFD model is useful for authors because this work, together with previous studies on 90° and 60° ribs, aims to be exploited during parametric early design-phases of new heat exchangers.

Keywords: forced convection, heat exchanger, rectangular channel, ribbed surfaces, CFD validation

1. Introduction

Heat transfer enhancement is crucial in several industrial applications as well as in the transport sector. High performance heat exchangers are required for electronic components cooling especially associated to modern applications such as high performance CPUs cooling and electric motors or batteries cooling of electric vehicles [1] in addition to traditional applications like gas turbine and solar air heaters.

Analyzing the most recent reviews [2–4] in literature related to this broad field, it is possible to conclude that, as far as ribbed channels with conventional ribs are concerned, V-shaped and W-shaped ribs configurations are the most favourable for heat transfer enhancement. It has also been found that the optimum angle of attack is 60° for both configurations. Furthermore, heat transfer enhancement sees an increase if V-shaped ribs present gaps along them [5].

Over years, the authors have conducted many experimental measurements on heat transfer and fluid-dynamics regarding forced convection inside channel of high aspect ratio (AR) with ribs. We have tested conventional ribs varying pitch-to-side and blockage ratio, and all of these heat transfer and fluid-dynamic data have now been exploited for creating a reliable CFD model applicable for a wide range of ribs configurations that could be used as starting point for the early design phase of new heat exchangers. In this work we present a comparison between experimental and numerical results regarding high AR rectangular channel with 60° V-shaped ribs.

2. Model and method

2.1. Experimental setup

The experimental setup can be easily summarized as a rectangular channel with AR=10 in which air flows at Reynolds between 700 and 8000. The air encounters ribs on the upper and lower channel walls which operate at a constant temperature, while ribs are adiabatic as the lateral walls. In this specific case air flows at Reynolds number of 7200 and ribs are V-shaped with an angle of attack of 60° degree (with respect to the main stream). They have been called V_{down} and V_{up} depending on what is the direction the apexes of the ribs: V_{down} ribs have apexes pointing towards the flow direction, while V_{up} ribs have apexes pointing forward the flow direction. In this study ribs have a squared cross section with e = s = 4 mm, they are parallel and in line as can be seen in figure 1. Moreover, we have considered for both configuration three different adimensional pitch-to-side ratio (p/s): 10, 20, 40, namely P10S4, P20S4, P40S4. For further details and information on measurements, data processing and error analysis we suggest you refer to [6], while table 1 presents a summary of the geometric parameters.

2.2. Numerical procedure

The computational domain is a 3D rectangular channel made up of three different solids. These are merged to create a test section, an inlet region and an outlet region. This approach helps to mitigate issues associated with reverse flow, particularly common in geometries featuring internal turbulators. Air, treated as an incompressible fluid, is the working fluid and is operated at Reynolds 7200 providing a turbulent flow regime.

The air enters the domain at $T_{in} = 296K$ or $T_{in} = 300K$ depending on the configuration considered. The temperature profile develops along the channel while the velocity profile at the inlet is fully developed, obtained in previous simulations at the same Reynolds number in an identical channel but without ribs. At the outlet $P_{out} = 0$ is applied. Walls are subject to a no-slip velocity condition, all surfaces −upper, lower, and lateral− operated at constant temperature $T_{wall} = 313.15K$. This condition does not apply to ribs walls, they are all considered adiabatic, with $q = 0$ imposed. This distinction is made to emphasize the beneficial fluid-dynamic effects of the ribs, dissociating them from potential fin-like behaviors. Also in the experimental setup they

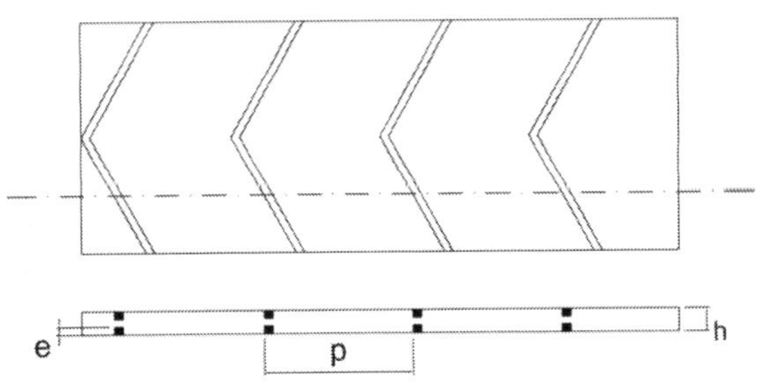

Figure 1. V_{up}, $\alpha = 60°$, in line.

Table 1. Channel geometric parameters.

	Value [m]
Height, h=H/2	0.012
Width, w=W/2	0.12
Length, L	0.84
Hydraulic diameter, Dh	0.02128
Ribs height, ribs width, e = s	0.004

have very low thermal conductivity, since they are made in wood and/or Plexiglass. For this work we applied symmetries along length and spanwise direction, managing to work with a quarter of the domain. The boundary conditions presented are consistent with the experimental setup, with the exception of the lateral walls. In the numerical setup they are considered at constant temperature, whereas in the experimental setup they are almost adiabatic. This difference is based on prior studies [7], which demonstrate that the heating of the lateral walls at constant temperature provides the most appropriate representation of the experimental configuration within the numerical framework.

Numerical simulations were performed exploiting the Ansys Fluent 2023 R1 software platform. The flow field is modelled using RANS, with the $k - \omega -$ SST turbulence model. The SIMPLEC algorithm was chosen among the others after some comparative analysis from earlier tests, exhibiting better and quicker convergence. We carried out a sensitivity analysis to be sure of proposing a grid-independent solution, by examining variations along all three directions. Notably, we observed that the influence of the mesh is more significant along channel height (y). In summary, after testing structured grids with consistent topologies varying the number of elements, a grid with approximately $80 \cdot 10^6$ elements was selected for the entire channel, meaning approximately $20 \cdot 10^6$ elements for the considered computational domain, reduced thanks to symmetries, but the exact number of elements may vary depending on the specific channel configuration as the number of ribs varies depending on the pitch-to-side ratio. Applying a medium-quality structured mesh resulted in an average error of about 1% over the analyzed parameters, but in no case exceeded 4%, with respect to a fine one. RANS simulations were stopped when the continuity residual fell below $1 \cdot 10^{-5}$ to ensure that each simulation did not exceed 96 hours of computation. For further information regarding the numerical model and the sensitivity analysis, we recommend [8].

3. Results

As mentioned above, the forced convection of air inside the channels analysed takes place at Re 7200. Reynolds number selection is primarily influenced by two factors. Firstly, it represents the upper limit of Reynolds numbers tested thus far with the current experimental setup. Secondly, in line with previous analyses [7], it has been found that the accuracy of the RANS model $k - \omega -$ SST improves with increasing Reynolds numbers. Numerical outcomes related to heat transfer properties have been compared with experimental measurements for the selected configurations.

3.1. Bulk temperature and heat exchanged

Before proceeding with the analysis of the thermal characteristics, we would like to have a brief indication of how effectively our numerical model is able to represent the experimental setup. Since at the experimental level the directly measured quantity is the outlet temperature, it is possible to compare it with the bulk temperature obtained numerically at the outlet section T_{out}. The error is very small for each configuration considered as shown in figure 2. A quantity that is more representative of the problem, could be the thermal power Q. The heat exchanged by the whole channel is calculated as Q in equation 1 and the errors remains below 5% for all the channel configurations except P40S4 V_{donw}, figure 3.

$$Q = \dot{m}c_p(T_{out} - T_{in}) \tag{1}$$

Both figures show, in addition to the experimental and numerical values, the associated experimental uncertainties, represented by means of error bars. In particular, the temperature was measured with a accuracy of $0.05K$, while the uncertainty on the heat exchanged was calculated using the Moffat method [9] and is equal to 2.3%.

3.2. Convective heat transfer coefficient (HTC)

Integral results for heat transfer characteristics are expressed by means of the mean convective HTC computed via $\Delta T_{m,ln}$ when a temperature profile can be considered developed inside a channel with an imposed wall temperature. Then, the mean convective HTC $h_{m,ln}$ can be computed from the energy balance as in equation 2.

$$h_{m,ln} = \dot{m}c_p(T_{out} - T_{in})/(A\Delta T_{m,ln}) \tag{2}$$

The comparison between the experimentally determined mean convective HTC and those calculated numerically, is shown in figure 4. The values obtained by CFD underestimate the experimental ones except for P10S4 for both V_{down} and V_{up}.

The discrepancies can be attributed to factors arising from both the application of the RANS model and the assumption of fully developed temperature profile inside the channels, which may not be achieved. Numerical results are consistent with the experimental findings, with discrepancies remaining below 10% for all the configurations except for P40S4 V_{down} that

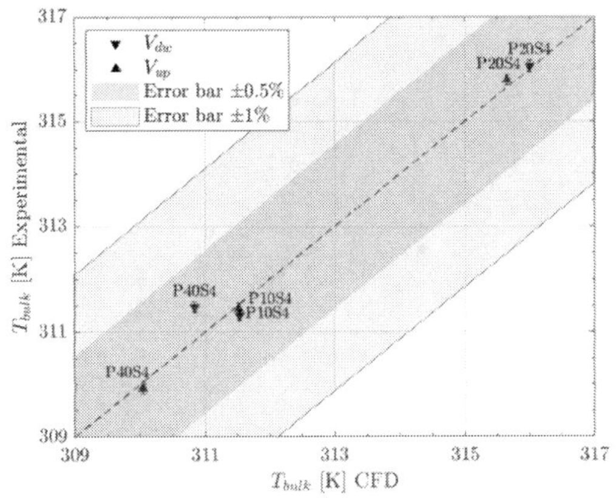

Figure 2. T_{bulk}, comparison between experimental and CFD data.

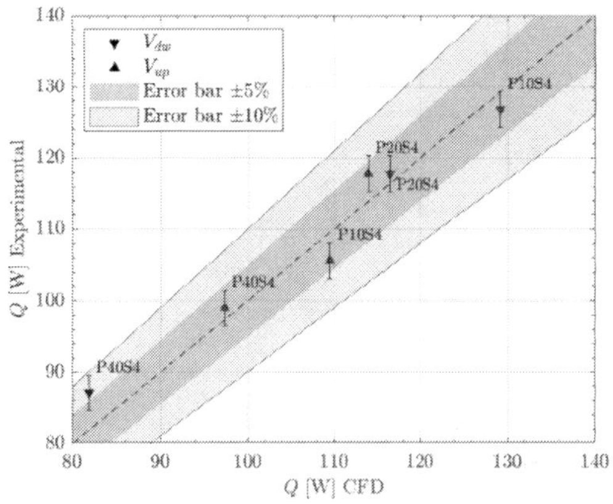

Figure 3. Q, comparison between experimental and CFD data.

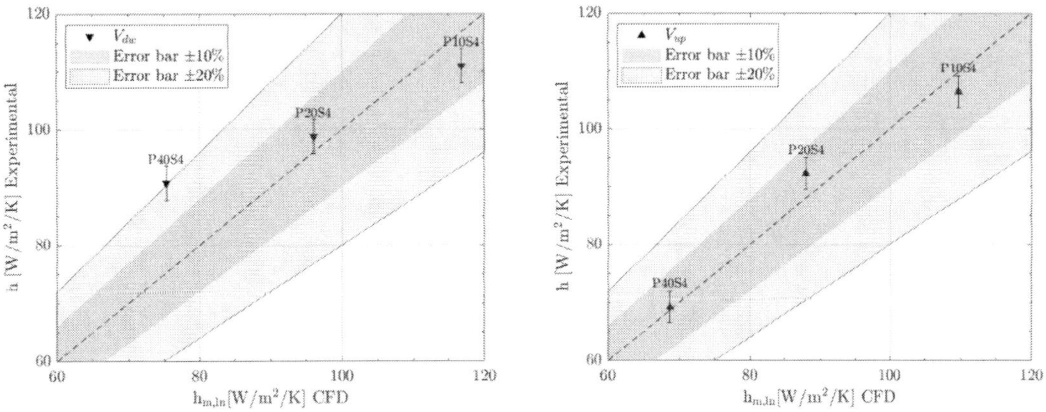

(a) V-down configuration.

(b) V-up configuration.

Figure 4. $h_{m,ln}$, comparison between experimental and CFD data.

reaches 20%. The capability of the RANS model to capture the associated flow and heat transfer phenomena is quite good. It's worth noting that the uncertainty associated with the experimental convective HTC is 3%.

To see how effective the developed temperature field assumption is, we can investigate the local convective HTC.

The local convective HTC is computed as h_z, equation 3.

$$h_z = q_z/(T_w - T_{b,z}) \tag{3}$$

T_b and the heat flux q_z have been evaluated as mean section variables and mean line variables, in that order, on surfaces and lines built along the channel length, equidistant from each other, and

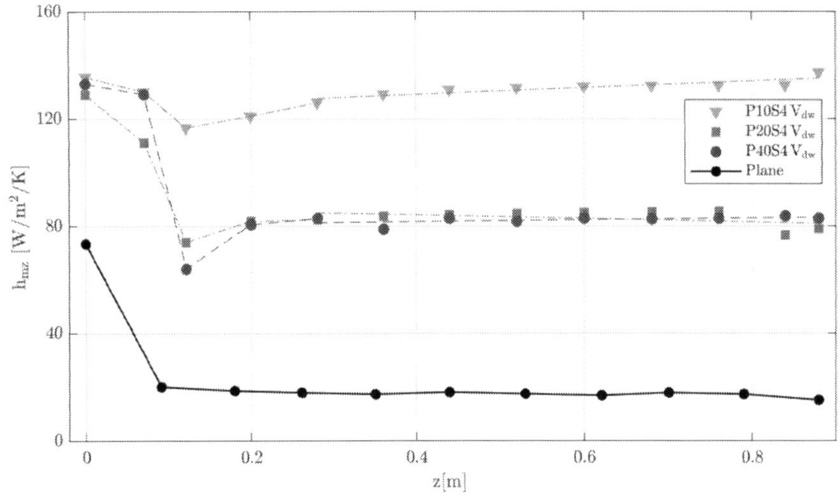

(a) V-down configuration.

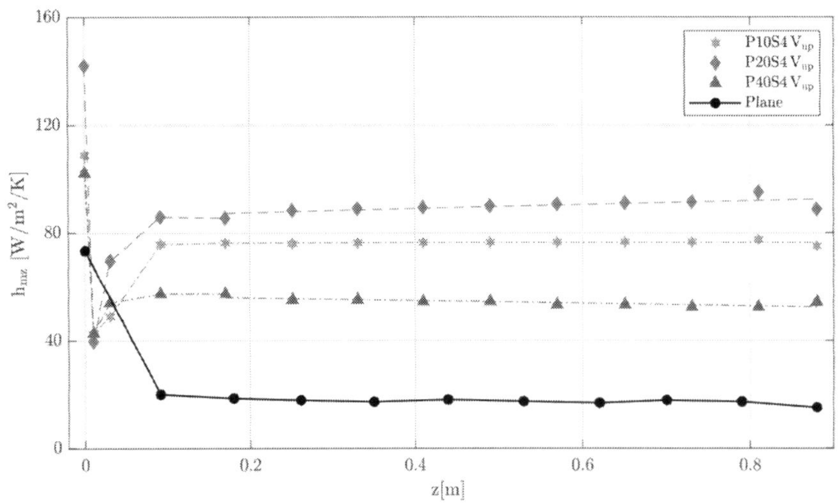

(b) V-up configuration.

Figure 5. Trend of the local convective HTC

in a number that ensures that the convective HTC is no longer affected by their numbers. Then each contribution of the local convective HTC h_z has been weighted through the heat transfer area ΔA_z of influence, and presented in figure 5 as h_{mz}. The figure illustrates the trend of the local convective HTC, h_{mz}, plotted against the channel length, z. Additionally, the distribution of h_z for a plane channel is included for comparison. The presence of the ribs brings a substantial change in the evolution of the local convective HTC, both in terms of the distribution of h_z along z and the value assumed (please note the different values of the y-axis). However, the trends of h_z in the case of ribbed channels present similar pattern, after taking into account the inlet effects the coefficient stabilizes around a value and the distribution becomes almost flat representing a state of thermally developed flow. Only for the V_{down} P10S4 configuration the distributions seem to deviate, probably because this condition is not fully achieved or because the RANS model is not sensitive enough to the complexity of the case.

We want to propose in detail the distribution along the channel length of h_z for the P40S4 V_{up} configuration thanks to figure 6. The local convective heat transfer coefficient increases significantly immediately after the rib until it gradually decreases as the effect of the obstacle decreases. We would also like to point out that after the first two modules, the distribution of h_z between two ribs follows the same shape, i.e. it increases, stabilises, decreases and then increases again, and the values tend to repeat themselves the same at the same distance. In order to increase heat transfer, this shows how tighter ribs are more effective.

At this point it is interesting to propose the actual convective HTC, h_m (equation 4), built considering each contribution of the local convective HTC h_z weighted by the affected heat transfer area ΔA_z. The actual convective HTC offers more exhaustive evaluation compared to $h_{m,ln}$, as it takes into account not only the impact of the channel entrance but also the variations in h_z along the channel.

$$h_m = \frac{1}{\sum_{z=1}^{n_{sections}} \Delta \ A_z} \left(\sum_{z=1}^{n_{sections}} h_z A_z \right) \tag{4}$$

To get an indication of how much the use of $h_{m,ln}$ introduces an error compared to the use of the actual h_m, we propose to compare them in a portion of the channel domain that does not take into account the inlet and outlet effects, starting the evaluation from $z = 0.2m$ until just before the outlet section. Figure 7 shows these data, it can be seen that the error remains

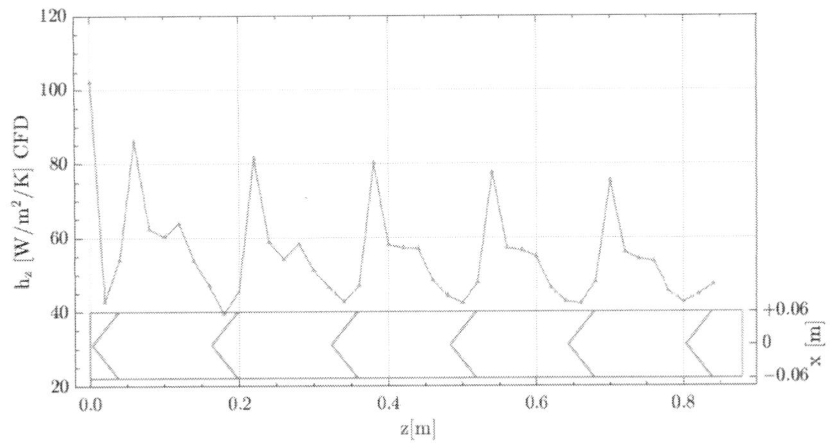

Figure 6. Detailed trend of the local convective HTC h_z for P40S4 V_{up} channel configuration.

always below 20%, the worst behavior is of those configurations that, looking at the trend of h_z in figure 5, show a distribution that is not yet flat but slightly increasing like P10S4 both V_{up} and V_{down}.

4. Discussion and Conclusions

A CFD model for a high aspect ratio rectangular channel with ribs on the upper and lower walls in V-down and V-up configuration has been presented with this work. The RANS $k-\omega-$SST turbulence model has been used, which showed good agreement by comparing the main heat transfer characteristics. Besides, this study confirms literature findings that V_{down} ribs provide better heat transfer enhancement than V_{up}, combined with the fact that it shows us how the closer the ribs are, the greater the increase in heat transfer. Thanks to the results obtained, it is also possible to use this CFD RANS model for different dimensionless pitch-to-side ratios, included in the analyzed range, at an early stage of parametric studies. This work is, actually, a further step of a more comprehensive work. Over time we have built CFD RANS models for different configurations, so that we have managed to have an efficient CFD model for ribs at 90°, 60° tilted and 60° V_{down} and V_{up}, the most popular in the literature. Starting from these CFD models, the idea is to find by an optimization process innovative geometries according to new design methods based on additive manufacturing that can further increase the heat transfer. Performance of these geometries will be first numerically tested by RANS model, in order to select the most promising once that will be further investigated with more accurate CFD simulations - such as LES -, and eventually experimentally. Now that the authors have found that the flow becomes thermally fully developed for most of the geometries analysed, this will allow the use of a periodic domain. Thus, by reducing the domain, we can increase the complexity of the simulation by dealing with the LES approach, for instance.

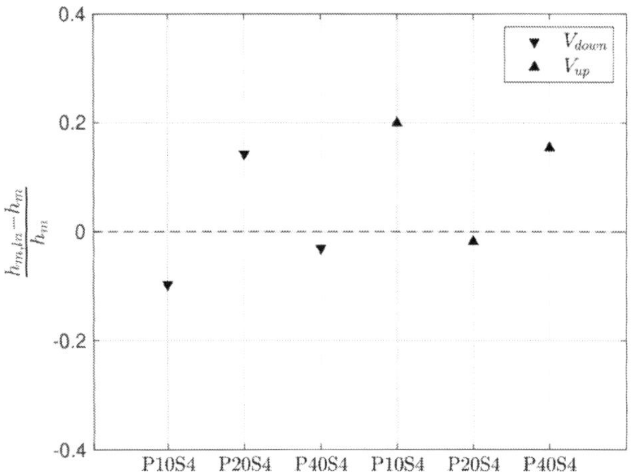

Figure 7. Comparison of convective HTCs computed from CFD results.

Acknowledgments

This work has been partially funded by means of the EU - Next GenerationEU - Project 2022SJP2A5 - CUP D53D23004040006 after the competitive MUR Call n.104 - PRIN 2022.

Journal of Physics: Conference Series **2940** (2025) 012002 doi:10.1088/1742-6596/2940/1/012002

References

[1] Wang X, Li B, Gerada D, Huang K, Stone I, Worrall S and Yan Y 2022 *Appl. Therm. Eng.* **201** 117758
[2] Sharma S K and Kalamkar V R 2015 *Renew. Sustain. Energy Rev.* **41** 413–435
[3] Alam T and Kim M H 2017 *Renew. Sustain. Energy Rev.* **69** 387–400
[4] Singh I and Singh S 2018 *Renew. Sustain. Energy Rev.* **92** 405–425
[5] Jain S K, Agrawal G D and Misra R 2019 *Heat Mass Transfer* **55** 3369–3412
[6] Fustinoni D, Gramazio P, Colombo L P and Niro A 2014 *Proc. 15th International Heat Transfer Conference*
[7] Corti M, Gramazio P, Fustinoni D, Vitali L and Niro A 2023 *J. Phys.: Conf. Ser.* **2509** 012009
[8] Corti M, Gramazio P, Fustinoni D and Niro A 2024 *J. Phys.: Conf. Ser.* **2685** 012001
[9] Moffat R J 1988 *Describing the uncertainties in experimental results* vol 1 (Experimental Thermal and Fluid Science)

Journal of Physics: Conference Series **2940** (2025) 012003 doi:10.1088/1742-6596/2940/1/012003

Computational fluid dynamics simulation of the homogeneous flow regime in a large-scale bubble column

N Varallo*, **G Besagni, R Mereu and F Inzoli**

Politecnico di Milano, Department of Energy, Via Lambruschini 4a, 20156 Milano, Italy

* Corresponding author: nicolo.varallo@polimi.it

Abstract. Bubble column reactors are widely used in various industrial applications, yet their characterisation at both global and local scales remains highly complex. This study presents a CFD Eulerian multi-fluid approach to model the hydrodynamics of large-scale bubble columns, focusing on the homogeneous flow regime. Transient 3D simulations were performed using the commercial software ANSYS Fluent 2023 R2, assuming a fixed mono- or poly-dispersed bubble size distribution to represent the dispersed gas phase. The results were thoroughly analysed and the discrepancies between the numerical results and the experimental data were discussed, highlighting areas for future improvement.

1. Introduction

Bubble columns are gas-liquid reactors used in several industrial applications. They offer excellent heat and mass transfer between phases and low operating and maintenance costs due to the lack of moving parts. For example, they are used in the Fischer-Tropsch process to produce liquid hydrocarbons and in the LOPROX process to improve the biodegradability of highly contaminated products.

A basic bubble column setup consists of a vertical container with a gas distributor at the bottom that disperses the gas phase into bubbles or coalescence-induced structures. The liquid phase can flow either co-currently or counter-currently with the rising bubbles. In batch mode, the bubble column operates with bubbles rising through a stationary liquid phase.

Although the design is simple, bubble columns are characterized by a complex fluid dynamics resulting in different flow regimes. The fluid dynamics of large-scale bubble columns can be described by six distinct flow regimes. As the gas flow rate increases, the following regimes are observed: (1) mono- and (2) poly-dispersed homogeneous flow regimes, (3) transition flow regime without and with (4) coalescence-induced structures, (5) pseudo- and pure (6) heterogeneous flow regimes (Figure 1).

The homogeneous flow regime occurs at low superficial gas velocities where bubbles do not coalesce. Within this regime, the flow can be further classified into two types based on the bubble size distribution (BSD): the monodispersed homogeneous flow regime and the poly-dispersed homogeneous flow regime.

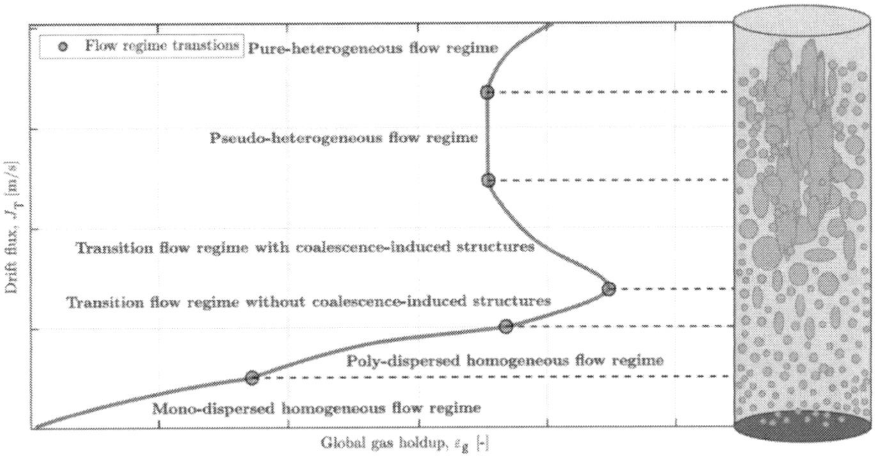

Figure 1. Operating curve of the bubble column.

The transition flow regime without coalescence-induced structures arises from unstable coalescence-induced structures, which cause gas holdup oscillations and are linked to hydrodynamics instabilities. As the gas flow rate increases, these unstable structures evolve into stable coalescence-induced structures, marking the transition to the flow regime with coalescence-induced structures. However, while these structures are visible in the bubble column, they are not fully developed. As the gas flow rate continues to rise, the gas holdup decreases, indicating a mismatch between higher gas flow rates and the growth of coalescence-induced structures.

When the gas flow rate increases further, the pseudo-heterogeneous flow regime emerges. This regime is characterized by a balance between the increase in the gas flow rate and the formation of coalescence-induced structures. At even higher superficial gas velocities, the flow enters a fully heterogeneous regime, where higher rates of coalescence and breakage lead to a broader range of bubble sizes.

This paper focuses on the homogeneous flow regimes and is structured as follows. Section 2 discusses the experimental benchmark used for model validation, while Section 3 details the model and numerical settings. Section 4 compares the numerical results with experimental data, and the paper concludes with a summary and suggestions for future work.

2. Experimental benchmark

Experimental data from Besagni [3] were used as a reference for this study. The experimental setup consists of an unpressurised vertical plexiglass tube with a height of 5.3 m and an internal diameter of 0.24 m. The column was equipped with a perforated plate sparger with 581 holes of 0.5 mm diameter. To assess the global gas holdup, bed expansion was measured by determining the height of the liquid-free surface when air was introduced into the column.

A dual fiber optical probe system was used to measure local fluid dynamic properties, specifically the local void fraction. The optical probe was inserted into the flow through an access port in the fully developed flow region.

The bubble size distribution at different superficial gas velocities were obtained through image analysis.

Journal of Physics: Conference Series **2940** (2025) 012003 doi:10.1088/1742-6596/2940/1/012003

3. The numerical model

3.1. Governing equations

In this study, the Eulerian two-fluid approach was employed. For an isothermal flow without mass transfer, the continuity and momentum equations for the k-phase are given as follows.

$$\frac{\partial}{\partial t}(\alpha_k \rho_k) + \nabla \cdot (\alpha_k \rho_k \boldsymbol{u}_k) = 0 \quad (1)$$

$$\frac{\partial}{\partial t}(\alpha_k \rho_k \boldsymbol{u}_k) + \nabla \cdot (\alpha_k \rho_k \boldsymbol{u}_k \boldsymbol{u}_k) = -\alpha_k \nabla p + \nabla \cdot (\alpha_k \bar{\tau}_k) + \alpha_k \rho_k \boldsymbol{g} + \boldsymbol{M}_{I,k} \quad (2)$$

The term on the right-hand side of Equation (2) accounts for the pressure gradient, the stresses (including Reynolds and viscous stresses), the body forces, and the momentum exchange between the phases.

3.2. Momentum exchange between the phases

The interactions between the continuous and dispersed phases are included as a source term in the momentum equation. The last term of Equation (2) incorporates various physical mechanisms, including drag, lift, turbulent dispersion, wall lubrication, and virtual mass forces.

$$\boldsymbol{M}_{I,k} = \boldsymbol{F}_{D,k} + \boldsymbol{F}_{L,k} + \boldsymbol{F}_{TD,k} + \boldsymbol{F}_{WL,k} + \boldsymbol{F}_{VM,k} \quad (3)$$

The closure models applied in this study are provided in Table 1. Special emphasis was placed on the lift force, with a comparison of various lift coefficient correlations (Figure 2), which predict different critical bubble diameters at which the lift coefficient changes sign.

Table 1. Closure relations used in the modelling of the momentum exchange between the phases.

Force	Correlation	Notes
Drag	Tomiyama [4]	
Lift	Tomiyama [5]	
	Ziegenhein [6]	
	Hessenkemper [7]	
Turbulent dispersion	Burns [8]	Neglected in the poly-dispersed homogeneous flow regime
Wall lubrication	Antal [9]	
Virtual mass	Neglected	

3.3. Turbulence modelling

In the homogeneous flow regime, due to the low density and small scale of the dispersed gas, it is sufficient to consider turbulence only in the continuous phase [10]. The k-ω SST model was used to estimate the Reynolds stresses using the Production Kato-Launder modification option. The model constants were set to their single-phase values. Bubble induced turbulence was not considered in this study.

IOP Publishing
doi:10.1088/1742-6596/2940/1/012003

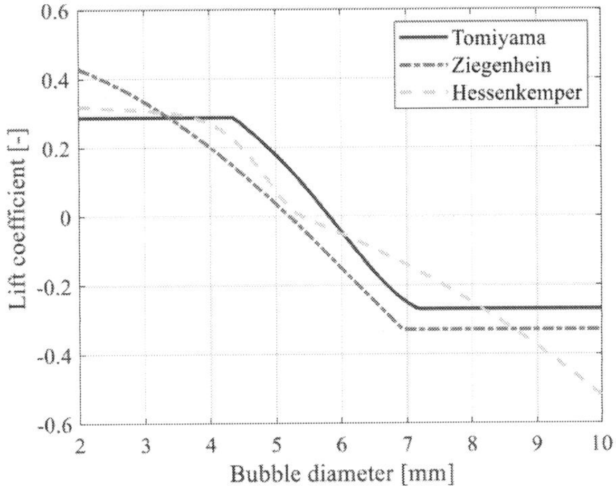

Figure 2. Evolution of the lift coefficient with the bubble diameter.

3.4. Numerical settings and boundary conditions

Transient 3D simulations were performed using the commercial software ANSYS Fluent 2023 R2 on a structured grid of 260,000 elements. The results were averaged over 60 seconds of flow time to ensure they were independent of initial transient. The time step was set to 0.001 to keep the Courant-Friedrichs-Lewy number below 1. The Phase-Coupled SIMPLE (PC-SIMPLE) algorithm was used for pressure-velocity coupling, and a second-order bounded Euler implicit scheme was used for temporal discretization.

For spatial representation, the least squares cell-based formulation and the PRESTO! discretization scheme were used for gradients and pressure, respectively. Due to the transient nature of the flow in bubble columns, higher order schemes were required, so the QUICK method (third order accuracy) was used for both the momentum and volume fraction equations, while the second order upwind scheme was used for the turbulent quantities. A uniform inlet was employed to model the gas sparger, with the gas velocity specified at the column inlet. The turbulent quantities were set based on Kawase [11]. A degassing boundary condition was applied at the column outlet. Table 2 provides a summary of the cases examined.

Table 2. Cases studied.

Case	Superficial gas velocity [cm/s]	Flow regime
C1	0.38	Mono-dispersed
C2	3.44	Poly-dispersed

3.5. Phases properties

A variable air density was considered due to the hydrostatic head. The air density varied by ±14% from a fixed value calculated at the centre of the column, with the air density at the bottom and top of the column equals to 1.6 kg/m³ and 1.2 kg/m³, respectively. Water properties were assumed to be at ambient temperature and pressure.

3.6. Dispersed phase modelling

This study focuses on the homogeneous flow regimes, and depending on the specific flow regime being analyzed, either a fixed mono-dispersed or poly-dispersed approach is used. The mono-dispersed approach assumes a constant, single bubble diameter, making it appropriate for simulating the mono-dispersed homogeneous flow regime.

Journal of Physics: Conference Series **2940** (2025) 012003 doi:10.1088/1742-6596/2940/1/012003

In contrast, the fixed poly-dispersed approach incorporates three velocity groups for the gas phase. This allows the model to account for the different dynamics of small and large bubbles. The three bubble classes implemented are:

- the *positive-lift* class refers to bubbles with a diameter smaller than the critical diameter (d_{cr}), at which the lift force changes sign, resulting in a positive lift coefficient.
- the *zero-lift* class consists of bubbles with a diameter approximately equal to the critical diameter (d_{cr}), resulting in a zero lift coefficient.
- the *negative-lift* class includes bubbles with a diameter larger than the critical diameter (d_{cr}), resulting in a negative lift coefficient.

The bubble diameter for each class, along with the mass fraction of small, medium, and large bubbles, were determined based on the experimental data.

4. Results

Table 3 presents a comparison of the experimental and predicted global gas holdup. This parameter is a crucial flow property, as it governs the mean residence time of the dispersed gas phase within the column. In addition, when combined with the bubble size distribution, it determines the interfacial area for heat and mass transfer rates. Consequently, accurate prediction of the global gas holdup is essential for reactor design, optimization and scale-up processes.

Table 3. Cases studied.

Case	$\varepsilon_{g,EXP}$ [%]	$\varepsilon_{g,CFD}$ [%]	Error [%]
C1-Tomiyama	1.57	1.49	-5.59
C1-Ziegenhein	1.57	1.50	-4.60
C1-Hessenkemper	1.57	1.49	-5.02
C2- Tomiyama	16.26	15.26	-6.11
C2-Hessenkemper	16.26	14.57	-10.37

The proposed numerical model provides a reliable estimation of the global gas holdup. Specifically, the model slightly underestimates the value, with a maximum error of -10.37%. This underestimation can be attributed to an underestimation of the drag coefficient, which leads to a reduced residence time of bubbles in the column and, consequently, a lower global gas holdup.

For case study C1, the differences between the various lift coefficient correlations are negligible, with the error consistently staying below 10%.

Figure 3 illustrates the comparison between the numerical and experimental local gas holdup profiles. Although the flow is in the mono-dispersed homogeneous regime, the experimental profile is centre-peaked, while the numerical profile is wall-peaked. This discrepancy is mainly due to air maldistribution at the gas sparger.

At low superficial gas velocities, air flows primarily through the central sparger holes, not through all the holes, which affects the local gas volume fraction profiles. This issue is common at low volumetric flow rates, where the overpressure is insufficient to activate all holes equally. Since the sparger maldistribution could not be quantified precisely, it was not incorporated into the numerical model but should be considered when interpreting the numerical results.

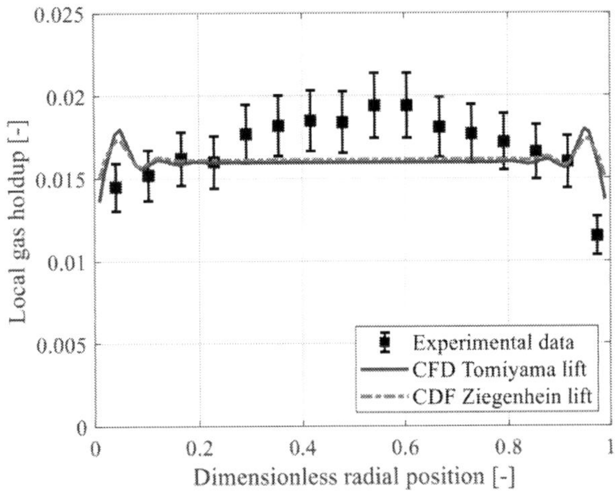

Figure 3. Comparison between experimental and numerical local gas holdup profiles. Case C1.

A comparison of different lift coefficient correlations reveals that the Hessenkemper model produces identical results to the Ziegenhein correlation, which are omitted from the figure for clarity. As depicted in Figure 3, both correlations predict the same lift coefficient for the bubble diameter considered in this study ($d_b = 3.2$ mm). The Tomiyama correlation, on the other hand, predicts slightly lower local gas holdup at the column centre, while showing more pronounced peaks near the walls. Nonetheless, no significant differences were observed among the three lift coefficient correlations within the monodispersed homogeneous flow regime.

In the poly-dispersed homogeneous flow regime, the results for the local void fraction profile are shown in Figure 4, with comparisons made between the Tomiyama and Hessenkemper lift coefficient correlations. The numerical model does not accurately predict the local void fraction profile near the wall, as it overestimates the void fraction near the vertical wall, resulting in a wall-peaked profile. This suggests an over-estimation of small bubbles compared to the experimental data. Small bubbles are evenly distributed throughout the column cross section, reducing liquid recirculation and increasing the gas volume fraction. By increasing the number of small bubbles while keeping the number of large bubbles constant, the local gas holdup profiles can shift from centre peak to flat to wall peak.

At the column centre, the numerical results align perfectly with the experimental data. A comparison between the Tomiyama and Hessenkemper lift coefficient correlations shows that both yield similar outcomes, with the Hessenkemper correlation demonstrating slightly better performance in the near-wall region.

The discrepancies between the numerical and experimental results can be attributed to several factors. One important aspect is the influence of neighbouring bubbles on the lateral motion of bubbles. Existing lift coefficient correlations are typically developed for isolated bubbles rising in a quiescent liquid. However, as numerous studies have shown, these correlations lose accuracy when bubbles interact with other bubbles in proximity. While the influence of neighbouring bubbles on the drag force has been extensively addressed in the literature by introducing a swarm factor related to local or global gas volume fractions, the swarm effect on the buoyancy force remains unexplored in experimental studies.

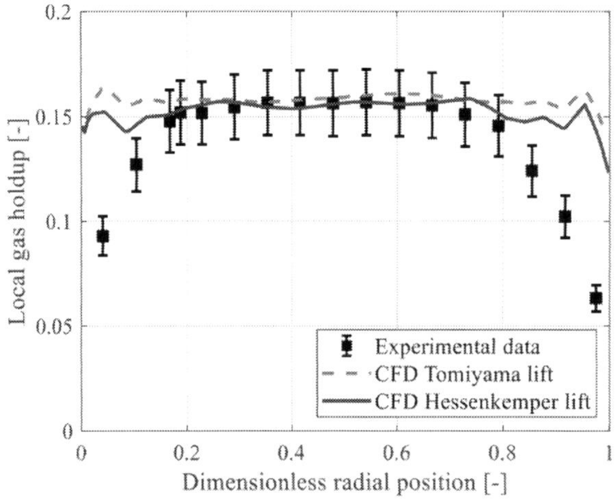

Figure 4. Comparison between experimental and numerical local gas holdup profiles. Case C2

5. Conclusions

This study involved CFD simulations of a large-scale bubble column operating in both the mono- and poly-dispersed homogeneous flow regimes. The numerical results were in good agreement with the experimental data for global gas holdup. However, the CFD model did not accurately predict the near-wall local gas holdup. In the mono-dispersed homogeneous flow regime, air maldistribution at the sparger caused air injection to occur only at the centre of the column, leading to centre-peaked local gas holdup profile. In the poly-dispersed homogeneous flow regime, the issue may be due to the lack of a swarm effect on the lift force, which should be incorporated into the numerical model. Future experimental and numerical investigations should focus on the lateral motion of bubble swarms to develop correlations for momentum exchange between the phases that account for the presence of other bubbles.

6. References

[1] Rollbusch P, Bothe M, Becker M, Ludwig M, Grünewald M, Schlüter M and Franke R 2015 Chem. Eng. Sci. **126** 660-678.
[2] Besagni G 2021 *Chin. J. Chem.* Eng. **40** 48-52.
[3] Besagni G 2021 *Int. J. Multiph. Flow* **135** 103510.
[4] Tomiyama A 1998 Multiph. Sci. Techenol. **10** 369-405.
[5] Tomiyama A, Tamai H, Zun S and Hosokawa 2002 *Chem. Eng. Sci.* **57** 1849-1858.
[6] Ziegenhein T, Tomiyama A and Lucas D 2018 *Int. J. Multiph. Flow* 108 11-24.
[7] Hessenkemper H, Ziegenhein T, Rzehak R, Lucas D and Tomiyama A 2021 *Int. J. Multiph. Flow* **138** 103587.
[8] Burns A D, Frank T, Hamill I, Shi J M 2004 *5th international conference on multiphase flow ICMF* 4 1-17.
[9] Antal S P, Lahey Jr and Flaherty J E 1991 *Int. J. Multiph. Flow* **5** 635-652.
[10] Rzehak R and Krepper E 2013 *Nucl. Eng. Des.* **265** 701-711.
[11] Kawase Y and Moo-Young M 1989 *Chem. Eng. J.* **40** 55-58

Data-driven thermal characterization of a 1-D model for sensible stratified thermal energy storage

A V Anacreonte[1,2,*], F Capolupo[1,2], R Russo[2], R Vitobello[3], and M Musto[1,2].

[1] Department of Industrial Engineering, University of Naples Federico II, P.le Tecchio, 80, Naples, Italy
[2] Institute of Applied Sciences and Intelligent Systems, National Research Council of Italy, Via Pietro Castellino 111, Naples, Italy
[3] TVP Solar SA, 10 rue du Pré-de-la-Fontaine, Satigny, 1242, Switzerland

* Corresponding author: vitalianoalessandro.anacreonte@unina.it

Abstract. Thermal Energy Storage (TES) facilities have a wide range of applications in thermal networks. They enable adaptable functionality by effectively and timely managing the supply and demand of thermal energy. This practice offers both environmental and economic benefits, such as reduced CO_2 emissions, lower energy generation costs, and decreased operational expenses. Although there are various TES technologies available in the market with higher energy density, such as those based on latent heat and phase change materials, hot water TES units are predominantly used due to their cost-effectiveness, simplicity, and favourable thermal properties. Although many two-dimensional and three-dimensional simulation approaches have been employed to assess the fluid dynamic behaviour of TES components, there has been growing interest in one-dimensional models, as they are better suited for time-dependent, plant-scale optimization studies. However, 1-D models often sacrifice accuracy for computational efficiency, leading many researchers to focus on improving their precision. In this study, field data acquired in an industrial solar thermal power plant are employed to validate and thermally characterize a sensible stratified thermal energy storage using pure water as Heat Transfer Fluid (HTF). This approach aims to bridge the gap between qualitative 1-D modelling and practical applications. The results demonstrate the advantages of data-based approaches while also highlighting some of the inherent limitations of 1-D modelling methods.

1. Introduction

Thermal Energy Storage (TES) systems play a crucial role in thermal networks, offering versatile functionality by efficiently managing the supply and demand of thermal energy. This practice provides environmental and economic benefits, including reduced CO_2 emissions, lower energy production costs and decreased operating expenses [1]. Despite the existence of various TES technologies with higher energy densities, such as those based on latent heat and phase change materials, hot water TES units remain prevalent due to their cost-effectiveness, simplicity, and favourable thermal characteristics [2]. Among sensible thermal energy storage systems, stratified thermal energy storage (SSTES) units have gained popularity since they generally reduce construction costs and are more space-efficient when compared to two-tank solutions [3]. Stratified stores, also known as thermocline thermal storages, take advantage of the temperature-induced difference in density of the stored fluid, ensuring minimal mixing between the hot and cold regions of the tank if the charging (or discharging) process is properly carried out [4]. Indeed, the separation region between the hot and cold zone of the storage, which entirely includes the steep temperature gradient, is known as thermocline. Several factors contribute to the destratification process, in which the thermocline thickens, including internal diffusion between adjacent fluid layers, convection with the surroundings, and axial conduction through the storage walls [5].

1.1. State of art of stratified thermal energy storage modelling

The numerical modelling of a thermocline thermal storage system is quite challenging. primary requirement is to accurately represent the thermocline, to correctly estimate the thickness of the region in which the steep temperature gradient is present. Indeed, the most common performance indicators for a stratified TES mainly account for this parameter [6]. However, when simulating complex energy networks, such as solar thermal power plants, the integration of numerically expensive TES models leads to unmanageable computational times. Therefore, it is crucial to optimize the trade-off between computational resources and the degree of accuracy desired for the modelling of thermocline store.

Over the past decades, several approaches to modelling stratified TES have been proposed in literature [7]. The more comprehensive models are developed in 3-D, where mass, momentum, and energy balance equations are solved on a 3-dimensional grid. Such models are designed to accurately represent the full range of phenomena occurring in the storage and are extremely computationally expensive. Therefore, 3-D models are usually employed to numerically test and optimize the design of either the store or the diffuser [8–9], or to study complex interdependent phenomena, such as transient cooling induced by heat losses through the mantle, as in [10]. On the other hand, 2-dimensional models retain a good degree of accuracy while being less computationally expensive, since most of the relevant phenomena that lead to destratification can be modelled in two dimensions with good approximation [11]. This is particularly true when the geometry of the store allows for an axisymmetric reduction, such as in [12–14]. Moreover, multi-dimensional models are often used to validate simplified models [15], which is also the case for the current study.

When studies involve either large scale energy network or optimization processes, 3-dimensional and 2-dimensional models carry an unbearable computational burden. In such cases, 1-dimensional models are preferable, although less accurate [16]. Nevertheless, it has been shown that assuming a uniform temperature along the radial axis can still result in a good degree of approximation with experimental data, especially in the case of cylindrical tanks with inlet and outlet sections located on the axis of symmetry [17], such as in the current study.

The goal of this research is to show how data-driven modelling can enhance the accuracy of 1-dimensional model for thermocline storages. To accomplish this, experimental data are collected from an industrial stratified storage that serves as heat depository in a solar thermal power plant and are used to thermally characterize and validate the numerical model. To characterize phenomena that could not be experimentally recorded, a previously validated 2-D model is deployed [13]. The results demonstrate the advantages of data-driven approaches while highlighting some of the limitations inherently associated with 1-D modelling methods.

2. Mathematical and numerical modelling

2.1. Reference system

The reference system used to characterize and validate the model is shown in figure 1. Data are retrieved every minute for one year, from November 2022 to October 2023. The system is located in Switzerland and is mainly composed by a thermocline storage used as a buffer tank in an industrial solar thermal power plant.

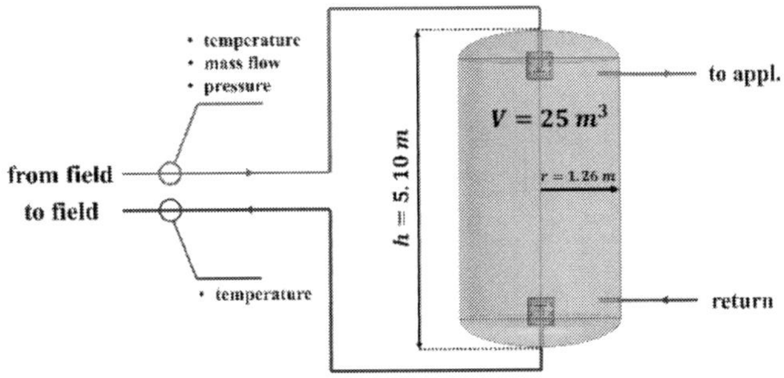

Figure 1. Schematization of the reference system, consisting of a thermocline storage (center) and connections to a production circuit (left) and a consumption circuit (right).

2.2. Mathematical formulation and numerical modelling

To develop a 1-dimensional schematization of a thermal energy storage system using water as HTF, a multi-node model is adopted. Equation (1), which describes the energy balance in the storage tank along an ascending vertical axis z, is discretized using a finite volume method, considering *N* layers of height Δz (figure 2).

$$\rho c_p A^S \frac{\partial T(z,t)}{\partial t} + \dot{m} c_p \frac{\partial T(z,t)}{\partial z} = A^S k \frac{\partial^2 T(z,t)}{\partial z^2} + U\,P\left(T_{amb}(t) - T(z,t)\right) \quad (1)$$

From left to right, the first term is the energy accumulation, the second term accounts for the enthalpy fluxes due to charge (or discharge) of the storage, the third term models diffusion phenomena occurring between the layers, and the last term represents the heat losses to the environment. In the equation, ρ ($kg\ m^{-3}$) is the density of pure water, c_p ($J\ kg^{-1}\ K^{-1}$) is the specific heat capacity, k ($W\ m^{-1}\ K^{-1}$) is the thermal conductivity, U ($W\ m^{-2}\ K^{-1}$) is the thermal conductance of the insulation layer, A^S (m^2) is the cross-sectional area of the storage, P (m) is the tank perimeter, and $\dot{m}$ ($kg\ s^{-1}$) is the mass flux either mass flux either entering, leaving, or both. In fact, the boundary conditions at the inlet and outlet sections of the storage depend on its operating conditions:

- Charge phase: $\left.\frac{\partial T(z,t)}{\partial z}\right|_{z=0} = 0; T|_{z=H} = T_{inlet}$;
- Discharge phase: $T|_{z=0} = T_{return}$; $\left.\frac{\partial T(z,t)}{\partial z}\right|_{z=H} = 0$;
- Idle time: $\left.\frac{\partial T(z,t)}{\partial z}\right|_{z=0} = 0; \left.\frac{\partial T(z,t)}{\partial z}\right|_{z=H} = 0$.

Where T_{inlet} and T_{return} are the temperatures of the inlet fluid and the return fluid, respectively, and $z=0$ and $z=H$ refer to the bottom and the top of the tank, respectively, since H (m) is the total height of the storage. It should be noted that, while idle phases are inherently defined as stand-alone, the same does not apply to charge and discharge phases, which can happen simultaneously. The discretization is accomplished using a finite volume method with a 1st order discretization in space and an implicit time formulation. After discretization, equation (1) transforms into equations (2) and (4) for the outlet and inlet nodes, respectively, and into equation (3) for the internal nodes:

$$U\,A^L(T_{amb} - T_1) + \frac{k_{eff}\,A^S}{\Delta z}(T_2 - T_1) + \dot{m}_c c_p (T_2 - T_1) + \dot{m}_d c_p (T_{return} - T_1) = \frac{\rho c_p A^S \Delta z}{\Delta t}(T_1^{t+1} - T_1^t) \quad (2)$$

$$UA^L(T_{amb} - T_i) + \frac{k_{eff}\,A^S}{\Delta z}(T_{i-1} - 2T_i + T_{i+1}) + \dot{m}_c c_p (T_{i+1} - T_i) + \dot{m}_d c_p (T_{i-1} - T_i) = \frac{\rho c_p A^S \Delta z}{\Delta t}(T_i^{t+1} - T_i^t) \quad (3)$$

$$U\,A^L(T_{amb}-T_N)+\frac{k_{eff}\,A^S}{\Delta z}(T_N-T_{N-1})+\dot{m}_c c_p(T_{inlet}-T_N)+\dot{m}_d c_p(T_N-T_{N-1})=\frac{\rho c_p A^S \Delta z}{\Delta t}(T_N^{t+1}-T_N^t) \tag{4}$$

where Δz (m) is the height of the single layer, $\dot{m}_c$ ($kg\ s^{-1}$) is the incoming mass flux during charging, $\dot{m}_d$ ($kg\ s^{-1}$) is the incoming mass flux in discharge phases, A^L (m^2) is the lateral surface area of the single layer, Δt (s) is the imposed time step of the model, and k_{eff} ($W\ m^{-1}\ K^{-1}$) is the effective thermal conductivity of material composing the layers, i.e., a weighted average of the thermal conductivity of pure water and stainless steel, the latter being the shell material (equation (5)). Although the shell cross-sectional surface area is significantly smaller than that of the water, given the stainless steel's higher thermal conductivity, this term cannot be neglected [18]. Moreover, it has been shown that the effect of axial heat conduction is particularly relevant in models that also account for idle phases [19]. However, both the shell thickness and the thermal conductance of the external insulation layer are unknown parameters, thus needing to be correctly determined to obtain a more precise reproduction of the system's real behaviour. The details of the determination process for these thermophysical quantities are provided in section (3.1). It should be noted that, in the left side of the previously presented equations, the time dependency of the variables is omitted for conciseness.

$$k_{eff}=k_f+k_s\left(\frac{r_{out}^2-r_{int}^2}{r_{out}^2}\right) \tag{5}$$

In which, r_{out} (m) is the radius of the storage including the shell thickness, excluding the outer insulation layer, r_{int} (m) is the radius which only includes the fluid dominium, while k_f ($W\,m^{-1}\,K^{-1}$) and k_s ($W\ m^{-1}\ K^{-1}$) are the thermal conductivities of the fluid (water) and solid (stainless steel) materials, respectively.

By dividing equation (3) for ($k_{eff}\ A^S\ \Delta z^{-1}$), the latter being the multiplier of the algebraic sum of adjacent nodal temperatures, equation (6) is derived.

$$(\mathrm{T}_{i-1}-2\mathrm{T}_i+\mathrm{T}_{i+1})+\frac{\dot{m}_c\,c_p\,\Delta z}{k_{eff}\,A^S}(T_{i+1}-T_i)+\frac{\dot{m}_d\,c_p\,\Delta z}{k_{eff}\,A^S}(T_{i-1}-T_i)+\frac{UA^L\Delta z}{k_{eff}A^S}(T_{amb}-T_i)=\frac{\rho c_p(\Delta z)^2}{\Delta t\,k_{eff}}(T_i^{t+1}-T_i^t) \tag{6}$$

Furthermore, the coefficients in equation (6) can be combined in numerical dimensionless parameters:

- $\frac{\dot{m}_c\,c_p\,\Delta z}{k_{eff}\,A^S}=\frac{v_c\,\rho\,c_p\,\Delta z}{k_{eff}}=\frac{v_c\,\Delta z}{\alpha}=Pe_z^c$
- $\frac{\dot{m}_d\,c_p\,\Delta z}{k_{eff}\,A^S}=\frac{v_d\,\rho\,c_p\,\Delta z}{k_{eff}}=\frac{v_d\,\Delta z}{\alpha}=Pe_z^d$
- $\frac{UA^L\Delta z}{k_{eff}A^S}=\frac{U\,\Delta z}{k_{eff}}=Bi_z$
- $\frac{\rho\,c_p\,(\Delta z)^2}{\Delta t\,k_{eff}}=\frac{(\Delta z)^2}{\alpha\,\Delta t}=(Fo_z)^{-1}$

Where Pe_z^c and Pe_z^d are the numerical Péclet number for the charge and discharge phases, respectively, Bi_z is the numerical Biot number, and Fo_z is the numerical Fourier number. In the previous correlations, α ($m^2\ s^{-1}$) is the weighted thermal diffusivity of water and steel, v_c ($m\ s^{-1}$) is the inlet velocity of the fluid during charge phases, and v_d ($m\,s^{-1}$) is the inlet velocity of the fluid during discharge phases. By replacing the aforementioned numerical dimensionless parameters in equation (6) and by highlighting the nodal temperature for the instant $(t+1)$, equation (7) can be derived.

$$T_i^{t+1}=T_i^t+Fo_z[(Pe_z^c+1)T_{i+1}+(Pe_z^d+1)T_{i-1}-(Pe_z^c+Pe_z^d+Bi_z+2)T_i+Bi_zT_{amb}] \tag{7}$$

The same operations can be applied to equation (2) and equation (4) to obtain a similar correlation which is valid for the inlet and outlet sections of the thermal storage, respectively. The model is developed under the assumption of thermal equilibrium, for each layer, between the store fluid and the wall. This assumption is supported by the presence of a large heat transfer coefficient between the fluid and the wall, the large conductivity of the metallic material, and the relatively small thickness of the latter. The assumed values of these thermophysical properties are constant and are evaluated at 50 °C, the latter being the temperature that minimizes the overall error induced by the assumption, given the range in which the system operates. The assumed values of the thermophysical are summarized in table 1.

Table 1. Thermophysical properties of both water (HTF) and stainless steel (wall), assumed as constant, used in the model.

water (50 °C)	value	unit
ρ	988.04	$kg\ m^{-3}$
c_p	4.182	$kJ\ kg^{-1}\ K^{-1}$
k_f	0.641	$W\ m^{-1}\ K^{-1}$
stainless steel (50 °C)	**value**	**unit**
k_s	15.00	$W\ m^{-1}\ K^{-1}$

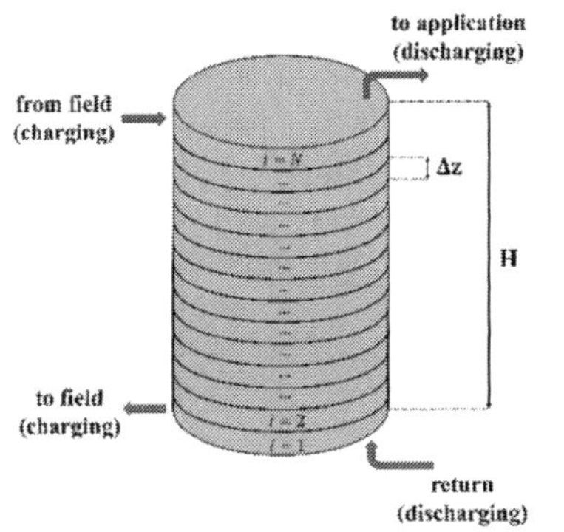

Figure 2. Finite volume method discretization.

2.3. Grid density and numerical diffusion

Numerical 1-D models inherently assume a uniform temperature in each layer of fluid, effectively implying infinite thermal diffusivity. This assumption can lead to a phenomenon known as numerical diffusion [20], where the temperature profile along the vertical axis of the storage becomes artificially smoothed. Additionally, numerical diffusion is highly sensitive to both the number of nodes (thus, the selected spatial step Δz) and the selected time step. This is shown in figure 3(a) and figure 3(b), for different numbers of nodes and time step values. Powell et al. [20] tested configurations with up to 10,000 layers but found that even this did not yield a fully converged solution, suggesting that an even greater number of nodes may be required to improve accuracy. However, computational time increases exponentially with the number of nodes and is inversely proportional to the time step. Therefore, a trade-off between these parameters must be made. For this study, a number of nodes (*N*) equal to 500 was selected, with a fixed time step (Δt) of 60 seconds.

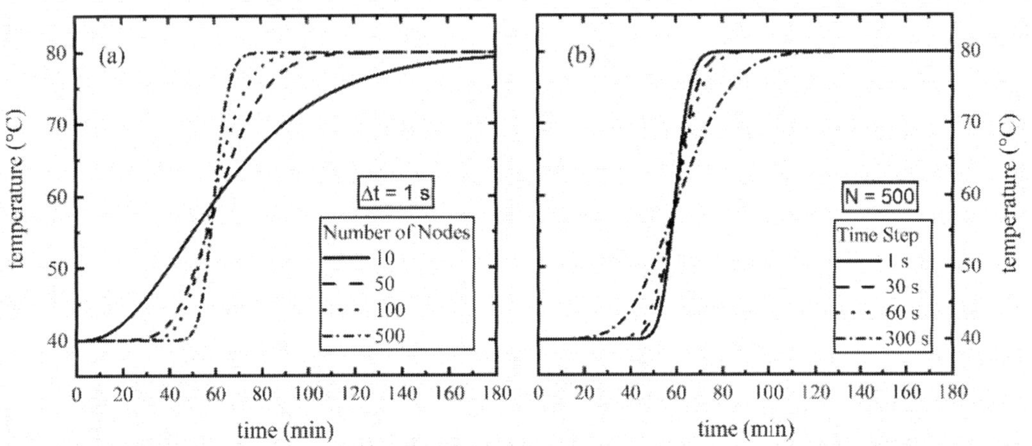

Figure 3. Temperature profiles for different values of **(a)** number of nodes and **(b)** value of time step. In each simulation, the storage starts at a uniform temperature of 40 °C, with an ambient temperature of 20 °C. The inlet mass flux is 2 $kg\ s^{-1}$.

2.4. *Buoyancy effect*

A critical aspect of modeling thermocline storage systems is accounting for mixing phenomena caused by the presence of two adjacent layers, with a colder fluid on top and a hotter fluid below. This mixing can occur for two primary reasons: (1) temperature inversion, where the initially hotter upper layer loses heat more rapidly to the environment, or (2) the inlet fluid being slightly cooler than the fluid already present in the top section of the tank, a common scenario in solar thermal power plants. However, in 1-dimensional models, buoyancy effects cannot be directly captured because the momentum equation is not solved.

Several numerical techniques have been proposed in the literature to simulate natural convection in 1-D models. The simplest approach involves using iterative procedures that reorganize the layers when a temperature inversion is detected. While this method is computationally efficient, it can significantly overestimate the amount of usable energy stored in the tank [21]. Other common methods include homogenizing the temperature around the inversion [22] or applying a turbulent diffusion coefficient that increases only when a temperature inversion is detected [23]. In this study, the max function approach proposed by [24] is employed.

3. Data-driven approach for the determination of thermophysical parameters

One limitation of the modelling process demonstrated so far is the need for precise knowledge of the thermophysical and geometrical properties of the system to be reproduced. While many of these properties were available, such as the temperature inside and outside the store, flow variables and others, it was not possible to obtain information on the thickness of the stainless-steel layer and the total thermal conductivity of the insulation layer. For the latter, however, both the thickness (250 mm) and the material (rock wool) are known. Nevertheless, relying solely on data sheet values for rock wool insulation panels can be misleading, as it has been shown how this type of material tends to deteriorate and lose some of its mechanical and thermal properties, mainly due to compression and moisture ingress [25, 27]. For these reasons, a data-driven approach is proposed in this study to reduce the error resulting from the incorrect estimation of these quantities.

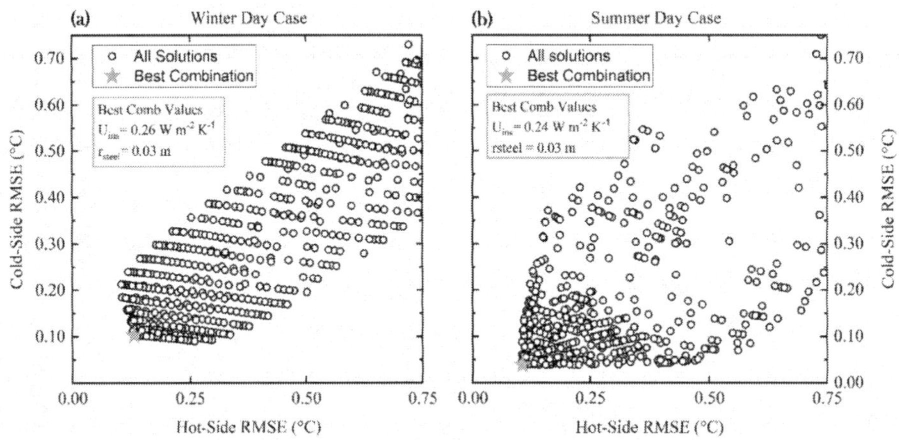

Figure 4. Data-driven optimization approach to reduce the modelling errors due to unknown thermal and geometrical characteristics of the reproduced system. The comparison is carried out both in **(a)** winter and **(b)** summer to minimize the error throughout the entire year.

To accurately quantify both the shell thickness and the insulation layer conductance, an optimization routine is employed. The first step of the process consists in the selection of two idle phases, in order to not account for the thermal perturbations deriving from either charging or discharging the store. The idle phases are chosen under sufficiently distinct ambient conditions, i.e., one during wintertime and the other during summer. Then, a multi-objective optimization algorithm is deployed. Realistic ranges of variation are chosen for both the decisional variables: the stainless-steel layer thickness (r_{steel}) is varied in the range of 10 *mm* – 250 *mm* with a step of 10 *mm*, while the conductance is varied in the range of 0.01 $W\,m^{-1}K^{-1}$ – 0.5 $W\,m^{-1}K^{-1}$, with a step size of 0.01 $W\,m^{-1}K^{-1}$. The objective functions are the Root Mean Square Errors (RMSE) (°C) evaluated comparing the dissipation-induced cooling curve of the real system and the one predicted by the model, for both the hot side (T_{hot}) and the cold side (T_{cold}) of the storage. This approach is chosen because the different volumes of water dissipate heat through the mantle and the insulation differently, since the ΔT with ambient temperature is up to 20 °C larger in one of the two cases. Therefore, the values obtained through the optimization process need to be able to reconcile both the hot and cold sides of the storage. The results of the investigation are presented in figure 4. In both the winter and the summer case, the value of r_{steel} that most accurately modelled the real system's behaviour converges to 30 *mm*. Additionally, in terms of thermal conductance, both optimization routines converged to very similar values, namely 0.26 $W\,m^{-1}K^{-1}$ and 0.24 $W\,m^{-1}K^{-1}$ for the winter and summer day, respectively. For this reason, a shell thickness of 30 *mm* is imposed, while the thermal conductance of the insulation layer is set to 0.25 $W\,m^{-1}K^{-1}$, a reasonable compromise between the winter and summer case results.

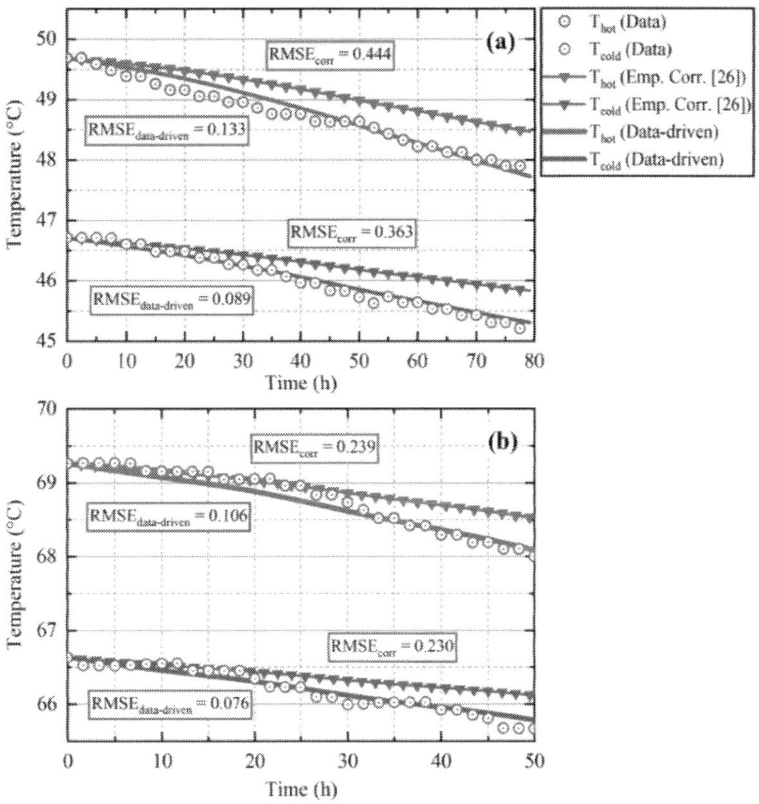

Figure 5. Comparison between results obtained using a correlation-based approach and a data-driven approach for the evaluation of the thermal conductance of the insulation layer of a thermocline storage during **(a)** a fall idle phase and **(b)** a springtime idle phase.

To validate the results obtained from the data-driven approach, the model is used to simulate a different set of idle phases, that occurred during fall and spring (figure 5). The results are also compared with those obtained using the same model, but employing an empirical correlation, described in [26], for evaluating of the averaged convective heat transfer coefficient, as per equation (8):

$$\overline{Nu} = 0.3 + \frac{0.62\,Re^{1/2}Pr^{1/3}}{[1 + (0.4/Pr)^{2/3}]^{1/4}}\left[1 + \left(\frac{Re}{282{,}000}\right)^{5/8}\right]^{4/5} \tag{8}$$

in which the Reynolds number is evaluated using the annual average wind velocity (4.4 $m\ s^{-1}$) for the area, while using the annual average temperature (9.5 °C) for the evaluation of the thermophysical properties of the air. On the other hand, the properties of the insulation layer were assumed to correspond solely to those of the rock wool, hence neglecting the outer thin metal layer. The datasheet value of thermal conductivity of the rockwool is equal to 0.037 $W\ m^{-1}\ K^{-1}$.

The values obtained using the data-driven approach result in a reduction of error in each simulated case, for both the cold and the hot side of the storage. While the correlation-based model consistently overestimates the insulating capability of the rock wool layer, the data-driven approach accounts for the performance degradation of the exterior insulation, reducing the RMSE values by up to four times. Furthermore, in the fall idle phase (figure 5(a)), over the course of 80 hours the correlation-based model accumulates a temperature difference of 0.83 °C on the hot side of the storage and 0.62 °C on the cold side. When these differences are integrated over a year, the model could significantly overestimate the amount of useful energy stored. This behaviour is also evident in the spring case (figure 5(b)), though less pronounced due to the higher ambient temperature, which mitigates heat dissipation.

4. Conclusions

This study examines the potential of data-driven methodologies for modelling thermal energy storage in one dimension. After the base model was characterized, a multi-objective optimization routine was deployed to determine the unknowns of the reference system. This approach has proven to be a valuable tool for significantly increasing the precision of low-dimensional models and more accurately predicting temperature evolution over long periods of component operation. The deviation from experimental data was reduced in each simulated scenario, and the model consistently predicted the outcome temperatures of idle phases, aiding in the accurate evaluation of useful storable energy.

Moreover, the same data-driven methodology can be used to estimate potential declines in the insulation capabilities of storage systems that have been operational for several years, enabling the degree of degradation within the insulating layer to be assessed.

5. References

[1] Jenkins N 2021 *Microgrids and local energy systems* (London: IntechOpen).
[2] Han YM, Wang RZ, and Dai YJ 2009 *Renew. Sustain. Energy Rev.* **13**(5) 1014.
[3] He Z, Qian Y, Xu C, Yang L, and Du X 2019 *Energy Procedia* **158** 4471.
[4] Sliwinski BJ, Mech AR, and Shih TS 1978 *6th International Heat Transfer Conf.*
[5] Stamps DW and Clark JA 1992 *Int. J. Heat Mass Transf.* **35**(3) 727.
[6] Haller MY, Cruickshank CA, Streicher W, Harrison SJ, Andersen E, and Furbo S 2009 *Sol. Energy* **83**(10) 1847.
[7] Chekifi T and Boukraa M 2023 *J. Energy Storage* **68** 107893.
[8] Hosseinnia SM, Akbari H, and Sorin M 2021 *J. Energy Storage* **40** 102682.
[9] Cònsul R, Rodríguez I, Pérez-Segarra CD, and Soria M 2004 *Sol. Energy* **77**(2) 179.
[10] Paing S, Anderson T, and Nates R 2019 *6th Asia-Pacific Solar Research Conf.*
[11] van Berkel J, Rindt CCM, and van Steenhoven AA 2002 *Int. J. Heat Mass Transf.* **45**(2) 343.
[12] Liang H, Liu L, Zhong Z, Gan Y, Wu JY, and Niu J 2022 *Appl. Energy* **310** 118526.
[13] Anacreonte AV, Musto M, Bianco N, Vitobello R, and Russo R 2024 *J. Phys. Conf. Ser.* **2685**(1) 012005.
[14] Kocijel L, Mrzljak V, and Glažar V 2020 *Energy* **194** 116878.
[15] De Césaro Oliveski R, Krenzinger A, and Vielmo HA 2003 *Sol. Energy* **75**(2) 121.
[16] Untrau A, Sochard S, Marias F, Reneaume JM, Le Roux GAC, and Serra S 2023 *Appl. Energy* **333** 120614.
[17] Zachár A 2020 *Sol. Energy* **195** 239.
[18] Shyu RJ, Lin JY, and Fang LJ 1989 *J. Sol. Energy Eng.* **111**(1) 54.
[19] Nelson JEB, Balakrishnan AR, and Srinivasa Murthy S 1999 *Appl. Therm. Eng.* **19**(1) 89.
[20] Powell KM and Edgar TF 2013 *Energy Convers. Manag.* **76** 865.
[21] Franke R 1997 *Sol. Energy* **60**(3-4) 171.
[22] Kleinbach EM, Beckman WA, and Klein SA 1993 *Sol. Energy* **50**(2) 155.
[23] Hawlader MNA, Bong TY, and Lee TS 2007 *Int. J. Sol. Energy* **6**(2) 119.
[24] Lago J, De Riddler F, Mazairac W, and De Schutter B 2019 *Appl. Energy* **248** 640.
[25] Nagy B, Simon TK, and Nemes R 2020 *J. Therm. Anal. Calorim.* **139**(1) 169.
[26] Bergman TL, Lavine AS, Incropera FP, and Dewitt DP 2011 *Fundamentals of Heat and Mass Transfer. Seventh Edition.* (Hoboken:Wiley) p 458.
[27] Abdou A and Budaiwi I 2013 *Constr. Build. Mater.* **43** 533.

Journal of Physics: Conference Series **2940** (2025) 012005 doi:10.1088/1742-6596/2940/1/012005

Electro-thermal model of a pouch cell battery considering thermal runaway

F Piccirillo[1,*], F Scognamiglio[1], M Iasiello[1], and W K S Chiu[2]

[1] Dipartimento di Ingegneria Industriale, Università degli Studi di Napoli Federico II, Piazzale Tecchio 80, 80125, Napoli, Italy
[2] Department of Mechanical Engineering, Univerity of Connecticut, 191 Auditorium Road, 06269, Storrs (CT), USA

* Corresponding author: francesco.piccirillo@unina.it

Abstract. Li-ion batteries are increasingly expanding their fields of application: automotive, electronic devices, and residential are just a few demonstrations of that. One of the most critical problems with batteries is overheating, which can lead to thermal runaway and a battery fire. For these reasons, the need for an accurate multiphysics model is growing and consequently an electro-thermal model of a pouch cell battery, including thermal runaway reactions, is proposed. It involves a two-dimensional electrical model for the cell potential distribution, and a three-dimensional thermal model to describe the temperature field. The two models are coupled because parameters such as the overpotential resistance and the thermal runaway reactions are temperature-dependent. The electrical model is based on the charge conservation theory, coupled with the ohmic law equations. The thermal model is governed by the heat diffusion equation with total heat generation calculated as the sum of three different parts: ohmic, irreversible and reversible, while thermal runaway occurrence is governed by an Arrhenius equation. After a comparison with literature data, numerical results in terms of temperature are shown, demonstrating a significant temperature rise during the thermal runaway process.

Nomenclature

Latin Symbols		τ	Time constant (s)
C	Capacity (Ah)	φ	Electric potential (V)
c_p	Heat capacity at constant pressure (J/kgK)	**Subscripts**	
d	Thickness (m)	*amb*	Ambient
h	Convective heat transfer coefficient (W/m^{2}K)	*av*	Average
I	Current (A)	*c*	Casing
J	Transversal current density (A/m^2)	*cc*	Current collector
k	Thermal conductivity (W/mK)	*e*	Electrode
N	Unit assembly	*G*	Global
R_A	Specific overpotential resistance (Ωm^2)	*irr*	Irreversible
Q	Heat generation (W/m^3)	*n*	negative
t	Time (s)	*ohm*	Ohmic
T	Temperature (K)	*p*	positive
U	Open-circuit potential (V)	*rev*	Reversible
Greek Symbols		*s*	Separator
ρ	Density (kg/m^3)	*tot*	Total
σ	Electrical conductivity (S/m)	*UA*	Unit assembly

1. Introduction

The battery thermal management problem is becoming increasingly central in the energy transition context. Estimating the temperature field and managing it allows one to improve the performance of battery-associated energy systems (vehicles, planes, electronic devices, and so on). Different models have been studied to correctly calculate the thermal generation of batteries. Thermal-electrochemical models calculate the heat generation from the chemical reactions occurring within the battery. Pseudo two-dimensional model (P2D) is one of the most utilized electrochemical models, and is based on Fuller, Doyle and Newman theory [1]. Different optimized versions have been studied, Gao et al. [2] proposed a control-oriented model, reducing computational time under dynamic-load simulations and providing better results in terms of temperature accuracy. Most of the electro-thermal models are based on Bernardi's theory [3], [4]. They simulate the chemical behavior with only electrical parameters, including the entropy coefficient, related to intercalation phenomena of the chemical reactions. The most important problems related to batteries are the aging and thermal runaway issues. Both are associated with overtemperature problems; indeed, it is important to study a proper thermal management system [5] in order to avoid these concerns. For this reason, a coupled electro-thermal model of a pouch cell battery is analyzed, studying its behavior during the thermal runaway process, through Arrhenius laws, due to overheating.

1.1. Thermal runaway

Thermal runaway happens when there is uncontrolled increase in temperature due to electric, thermal, or mechanical abuse. Thermal runaway processes involve several key steps:

- Initial decomposition of the solid electrolyte interface (SEI). It occurs at temperatures ranging from 90 to 120 °C.
- When SEI decomposes, it will not protect graphite, and the lithium intercalated in the anode will react with the electrolyte solvents above 120 °C.
- The separator melts at around 135 °C, causing the electrodes to short-circuit, breaking down the cathode at 170 °C.
- Electrolyte decomposes at 200 °C, emitting flammable gases.

The most serious scenario is thermal runaway cascading, in which the failure of a single cell triggers the failure of several neighboring cells, thus disrupting the entire battery pack. Experimental setups are constructed [6] and different thermal runaway models are analyzed [7], [8], also considering different cathode materials [9].

2. Methodology

The pouch cell electrical and thermal performance has been investigated through a coupled model. The aim is to evaluate the heat rate production from a two-dimensional electrical model, in order to predict the battery temperature field through a three-dimensional thermal model. A two-dimensional model is necessary for a pouch cell to properly characterize the potential gradients, which generate non-negligible heat generation as it relates to hotspot formation, whereas a three-dimensional model is necessary because of the lower thermal conductivity of the battery along the thickness direction compared with the other directions, which results in a marked temperature gradient. The behavior during thermal runaway was also simulated, by describing the total heat generation via Arrhenius equations, produced once specified temperature thresholds are exceeded. The battery analyzed is a 20 Ah Li-ion pouch cell with an NMC cathode (figure 1), the ePLB C020 from EIG [10]. It consists of a positive tab, a negative tab, an external case, and a core, composed of $N = 18$ unit assemblies in parallel between them. Each unit assembly is composed of a positive/negative current collector, two positive/negative electrodes, and two separators, with the core thickness $d_{core} = N(d_{cc,p} + 2d_{e,p} + 2d_s + 2d_{e,n} + d_{cc,n})$. All the layers are porous except for the current collectors, and they are inserted in a concentrated electrolyte liquid. The gap between the tabs and the case W_g is 23.5 mm and the case lengths D_1, D_2, and D_3 are, respectively, 15 mm, 11 mm, and 6 mm. Battery characteristics are taken from [10] while the geometric parameters and thermal characteristics of the components are listed in table 1.

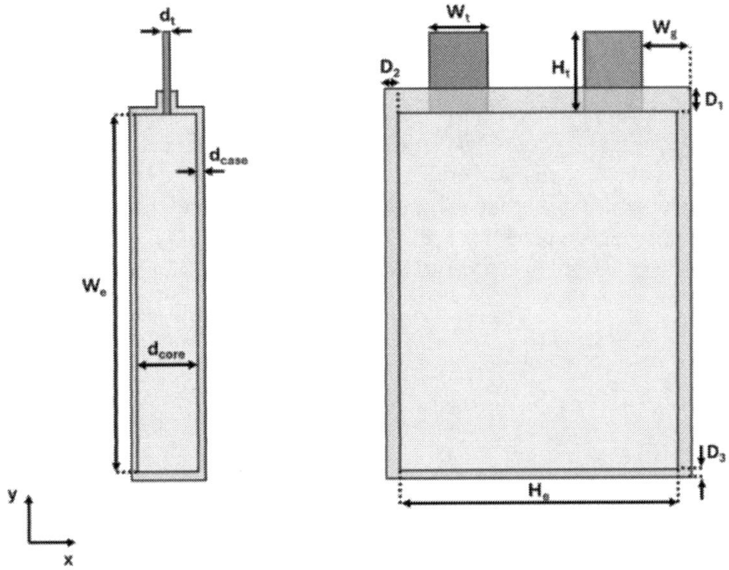

Figure 1. NMC pouch cell analyzed

Table 1. Geometric parameters and thermal characteristics

	Material	W_i [mm]	H_i [mm]	d_i [mm]	σ_i [MS/m]	ρ [kg/m^3]	c_p [J/kgK]	k [W/mK]
Positive current collector	Aluminum	125	195	21	37.8	2702	903	238
Negative current collector	Copper	125	195	12	59.6	8933	385	398
Positive electrode	NMC	125	195	70 x 10^{-3}	/	2895	1270	1.58
Negative electrode	Graphite	125	195	79 x 10^{-3}	/	1555	1437	1.04
Separator	Polymer	125	195	25 x 10^{-3}	/	1017	1978	0.34
Positive tab	Aluminum	30	36	0.3	37.8	2702	903	238
Negative tab	Copper	30	36	0.3	59.6	8933	385	398
Casing	Aluminum-laminated	/	/	16 x 10^{-2}	/	1150	1900	55.1

Different models have been developed based on this battery. Taheri et al. developed an electro-thermal model, without considering the generation due to potential gradients at the current collectors [11]; while Goutam et al. developed an electrothermal model similar to the one proposed, without simulating the thermal runaway process [12].

2.1. *Electrical model*

A two-dimensional electrical model, based only on the current collectors, is used to evaluate the potential fields, because their gradient leads to a concentrated heat generation which cannot be estimated without analyzing in at least two dimensions. A three-dimensional model is not examined because the potential distribution is assumed to be two-dimensional, as the thickness of the unit assembly layers is negligible compared to the width and length dimensions. The charge conservation theory, coupled with Ohm's law, is described in equations (1) and (2).

$$\sigma_{cc,i}\left[\frac{\partial}{\partial x}\left(\frac{\partial \varphi_i}{\partial x}\right)+\frac{\partial}{\partial y}\left(\frac{\partial \varphi_i}{\partial y}\right)\right]+\frac{n_i J}{d_{cc,i}}=0 \qquad i=p,n \tag{1}$$

$$J=\frac{U-(\varphi_p-\varphi_n)}{R_A} \tag{2}$$

Where n_i is the normal unit vector on each electrode surface, with $n_p = \{0,0,-1\}$ and $n_n = \{0,0,1\}$. The specific overpotential resistance R_A is the product of the unit assembly resistance and the area perpendicular to the transverse current density (current collector surface). The unit assembly resistance is N-times greater than the pouch cell global resistance, because all the unit assemblies are in parallel with each other (equation (3)). Pouch cell global resistance R_G is calculated by an equivalent circuit model (ECM). This is developed to synthesize the battery electrical functioning and a second-order Thévenin model is used with a resistance R_0 in series with two RC elements on each point of the current collectors, where the capacitive behavior is assessed by a time value τ. The global resistance value, unit assembly resistance and specific overpotential resistance are obtained through equations (3-5).

$$R_G = \left[R_0 + R_1\left(1 - e^{-\frac{t}{\tau_1}}\right) + R_2\left(1 - e^{-\frac{t}{\tau_2}}\right)\right] \tag{3}$$

$$R_{UA} = R_G \cdot N \tag{4}$$

$$R_A = R_{UA} \cdot A_{cc} \tag{5}$$

Both the open-circuit potential U and the global resistance R_G components are described by fittings made on experimental results, taken from [13], as a function of the state of charge (SOC) and temperature T. The equation for R_G is valid across the entire SOC range (0–1) and within the battery's operating temperature range (-30°C to 60°C). The SOC evolution is described by equation (6), which accounts for its dependence on the applied current I over time and the battery capacitance C, in Ah.

$$SOC(t) = SOC(t_0) - \frac{1}{3600C}\int_{t_0}^{t} I(t)dt \tag{6}$$

Finally, the boundary conditions are defined as in equations (7-9). Equation (7) is a Neumann condition on the positive current collector, with an in-plane current I_p exiting (discharge) or entering (charge) at the interface between the positive current collector and the positive tab. The current I is positive during discharge and negative during charge. Equation (8) is a Dirichlet condition on the negative current collector for the potential distribution reference, this is applied at the interface between the negative current collector and the negative tab. Equation (9) is a Neumann condition for the rest of the boundaries that are electrically insulated. Based on the results of the electrical model, it is possible to calculate the battery heat generation rate for the thermal model.

$$I_{cc} = -\sigma_{cc,p}\nabla\varphi_p = \frac{I}{N \cdot W_{tab} \cdot d_{cc,p}} \tag{7}$$

$$\varphi_n = 0 \tag{8}$$

$$\nabla\varphi_i = 0 \tag{9}$$

2.2. Thermal model

The three-dimensional thermal model allows the calculation of the battery temperature field. Four domains are analyzed: battery core, casing, positive tab, and negative tab. To avoid examining a very complex model, the core components are summarized into a single object, considering equivalent density, heat capacity and thermal conductivity, calculated as in [7]. Three dimensions are required as the temperature variation along the thickness is not negligible. The model is governed by the differential

Journal of Physics: Conference Series **2940** (2025) 012005 doi:10.1088/1742-6596/2940/1/012005

equation of conduction with total heat generation Q_{tot} calculated as the sum of three different parts in several domains, described in equations (10-15). Q_{irr} is the irreversible heat generation due to Joule effect, Q_{rev} represents the heat produced because of entropy changes related to intercalation phenomena of the chemical reactions, and it could be endothermic or exothermic depending on the SOC and on the charge/discharge mode; then Q_{ohm} is generated by the ohmic resistance due to the potential gradients. The heat generation behavior is different on each cell component: Q_{ohm} is generated only in the current collectors, while Q_{irr} and Q_{rev} in all the battery core. For the battery tabs, the heat generation is due to Q_{irr} and the casing is not a heat-generating component. All the heat generation terms are space- and time-dependent because they are function of temperature T, transversal current density J, and overpotential resistance R_A. The thermal conductivity k is direction-dependent [7].

$$\rho c_p \frac{\partial T}{\partial t} = k_{in}\frac{\partial^2 T}{\partial^2 x} + k_{in}\frac{\partial^2 T}{\partial^2 y} + k_{th}\frac{\partial^2 T}{\partial^2 z} + Q_{tot} \tag{10}$$

$$Q_{tot,core} = Q_{irr,core} + Q_{rev,core} + Q_{ohm,core} \tag{11}$$

$$Q_{irr,core} = \frac{J^2 \cdot R_A}{d_{core}} \tag{12}$$

$$Q_{ohm,\,core} = \frac{\sigma_{cc,p}(\nabla\varphi_p)^2}{d_{core}} d_{cc,p} + \frac{\sigma_{cc,n}(\nabla\varphi_n)^2}{d_{core}} d_{cc,n} \tag{13}$$

$$Q_{rev,core} = \frac{J \cdot T}{d_{core}}\frac{dU}{dT} \tag{14}$$

$$Q_{irr,tab,i} = \left(\frac{I}{A_{tab,i}}\right)^2 \frac{1}{\sigma_{tab,i}} \tag{15}$$

Where dU/dT is the entropic coefficient, and it is a SOC function. Its trend is summarized in a fitting equation based on experimental analysis, taken from [13]. Boundary condition for the thermal model is a convective heat dissipation through all battery surfaces.

$$Q_{conv} = h \cdot (T - T_{amb}) \tag{16}$$

The core heat generation is calculated through the electric model, then is transferred to the three-dimensional thermal model assuming that total heat is uniform along the thickness. Temperature field is calculated through the thermal model and then, in order to evaluate the temperature dependence of some parameters in the electrical model, it is transferred by taking an average along thickness in the z direction, as in equation (17).

$$T_{2D} = \frac{1}{d_{core}} \int_0^{d_{core}} T_{3D} dz \tag{17}$$

2.3. *Thermal runaway model*

The battery during the thermal runaway process is described by Arrhenius law, adopting the constant fuel model. It is divided into 4 stages, occurring at different process average onset temperatures, corresponding to the component decompositions within the battery, which in turn cause heat generations:

- SEI decomposition (90 °C < T_{av} < 120 °C)
- Anode decomposition (T_{av} > 120 °C)
- Cathode decomposition (T_{av} > 170 °C)

- Electrolyte decomposition (T_{av} > 200 °C)

Heat generation is governed by equation (18), and all the parameters are described in table 2 and are taken from [9] for an NMC cathode.

$$Q_{TR,i} = q_i A_i c_{i,o} \exp\left(-\frac{E_{a,i}}{RT}\right) \quad (18)$$

Table 2. Parameters for thermal runaway equations

	Density ρ_i [kg/m^3]	Heat of reaction q_i [J/kg]	Frequency factor A_i [1/s]	Activation energy $E_{a,I}$ [J/mol]	Concentration $C_{i,0}$
SEI decomposition	1.7×10^3	2.57×10^5	1.67×10^{15}	1.35×10^5	0.15
Anode decomposition	1.7×10^3	1.71×10^6	2.5×10^{13}	1.35×10^5	0.75
Cathode decomposition	1.29×10^3	7.9×10^5	2.25×10^{14}	1.274×10^3	0.04
Electrolyte decomposition	5×10^2	1.55×10^5	5.14×10^{25}	$2.74\ 10^5$	1

3. Results

Validation on the electro-thermal model, without considering thermal runaway reactions, is performed on the surface average temperature of the battery. A continuous *5C* and *2C* discharge, starting from the maximum charge, is simulated with a different heat transfer coefficient between core and tabs (h_{core} = *17.5 W/m²K* and h_{tab} = *12.5 W/m²K*) and an external temperature T_{amb} of 25 °C. The simulations results are compared with the experimental results from [12]. In figure 2 (a), the average surface temperature trends as a function of depth of discharge (DOD), are depicted for different discharges, obtaining a maximum error of 3.6 °C, and 2 °C for a *5C* and a *2C* discharge, respectively. In figure 2 (b), the temperature field for a *5C* discharge at the end of discharge is showed. The maximum temperature is observed at the core, and there is a marked temperature difference between the two terminals (3 °C); this is due to the different electrical conductivity σ, which results in higher heat generation Q_{irr} for the positive terminal, being made of aluminium. Two different meshes were used, one for the electrical model (1300 elements) and one for the thermal (76000 elements), where both timestep and grid consistency were verified with steps and grids doubled and halved.

Then a dynamic current is applied to the battery with a 5C charge/discharge cycle (500 s square cycle, with continuous second derivative and a transition zone of 30%, initial SOC = 100%) with reduced heat exchange with ambient (h = *2-5 W/m²K*) and an ambient temperature T_{amb} of 25 °C in order to simulate a battery overheating that leads to a thermal runaway. The average external battery temperature is shown in figure 3 (a). Once the battery reaches 90 °C, corresponding to the decomposition of SEI, the runaway equations are introduced. The curve tends to drop slightly at some places, as they correspond to the charge/discharge mode change phases. During the decomposition phase of the SEI, relevant differences are not noticed. By the time the anode decomposition temperature is reached, the temperature derivative dT/dt over time grows a lot until it tends to infinity. For this reason, the computational analysis is stopped. The trends differ between the two convective heat transfer coefficients considered. With an higher h, the battery tends to heat up slowly as it is able to exchange heat with the external environment more effectively, this results in an increase of about 15 minutes in the time before the SEI decomposition temperature is reached. It is important to design an efficient thermal management system also to increase the time before the thermal runaway process begins. Finally, figure 3 (b) depicts the surface temperature plot at the end of simulation for h = *2 W/m²K*, the battery core is almost at the same temperature as thermal runaway initiates. For this reason, the tabs are less heated than the core (about 20 and 15 °C for the negative and positive tabs, respectively) since thermal runaway was observed to occur only inside the battery.

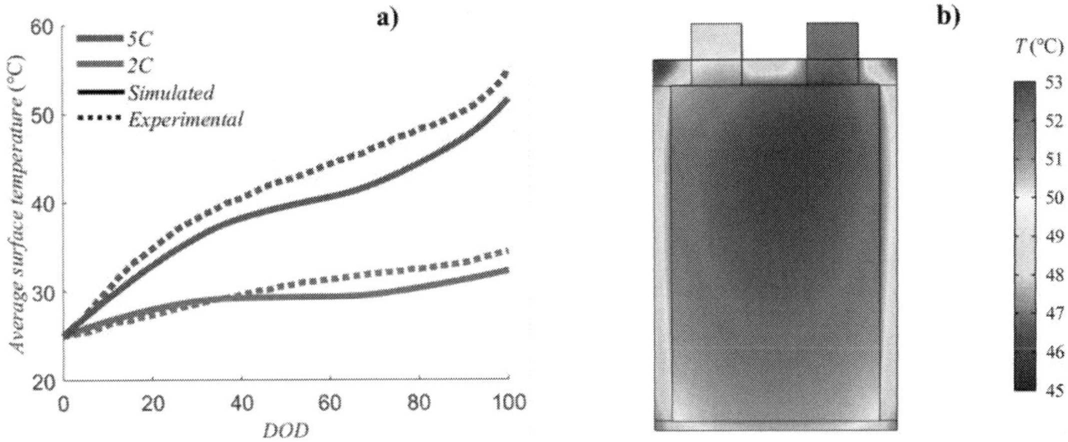

Figure 2. Surface temperature trends as a function of depth of discharge (a), and temperature field for a 5C discharge at the end of discharge (b)

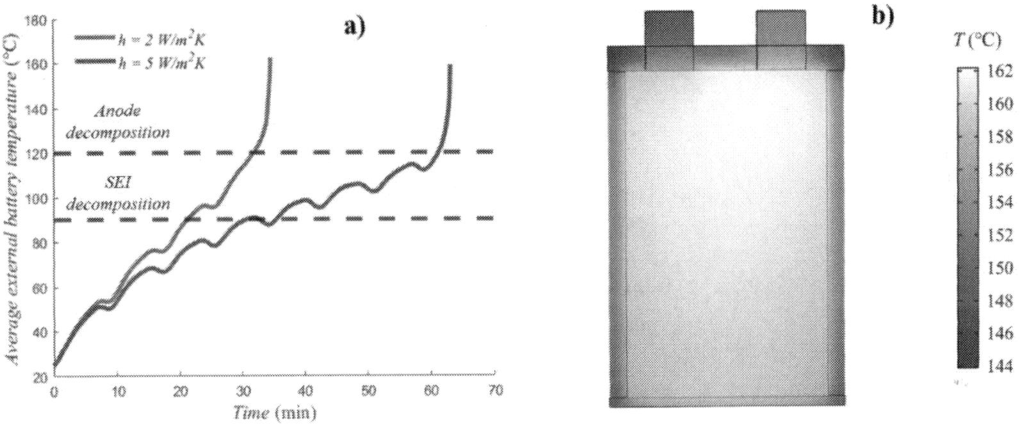

Figure 3. Average external battery temperature as a function of time during thermal runaway process (a), and battery surface temperature field at simulation end (b)

4. Conclusions

A coupled electro-thermal model of a pouch cell battery was analyzed. The model is two-dimensional for the electrical part, based on charge conservation theory, which allows the computation of the potential fields on the current collectors and the heat generated by the battery. The three-dimensional thermal model calculates the temperature field at each timestep. The equations describing the thermal runaway process are also modeled through the Arrhenius laws, whose parameters vary as a function of the temperature reached by the battery and therefore the degradation state. A validation based on average surface temperature against experimental data from [12], was investigated at two different discharge currents (2C and 5C), showing very similar trends between them. The temperature field at the end of a 5C discharge is also shown, noting that the maximum temperature is at the battery core. The poles show different temperatures (about 3 °C), because of the difference in thermal conductivity σ, resulting in different heat generation $Q_{irr,tab,i}$. A charge/discharge cycle is applied to the battery to simulate the thermal runaway process due to overheating. Low heat transfer coefficients h are used in order to emulate a lower heat exchange rate with ambient. After the battery achieves the SEI decomposition temperature, the temperature trend does not appear to vary much. However, once the anode decomposition temperature is exceeded, the temperature increases rapidly to infinity. For this reason, it may be difficult to be able to cool the battery after anode decomposition initiates. A higher convective

heat transfer coefficient *h* allows for a more gradual rise in temperature resulting in an increase of about 15 minutes in reaching the decomposition temperature of the SEI in the case of 5 W/m^2K instead of 2 W/m^2K. This suggests how important it is to properly design a battery thermal management system to increase the time before runaway is reached.
As possible future developments, a transition from a single battery to a battery pack is expected. In this way, different thermal management systems could be simulated to study the best solutions to avoid the thermal runaway phenomenon. A propagation model for the thermal runaway phenomenon in a battery pack is also under development.

5. References

[1] T. F. Fuller, M. Doyle, and J. Newman, "Simulation and Optimization of the Dual Lithium Ion Insertion Cell," *J Electrochem Soc*, vol. 141, no. 1, pp. 1–10, Jan. 1994, doi: 10.1149/1.2054684.
[2] Y. Gao, C. Zhu, X. Zhang, and B. Guo, "Implementation and evaluation of a practical electrochemical- thermal model of lithium-ion batteries for EV battery management system," *Energy*, vol. 221, p. 119688, Apr. 2021.
[3] D. Bernardi, E. Pawlikowski, and J. Newman, "A General Energy Balance for Battery Systems," *J Electrochem Soc*, vol. 132, no. 1, pp. 5–12, Jan. 1985.
[4] Y. wei Pan *et al.*, "A computational multi-node electro-thermal model for large prismatic lithium-ion batteries," *J Power Sources*, vol. 459, p. 228070, May 2020.
[5] W. Zichen and D. Changqing, "A comprehensive review on thermal management systems for power lithium-ion batteries," *Renewable and Sustainable Energy Reviews*, vol. 139, p. 110685, Apr. 2021.
[6] A. V. Shelkea *et al.*, "Combined numerical and experimental studies of 21700 lithium-ion battery thermal runaway induced by different thermal abuse," *Int J Heat Mass Transf*, vol. 194, p. 123099, Sep. 2022.
[7] M. Cianciullo, G. Vilardi, B. Mazzarotta, and R. Bubbico, "Simulation of the Thermal Runaway Onset in Li-Ion Cells—Influence of Cathode Materials and Operating Conditions," *Energies (Basel)*, vol. 15, no. 11, p. 4169.
[8] A. Melcher, C. Ziebert, M. Rohde, and H. Seifert, "Modeling and Simulation of the Thermal Runaway Behavior of Cylindrical Li-Ion Cells—Computing of Critical Parameters," *Energies (Basel)*, vol. 9, no. 4, p. 292, Apr. 2016.
[9] V. Talele *et al.*, "Computational modelling and statistical evaluation of thermal runaway safety regime response on lithium-ion battery with different cathodic chemistry and varying ambient condition," *International Communications in Heat and Mass Transfer*, vol. 146, p. 106907, Jul. 2023.
[10] Datasheet of ePLB C020 by EIG (Energy Innovation Group) at: http://www.liionbms.com/pdf/eig/ePLBC.pdf (accessed on 20th May 2024).
[11] Taheri, P., Yazdanpour, M., & Bahrami, M. (2013). Transient three-dimensional thermal model for batteries with thin electrodes. Journal of Power Sources, 243, 280-289.
[12] S. Goutam *et al.*, "Three-dimensional electro-thermal model of li-ion pouch cell: Analysis and comparison of cell design factors and model assumptions," *Appl Therm Eng*, vol. 126, pp. 796–808, Nov. 2017.
[13] F. Piccirillo, N. Bianco, L. P. Di Noia, P. Guerriero, and M. Iasiello, "Analysis of the Thermal Behavior of Li-Ion Pouch Battery Cell - Part I: Finite Element Simulations Including the Entropic Coefficient," in *2022 28th International Workshop on Thermal Investigations of ICs and Systems (THERMINIC)*, IEEE, Sep. 2022, pp. 1–6.

Journal of Physics: Conference Series **2940** (2025) 012006 doi:10.1088/1742-6596/2940/1/012006

Evaporation of arrays of single-component drops on hydrophobic and hydrophilic surfaces

S Tonini* and G E Cossali

Department of Engineering and Applied Sciences, University of Bergamo, Italy

* Corresponding author: simona.tonini@unibg.it

Abstract. The paper focuses on modelling the evaporation of arrays of single-component identical sessile droplets, a phenomenon encountered in various applications. An analytical approach is used and developed to treat ordered and scattered sessile drop arrays of any numerosity. An approximate analytical solution for the vapour field and evaporation fluxes from general arrays of sessile drops on any type of substrate (hydrophobic or hydrophilic) is found by a proper superposition of analytical solutions of single drop fields. Comparison with accurate numerical solutions and available experimental results validates the approach. An analysis of ordered and scattered square arrays from 4 to 49 drops on substrates with contact angles from 30° to 150°, is reported and used to study the dependence of the evaporation screening coefficients on array patterns.

1. Introduction

The interest of the research community in the topic of the evaporation of groups of sessile or pendant drops is quite recent compared to that of the evaporation of single sessile drops, as witnessed by the huge amount of analytical, experimental, and numerical investigations on the latter. Recent reviews on the topic [1-3] motivate this interest by the recognised importance in many applied and research fields. The many coexistent phenomena occurring in sessile and pendant drop evaporation, such as thermal Marangoni effects [4], contact line instabilities [5] and drop shape effect [6, 7], and the added complexity due to different modes of contact line motion (see, for example [6, 8, 9]) make this research topic scientifically challenging. The interest for applications is also broad, and some examples are micro/nano fabrication [10], DNA/RNA micro-arrays deposition on solid substrates [11] and inkjet printing [12] (see also [2, 13] for descriptions of many other applications). The study of multiple sessile drops evaporation is further complicated by the interaction of the evaporation fluxes from each drop, causing the so-called "shielding" effect: The increase of the vapour concentration due to the evaporation of neighbouring droplets decreases the evaporation rate of each droplet. New phenomena were also detected, mainly related to drop spacing, such as the inception of instability [14], modification of evaporation modes [15, 16] and drop deformation [17]. The classical analytical approach to studying evaporating sessile drop arrays is based on the well-supported assumption that the phenomenon is diffusion-limited, as for single drop evaporation, allowing a reduction of the complexity of the analysis. Since the interest in this topic is relatively recent, the number of analytical approaches to model the phenomenon is still limited. An analytical model for the evaporation of groups of thin drops (i.e. drops highly deformed by gravity) with a circular basis has been proposed in a recent study [18], and the comparison with measurements on ordered and scattered sessile drop arrays on hydrophilic surfaces [19] yielded positive matches. An analytical model for the evaporation of pairs and triplets of sessile drops on a line, for both hydrophilic and hydrophobic substrates, was proposed in [20]. A satisfactory agreement with the predictions of [18] was found for the case of very low contact angle, and a comparison with the experimental data of [13, 15, 21] yielded a satisfactory agreement. This paper proposes a relatively simple analytical model for the evaporation of sessile and pendant drop arrays of any numerosity and pattern and for any contact angle. The model is based on the superposition of the analytical solutions for single drop evaporation, considering the effect of Stefan flow and temperature-dependent thermophysical properties (by the approach developed in [23]).

2. Mathematical model

Sessile drop evaporation is a diffusion-controlled phenomenon that can be modelled assuming quasi-steady diffusion in the gas mixture. Steady-state species and energy conservation equations in the gas mixture can be written (neglecting viscous dissipation in the energy equation, as well as other minor terms assumed irrelevant for the present analysis) as follows:

$$\nabla_j N_j^{(p)} = 0; \qquad \rho U_j \nabla_j \left(c_p^{(1)} T\right) = \nabla_j \left(k_{mix} \nabla_j T\right) \tag{1}$$

where p=1 for the vapour species and $p = 0$ for the ambient gas, $N_j^{(p)} = N_j^{(T)} y^{(p)} - cD_{10} \nabla_j y^{(p)}$ are the molar fluxes, $N_j^{(T)} = N_j^{(1)} + N_j^{(0)}$ is the total molar flux, c is the molar density, D_{10} is the binary diffusion coefficient, $y^{(p)}$ is the molar fraction of the p-component, k_{mix} is the gaseous mixture conductivity, $c_p^{(1)}$ is the vapour specific heat capacity, ρ is the mass density of the mixture and U_j is the mass average velocity (that accounts for the Stefan flow, while external convection is neglected). It can be shown [23, 24] that the analytical solution of Equations (1) can be found, even when the temperature dependence of the gas mixture thermophysical properties are considered, once the solution of the Laplace problem

$$\nabla^2 \Phi = 0; \qquad \Phi(\text{surface}) = 1, \; \Phi(\text{infinity}) = 0 \tag{2}$$

is known for the particular geometry of the liquid surface, even for multiple drops. The vapour and the sensible heat fluxes can then be found as

$$n_j^{(1)} = Mm_j^{(1)} N_j^{(1)} = Mm_j^{(1)} f_n^{(1)} \nabla_j \Phi; \qquad q_j = f_T \nabla_j \Phi \tag{3}$$

where $Mm^{(1)}$ is the vapour molar mass, $f_n^{(1)}$ and f_T are constants that depend on the boundary conditions and on the gas mixture properties (see again [23, 24] for details) and evaporation and heat rates can be found by integrating the corresponding fluxes over the liquid surface. The analytical solution to the Laplace problem (Equation 2) for the geometry of a single sessile drop was first proposed by [22] under the form

$$\Phi = \int_0^\infty P_{i\tau-1/2}(\cosh \xi) \frac{\cosh\left[(\pi - \psi_0)\tau\right] \cosh(\psi \tau)}{\cosh(\pi \tau) \cosh(\psi_0 \tau)} d\tau, \tag{4}$$

where $P_{i\tau-1/2}(x)$ are the Legendre functions of first kind with complex index, which are sometimes called conical or Mehler functions [25]. In Equation (4), ξ and ψ are the toroidal coordinates (see [24]) and the equation $\psi = \psi_0$ represents a spherical cap, which is the shape of a sessile and pendant drop of a size smaller than the capillary length, $L_c = (\sigma/\rho g)^{1/2}$. The quantity ψ_0 is related to the contact angle by $\theta_c = \pi - \psi_0$. This solution was widely used to model the evaporation of sessile drops [26, 27] for constant thermophysical properties and nil Stefan flow. Later it was shown [23] that the same function Φ can be used as an auxiliary function to solve the problem of the quasi-steady evaporation of a sessile drop with non-nil Stefan flow and temperature dependence of the thermophysical properties. In [20], a method was used to model the evaporation of pairs and triplets of sessile drops, and here it is generalized for arrays of sessile drops of any pattern.

2.1. The array of sessile drops

The structure of an array of identical sessile drops can be defined by the substrate coordinates of each sessile drop centre (i.e. the centre of the circular drop base, see Figure 1) given as (x_k,y_k). The analytical solution of the Laplace problem (Equation 2) for a drop having (x_k,y_k) as coordinates of its centre can be found again from Equation (4) as $\Phi_k(x,y,z)=\Phi(x-x_k,y-y_k,z)$. It is easy to show that the function obtained as a linear combination of all the solutions $\Phi_k(x,y,z)$ for all the N_d drops in the array

$$\Phi_T(x,y,z)=\sum_{k=1}^{N_d}\gamma^{(k)}\Phi_k(x,y,z), \tag{5}$$

is also a solution to the Laplace equation, and vanishes at infinity. The Dirichlet condition Φ_T=1 on the liquid interface can be used to find the coefficients $\gamma^{(k)}$. Following the approach suggested in [20], the coefficients $\gamma^{(k)}$ can be chosen in such a way that Φ_T=1 on the drop apexes. This method yields an approximate solution to the Laplace problem, which was shown to be sufficiently accurate for drop pairs and triplets, even for a drop distance equal to twice the drop diameter: The relative error on the calculated evaporation rates (compared to an accurate numerical solution) was better than 7×10^{-3}. Since the coordinates of the p-th drop apex are (x_p,y_p,z_0+R_d), where z_0 is the z-coordinate of the drop centre (Figure 1), the condition $\Phi_T(x_p,y_p,z_0+R_d)=1$ yields the following system of N_d linear equations:

$$1=\sum_{k=1}^{N_d}\Phi_{pk}\gamma^{(k)}, \tag{6}$$

where $\Phi_{pk}=\Phi\ (x_p-x_k,y_p-y_k,z_0+R_d)$ is a symmetric matrix with unitary diagonal elements (since $\Phi(0,0,z_0+R_d)$=1). The evaporation rate of each drop in the array can be calculated as

$$m_{ev}^{(k)}=f_n^{(1)}\int_{\Omega_k}\nabla_n\Phi_T dS, \tag{7}$$

where Ω_k is a surface that surrounds only the k-th drop. It can be demonstrated (see [20]) that the coefficients $\gamma^{(k)}$ are always lower than 1, and they are equal to the ratio between the evaporation rate of the k-th drop in the array and that of the same drop evaporating alone; that is

$$\gamma^{(k)}=\frac{\int_{\Omega_k}\nabla_n\Phi_T dS}{\int_{\Omega_k}\nabla_n\Phi_k dS}=\frac{m_{ev}^{(k)}}{m_{ev,is}^{(k)}} \tag{8}$$

The coefficients $\gamma^{(k)}$ are then the so-called *screening coefficients*, and they can be used to quantify the effect of the presence of the other drops on the evaporation characteristics of each drop in the array.

2.2. Validation of the model

The accuracy of the model was tested against accurate numerical solutions. The commercial code COMSOL Multiphysics® was used to solve the problem set by Equation (2) for multiple sessile drops with different contact angles. A 3D axis-symmetric unstructured grid with triangular tessellation of about 2.2×10^5 cells was used for the finite element discretisation approach, setting the computational domain as a quarter of a sphere (see Figure 1). The size of the domain (the radius R_∞) was chosen to be equal to 15 times the minimum distance between the drops, L (see also [20], [28]). An iterative procedure was used to mitigate the problem set by the finiteness of the domain, by substituting the Dirichlet condition at infinity with an equivalent condition on the outer boundary, and tests were performed for many drop shapes and conditions. Grid independence analysis was performed (see again [20], [28]). The integration of the fluxes was performed both on the liquid surface and on a spherical surface containing the drop as a double check. To assess the accuracy of the numerical model, the numerical calculations were first compared with exact analytical solutions for single drops on hydrophilic and hydrophobic substrates and for twin sessile drops with θ_c= 90°, obtaining a numerical relative error on the evaporation rates always lower than 0.23%. The numerical model was then used as a benchmark to validate the analytical model for various arrays of droplets.

Journal of Physics: Conference Series **2940** (2025) 012006 doi:10.1088/1742-6596/2940/1/012006

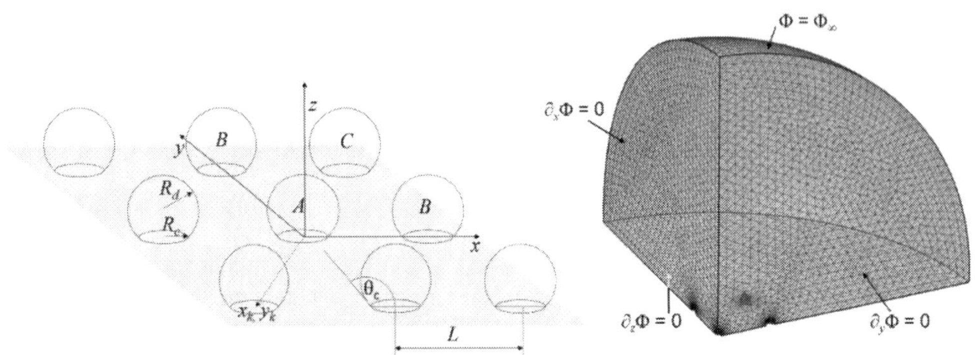

Figure 1. (Left) Sketch of a 3 × 3 drop array with the definition of the drop coordinates, radii and contact angle; (right) sample of the computational grid with the imposed boundary conditions.

Ordered arrays with four and nine droplets were analysed, varying the minimum drop distance from $3.8R_d$ up to $11.4R_d$, with contact angles equal to 60°, 90° and 120°. Figure 2 reports the screening coefficient of one of the four identical drops in the 2 × 2 array and of three of the nine drops of the 3 × 3 array (refer to Figure 1 for a schematic of the 3×3 array). The innermost drop (drop A) is the most screened one, while the drop C on the corner is the least affected by the presence of the neighbouring drops. The comparison between the numerical and analytical model predictions shows very good accuracy, with a maximum difference of about 1.0×10^{-2}, 3.1×10^{-3}, and 5.9×10^{-3}, for drops A, B and C, respectively (see Figure 1), while the average differences are equal to 3.4×10^{-3}, 1.4×10^{-3} and 2.2×10^{-3}, for drops A, B and C, respectively.

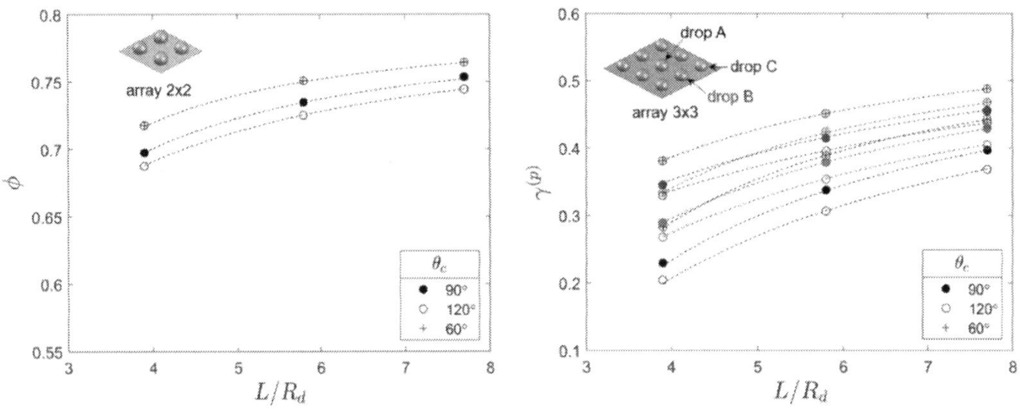

Figure 2. Screening coefficients as a function of the drop non-dimensional distance L/R_d for (left) 2 × 2 and (right) 3 × 3 ordered drop arrays on different solid substrates. The dashed curves correspond to the results from the analytical model, while the symbols refer to the numerical model calculations.

Figure 3 (left) reports a summary of the comparison between the values of the screening coefficient predicted by the model and obtained from experimental measurements, for two and three drops on a line (see [20] for details). In [15], measurements on the evaporation history of pairs of deionized water sessile drops on hydrophobic substrates (polydimethylsiloxane, PDMS) in air at ambient conditions and initial contact angle of 115° were reported for three drop interdistances L. Since the non-dimensional value $L/2R_d$ changes due to the change in drop size, Figure 3 (left) shows the model predictions for both initial and final contact angles (lines) and the experimental values (open triangles). Considering the reported variability of the measured data and the simplifying assumptions of the proposed model, this comparison may be considered satisfactory.

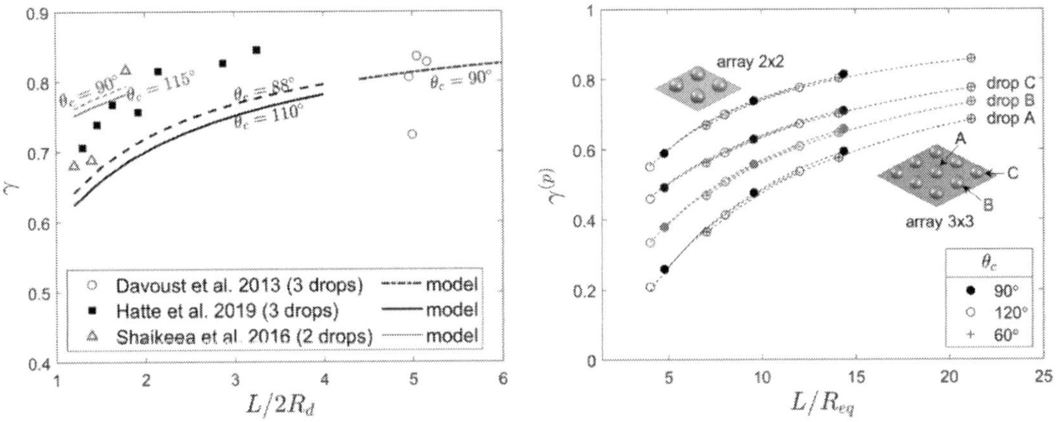

Figure 3. (Left) Experimental and model predictions of the screening coefficients vs. non-dimensional distance $L/(2R_d)$, for 3-drop linear array [21], [13] and drop pairs [15]. (Right) Screening coefficients vs drop non-dimensional distance L/R_{eq}, for 2×2 and 3×3 ordered drop arrays on different substrates.

Measurements on the evaporation characteristics of the central drop in a triplet of sessile drops on PDMS with an initial contact angle of $\theta_c = 110° \pm 2°$ were reported in [13]. The angle was observed to decrease, reaching about 88° at the end of evaporation. Figure 3 (left) shows the experimental values (solid squares) and those predicted by the model (lines) for both initial and final contact angles. Despite an evident underestimation of the values of the screening coefficient by the model, the results of the comparison can be considered acceptable. Finally, the experiments reported in [21], where the contact angle was kept constant at 90° by electrowetting methods, were used to calculate the screening coefficient for the central drop of a linear sessile drop triplet. The data (open circles) reported in Figure 3 (left) also fit the predictions (line) in this case. Despite the observed discrepancies, the comparison of the analytical prediction to the experimental values can be considered satisfactory overall.

3. Results and discussion

The vapour field around an ensemble of drops is the result of an interaction of the evaporating fluxes from each drop that reduces the total evaporation rate with respect to non-interacting drops of the same size and under the same conditions. The screening coefficient of each drop, $\gamma^{(p)}$, does not depend on the thermophysical properties, the evaporating conditions, and the drop size, but it depends on the wettability, drop separation, numerosity of the array and position inside the array. Figure 4a shows the screening coefficients of each drop in an ordered array of nine drops, as calculated by applying the present analytical model, and it is seen that the drop in the innermost position has the smallest screening coefficient, while those on the corners, which are less screened by the other ones, have the largest values. This also happens for non-ordered arrays; that is arrays where the drops are not positioned on a square grid, as seen in Figure 4b. The screening coefficients also depend on the compactness of the array. Figures 4c and 4d show the screening coefficients of arrays of the same numerosity (N_{dr}=16) on the same substrate, but with different drop interdistances. The screening coefficients are larger for the less compact array. To quantify these observations, the minimum value γ_{min}, the maximum value $\gamma_{\max}$, and the average value γ_{ave} of the screening coefficients in square arrays (i.e. arrays made by $n \times n$ drops) were calculated for different arrays numerosity and substrate wettability. As observed in [20], for drop pairs and triplets, the use of a non-dimensional distance, defined as the ratio between the drop interdistance and the volume equivalent radius, $R_{eq} = (3V/4\pi)^{1/3}$, is useful to collapse results from linear arrays of different compactness. As an example, the results of Figure 2 are reported in Figure 3b using the quantity L/R_{eq} instead of L/R_d. It is evident that there must exist a relationship between the drop screening coefficient and the drop non-dimensional distance L/R_{eq}, almost independent of the surface wettability. In the following, the non-dimensional distance from the two closest drops in the ordered array (see Figure 1) will then be defined as $\hat{L} = L/R_{eq}$ and used as a measure of the array compactness. The value of the average screening coefficient increases by increasing L, as observed above, but using $\hat{L} = L/R_{eq}$ the results from different contact angles collapse almost on a single curve for any given array numerosity, N_{dr}, a result that cannot be obtained using other definitions of compactness.

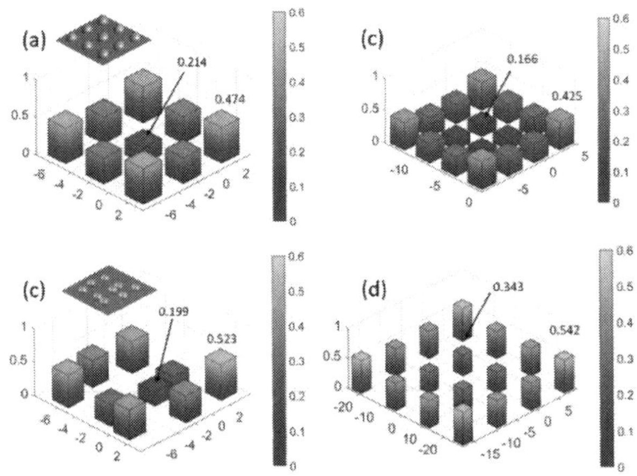

Figure 4. Values of $\gamma^{(P)}$ for substrates with θ_c = 120° for (a) a 3 × 3 square array with $L = 4R_d$, (b) a 9-drop scattered array with average L_{ave}=4R_d; (c) a 4 × 4 square array with $L = 4R_d$, and (d) a 4 × 4 square array with L= 8R_d. The minimum and maximum values of $\gamma^{(P)}$ are reported on the images.

Figure 5 illustrates the results obtained for four values of the compactness parameter, and six values of array numerosity from N_{dr}=4 (2×2 arrays) to N_{dr}=49 (7×7 arrays).

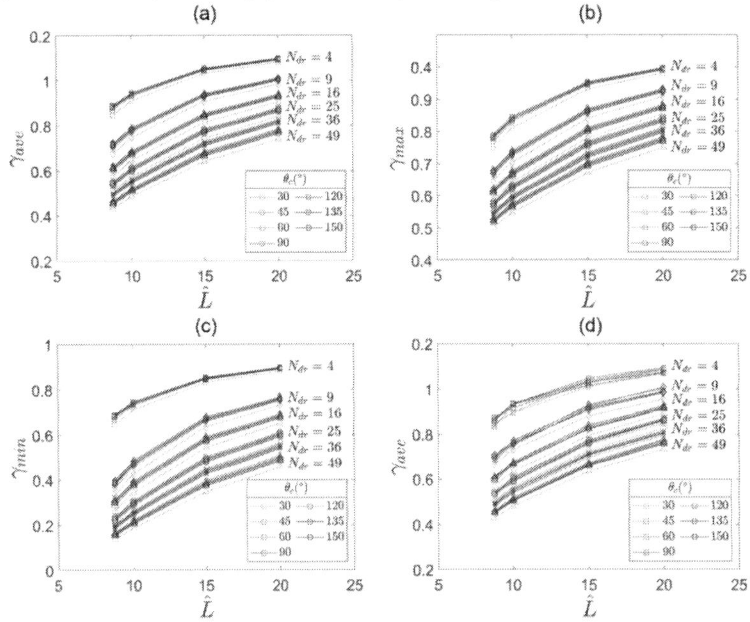

Figure 5. Screening coefficients for ordered (a: average value, b: minimum value, c: maximum value) and scattered (d: average value) drop arrays as a function of the drop non-dimensional distance. Drop numerosity from 4 to 49. Contact angle θ_c from 30° to 150°.

An increase in N_{dr} causes a decrease in the screening coefficients. A similar behaviour can be observed on the values of minimum and maximum screening coefficients in the drop array, also reported on Figure 5(b-c). A larger number of drops with the same interdistance causes a larger screening on the evaporation of the innermost drops while also decreasing the value of $\gamma^{(k)}$ on the outer drops since those drops are also screened by a larger number of drops into the array. The minimum value of the screening coefficient can reach, for arrays of larger numerosity, relatively low values (Figure 5b). Again, the use of $\hat{L} = L/R_{eq}$ collapses the results for different substrate wettability on almost single curves, and a decrease of the compactness (increase of L) causes an increase of γ_{max} and γ_{min}, independent of the substrate wettability. To analyse the effect of an irregular distribution of drops in an array, scattered arrays were considered. From each ordered array, a scattered array was generated by randomly shifting each drop, with an average shift amplitude of about 50% of the structured array interdistance (see Figure 4b). Care was taken to exclude all shifts that cause an overlapping of drops. The values of the screening coefficients of each drop were evaluated, for a range of contact angles from θ_c=30° to θ_c=150°, numerosity from 4 to 49 drops and different average drop interdistances. Since these arrays are not ordered, a new definition of the compactness is needed. In [29], the value of the variance of drop coordinates, defined as

$$\sigma_c^2 = \sum_{p=1}^{N_{dr}} \left| x_j^{(p)} - \bar{x}_j \right| \Big/ N_{dr}, \tag{9}$$

was used to define the average interdistance among floating drops. Following that path, in the present investigation, the compactness of the scattered arrays was defined as

$$\hat{L} = \frac{\sigma_c}{R_{eq}} \frac{\sqrt{6}}{(N_{dr}-1)^{1/2}}, \tag{10}$$

based on the observation that for regular $n \times n = N_{dr}$ square arrays the value of σ_c can be directly calculated, yielding

$$\sigma_c = L\left(n^2-1\right)^{1/2} \Big/ \sqrt{6}, \tag{11}$$

and then the definition (Equation 10) generalises this result. The values of γ_{ave} for scattered arrays on the same range of numerosity and substrate wettability are shown in Figure 5d. Using the compactness defined by Equation (10) almost collapses all the results on single curves, one for each value of N_{dr}. A comparison of the values of γ_{ave} for ordered and scattered arrays with the same compactness, numerosity, and contact angle is reported in Figure 6a. The average scattering of the data along the bisector is better than 1.6%.

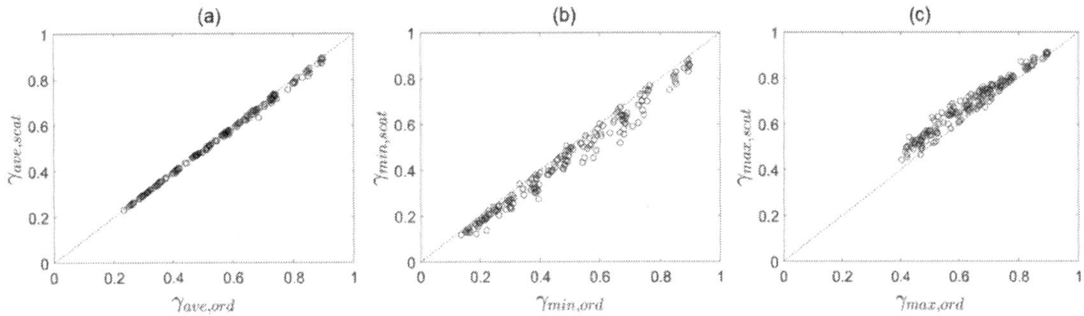

Figure 6. Comparison of the (a) average, (b) minimum, and (c) maximum values of the screening coefficient for square ordered and scattered drop arrays.

The values of minimum and maximum values of the screening coefficient in the array were also evaluated and the comparison between ordered and scattered arrays is shown in Figure 6(b-c). The data are more scattered around the bisector, as expected. The average discrepancy equals about 6.5% and 5.5% for γ_{min} and γ_{max} respectively. Moreover, the inequalities $\gamma_{min,scat} < \gamma_{min,ord}$ and $\gamma_{max,scat} > \gamma_{max,ord}$ are expected and observed. This shows that the ordered array is the one where the non-uniformity of the screening coefficient distribution is the lowest.

4. Conclusions

An analytical model for predicting the evaporation characteristics of arrays of identical sessile and pendant drops with general disposition on substrates of any wettability was developed. The application to square ordered and scattered arrays was analysed in detail. The screening coefficient was used to quantify the effect of the other drops on the evaporation characteristics of each drop in the array. The model compared well with accurate numerical solutions of the same problem, and it was validated with available experimental measurements on linear arrays. The non-dimensionalisation of the drop distance based on the drop equivalent radius allows one to compact the results from substrates of different wettability (θ_c=30°-150°). The array numerosity (from 4 to 49) was found to decrease the average screening coefficient and its minimum and maximum values, almost independently of the substrate wettability. The definition of a drop interdistance based on the variance of the drop coordinates on the substrate allows one to find a strong similarity between the evaporation characteristics of ordered and scattered arrays. In principle, the model can be extended to non-square arrays, but a different definition of the average drop interdistance may be necessary.

5. References

[1] Erbil H Y 2012 *Adv. Colloid Interfac.* **170** 1-2 67-86.
[2] Brutin D and Starov V 2018 *Chem. Soc. Rev.* **47** 2 558-585.
[3] Gurrala P, Balusamy S, Banerjee S and Sahu K C, 2021 *Fluid Dyn. Mater. Proc.* **17** 2 253-284.
[4] van Gaalen R T, Diddens C, Wijshoff H M A and Kuerten J G M, 2021 *J. Colloid Inter. Sci.* **584** 622-633.
[5] Murisic N and Kondic L 2011 *J. Fluid Mech.* **679** 219-246.
[6] Van Der Heijden T W, Darhuber A A and Van Der Schoot P 2018 *Langmuir* **34** 41 12471-12481.
[7] Tonini S and Cossali G E 2024 *Physics of Fluids* **36** 2 027131.
[8] Picknett R G and Bexon R 1977 *J. Colloid Inter. Sci.* **61** 2 336-350.
[9] Shanahan M E R 1995 *Langmuir* **11** 3 1041-1043.
[10] Xia D and Brueck S R J 2008 *Nano Letters* **8** 9 2819-2824.
[11] Chopra M, Li L, Hu H, Burns M A and Larson R G 2003 *J. Rheol.* **47** 5 1111-1132.
[12] Park J and Moon J 2006 *Langmuir* **22** 8 3506-3513.
[13] Hatte S, Pandey K, Chakraborty S and Basu S 2019 *J. Fluid Mech.* **866** 61-81.
[14] Schaefle C, Bechinger C, Rinn B, David C and Leiderer P 1999 *Phys. Rev. Lett.* **83** 25 5302-5305
[15] Shaikeea A J D and Basu S 2016 *Langmuir* **32** 1309-1318
[16] Carrier O, Shahidzadeh-Bonn N, Zargar R, Aytouna M, Habibi M, Eggers J and Bonn D 2016 *J. Fluid Mech.* **798** 774-786.
[17] Khilifi D, Foudhil W, Fahem K, Harmand S and Ben J S 2019 *Therm. Sci.* **23** 2 B 1105-1114.
[18] Wray A W, Duffy B R and Wilson S K 2020 *J. Fluid Mech.* **884.**
[19] Edwards A M, Cater J, Kilbride J J, Le Minter P, Brown, C V, Fairhurst D J and Ouali F F 2021 *App. Phys. Lett.* **119** 15 151601.
[20] Tonini S and Cossali G E 2022 *Phys. Rev. E* **105** 054803.
[21] Davoust L. and Theisen J 2013 *Sensors and Actuators B:Chemical* **189** 157-164,
[22] Lebedev N N 1972 *Special Functions and Their Applications* (Dover Publications).
[23] Cossali G E and Tonini S 2019 *Int. Jour. Heat Mass Tran.* **138** 1166-1177.
[24] Cossali G E and Tonini S 2021 *Drop Heating and Evaporation: Analytical Solutions in Curvilinear Coordinate Systems* (Springer).
[25] Olver F W J, Lozier D W, Boisvert R F and Clark C W, Eds. 2010 *NIST Handbook of Mathematical Functions* (Cambridge University Press).
[26] Popov Y 2005 *Phys. Rev. E* **71** 036313.
[27] Chini S F, Amirfazli A 2017 *Adv. Colloid Interfac.* **243** 121-128.
[28] Tonini S and Cossali G E 2023 *Phys. Fluids* **35** 3.
[29] Cossali G E and Tonini S 2020 *Int. J. of Heat Mass Transfer* **162** 120315.

Journal of Physics: Conference Series **2940** (2025) 012007 doi:10.1088/1742-6596/2940/1/012007

Influence of thermal buoyancy on heat transfer in spacer-filled channels for Membrane Distillation

M Ciofalo, N Cancilla*, A Cipollina, A Tamburini and G Micale

Dipartimento di Ingegneria, Università degli Studi di Palermo, Viale delle Scienze Ed. 6, 90128 Palermo, Italy

* Corresponding author: nunzio.cancilla@unipa.it

Abstract. Numerical results are discussed for the flow in a horizontal plane channel filled with a novel sphere-rod spacer and exchanging heat from both the top and the bottom sides. Direct Numerical Simulations (DNS) are compared with RANS ones based on different turbulence models in the Reynolds number range 100~2000. Preliminary comparisons for non-buoyant flow show that models using wall functions perform poorly, grossly overpredicting Nusselt numbers, while ω-based models resolving the viscous-conductive sublayer all yield satisfactory results. In the presence of buoyancy, simulations using either DNS or the k-ω model yield a thermal asymmetry between top and bottom wall, confirmed by experiments and related to the stable or unstable thermal stratification occurring in the lower and upper layers of the channel. The asymmetry, large at low Re, becomes negligible for Re≥1000. The Spalart-Allmaras model yields satisfactory results in the absence of buoyancy but grossly overpredicts Nu in buoyant flows.

1. Introduction

In Membrane Distillation (MD) two fluids at different temperatures flow in two channels, feed (hot) and permeate (cold), separated by a hydrophobic membrane. The temperature difference between the fluids creates a trans-membrane vapour pressure difference among the channels, which is the driving force for the separation. The membrane allows the passage of vapour from the feed to the permeate, where it condenses, while rejecting the liquid and the non-volatile components (e.g., salts).

In MD, spacers are commonly employed both to support the membranes and to promote mixing, thus minimizing temperature polarization. Their optimum design requires a knowledge of the distribution of the local heat transfer coefficient and of its dependence on Reynolds number, spacer geometry and orientation between spacers and flow. In previous work [1,2,3] some of the present authors used a combination of experimental measurements (based on Thermochromic Liquid Crystals and digital image processing) and numerical simulations (including direct numerical simulations of turbulence) to study the distribution of the local heat transfer coefficient and the overall performances of different spacer configurations. Comparative results guided the choice of the best turbulence models in the Reynolds number range expected in MD (~200-2000) [4].

Recently, first Koutsou and Karabelas [5] and later Chong *et al.* [6] investigated by CFD a novel spacer geometry for mixing promotion (Figure 1). It consists of spheres connected by cylindrical rods so that the membrane-spacer contact is limited to isolated spots. Preliminary, TLC-based, experimental results were obtained for these spacers by some of the present authors [7].

Here, the results of direct numerical simulations (DNS) and RANS simulations using different turbulence models are presented for this spacer geometry under laminar to early turbulent flow conditions (Re≈100-2000). Two-side heat transfer is assumed. The influence of thermal buoyancy on heat transfer is also assessed.

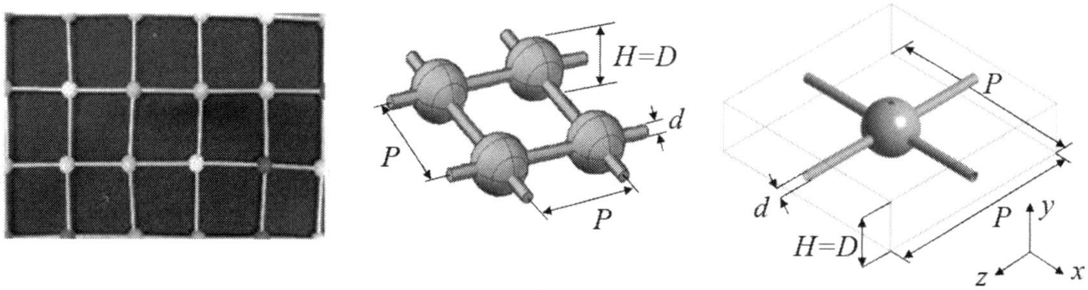

(a) (b) (c)

Figure 1. Sphere-rod spacers. (a) real spacer; (b) Sketch of the geometry; (c) computational domain. Nomenclature: H = channel height; D = sphere diameter; d = filament diameter; P = pitch.

2. Models and methods

2.1 Computational domain, governing equations and boundary conditions

The computational domain, representing the repetitive unit cell of a streamwise and spanwise indefinite array, is shown in Figure 1(c). The direction of the axes is indicated. The membranes are assumed to lie flat and horizontal, i.e. orthogonal to the gravity vector, as in the experiments [7].

Simulations were carried out under the assumption of fully developed flow and thermal field using the unit-cell approach, described in detail in previous work [2] ("Supplementary Material" section). The variables simulated were the velocity **u** and the periodic components p of pressure and T of temperature, so that translational periodicity could be imposed to the flow and thermal fields between opposite boundaries of the computational domains.

The governing equations were the continuity, momentum and energy ones for a Newtonian incompressible fluid with the Boussinesq approximation for buoyancy. In Cartesian tensor notation:

$$\frac{\partial u_j}{\partial x_j}=0 \tag{1}$$

$$\frac{\partial \rho u_i}{\partial t}+\frac{\partial \rho u_j u_i}{\partial x_j}=-\frac{\partial p}{\partial x_i}+\frac{\partial}{\partial x_j}\mu\frac{\partial u_i}{\partial x_j}-\rho g_i\beta\left(T-T_0\right)+G_i \tag{2}$$

$$\frac{\partial \rho c_p T}{\partial t}+\frac{\partial \rho c_p u_j T}{\partial x_j}=\frac{\partial}{\partial x_j}\lambda\frac{\partial T}{\partial x_j}+S \tag{3}$$

where ρ, μ, λ and c_p are density, dynamic viscosity, thermal conductivity and specific heat; g_i is the i-th component of the gravity acceleration; β is the thermal expansion coefficient; T_0 is a reference temperature; G_i is a forcing term (driving pressure gradient) balancing the large-scale pressure loss; and S is a volumetric heat source term balancing the large-scale enthalpy loss from the hot fluid. A detailed analysis of the governing equations [2] shows that S must be expressed as $(\langle q''_w\rangle u_s)/(HU_s)$, in which $\langle q''_w\rangle$ is the area average of the heat flux q''_w exiting the channel from the walls, u_s is the local velocity component along the direction **s** of the forcing term $\mathbf{G}=\{G_i\}$, U_s is its average over the total volume (*superficial* velocity) and H is the channel height.

In the present simulations the **s** direction lay along the x axis. The forcing term G_i was imposed and dynamically adjusted so as to obtain a prescribed Reynolds number.

The no slip condition was imposed to the velocity field on all solid walls. In regard to thermal boundary conditions, the spacer surface was assumed to be adiabatic while on the upper and lower wall (simulating the membranes) Robin (third type) conditions were imposed:

$$T_w=T_c+rq''_w \tag{4}$$

where T_w is the wall temperature, T_c is the temperature of the external cold fluid (the condensate in a MD module) and r is an interposed thermal resistance, inclusive of the conductive resistance of the wall (membrane) and of the advective resistance on the cold channel side, supposed known.

The fluid properties were set to those of water at 39°C [9]. These values are immaterial once results are expressed in dimensionless form. Simulations were performed by the Ansys-CFX® code [8].

2.2 Main definitions

The Reynolds number Re is defined here as:

$$\mathrm{Re}=\frac{\rho U_s d_{eq}}{\mu} \tag{5}$$

where $d_{eq}=2H$ is the hydraulic diameter of the channel. The local Nusselt number Nu is defined as:

$$\mathrm{Nu}=\frac{q''_w}{\lambda\left(T_w-T_h\right)} \tag{6}$$

where T_h is the bulk temperature of the hot fluid. The mean Nusselt number $\langle\mathrm{Nu}\rangle$ is computed by substituting surface averages $\langle q''_w\rangle$ and $\langle T_w\rangle$ for q''_w and T_w. Partial averages can be computed over the top or the bottom wall to obtain the corresponding Nu.

2.3 Computational details

The most demanding case regards the direct numerical simulations (DNS). The Kolmogorov scale of the smallest energy carrying eddies is $\Lambda_K/H=(\nu^3/\varepsilon)^{1/4}$ in which ε is the rate of turbulence energy dissipation per unit mass and $\nu=\mu/\rho$ is the kinematic viscosity. In the present plane channel geometry, identifying ε with the total mechanical energy dissipation, it follows from the above definitions that:

$$\Lambda_K=2.378\,f^{-1/4}\,\mathrm{Re}^{-3/4}\,H \tag{7}$$

where $f=|\mathbf{G}|(4H/\rho U_s^2)$ is the Darcy friction coefficient. For the highest Reynolds number simulated by DNS (Re≈2000), preliminary simulations indicate $f\approx0.5$. Equation (7) yields $\Lambda_K\approx10^{-2}H$. Therefore, an isotropic grid ($\Delta x=\Delta y=\Delta z=\Delta$) with 90 finite volumes along the cross-stream direction y and 360 along the x and z directions, yielding $\Delta/\Lambda_K\approx1$, was judged adequate. The total number of the finite volumes was ~11.6 million. With this choice the viscous-conductive sublayer $y^+\leq11$ on the thermally active walls was resolved by ~10 volumes.

For the same case (DNS at Re≈2000) a time step of 10^{-3} s was adopted, corresponding to a RMS Courant number $\Delta t/(u\Delta x)$ of ~0.5. The simulations were protracted for 20 s, corresponding to ~30 wash-through times (P/U_s); the last 12 s were used to compute averages and statistics. The computing time was of the order of $1.3\cdot10^5$ core-hours. The same Courant number and number of wash-through times were adopted at different Reynolds numbers. RANS simulations were run in steady-state mode with grids having from ~2.5 to ~6 million finite volumes. The symmetry properties of the geometry were not exploited to simplify the computational domain. Based on previous computational experience [2,3], grid independence was amply satisfied in all cases.

2.4 Turbulence models

Most of the turbulence models which were compared can be grouped into families according to two criteria. The former criterion distinguishes first order eddy viscosity/eddy diffusivity models (k-ε [9], k-ω [11], BSL k-ω [12], SST [12]) from second order Reynolds stress/Reynolds flux models (LRR RS [13], SSG RS [14], ω RS [8]). The latter criterion distinguishes ω-based models, in which the viscous/conductive sublayer is explicitly resolved by the computational grid (k-ω, BSL k-ω, SST, ω RS) from k-based models, in which the sublayer is not explicitly resolved and wall functions are used (k-ε, LRR RS, SSG RS). A further model considered was the one-equation Spalart-Allmaras model [15] in which a transport equation for the eddy viscosity is solved. Space does not allow here a description of the above mentioned models, but they are all well documented in the cited references.

3. Results

3.1 Preliminary selection of turbulence models in the absence of buoyancy

Preliminarily, all models were compared with DNS for a Reynolds number of ~2000 in the absence of buoyancy (i.e., for β=0). Results are summarized in Figure 2, which shows maps of the local Nusselt number. The mean Nusselt number over a wall (indifferently top or bottom) is indicated.

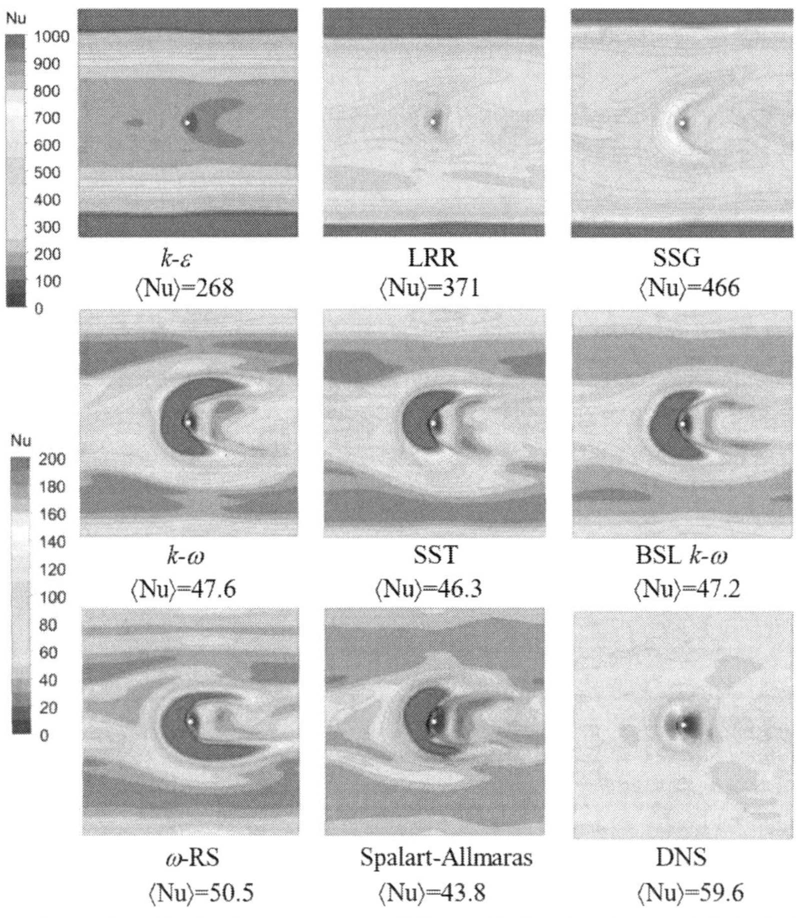

Figure 2. Nusselt number distributions over one of the walls for the nine models tested (Re≈2000, no buoyancy, flow from left to right). DNS results are time averaged.

It is clear from these results that models using wall functions, either of first (k-ε) or of second order (LRR, SSG), are totally inadequate: they overpredict ⟨Nu⟩ by 4-8 times with respect to DNS and yield a wrong distribution of the local Nusselt number. All ω-based models resolving the viscous-conductive sublayer yield acceptable and very similar results, with computed values of ⟨Nu⟩ from 78% to 85% of the DNS prediction. Even the one-equation Spalart-Allmaras model yields a satisfactory value of ⟨Nu⟩, only slightly below the ω-based models, and a similar spatial distribution of Nu.

3.2 *Influence of thermal buoyancy*

On the basis of the above comparisons, all following simulations were limited to three models: DNS, k-ω (representative, with only ±4% differences in ⟨Nu⟩ and utterly negligible differences in flow quantities, of all ω-based models) and Spalart-Allmaras (as a one-of-a-kind, eddy viscosity transport-based, model). All subsequent simulations were performed with buoyancy on, with the explicit purpose of assessing the influence of natural convection on heat transfer coefficients and its dependence on the Reynolds number in a spacer-filled channel geometry.

Note that, in a simple "void" (i.e., spacerless) horizontal plane channel, thermal buoyancy causes an asymmetry between the Nu distributions on the top and bottom walls even in the presence of perfectly symmetrical thermal boundary conditions. This is illustrated in Figure 3. Graph (a) is a sketch of the temperature distribution across a hot channel surrounded by two identical cold channels, as occurs in direct contact membrane distillation (DCMD) stacks and in the test section used in the most recent of our experiments [7]. Stable thermal stratification $\partial T/\partial y>0$ (expected to be associated with laminarized flow and low heat transfer rates) will occur in the bottom half of the hot channel, unstable stratification $\partial T/\partial y<0$ (expected to be associated with buoyancy-promoted turbulence and high heat transfer rates) in the top half. Graph (b) is a plot of the experimentally measured ⟨Nu⟩ on the top and bottom walls as functions of Re for a temperature difference of ~20°C between the hot and cold channels [7]. Consistently with the above remarks on thermal stratification, $\langle Nu\rangle_{top}$ largely exceeds $\langle Nu\rangle_{bottom}$ at low Re, while the two values become almost identical for Re>~1000.

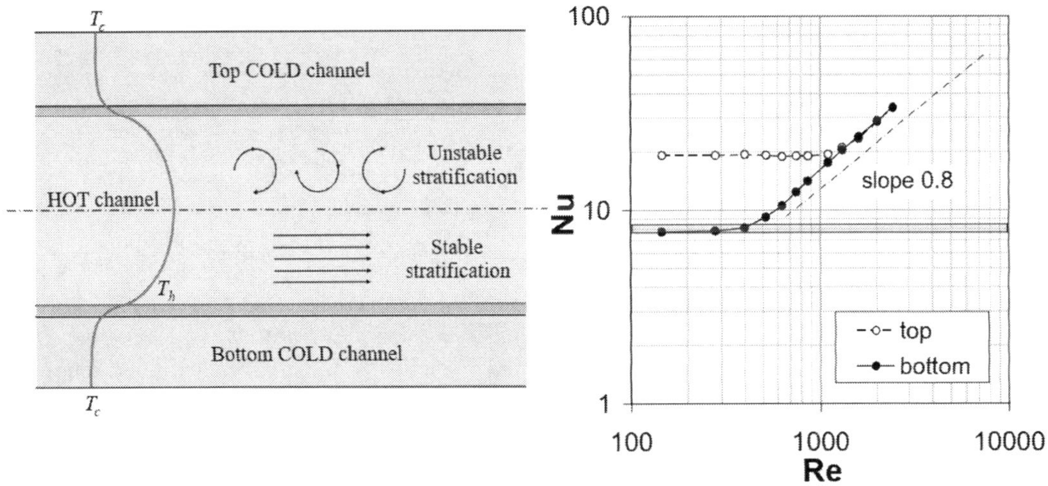

(a) (b)

Figure 3. Influence of buoyancy in a "void" (spacerless) horizontal plane channel. (a) Sketch of the temperature distribution across the channel, showing stable and unstable thermal stratification regions; (b) experimental ⟨Nu⟩ on the top and bottom wall as a function of Re.

An *a priori* estimate of the importance of buoyancy can be made on the basis of the Richardson number $Ri=Ra/Re^2$. Here, Re is the Reynolds number as defined by Eq. (5). Ra is the Rayleigh number which, for the configuration in Figure 3(a), can be computed for a length scale $H/2$ as:

$$\mathrm{Ra} = \frac{g\beta\left(T_h - T_w\right)(H/2)^3}{\nu\alpha} \tag{8}$$

where $\alpha=\lambda/\rho c_p$ is the thermal diffusivity of the fluid. Taking account of the thermal boundary condition in Eq. (4) and of the definition of Nu in Eq. (6), the temperature difference T_h-T_w can be expressed as $a/(1+b\mathrm{Nu})$, with the constants a and b depending on the values chosen for r and T_c in Eq. (4). A common criterion is that heat transfer is dominated by buoyancy for Ri>~0.25.

Identifying Nu with $\langle \mathrm{Nu}\rangle_{top}$ as computed by the k-ω model, the behaviour of Ri as a function of Re is that shown in Figure 4.

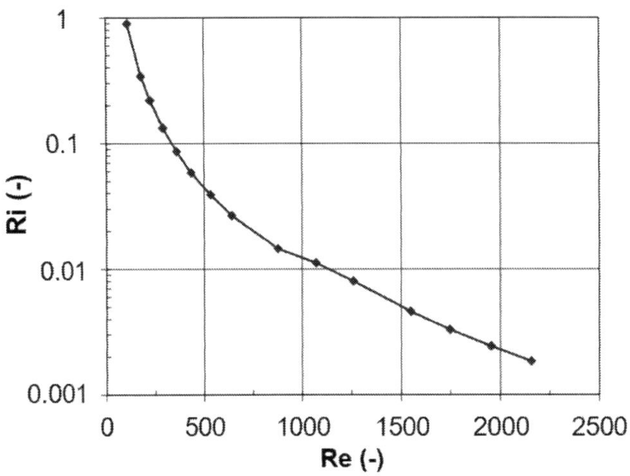

Figure 4. Richardson number as a function of the Reynolds number for a sphere-rod spacer-filled channel.

It can be observed that Ri is almost 1 at the lowest Re investigated (~100) but decreases rapidly with Re and becomes utterly negligible at Re>~1000. Therefore, one may expect the influence of buoyancy to be significant only at low Reynolds number, especially Re<300, and to practically vanish for Re>1000. The void-channel results in Figure 3(b) confirm this estimate.

Considering now spacer-filled channels, mean Nusselt numbers on the top and the bottom wall predicted by different turbulence models are reported in Figure 5 as functions of the Reynolds number. Experimental results in [7] are also shown.

DNS results behave like the experimental data for a void channel reported in Figure 3(b): $\langle \mathrm{Nu}\rangle_{top}$ largely exceeds $\langle \mathrm{Nu}\rangle_{bottom}$ at low Re, while the two values practically coincide and exhibit a roughly 0.8 power-law trend for Re≥~1000. The k-ω model (used here with buoyancy-related turbulence production terms activated in the k-transport equation but turned off in the ω-transport equation, as suggested in [16]) behaves similarly to DNS, although top and bottom ⟨Nu⟩ values do not quite collapse on a single trend (as in DNS) up to Re≈4000. In the presence of buoyancy, the Spalart-Allmaras models becomes grossly inadequate, overpredicting ⟨Nu⟩ by 6-20 times with respect to DNS on *both* walls.

Experimental results in [7] are qualitatively similar to DNS predictions, in that $\langle \mathrm{Nu}\rangle_{top}$ exceeds $\langle \mathrm{Nu}\rangle_{bottom}$ at low Re, while the two values become practically coincident for Re≥~1000. However, the experimental top-bottom asymmetry is lower than that exhibited by both DNS and k-ω predictions; more precisely, experimental, DNS and k-ω results are very close to one another on the bottom wall, but differ significantly on the top wall. A careful analysis of the differences between the experimental setup and the computational assumptions evidences that in the experiments the fluid enters the test section with significant inlet turbulence levels, caused by the upstream circuit including expansions and direction changes. On the contrary, all simulations are conducted for fully developed flow and thermal conditions, and turbulence levels are the equilibrium ones between production (by shear, obstacles and buoyancy) and dissipation. Further investigations will be necessary to clarify this issue.

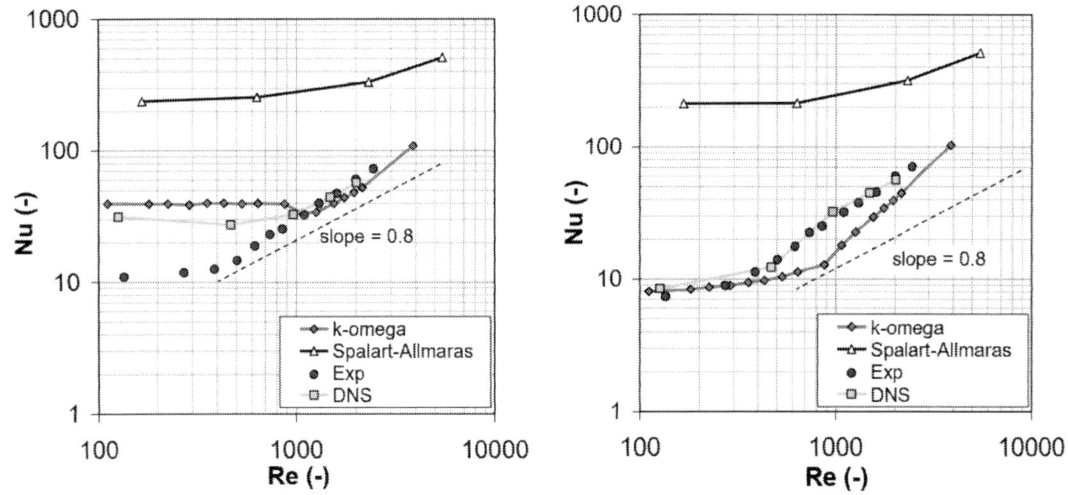

(a) (b)

Figure 5. ⟨Nu⟩ as a function of Re for the sphere-rod spacer geometry of Figure 1 and different turbulence models. Experimental results [7] are also reported. (a) Top wall; (b) bottom wall.

Figure 6 reports maps of the turbulent kinetic energy k (a, c) and of the time-averaged temperature (b, d) in the mid xy plane predicted by DNS for Re≈130 (a, b) and Re≈1000 (c, d).

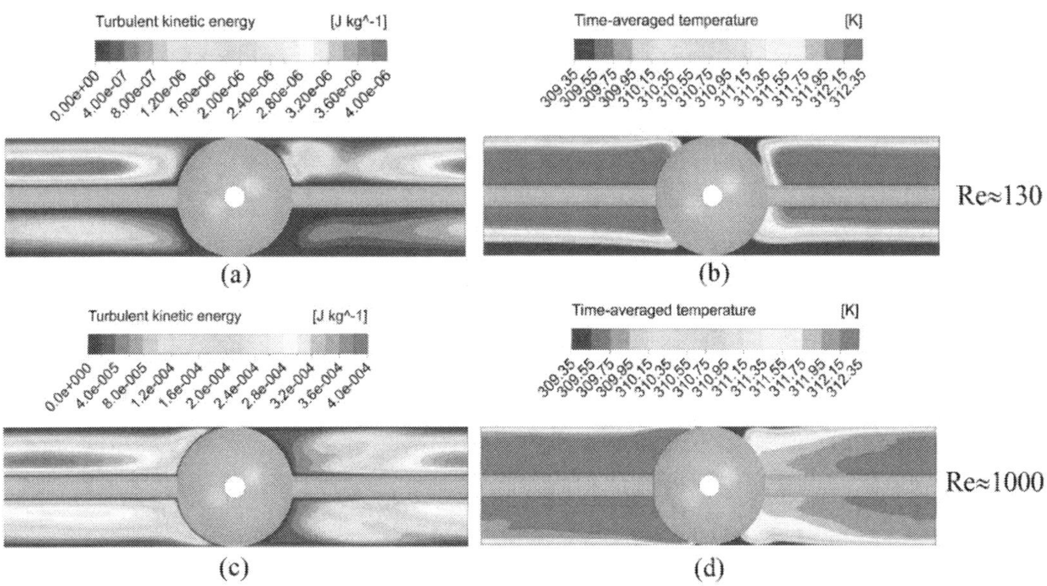

Figure 6. Turbulent kinetic energy k (a, c) and time-averaged temperature (b, d) in the mid xy plane predicted by DNS. (a, b) Re≈130; (c, d) Re≈1000. Flow is from left to right.

Consistently with the previous remarks on the Richardson number, the top-bottom asymmetry of both the k and the T distributions is much higher at the lower Reynolds number (Ri≈0.7), while it becomes barely noticeable at the higher Re (Ri≈0.01). Notably, values of k, once scaled by Re^2 (i.e., by U_s^2) are similar in the two cases, indicating that turbulence is sustained even at a Reynolds number as low as ~130 thanks to the contribution of production by buoyancy (which is almost independent of Re).

4. Conclusions

Numerical results were obtained for buoyant and non-buoyant flow in a horizontal plane channel filled with a novel sphere-rod spacer and exchanging heat from both the top and the bottom sides, representative of membrane distillation units.

Preliminary simulations were conducted setting buoyancy off. In this case, in comparison with Direct Numerical Simulations, RANS models of both first and second order using wall functions (k-ε, SSG RS, LRR RS) performed poorly, grossly overpredicting Nusselt numbers, while ω-based models resolving the viscous-conductive sublayer (k-ω, BSL k-ω, SST, ω RS) all yielded satisfactory results with small differences between specific models.

In the presence of buoyancy, DNS runs predicted significant turbulence levels, sustained by buoyancy production, even at very low Reynolds numbers (e.g. ~100). ω-based RANS models failed to predict turbulence at Re<~1000, yielding instead velocity fluctuations in the iteration space, which are unphysical in steady-state simulations. Both DNS and k-ω simulations yielded a thermal asymmetry between top and bottom wall, confirmed also by experimental results for both spacerless and spacer-filled channels. This asymmetry is explained by the stable or unstable thermal stratification occurring in the lower and upper layers of the channel, respectively; it is large at low Re but becomes negligible for Re>~1000, consistently with estimates based on the Richardson number.

The Spalart-Allmaras eddy viscosity transport model exhibited an anomalous behaviour, yielding satisfactory results in the absence of buoyancy but a gross overprediction of Nusselt numbers in buoyant flows.

5. References

[1] Tamburini A, Pitò P, Cipollina A, Micale G and Ciofalo M 2013 *J. Membr. Sci.* **447** 260
[2] Tamburini A, Renda M, Cipollina A, Micale G and Ciofalo M 2016 *Int. J. Heat Mass Transf.* **93** 1190
[3] Ponzio F N, Tamburini A, Cipollina A, Micale G and Ciofalo M 2017 *Int. J. Heat Mass Transf.* **104** 163
[4] Ciofalo M, Di Liberto M, La Cerva M and Tamburini A 2019 *Int. J. Therm. Sci.* **145** 106040
[5] Koutsou C P and Karabelas A J 2015 *J. Memb. Sci.* **488** 129
[6] Chong Y K, Liang Y Y, Lau W J and Fimbres Weihs G A 2022 *Int. J. Heat Mass Transf.* **191** 122819
[7] Cancilla N, Tamburini A, Tarantino A, Visconti S and Ciofalo M 2022 *Membranes* **12** 1029
[8] ANSYS 2018 *ANSYS-CFX Reference Guide Release 18.2*
[9] Launder B E and Spalding D B 1974 *Comput. Methods Appl. Mech. Eng.* **3** 269
[10] Yakhot V, Orszag S A, Thangam S, Gatski T B and Speziale C G 1992 *Phys. Fluids A* **4** 1510
[11] Wilcox D C 1988 *AIAA J.* **26** 1299
[12] Menter F R 1994 *AIAA J.* **32** 1598
[13] Launder B E, Reece G J and Rodi W 1975 *J. Fluid Mech.* **68** 537
[14] Speziale C G, Sarkar S and Gatski T B 1991 *J. Fluid Mech.* **277** 245
[15] Spalart P R and Allmaras S R 1994 *La Recherche Aérospatiale* **1** 5
[16] Ciofalo M 2021 *Thermofluid Dynamics of Turbulent Flows – Fundamentals and Modelling* Springer, UNIPA-Springer Series

Acknowledgments
This work was partly performed within the Sol2H2O project (European Twinning for research in Solar Energy to Water Production and Treatment Technologies), funded by European Union under Horizon Europe research & innovation programme, grant agreement No: 101079305.

Journal of Physics: Conference Series **2940** (2025) 012008 doi:10.1088/1742-6596/2940/1/012008

Lumped Parameter Modelling of Internal Temperature Dynamics in Multichip Power Modules

Giovanni Alì[1], Daniela Cavallaro[2], Alberto Fichera[1], Marta Lo Faro[1], Stefano Orlando[2], Arturo Pagano[1*], Rosaria Volpe[1]

[1] Department of Electrical, Electronics and Computer Engineering, University of Catania, Catania, Italy
[2] STMicroelectronics, Catania, Italy

*E-mail: arturo.pagano@unict.it

Abstract. This study assesses a lumped parameter modelling strategy for the fast response, layer-resolved prediction of the thermal field within the die of a selected power module device. A benchmark 1D lumped parameter model was initially defined, providing a simple description of the thermal field through the identification of a low number of lumped parameters but retaining information only on the junction temperature. A refined analytical model was therefore developed by adding the internal temperature nodes at the interfaces between the device layers. Thus, with a moderate increase of the lumped parameters, the model was able to predict the thermal behaviour of the inner layers of the die, which is the fundamental information required by reliability analysis. The set of model parameters were determined from the step-response thermal impedances of the power module, simulated by means of a numerical model purposely assessed under a boundary condition of forced liquid cooling. The lumped parameter identification consisted in a curve fitting procedure calculating the couples of thermal resistance and conductance. Both models were implemented in Matlab-Simulink, and simulated in order to predict the thermal response of the device to conditions of both step-wise and alternating current power inputs.

1. Introduction

The increasing demand of efficient and compact power electronic systems is boosting the transition from conventional silicon devices, such as silicon MOSFET and IGBT, to wide bandgap (WBG) solution in silicon carbide (SiC), such as, in particular, SiC MOSFET [1]. The latter, in fact, provides both a relevant reduction of the power losses and the high thermal conductivity and high temperature resistance of SiC, which sensibly facilitate the thermal management and the possibility to increase the maximum accepted power density [2]. Yet, high temperature reliability is still a main limitation to a wider spreading of SiC power modules in high power applications. In fact, the thermal profile expected in operation, i.e. in response to a specified duty cycle or mission profile, affects the performance and the reliability of the device in two different ways:

- the deterioration of the materials constituting the layers of the device as a consequence of either transient overloading or repeated long-term exposures to high temperature profiles, typically occurring as delamination or dielectric erosion phenomena or as gate oxide breakdown [3];

- the breakage caused by the wear-out of fragile elements, such as the bond-wire or the solder layers, subjected to the thermal fatigue [4].

Thermal fatigue is caused by cyclical mechanical stresses arising as a consequence of the different thermal expansion coefficients of the materials; thus, it is associated mainly to transient operations, such as in particular the device start-up and shut-down, but also to the thermal cycling shown by the device in response to the frequency of the feeding current during regime operations.

It emerges that the design for reliability [5] of SiC MOSFET devices requires prediction tools that must be computationally effective while properly accounting for the intrinsic time dependence of regime operations as well as of typical duty cycles [6].

Not all modelling approaches for the prediction of the temperature field are appropriate for dynamic simulations. FEM models provide accurate predictions but require high computational efforts and, above all, are excessively time-consuming for practical application in the prediction of the temperature dynamics with time-varying power inputs [7]. Models based on the direct use of experimental measurements, on the other hand, require the physical instrumentation of a reference power module, providing direct access to the required monitoring points; moreover, accurate dynamic predictions are possible only with fast response optical temperature probes, available for surface temperature detection but not for internal nodes; therefore, the use of experimental methods is practically limited just to the prediction of the junction temperature but not for that of the thermal distribution in the device layers [4]; parenthetically, a general concern is that relevant uncertainty may affect the generalization to other devices of the results achieved from the instrumented reference module.

The most effective approach for reliable and time-effective dynamical predictions is represented by lumped-parameter models. These are based on a network of thermal impedances, made up of resistance-capacitance coupled elements, able to simulate the heat flow path within the device and to predict the dynamics of the temperature at the nodes of interest, i.e. those at the layers' interfaces which are related to reliability issues.

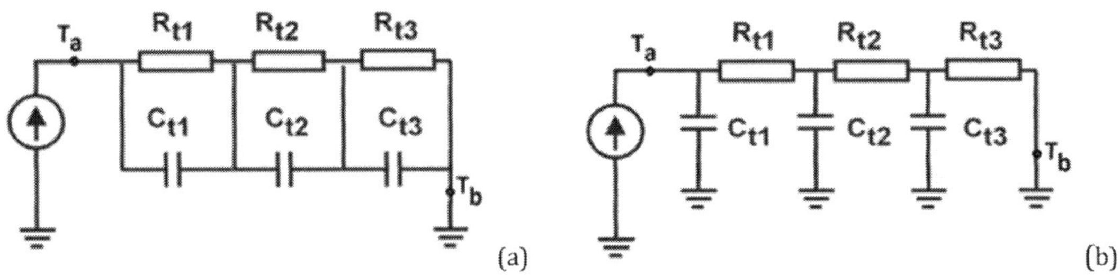

Figure 1. Lumped-parameter equivalent electric schemes of the thermal network between to temperature nodes T_a and T_b: (a) Foster network (b) Cauer network [9].

Two main approaches are used for the construction of a thermal network between two given temperature nodes, namely the Foster-type and the Cauer-type network schematized in figure 1 (a) and, respectively, (b) [8].

The thermal impedance between two temperature nodes T_a and T_b is defined as:

$$Z_{th}(t) = \frac{T_a(t) - T_b(t)}{P_d(t)} \tag{1}$$

with $P_d(t)$ instantaneous heat power crossing the branch of the network. Depending on the network order N_p, i.e. the number of poles of the impedance transfer function, Z_{th} analytically expresses the superposition of N_p thermal modes, each characterized by a time constant τ_n, produced in response to a step-wise variation of the applied P_d:

$$Z_{th}(t) = \sum_{n=1}^{N_p} R_{th,n} \cdot \left(1 - e^{-\frac{t}{\tau_n}}\right) \quad \text{with} \quad \tau_n = R_{th,n} \cdot C_{th,n} \tag{2}$$

Referring to the Foster network in figure 1 (a), the model order n and the parameters $R_{th,n}$ and $C_{th,n}$ are optimized in order to ensure a satisfactory curve fitting with the reference step-response of the device between the two nodes. While the intermediate nodes do not retain any physical meaning of intermediate temperatures, a main advantage of the approach is that the lumped parameters of the model can be easily extracted by either experimental measurements or by accurate numerical simulations of the step-response, provided that available data are satisfactorily resolved both in space and time. This explains why, being the junction and the case detectable boundary nodes, this approach is used for the junction-to-case Foster network experimentally derived by manufacturers and reported in the datasheets of commercial devices. Conversely, Foster networks between internal nodes are not easily obtained from experimental measurements but might be extracted from numerical simulations.

Referring to the Cauer network in figure 1 (b), each intermediate $R_{th,n} - C_{th,n}$ couple represents the thermal resistance and capacitance of each of the intermediate layers, calculated under the assumption of 1-D heat flow across the layer provided that the geometry and the thermophysical properties of the layers are properly assessed. Therefore, the internal nodes of the Cauer network retain the physical meaning of temperatures at the interface between layers [9]. Nonetheless, in practical applications, due to peripheral heat flows orthogonal to the main heat flow from junction to case, the validity of the assumption of 1-D heat flow is compromised; thus, the effective values of the $R_{th,n} - C_{th,n}$ couple of the Cauer network do not actually correspond to those theoretically calculated and the error introduced in the model predictions might be unacceptable. A viable solution to this problem is represented by the evaluation of the effective values of $R_{th,n} - C_{th,n}$ couples from the so-called Structure Function, SF [10]. As the SF is built up from the time constant spectrum of the junction-to-case Foster model of the reference thermal step-response, the SF-based Cauer model can be set to be more reliable.

The present study proposes the first step of assessment of a lumped-parameter thermal model of a SiC MOSFET multichip device, ACEPACK DRIVE, produced by STMicroelectronics for electric and hybrid vehicles. This is a three-phase inverter consisting of three modular legs, each made up of two switches formed by 8 dies topologically divided in two contiguous interacting quartets.

For the same power module device herein considered, a 1D thermal model was proposed in [11] providing the description of a quartet of die through a junction-to-case Foster network that accounts for the thermal coupling between dies but does not model the layers nodes beneath the dies.

The proposed lumped-parameter scheme is specifically designed for the 1-D thermal modelling of a single die with the aim of overcoming the limits of the basic junction-to-case 1-D Foster model, which is nonetheless assessed and used as a reference benchmark. With a marginal increase of the computational cost, the structure of the basic model is modified through the addition of the internal temperature nodes at the interfaces of the device layers. For both reported models, the lumped parameters were extracted by the numerical simulations of the step-response thermal impedances of the device by means of a curve fitting algorithm. A model with such a structure represents a tool for the dynamical prediction of the temperature profile within the device layers beneath the die and is, therefore, fundamental for properly addressing the issues involved in thermal fatigue analysis. Thus, scaling the thermal network to a quartet of dies and, thereafter, to the whole set of dies, represents the future development of the model herein discussed and aims at properly modelling the time-dependent temperature distribution within the device layers.

The approach proposed in the present study is similar to that reported in [3], [4], [10], [12], where lumped-parameter models were built for multidie devices based on either Foster or Cauer subnetworks connecting n temperature nodes of interest within the layers of the power module.

Reported results of the proposed model, in good agreement with the numerical simulations, outline its capability of predicting the dynamics of internal interface temperatures in response to an alternative current input.

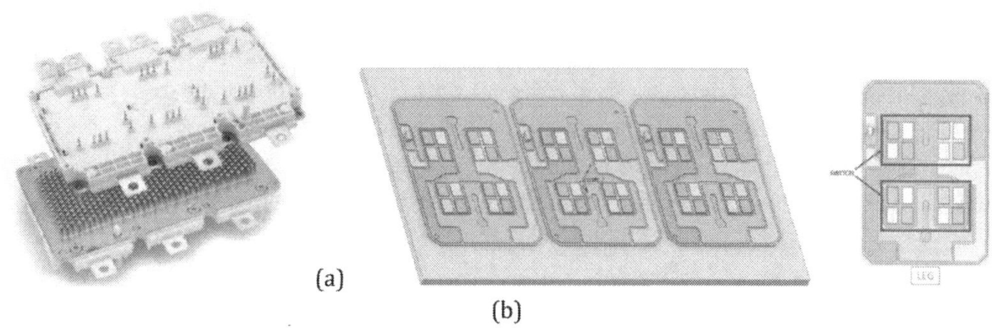

Figure 2. Schematic of ACEPACK DRIVE Power Module: (a) distribution of the three-phase legs; (b) detail of a leg, formed by a couple of switches with 8 diodes each.

2. ACEPACK DRIVE SiC MOSFET device

Silicon technology is indeed the most widely used power electronic devices but it is characterized by relative low switching velocity, affecting the power losses, and by reliability issues severely increasing with the growth of power density requirement for current applications in traction inverters. In fact, the automotive sector is leading the transition towards technologies based on novel semiconductor materials, such as silicon carbide (SiC) and gallium nitride (GaN). SiC MOSFET devices possess advantageous physical properties, such as admitted maximum junction temperatures up to 600°C, as compared to 150°C of silicon devices, and high thermal conductivity, which strongly enhances the heat dissipation efficiency, both substantially reducing the limitations on the maximum power density.

Further increases of the power density are achieved with power modules with a built-in heat sink, such as the SiC MOSFET ACEPACK DRIVE produced by STMicroelectronics, reported in figure 2 (a) and object of the modelling discussed in the present study. It is a traction inverter for electric and hybrid vehicles, in sizes ranging from 100 to over 300 kWe. Such high-power densities can be obtained by the direct connection of the heat sink and the baseplate through a Sn-Ag alloy solder, much more efficient than typical thermal interface material used for power modules. As shown in figure 2 (b), the three-phase device consists of 48 dies, divided into 3 legs (one for each of the phases) arranged on the baseplate; each leg consists of a pair of switches made up of 8 diodes. Figure 3 shows the vertical cross-section of a leg and the detail of the die connection to the common Cu layer shared by the diodes on the three legs.

3. Numerical model of ACEPACK DRIVE power module

The numerical modelling was developed in Ansys, through a virtual CAD model implemented in SpaceClaim and imported in Mechanical of Ansys. The constituting materials and the thermophysical properties of each of the layer schematized in figure 3 were assigned through the software library. These were considered isotropic and, with the exception of the sinter and solder layers, their temperature dependence was accounted during the simulation as they sensibly vary in the device operating range.

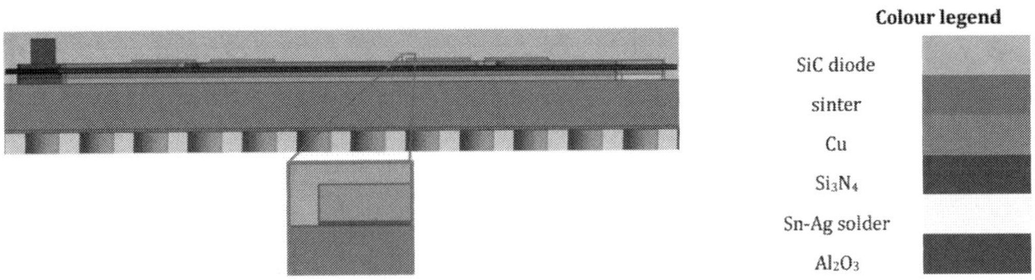

Figure 3. Cross-section of a leg of ACEPACK DRIVE with details of material layering.

In a previous numerical model, a uniform distribution was used for the elements of each layer. To achieve an accurate resolution of the local temperature profile along the device layers, the constraint on uniformity led to a mesh of 6,910,641 nodes and 1,517,380 elements, which was computationally very demanding. Therefore, without loss of accuracy, in the numerical model herein proposed a substantial reduction of the number of nodes and elements was achieved by adopting the edge sizing rule, so as to refine the mesh of the various layer with the desired accuracy. As shown in figure 4 (a), the edge refinement was performed optimizing the element size with respect to both the layer thickness and the local heat flow density, depending as a function of the distance from the dies, i.e. the active heat sources.

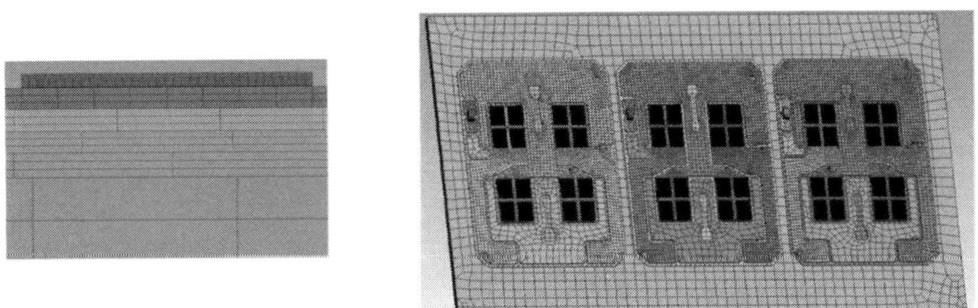

(a) (b)

Figure 4. Mesh details: (a) element distribution along the layer cross-section; (b) mesh refinement on the Cu layer under the switch of a leg (die region)

Moreover, to refine the distribution of the element, the MultiZone meshing tool was adopted for all layers, with the exception of the ceramic, solder and copper layers between them, where element distribution was already satisfactory. The MultiZone tool generates a pure hexahedral mesh, resulting in regions with structured mesh, but automatically switch on an unstructured mesh generation in those regions that are more difficult to capture.

Finally, a face-sizing rule was applied at the Cu layer under the footprint of the regions occupied by the dies, thickening the mesh to a value of 6·10-4 m, while the mesh of the remaining part of the copper layer was generated according to the general rule. This led to a smooth control of the variation of the mesh size from one layer to another. Figure 4 (b) reports a plan view of the final mesh, from which the main differences between the mesh of the layers beneath the heat sources are observable. In summary, the adopted meshing criteria led to a mesh consisting of 5,300,657 nodes and 1,073,916 elements, i.e. lower of about 25% with respect to the previous.

The setting of the boundary conditions was based on the specific requirement of achieving the maximum junction temperature at the die surfaces, $T_{j,max} = 175°C$, with a tolerance of 0.2°C. This is, in fact, the design limit of the device within which the manufacturer guarantees the performance and reliability of the power module. Thus, a set of preliminary numerical tests were performed in order to set the value of the power loss at the dies, P_{die}, leading the regime junction temperature at $T_{j,max}$. This was obtained for $P_{die} = 111\ W$, assuming uniformly distributed on each of the surfaces of the active dies.

For the scope of the present study, in order to avoid unnecessary CFD simulation of the coolant flow within the water-jacket, the boundary condition at the heat sink surface was simplified by assuming a constant heat transfer coefficient (HTC) dependent on the properties of the coolant and on the flow rate crossing the heatsink. For the simulation discussed in the present study it was always considered HTC=104 W/m2 °C with a constant coolant reference temperature equal to 65°C. Parenthetically, this HTC value had been previously evaluated by means of CFD simulations, performed considering a flow rate of 6 10-3 m3/min of a coolant solution of water and glycol at an inlet temperature of 65 °C. For all other surfaces, where no HTC value was assigned, adiabatic boundary conditions are automatically assigned by the numerical model's resolution algorithm, which is a simplified condition that is indeed acceptable in consideration of the robust thermal insulation determined by the protective gel that covers all the device surfaces. The numerical simulations were performed with the aim of determining the step response of the thermal field along the device layers caused by the activation of the dies, i.e. imposing a step change of the feeding power from $P_{die} = 0\ W$ to $P_{die} = 111\ W$. In consideration of the exponential time decay of the thermal step-response, the numbers and amplitudes of the simulation time steps were modulated, so as to achieve the proper resolution in a limited computational time.

3.1 Simulated thermal field

This section reports the description of the regime thermal field obtained at the extinction of the step response transitory under two conditions of interest:

i. step-wise simultaneous activation of all of the dies of the device with the maximum heat power $P_{die} = 111\ W$;

ii. ii. step-wise activation of a single die, selected on the central leg of the device, with the maximum heat power $P_{die} = 111\ W$.

Though numerical thermal maps for this case are not reported for the sake of conciseness, this simulation allowed to quantify: i. the time required for the extinction of the step response transitory, i.e. 13.5 s after the activation of the dies, ii. the maximum steady-state temperatures of about 175°C hottest die on each leg of the device, i.e. the cyan-colored dies of the bottom-right quartet of each leg in figure 2 (b), and iii. the substantial equivalence of the thermal field on the three legs.

The second simulated condition concerns the step response produced by the activation of a single die; among the possible dies, this was chosen by selecting the die of the central leg at which the maximum steady-state temperature was reached at the end of the numerical simulation discussed insofar, i.e. for the case of simultaneous stepwise activation of all dies with the maximum heat power $P_{die} = 111\ W$.

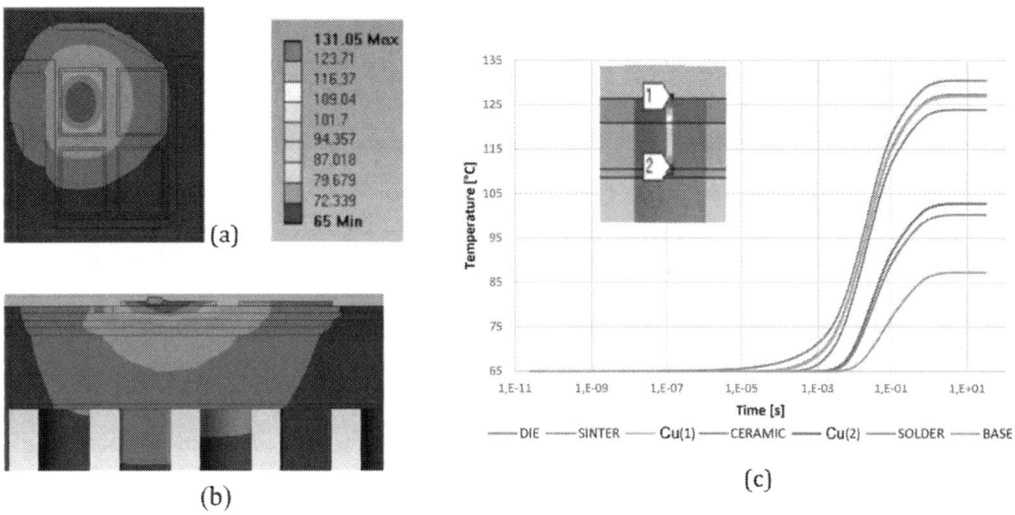

Figure 5. Step-response simulation with the activation of a single die in the central leg: (a)-(b) maps of the steady state thermal field reached at the end of the step response transitory, (c) step-response transitory of the internal temperatures at the interfaces between the layers.

Activating only this die with the same heat power, $P_{die} = 111\ W$, the highest temperature reached at steady state on the die surface, i.e. the maximum achived junction temperature, is 131.05 °C. Moreover, this value is attained at the simulation step 256, thus after 10.5 s since the die activation. Therefore, as expected, the maximum junction temperature is reached more rapidly and is lower of about 44°C than that obtained after the activation of all the dies, due to the peripheral heat flow along the layers. This clearly emerge from the thermal map reported in figure 5 (a), where an asymmetrical propagation is observed in the surrounding of the active die, as a consequence of the geometrical layout of the Cu layer that poses a limitation to the possibility to propagate heat towards the region on top and on the left of the die.

The temperature values at the interfaces between the layers are fundamental for the assessment of the lumped parameter thermal model discussed in the following paragraph. Therefore, the thermal distribution across the layers of the device, mapped in figure 5 (b), was extracted by the numerical model along a vertical path, originating in the point of the die where the maximum temperature is reached, and ending at the interface between the solder and the baseplate layers. In this way the transient step response of the temperatures at the layer interfaces reported in figure 5 (c) were evaluated and were subsequently used to derive the parameters of the model discussed in the next section.

4. 1-D lumped-parameter thermal models of a single die

The first proposed model recalls the canonical junction-to-case 1-D model and serves just as a reference benchmark. Worth mentioning that for the purpose of the present study, concerning a power module with a built-in heat sink, the reference 1-D case temperature is, in fact, substituted by the reference temperature of the coolant, T_{fl}, circulating in the water jacket of the device, which was assumed to be constant at the current state of the modelling activity. In this basic model, the junction temperature, $T_j(t)$, is the only prediction that can be withdrawn in order to characterize the thermal response of the device at the time varying power input to the die, $P_{die}(t)$. With this premises, a unique Foster network is required between the temperature nodes T_j and T_{fl}, which is identical to the one reported in figure 1 (a) for the generical nodes T_a and T_b. The numerical simulations discussed in the previous section for the step-wise activation die with a power input $P_{die} = 111\,W$ and a coolant temperature $T_{fl} = 65°C$ were used to calculate through equation (1) the overall numerical thermal impedance between the model nodes T_j and T_{fl}. Afterwards, the order $N_p$of the Foster model and the corresponding values of the N_p couples of lumped parameters $R_{th,n}$ and $C_{th,n}$ were assessed through a specific MATLAB® function that calculates the parameters of the Foster impedance curve providing the best fitting with the numerical impedance, for a given model order. Starting from the first order, the model order was increased until $N_p = 3$, for which the fitting was found to be satisfactory. For brevity, it is only mentioned that the third order parameters for the junction-to-case model and the corresponding Foster thermal impedance calculated by the fitting procedure were found to be comparable with the manufacturer expectations; moreover, the numerical and analytical predictions of the step response transitory of the junction temperature and of the junction-to-case thermal impedance were both found to be in good agreement.

The second 1-D model of the die was constructed in order to be able to provide the prediction of the internal temperatures at the interface of the layers of the die. It is important to highlight that knowing the temperatures of the intermediate layers is crucial to predicting which of them will be most critical and therefore will affected by failures more quickly and easily because of thermal fatigue. This goal was achieved by means of a novel thermal network, built as a sequence of Foster sub-models, one for each layer, connected at the interface nodes. The representation of this model in the Simulink-MATLAB® interface is reported in figure 6 (a), in which each a Foster block represents a sub-model corresponding to a layer of the die.

For the specified purpose, in the FEM simulation, monitoring points are selected at the center position of each die and on the surface of each layer at the same vertical position below each chip, and time series of temperature values are extracted. The thermal impedance curves for each layer are then calculated numerically, considering these temperature values at the top and at the bottom of each layer. Therefore, the impedances were calculated for each single layer, and then, the same procedure discussed for the first model was used for identifying the order of each sub-model and of the related set of $R_{th,n}$ and $C_{th,n}$ parameters.

Noticeably, in this case the thermal impedance of each Foster sub-model was calculated by applying the curve fitting procedure to the numerical thermal impedance calculated by the numerical time series on the interface surfaces, extracted as described in section 2. Again, for all the sub-models, the third order model was found to be the one that give a satisfactory fitting without an excessive increase of the number of parameters. Once the model parameters were identified, the model was simulated in order to compare its prediction with the available numerical results for the die step-response. Figure 6 (b) and (c) compare the numerical and analytical simulations of the step-response of the temperatures at the interfaces of the layers and of the thermal impedances of each layer, respectively.

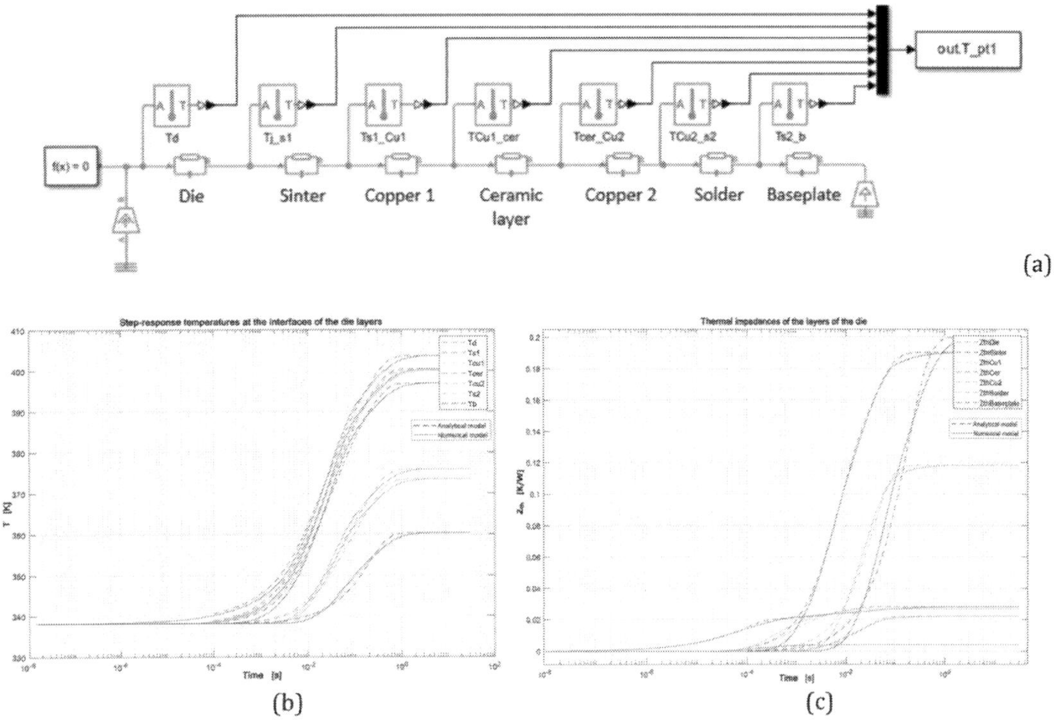

Figure 6. Mesh details: (a) element distribution along the layer cross-section; (b) mesh refinement on the Cu layer under the switch of a leg (die region); (c) planimetric view of the generated mesh.

As these preliminary results show satisfactory agreement, the analytical model was simulated under the condition of a pulsating power input, i.e. the time-varying input condition that corresponds to the alternative current that feeds the dies of the ACEPACK DRIVE module in actual operations. For comparison with previous results, figure 7 shows the predictions of the temperatures at the interfaces between the layers simulated under the condition of an alternative power input $P_{die}(t)$ of amplitude 111W and frequency 100 Hz. The straightforward advantage of the proposed lumped-parameter 1-D model is represented by its ability to produce, in a negligible simulation time, not only the dynamics of the junction temperature but also the thermal cycling of the interface temperatures, i.e. one of the main causes of thermal fatigue leading power electronics devices to failure.

Reported results show that, the proposed lumped-parameter model of a single die, though preliminary to the construction of the model of the whole multidie power module, demonstrates to reliably predict internal temperature dynamics with very limited computational effort; a main advantage is the very small number of parameters and the repeatable procedure that can be reproduced for other power modules in order to assess a device-specific tool for the prediction of the temperature dynamics and, thus, for thermal fatigue and reliability analysis for any type of mission profile.

5. Conclusion.

The study reports the definition of a refined 1-D lumped parameter models for the preliminary identification of the internal thermal impedances of the layer of a single die of ACEPACK DRIVE SiC MOSFET power module, produced by STMicroelectronics. The lumped parameters were extracted from the step response transient time series of the layer interface temperatures, simulated by means of an accurate numerical model purposely assessed. The model was shown to be able to produce a fast and computationally efficient prediction of the dynamics of the internal temperature distribution within the die, both for a step-wise variation of the power input and for an alternative power input, i.e. dynamical phenomena causing thermal cycling leading to failure.

The current development of the present study is oriented to the assessment of a 3-D lumped parameter model that encompasses, for each of the dies of the device, not only the 1-D model herein discussed but also the thermal mutual impedances that govern the peripheral heat transfer between confining dies.

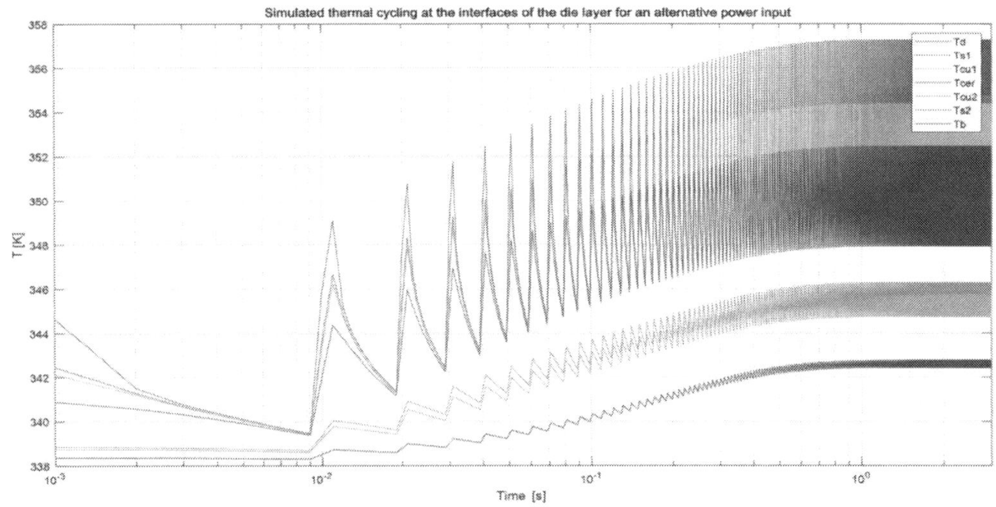

Figure 7. Predicted thermal cycling dynamics of the internal temperatures at the interfaces of the die layers, simulated for an alternative power input $P_{die}(t)$ of amplitude 111W and frequency 100 Hz.

References

[1] Mauromicale G, Raciti A, Rizzo SA, Susinni G, Abbatelli L, Buonomo S, Cavallaro D and Giuffrida V 2019 *Proc. 45th Annual Conference of the IEEE Industrial Electronics Society IECON*
[2] Zhu J, Kim H, Chen H, Erickson R and Maksimović D 2018 *Proc. IEEE Applied Power Electronics Conf. and Exp. – APEC*
[3] Van der Broeck CH, Lorenz RD and De Doncker RW 2019 *IEEE Trans. Power Electr.* **34 8** 7983
[4] Bahman AS, Ma K and Blaabjerg F 2015 *Proc. IEEE Appl. Power Electr. Conf. & Exp. - APEC*
[5] Kalker S, 2022 *IEEE J Emerging and Selected Topics in Power Electronics* **10 2** 1326
[6] Manzitto A *2020 Proc. Int Ex.& Conf. Power Electr. Intel. Motion, Ren. Energy and En. Manag.*
[7] Xu, 2023 *IEEE Transactions on Power Electronics* **38 5** 6522
[8] Bahman AS, Ma K and Blaabjerg F 2018 *IEEE Transactions on Power Electronics,* **24 4** 611
[9] Zhou Z, Holland PM and Igic P 2008 Proc. *26th Int. Conf. on Microelectronics MIEL 2008*
[10] Wang H et al. 2024 *IEEE Transactions on Power Electronics,* **39 5** 6225
[11] Allegra A et al. 2020 *2020 ELEKTRO*
[12] Yang X et al. 2023 *IEEE Journal of Emerging and Selected Topics in Power Electronics* **11 6** 5863

Modelling Phase Change Material Solidification/Melting for Accurate Thermal Energy Storage Design in Refrigeration Applications

A Ferraioli[1,*], O Carotenuto Ternullo[1], A Maiorino[1], F Giannetti[1] and C Aprea[1]

[1] Department of Industrial Engineering, University of Salerno, Via Giovanni Paolo II, 132, 84084 Fisciano (SA), Italy

[*] e-mail: alferraioli@unisa.it

Abstract. Phase Change Materials (PCMs) offer a promising solution for thermal energy storage (TES) in refrigeration, especially for critical applications like medicine and food preservation. This paper presents a new modelling approach to optimise PCM performance in cold storage systems. The research uses the Enthalpy-Porosity Model (EPM) to simulate PCM solidification and melting processes. This model offers a more accurate prediction than simpler methods and accounts for natural convection effects. The FreeFem++ toolbox and Galerkin method are employed for numerical solutions, enabling efficient parallel computing. By precisely modelling the phase change behaviour, the proposed method aids in designing high-efficiency and reliable cold storage systems with PCMs. This research contributes to advancements in PCM technology for sustainable refrigeration solutions.

1. Introduction

The European Climate Law, in July 2021, wrote down the goals set out in the former European Green Deal. According to the European Commission's decisions, greenhouse gas emissions must be reduced by at least 55% by 2030, and carbon neutrality for all EU countries is set to be achieved by 2050 [1].

Relying on renewable energies is not enough to fulfil such a challenging target. Efficiency enhancement of existing energy systems is essential as well.

Regarding this, the International Institute of Refrigeration (IIR) stated that cold industry technologies involving air cooling, food storage and vaccine preservation applications absorb 20% of electrical energy in the world, and this consumption is likely to redouble by 2050 due to global warming [2], [3]
.

This awareness led to introducing Phase Change Materials (PCM) as Thermal Energy Storage (TES) systems in refrigeration. These materials can absorb/release large amounts of latent heat during the melting/solidification, showing higher storage capacity than traditional TES systems based on sensible heat (SHTES).

When dealing with refrigeration purposes, two main topics must be taken on. The first one is the correct preservation of perishable goods, like food, vaccines, or medicines, while the second one is enhancing the energy efficiency of refrigerators. Latent Heat Thermal Energy Storage (LHTES) easily allows to meet both these requirements, rising as precious technology for future developments of cold chain. PCM coupling leads to better Coefficient Of Performances (COP) for conventional Vapour

Compression Refrigeration (VCR) and allows keeping chamber temperature steady at the prescribed value since thermal storage/release process will take place almost isothermally.

Designing an LHTES system for refrigeration purposes means retaining many input variables concerning PCM thermophysical properties, system geometry and, sometimes, even encapsulation solutions.

All these variables and many others strongly influence the thermodynamic performance of an LHTES system, both from energy and entropy point of view, and searching for their optimal configuration can't be done only through experiments because too many prototypes should be realised, leading to unacceptable design costs and time.

To accurately assess stored thermal energy and system autonomy, considering only mass and latent heat is not enough; phase propagation in time and space also needs to be taken into account, so that PCM-environment heat exchange is properly investigated. Numerical methods for phase change problems are needed for PCM modelling.

For this study, a simulation of the water melting/solidification cycle for refrigeration applications has been carried out, and the FreeFem++ PCM Toolbox [4] has been used. This program was initially tested considering Rayleigh number around 10^6 for water freezing and 10^8 for the melting of a generic material with linear density behaviour. In this research, more intense natural convection (Rayleigh≈10^9) has been successfully recreated during the ice-melting process. Such a stiff phenomenon required to set up very carefully numerical parameters for time and space integration to achieve method convergence and acceptable simulation cost. The chosen parameter values will be discussed in this paper.

2. Enthalpy-Porosity Model

To this day, the most accurate physical model describing phase change convection-diffusion problem (Stefan problem) is the Enthalpy-Porosity Model (EPM) [5], a powerful formulation based on single domain approach, enthalpy method and, finally, porous approximation for flow field across solid-liquid interface.

The dimensionless system of equations to be solved in both liquid and solid regions come from the incompressible Navier-Stokes equations, with Boussinesq approximation for buoyancy effects, and an enthalpy form for the energy equation [4]:

$$\underline{\nabla} \cdot \underline{u} = 0 \tag{1}$$

$$\frac{\partial \underline{u}}{\partial t} + (\underline{u} \cdot \underline{\nabla})\underline{u} + \underline{\nabla} p - f_B(\theta)\,\hat{\jmath} - \frac{1}{Re}\nabla^2 \underline{u} - A_{mushy}(\theta)\,\underline{u} = \underline{0} \tag{2}$$

$$\frac{\partial (C\theta)}{\partial t} + \underline{\nabla} \cdot (C\theta \underline{u}) - \underline{\nabla} \cdot \left(\frac{K}{RePr}\underline{\nabla}\theta\right) + \frac{\partial (CS)}{\partial t} = 0 \tag{3}$$

The buoyancy force $f_B(\theta)$ comes from Boussinesq approximation and allows to simulate natural convection in the liquid phase and it is defined according to density-temperature relation of the investigated material. Taking the buoyancy term into account is mandatory since, according to experimental and numerical data [4], [6] melting process is strongly influenced by convective heat transfer between phases, which deforms the interface shape and causes the transition to be faster. On the other hand, solidification takes place with almost no significant convection, similarly to a diffusion-only phenomenon.

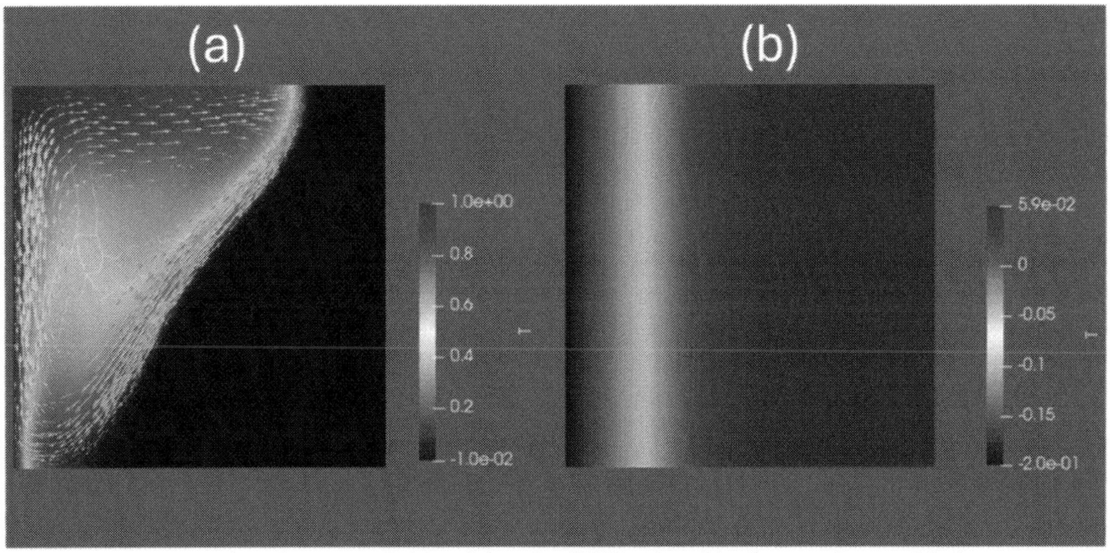

Figure 1. Comparison between melting (a) and solidification (b) of n-octadecane. Made on *FreeFem++* PCM Toolbox.

A Carman-Kozeny porous penalization is applied to velocity in equation (2). This A_{mushy} factor depends on dimensionless temperature θ through local liquid fraction L_f:

$$A_{mushy}(\theta) = -\frac{C_{CK}\left(1 - L_f(\theta)\right)^2}{L_f(\theta)^3 + b}, \qquad C_{CK} = 10^6, \qquad b = 10^{-6} \tag{4}$$

Where L_f is a Heaviside step function that is equal to 0 in solid phase and to 1 in liquid phase.

In equation (3), an enthalpy source term, which describes latent enthalpy level showing up after melting, includes the non-dimensional property S, which considers the latent heat of fusion [7]. It is a Heaviside step function that is zero in the solid phase while, in the liquid one, it is equal to the inverse of Stefan's number, defined as:

$$Ste = \frac{c_l \delta T}{h_{sl}} \tag{5}$$

Thanks to Carman-Kozeny penalization and latent enthalpy source term, the EPM can be defined as a "single-domain" or "fixed-domain" model, which means that balance equations can be coherently extended from liquid to solid phase. The advantage of this approach is that there is no need to explicitly track the phase interface and split the model in two parts. Solid-liquid interface is the $\theta = \theta_f$ isoline and so it is computed as part of the solution.

Of course, for a pure substance, properties like K, C, L_f and S should be step functions, having a discontinuity on θ_f. This can't be simulated by a numerical method, so the following smoothing function is adopted [7]:

$$F(\theta, a_s, \theta_s, R_s) = f_l + \frac{f_s - f_l}{2}\left\{1 + \tanh\left(a_s\left(\frac{\theta_s - \theta}{\varepsilon}\right)\right)\right\} \tag{6}$$

The prescribed function approximates a fictive ε wide finite mushy zone separating the two phases.

3. Numerical method

The numerical method proposed by PCM toolbox adopts two finite difference numerical schemes for time integration. The first one is the second-order Gear scheme, given by the following expression:

$$\frac{dU}{dt} \approx \frac{3U^{n+1} - 4U^{n} + U^{n-1}}{2\delta t} \tag{7}$$

The second one is the first-order Euler implicit scheme, only used for the first-time step. These schemes are applied both to velocity ($U = u$) and temperature fields ($U = \theta$) and they allow to calculate solution U^{n+1} at $t_{n+1} = (n + 1)\delta t$ by using the two previous known time states.

At each time step, Galerkin finite-element method is used for space integration. The computational domain is divided into a finite number of elements composed of nodes. Delaunay – Voronoi triangulation algorithm is employed by *FreeFem++* for mesh building. Therefore, original domain Ω is approximated to $\mathcal{T}_h$ mesh, made up of m triangular elements.

The solution will be approximated as a combination of polynomials associated with the finite elements and called shape functions. Approximation order will depend on the number of nodes chosen for the single element. Regarding this, PCM Toolbox implements Taylor hood combination, consisting of an interpolation that is quadratic for velocity (P_2 elements) and linear for pressure (P_1 elements), so Ladyzhenskaya-Babuska-Brezzi stability condition is also satisfied [8] elements.

Finally, the Galerkin method approximates equations (1)-(3) to a non-linear algebraic system for the current time, so the iterative Newton method must be applied. This means that nonlinear system is linearized and solved multiple times until convergence is reached. Then, solution is updated and Newton loop is performed again for next time step. This method is carried out for the entire time range prescribed for the analysis.

4. Resolution method

The *FreeFem++* PCM Toolbox is a code framework compiled in *FreeFem++*. It is based on the EPM and implements the previously described numerical method. It is a vey suitable tool for PCM design, mainly due to the great modelling freedom granted by *FreeFem++* environment that, similarly to a DNS, requires explicitly declaring the weak form of the investigated PDEs for Galerkin method application. This leads to freely model phase change terms in equations (1)-(3), which are Carman-Kozeny porous and enthalpy source terms. Smoothing function shape and width, building up the mushy region, can also be chosen by the user.

Another important feature of this code is the coupling with the *ffddm* package [9], bringing up domain decomposition methods for parallel computing, strongly enhancing computational efficiency. During computing, the original mesh is distributed into smaller overlapping meshes at each time step. Linearised algebraic system, coming from Galerkin Method at current Newton iteration, is also distributed. Then, each sub-system is solved with a preconditioned iterative method. The Restrictive Additive Schwartz (RAS) method is chosen for preconditioning, and then Generalised Minimal Residual (GMRES) is the applied Krylov solver.

Each sub-system is delivered to a different MPI process so that distributed FEM analysis is carried out in parallel by multiple cores of the computer. This allows to speed up the simulation up to four times concerning a sequential code [4].

Mesh adaptivity is also implemented with *adaptmesh* command, available in *FreeFem++* [10]. At each time step, the mesh is thickened where the highest gradients are found in the solution, and it is coarsened where the solution field is smoother. This feature allows for adequate approximate solutions always using the minimum number of nodes, reducing the computational cost for the simulation one more time.

Mesh adaptation efficiently tracks the liquid phase's solid-liquid interface and thermal and dynamic boundary layers.

5. Test case

PCM simulation can involve several applications related to cold chain to improve refrigerators performance. The test cases analysed in this paper are carried out to assess the melting-solidification process. The focus will be on water, which is a pure substance characterised by non-linear Gebhart-Mollendorf density law, with melting point at $0°C$. A rectangular computational domain with height $530\ mm$ and width $37.5\ mm$ is used for our analysis. The following two paragraphs discuss water melting and freezing and their post-process results.

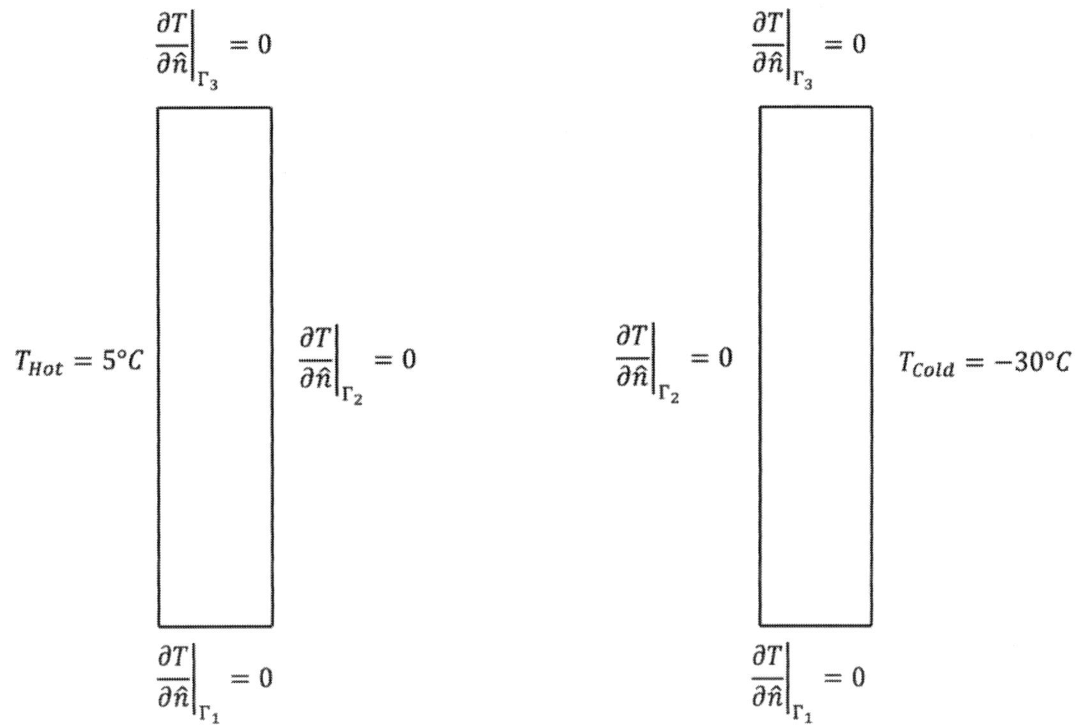

Figure 2. Boundary conditions for water melting and solidification

5.1. Water melting

The water melting test case is studied to evaluate phase propagation and quality change throughout the phenomenon. The melting process is a very robust and complex phenomenon to analyse. It is characterised by both convective heat transfer, due to the natural convection of the liquid phase, and diffusive one caused by conduction.

Starting from an initial condition in which water is totally solid $(L_f = 0)$ the boundary con our Stefan problem involves a homogeneous Neumann adiabatic condition on the two horizontal walls and the right vertical one, whereas a Dirichlet condition with an imposed temperature of $5°C$ set-up on the left vertical wall. Additionally, no-slip and no-permeability conditions are assigned to all four sides.

Since the main domain dimension is the height, the Rayleigh number becomes very challenging, and consequently, the Boussinesq term becomes significant. In this case Rayleigh number achieves $6.6313e9$. Stefan number, which represents the comparison between sensible and latent thermal inertia, is 0.0627; therefore, the melting process is completed in $t_{max} = 0.2$.

Adaptmesh algorithm, with minimum length of triangle edge $h_{min} = 1e - 4$, and time step equal to $\delta t = 1e - 4$ catch high gradients on solid-liquid interface and along the isotherm $T_m = 4°C$.

In Figure 3, the melting process is shown: solid-liquid interface shape is affected by two strong convection cells; therefore, when the propagation front gets in touch with the right wall, the heat transfer area between the two phases gets smaller, and the process slows down. Figure 4 shows a polynomial regression in which quadratic model well approximates the quality change (L_f) with 95% interval confidence.

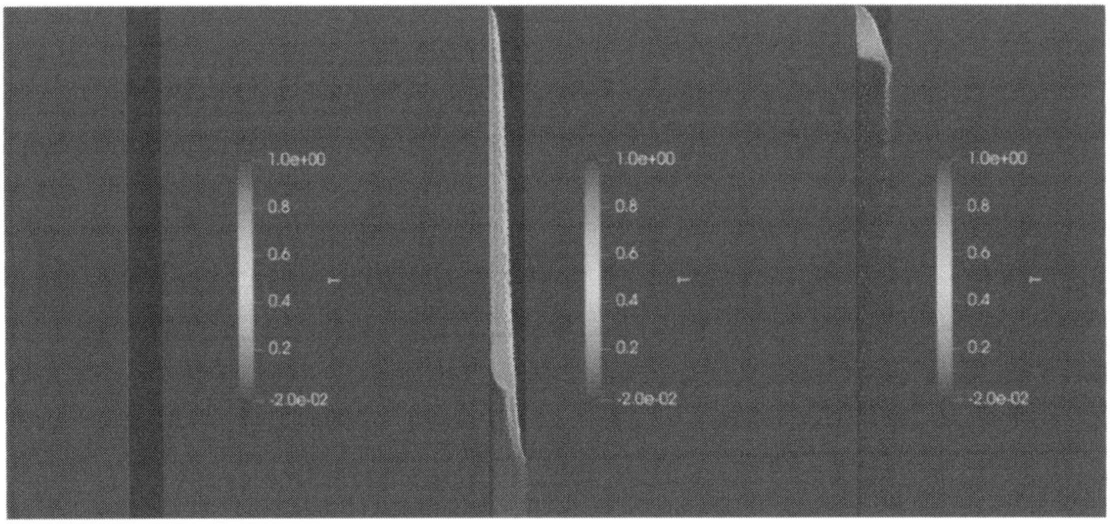

Figure 3. Melting in time and solid-liquid interface.

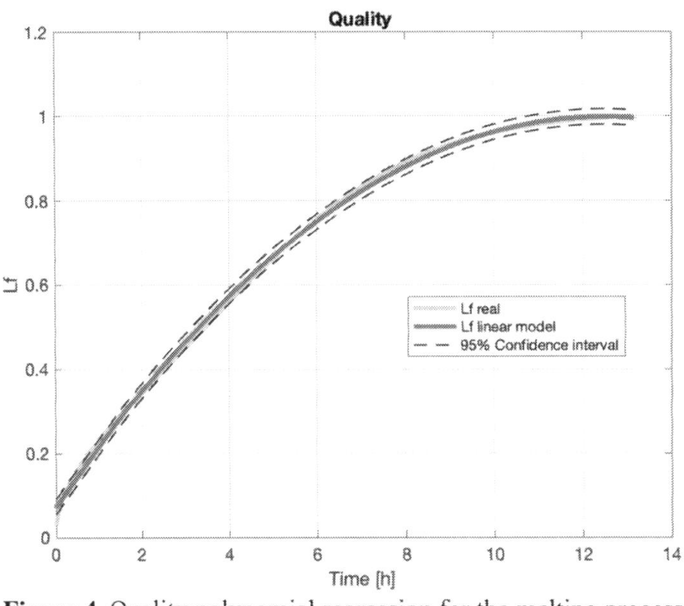

Figure 4. Quality polynomial regression for the melting process

5.2. Water freezing

Water freezing analysis is lighter than melting one because it is a pure diffusion phenomenon. Since there is no natural convection, the simulation parameters setting becomes more straightforward to manage.

Journal of Physics: Conference Series **2940** (2025) 012009 doi:10.1088/1742-6596/2940/1/012009

Time integration requires less computational efforts because both space and time grid are coarser. Boundary conditions are set up in this way: two horizontal and right walls are kept adiabatic, whereas, on the left side, there is a temperature Dirichlet condition of $-30°C$. No-slip and no-permeability conditions are assigned as well as done for previous study. Starting from an initial condition in which water is totally liquid ($L_f = 1$), convergence is achieved with an investigation time $t_{max} = 0.013$, a time step $\delta t = 1e-3$, one order of magnitude higher. Rayleigh and Stefan number are respectively $7.5746e8$ and 0.0075, whereas the minimum length of side triangle is kept at $h_{min} = 1e-4$.

In Figure 5 the solidification process is represented: the phase interface assumes a vertical shape thanks to the absence of natural convection.

Figure 6 shows a polynomial regression in which quadratic model well approximates the quality change (L_f) in the middle part of the curve with the same interval confidence of melting case, where the phenomenon is of the greatest interest.

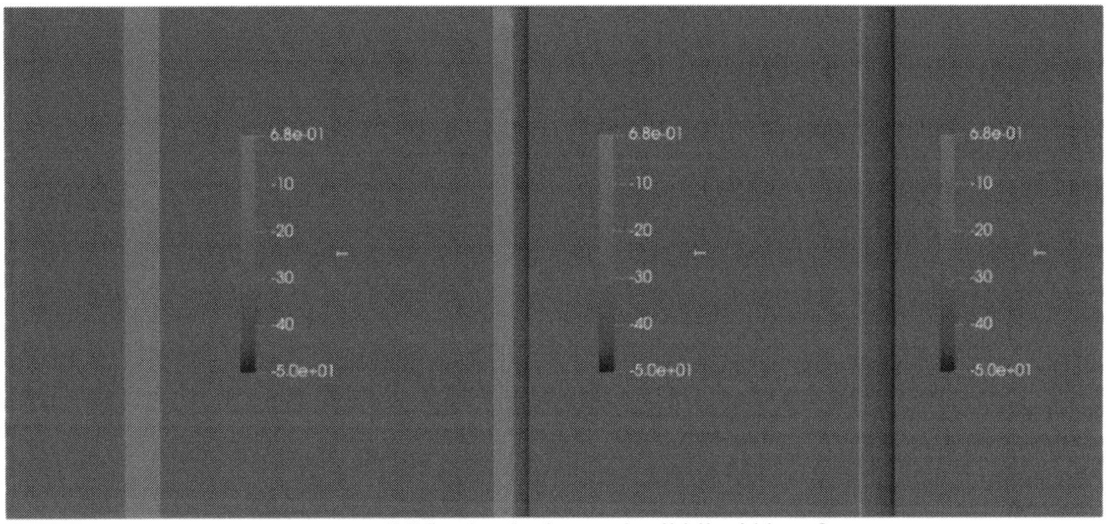

Figure 5. Solidification in time and solid-liquid interface.

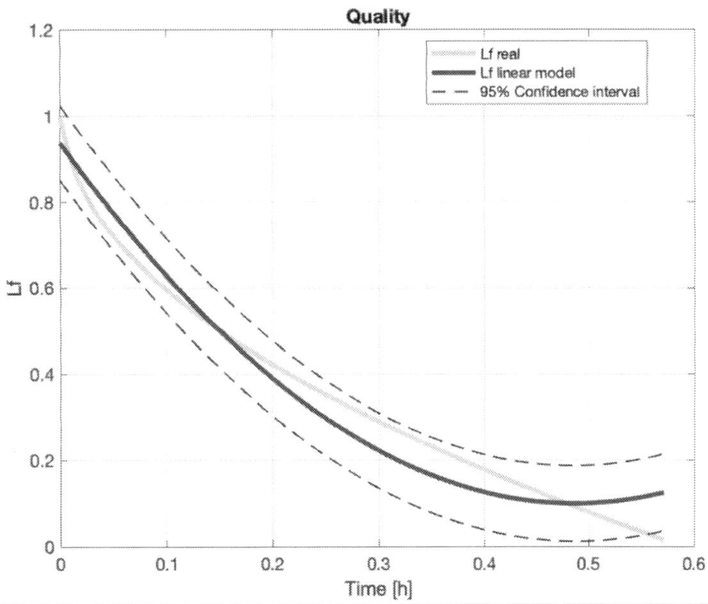

Figure 6. Quality polynomial regression for solidification process

6. Conclusions

The case studies have shown the several advantages of using a numerical method. Considering a very tall cavity has required to carefully setting up challenging parameters to achieve convergence and the parallel computation effectiveness has allowed to considerably reduce design time and costs.

As concerns melting and solidification processes, they are characterised by two different propagation fronts due to their intrinsic nature. Quality change is well approximated with quadratic polynomial regression; hence, this model could be used for a predictive control system operating on PCM charge/discharge.

This study underlines that the PCM toolbox is easy to test and adapt for several scenarios by changing substance properties, boundary conditions and geometry. However, this powerful tool allows to simulate the PCM behaviour under three assumptions: 2D computational domain, the possibility to treat only a pure substance, and the mushy zone modelling with an assigned width between two phases. The software FreeFem++ is preferred with respect to other commercial one because it is an open-source software and results very flexible in parameters management and numerical method tuning.

7. References

[1] K. Kulovesi, S. Oberthür, H. van Asselt, and A. Savaresi, "The European Climate Law: Strengthening EU Procedural Climate Governance?," *Journal of Environmental Law*, vol. 36, no. 1, pp. 23–42, Mar. 2024, doi: 10.1093/jel/eqad034.

[2] "The Role of Refrigeration in the Global Economy - 29th Informatory Note on Refrigeration Technologies," Nov. 2015. [Online]. Available: www.iifiir.org

[3] "THE ROLE OF REFRIGERATION IN THE GLOBAL ECONOMY 38 th Informatory Note on Refrigeration Technologies N° ISSN : en cours," Jun. 2019.

[4] G. Sadaka, A. Rakotondrandisa, P. H. Tournier, F. Luddens, C. Lothodé, and I. Danaila, "Parallel finite-element codes for the simulation of two-dimensional and three-dimensional solid–liquid phase-change systems with natural convection," *Comput Phys Commun*, vol. 257, Dec. 2020, doi: 10.1016/j.cpc.2020.107492.

[5] V. R. Voller, M. Cross, and N. C. Markatos, "An enthalpy method for convection/diffusion phase change," *Int J Numer Methods Eng*, vol. 24, no. 1, pp. 271–284, 1987, doi: 10.1002/nme.1620240119.

[6] A. Rakotondrandisa, I. Danaila, and L. Danaila, "Numerical modelling of a melting-solidification cycle of a phase-change material with complete or partial melting," 2018.

[7] I. Danaila, R. Moglan, F. Hecht, and S. Le Masson, "A Newton method with adaptive finite elements for solving phase-change problems with natural convection," *J Comput Phys*, vol. 274, pp. 826–840, Oct. 2014, doi: 10.1016/j.jcp.2014.06.036.

[8] "FONDAMENTI DI TERMOFLUIDODINAMICA COMPUTAZIONALE a cura di G. Comini, G. Croce ed E. Nobile," 2014.

[9] P.-H. 'Tournier, P. 'Jolivet, and F. 'Nataf, "FFDDM: FreeFem Domain Decomposition Method," https://doc.freefem.org/documentation/ffddm/index.html.

[10] F. Hecht, "New development in freefem+," *Journal of Numerical Mathematics*, vol. 20, no. 3–4, pp. 251–265, Dec. 2012, doi: 10.1515/jnum-2012-0013.

Journal of Physics: Conference Series **2940** (2025) 012010 doi:10.1088/1742-6596/2940/1/012010

Numerical simulation of film instability over a corrugated sheet

Nicola Suzzi and Giulio Croce

DPIA - Dipartimento Politecnico di Ingegneria e Architettura - Universitá di Udine - Via delle Scienze - 33100 - Udine (UD) - Italy

E-mail: nicola.suzzi@uniud.it, giulio.croce@uniud.it

Abstract. The evolution of a liquid layer flowing down a corrugated sheet driven by gravity, which is the characteristic configuration of structured packing, is numerically analysed via the solution of the governing lubrication equation, which reduces the 3D physical problem to a 2D mathematical problem. Disjoining pressure is used to model contact line dynamics and surface wettability, while full implementation of capillary pressure allows to investigate contact angles up to 60°. The effect of corrugation is introduced via the definition, in the governing lubrication equation, of non-uniform gravity acceleration. Furthermore, the additional capillary force, arising from variations in the solid surface curvature, is also implemented. Different corrugation geometries and flow conditions, the latter being defined by the reference film Capillary number, are investigated, looking for configurations that allow enhancement of the liquid-gas interface area at low liquid flow rate. Such an analysis provides a novel approach in the design of structured packing, widely used in carbon capture via chemical absorption.

1 Introduction

The evolution of a thin liquid layer is involved in a number of practical applications, such as the in-flight icing [1, 2] and the CO2 absorption through structured packing [3, 4, 5]. In the absorption process through structured packing, the gas solute flows up while a liquid solvent flows down the interspaces of a collection of corrugated sheets. A continuous film pattern enhances the interface area between liquid and gas; as a result, the mass transport process is incentivated. On the other hand, increasing the liquid flow rate over a certain threshold may reduce the void area, with a sudden increase of the pressure drop on the gas side (flooding condition). Simplified models usually assume a continuous film covering the solid wall of a flow channel [6, 7], resulting in higher values of the predicted absorption rate. A number of experimental analyses investigates both the break of an initially continuous film covering an inclined plate [8, 9] and the occurrence of finger instability when an advancing film front falling down an inclined plate is perturbed [10]. However, experiments do not allow to access the details of the flow pattern inside the structured packing. On the other hand, the numerical approach is a valuable option to predict the occurrence of film instability in complex geometries such as the one characterizing structured packing, with the eventual break of the continuous wetting layer or the formation of stationary rivulets. A liquid layer falling down an inclined plate bounded by lateral wall, which is the simplest configuration describing the hydrodynamics inside structured packing, was numerically investigated in [3, 5] using a fully 3D CFD analysis, which includes the Volume of Fluid (VoF) method to solve the liquid-gas interface. A large portion of a corrugated sheet was investigated in [4], using the VoF method incorporated in Ansys Fluent®. However, only few configurations were tested due to the large computational cost deriving from the fully 3D modelization. Lubrication theory, which allows to reduce the 3D physical problem of a thin

liquid layer flowing on a hydrophilic surface to a 2D mathematical problem, represents the main tool to predict the instability of a flowing film [11, 12, 13, 14]. The effect of the surface wettability was included in the lubrication framework in [15, 16, 17] through the disjoining pressure [18, 19]. However, such analyses are limited to low values of the contact angle, since the standard small slope approximation does not provide accurate results for $\theta \geq 30°$ [20]. It was proved in [21] that the implementation of the exact free surface curvature allows to investigate contact angles up to 60°, which falls in the range of structured packing applications. A vertical plate bounded by lateral walls, representative of the hydrodynamics inside structured packing, was numerically investigated in [21, 22] via the lubrication theory. Compared to the fully 3D modelization, it was proved in [23, 21] that the lubrication theory allows a great reduction of the computational cost, still providing accurate results for contact angles up to 60°.
Here, a liquid film flowing down a corrugated sheet, which is the basic element of structured packing, is numerically investigated via the lubrication theory. Following [21, 22, 24], the full implementation of the free surface curvature combined with the disjoining pressure model is used to investigate increased surface wettabilities, compared to the range of the standard formulation of the lubrication theory. Compared to previous literature analyses, a non-uniform gravity field is implemented in order to model the corrugated surface, providing a specific tool to study hydrodynamics in structured packing. Thus, the stationary liquid pattern is computed for different geometries and the influence of the corrugated geometry on the eventual dewetting (i.e. formation of stable rivulets) is discussed, since it is a crucial aspect for a proper design of the packing tower, used for absorption and distillation process.

2 Mathematical model

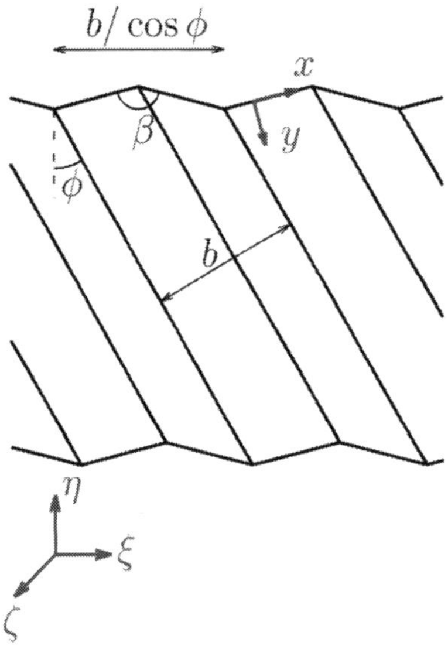

Figure 1: Geometry of a vertical corrugated sheet: crimp angle ϕ, corrugation angle β and corrugation length b. Reference Cartesian coordinate system $O\xi\eta\zeta$ and body-fitted orthogonal coordinate system Oxy laying on the solid surface.

Consider a thin film flowing down a vertical corrugated sheet like the one of Figure 1. Let h be the film thickness and $\mathbf{x} = (x, y)$ be a body-fitted orthogonal coordinate system laying on the corrugated sheet. The lubrication theory allows to integrate the continuity equation across the film thickness. Further assuming a low film Reynolds number gives a parabolic velocity profile across the film thickness, which can be averaged and substituted in the integrated continuity equation yielding the well-known lubrication equation,

$$\frac{\partial h}{\partial t} + \nabla \cdot \left(\frac{\rho\, g\, h^3}{3\,\mu}\, \tilde{\mathbf{g}} - \frac{\nabla p}{3\,\mu}\, h^3 \right) = 0,\ p = -2\,\sigma\,(\kappa + \kappa_s) - D + \rho\, g\, h\,(\hat{\mathbf{g}} \cdot \hat{\mathbf{n}}) \tag{1}$$

where $\tilde{\mathbf{g}}$ is projection of the gravity versor $\hat{\mathbf{g}}$ on the corrugated sheet, $\hat{\mathbf{n}}$ is the normal inward versor, $2\,\kappa_s$

is the solid surface curvature, $2\,\kappa$ is the free surface curvature and D is the disjoining pressure,

$$2\,\kappa = \frac{\frac{\partial^2 h}{\partial x^2}\left[1+\left(\frac{\partial h}{\partial y}\right)^2\right] + \frac{\partial^2 h}{\partial y^2}\left[1+\left(\frac{\partial h}{\partial x}\right)^2\right] - 2\,\frac{\partial h}{\partial x}\,\frac{\partial h}{\partial y}\,\frac{\partial^2 h}{\partial x \partial y}}{\left[1+\left(\frac{\partial h}{\partial x}\right)^2+\left(\frac{\partial h}{\partial y}\right)^2\right]^{3/2}} \tag{2}$$

$$D = B\left[\left(\frac{d}{h}\right)^n - \left(\frac{d}{h}\right)^m\right],\; B = \frac{\sigma}{d}\,\frac{(n-1)\,(m-1)}{n-m}\,(1-\cos\theta) \tag{3}$$

with $n > m > 1$. Compared to the standard formulation of the lubrication theory, the full curvature implementation replaces the small slope approximation, $2\,\kappa = \nabla^2 h$, allowing accurate investigation of contact angles up to 60° [21, 24], while the disjoining pressure models the surface wettability [15, 16]. The presence of a precursor film of thickness $d \ll h$ allows the modelization of moving contact lines. Since the body-fitted orthogonal coordinate system Oxy stands on a corrugated geometry, $\tilde{\mathbf{g}}$ and $\hat{\mathbf{n}} \cdot \hat{\mathbf{g}}$, i.e. the projections of the gravity versor $\hat{\mathbf{g}}$ along the corrugated surface and along its normal inward versor, depend on the position x, y. Introducing the following non-dimensional quantities,

$$\begin{gathered} H = \frac{h}{h_0},\; \delta = \frac{d}{h_0},\; \mathbf{X} = \frac{\mathbf{x}}{L_0},\; T = \frac{t}{(L_0/u_0)}, \\ P = \frac{p}{(\sigma\,h_0/{L_0}^2)},\; 2\,K = \frac{2\,\kappa}{(h_0/{L_0}^2)},\; \Pi = \frac{D}{(\sigma/h_0)} \end{gathered} \tag{4}$$

with h_0 being the undisturbed film thickness, u_0 being the undisturbed film velocity and L_0 being the scaling length,

$$u_0 = \frac{\rho\,g\,{h_0}^2}{3\,\mu},\; L_0 = \left(\frac{\sigma\,h_0}{\rho\,g}\right)^{1/3} \tag{5}$$

the governing lubrication equation can be recast in non-dimensional form,

$$\frac{\partial H}{\partial T} + \nabla\cdot\left(H^3\,\tilde{\mathbf{g}} - H^3\,\nabla P\right) = 0,\; P = -2\,(K+K_s) - (3\,\mathrm{Ca})^{-2/3}\,\Pi + (3\,\mathrm{Ca})^{1/3}\,H\,(\hat{\mathbf{g}}\cdot\hat{\mathbf{n}}) \tag{6}$$

with Ca being the film capillary number, $\mathrm{Ca} = \mu\,u_0/\sigma$.
An in-house code, previously developed in `FORTRAN` and validated with experimental evidence [23, 21, 24], is used to numerically solve Eq. (6) on an orthogonal, structured grid via finite volume method (FVM). A second order centered scheme such as the one of [25] was implemented for spatial discretization, while the Alternate Direction Implicit (ADI) approximate factorization of [26] was used for marching in time. Parallelization for a shared memory machine was achieved via the OpenMP interface.

3 Numerical setup

Figure 1 shows the reference geometry of a corrugated sheet, defined by: the crimp angle ϕ between the corrugation direction and the vertical one; the surface corrugation angle β; the corrugation length b, i.e. the distance between two consecutive peaks or valleys. The governing lubrication equation is solved on the body-fitted orthogonal coordinate system Oxy laying on the corrugated surface, with x denoting the horizontal direction. The projection of the gravity versor $\hat{\mathbf{g}}$ along the corrugated sheet surface is required in order to implement the hydrostatic pressure in Eq. (6),

$$\tilde{\mathbf{g}} = \begin{Bmatrix} \hat{\mathbf{g}}\cdot\hat{\mathbf{i}} \\ \hat{\mathbf{g}}\cdot\hat{\mathbf{j}} \end{Bmatrix},\; \hat{\mathbf{g}}\cdot\hat{\mathbf{i}} = 0,\; \hat{\mathbf{g}}\cdot\hat{\mathbf{j}} = \sqrt{1-(\hat{\mathbf{n}}\cdot\hat{\mathbf{g}})^2} \tag{7}$$

with $\hat{\mathbf{n}}$ being the normal inward versor to the corrugated sheet surface and $\hat{\mathbf{i}}, \hat{\mathbf{j}}$ being versors directed along x, y axes. Since the straightforward calculation of the substrate curvature, $2\,\kappa_s = -\nabla\cdot\hat{\mathbf{n}}$, would lead to a numerical discontinuity in correspondence of the sharp corrugations, a dummy smoothing radius of $r = b/15$, is used only for the implementation of the substrate curvature, required by the governing lubrication equation. The computational domain is composed of a portion of the corrugated sheet of non-dimensional size $L_X \times L_Y$. The liquid is injected through the topmost section, located at $Y = 0$, through which the undisturbed film thickness and film flux are imposed: $H = 1$, $\mathbf{Q}\cdot\hat{\mathbf{n}} = 1$, with $\hat{\mathbf{n}}$

denoting the normal to the inlet section and $\mathbf{Q} = H^3 \left(\tilde{\mathbf{g}} - \nabla P\right)$ being the non-dimensional film flux. A fully developed flow is imposed through the bottomest section, $Y = L_Y$, where $\nabla H \cdot \hat{\mathbf{n}} = 0$, $\nabla P \cdot \hat{\mathbf{n}} = 0$. Periodic boundary conditions are implemented through sections $X = 0, L_X$. Since the corrugated sheet geometry is recursive, L_X must be a multiple of the periodic length for a proper implementation of the periodic boundary conditions:

$$X_\pi = \sqrt{\left(\frac{b/L_0}{\cos\phi}\right)^2 + \left(\frac{b/L_0}{\tan(\beta/2)}\right)^2} \tag{8}$$

A dry domain is always initialized at $T = 0$. This means that the solid substrate is initially covered by the precursor film: $H(X,Y)|_{T=0} = \delta$. Following literature, the disjoining exponents are set to $n = 3$, $m = 2$, while the non-dimensional precursor film thickness is set to $\delta = 5 \times 10^{-2}$. After a grid independence analysis, it was verified that a spatial discretization step of ΔX, $\Delta Y \leq \delta$ is required to get accurate results. The computational domain dimensions are set to $L_X = 2X_\pi$ and $L_Y = 50$, with X_π being the periodic length of the corrugated geometry, evaluated along x direction via Eq. (8). The condition $L_X > \lambda_{cr}$ is verified for all the investigated geometries, with $\lambda_{cr} \simeq 7.86$ being the critical wavelength of the spanwise perturbations [16]: this allows to observe the finger instability phenomenon and, thus, to accurately predict the eventual dewetting of the domain, which defines the transition between continuous film and rivulet regime.

4 Result

Parametric computations are run at different corrugated sheet geometries and flow conditions, the latter being defined by the Capillary number and by the static contact angle. The stationary flow pattern is traced, in order to analyse possible transitions from a continuous film to an ensemble of rivulets. In particular, three corrugation angles are investigated, $\beta = 90°$, $113°$ and $143°$, while the non-dimensional corrugation length and the crimp angle are set to $b/L_0 = 12$ and $\phi = 45°$. The film capillary number is ranged in $10^{-2} \div 5 \times 10^{-2}$, while the static contact angle is set to $\theta = 45°$. The imposed parameters are consistent with real test cases involving CO2 absorption through structured packing: as an example, Ca of the order of 5×10^{-2} covers the typical application described in [5], while b/L_0 in the range [12, 20] were addressed in [4].

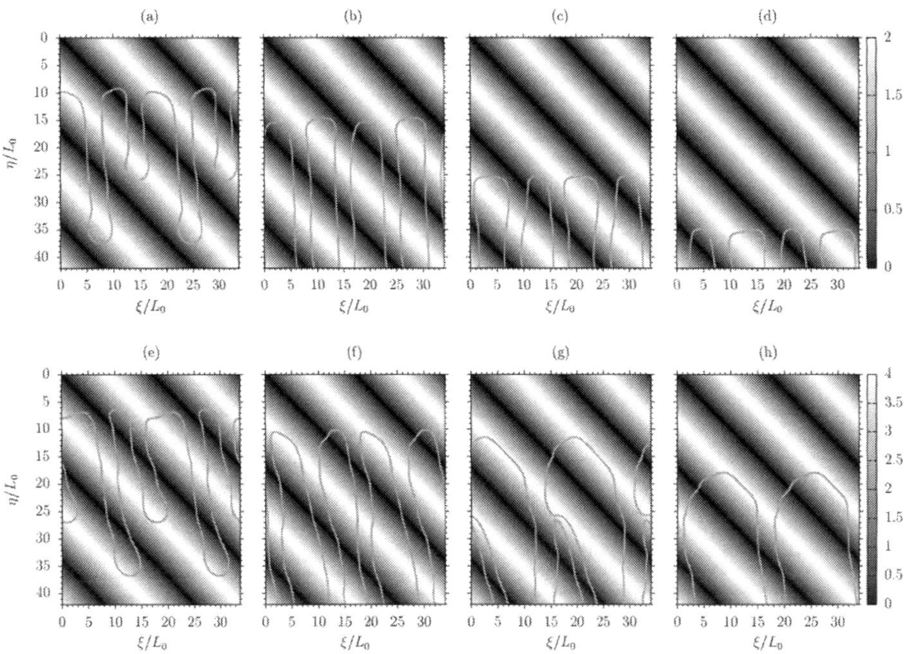

Figure 2: Contact line evolution at successive instants: $\beta = 143°$ (a-d); $\beta = 113°$ (e-h). $T = 20, 40, 80, 120$. $\theta = 45°$, $\mathrm{Ca} = 3 \times 10^{-2}$, $b/L_0 = 12$, $\phi = 45°$.

Figure 2 compares the moving contact line at successive instants, for the same $\mathrm{Ca} = 3 \times 10^{-2}$, $b/L_0 = 12$ and different corrugation angles: $\beta = 143°$, which corresponds to a corrugation height of $b/6$, and $\beta = 113°$, which corresponds to $b/3$. The stationary solution of both the configurations is characterized by a continuous film wetting the whole domain. As could be expected, the rivulets drain faster at lower corrugation height, meaning that the vertex between two neighbor rivulets moves downward with a higher velocity. Furthermore, the rivulets move essentially along the η direction when $\beta = 143°$, while they start to deviate, following the corrugation direction for $\beta = 113°$. This is due to the combined effect of gravity and surface tension. Indeed, the local substrate curvature induces additional capillary forces, that drive the liquid into the valleys of the corrugated sheet. A change in the number of rivulets is also reported: 4 rivulets are observed at high β, that is consistent with the one predicted via a stability analysis of a falling film subjected to spanwise perturbation such as the one of [16], $L_X/\lambda_{cr} = 4.4$ with $\lambda_{cr} \simeq 7.86$; 2 rivulets are observed at low β, meaning that film stability is influenced by the presence of the corrugations.

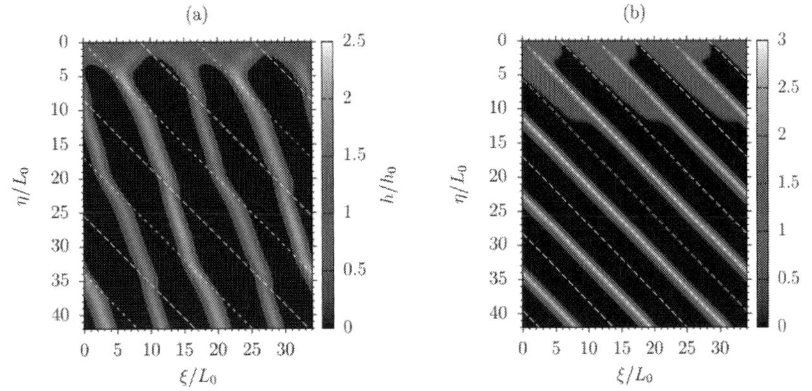

Figure 3: Stationary liquid layer distribution: $b/L_0 = 12$, $\theta = 45°$ (a); $b/L_0 = 8$, $\theta = 30°$ (b). $\mathrm{Ca} = 3 \times 10^{-}$, $\beta = 90°$, $\phi = 45°$. Dashed lines denotes valleys (short hatch) and peaks (long hatch) of the corrugated surface.

Indeed, when $\beta = 90°$, i.e. corrugation height equal to $b/2$, we observe a partial dewetting of the computational domain (i.e. receding front forms between two neighbor rivulets), covered by 4 stationary rivulets. The steady state solution for $\mathrm{Ca} = 3 \times 10^{-2}$, $b/L_0 = 12$ and $\beta = 90°$ is shown in Figure 3(a). The effect of the film Capillary number is shown in Figure 4, which refers to $\mathrm{Ca} = 4 \times 10^{-2}$, while the non-dimensional geometry is the same as Figure 3(a), i.e. $b/L_0 = 12$, $\beta = 90°$ and $\phi = 45°$: slightly increasing Ca, a change in the liquid flow pattern is observed, with the 4 rivulets, formed due to the perturbance induced by the corrugations, merging into to 2 rivulets, that slowly drain down the vertical corrugated sheet.

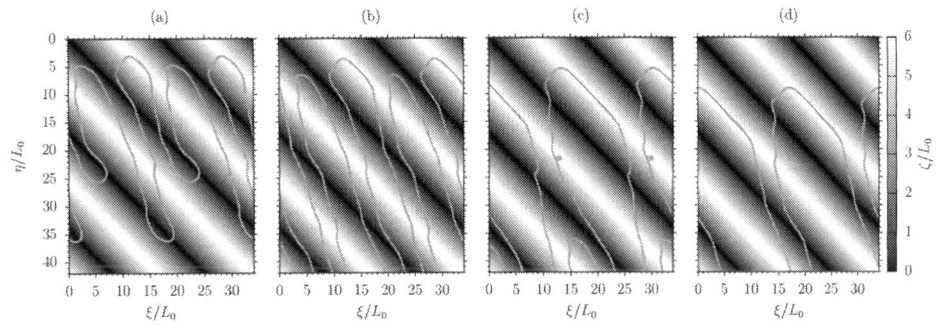

Figure 4: Contact line evolution at successive instants: $T = 20$ (a), $T = 40$ (b), $T = 80$ (c), $T = 120$ (d). $\theta = 45°$, $\mathrm{Ca} = 4 \times 10^{-2}$, $b/L_0 = 12$, $\beta = 90°$, $\phi = 45°$.

The flow pattern is traced as a function of both the Capillary number Ca and the corrugation angle β

in Table 1, which proves that the corrugation geometry affects the observed flow pattern: a continuous film, incentivated by choosing lower β, enhance the mass transfer process during the absorption process; however, lowering β decreases the corrugated sheet surface (i.e. the potential liquid-gas interface area) and may induce flooding occurrence, with an increase of the pressure drop and the possible obstruction of the gas flow. For Ca = 0.04, $\beta = 90°$, the vertices between two neighbor rivulets, formed due to finger instability, move downward with a small velocity $u/u_0 < 0.1$, see Figure 4, meaning that we are very close to the transition from film to rivulet regime

Ca \ β	90°	113°	143°
0.02	*	*	*
0.03	*	⋆	⋆
0.04	⋆	⋆	⋆
0.05	⋆	⋆	⋆

Table 1: Flow pattern map: the red asterisks denote a rivulet pattern; the blue stars denote a continuous film pattern. $\theta = 45°$, $b/L_0 = 12$, $\phi = 45°$.

An additional computation is run, in order to investigate the effect of changing the corrugation length b. Figure 3(b), referred to $b/L_0 = 8$ and $\theta = 30°$, shows a different behavior, if compared to the test case of Figure 3(a), even though a partial dewetting of the computational domain is observed in both the two cases. Indeed, the stationary rivulets follow the corrugation direction for the lower value of b, while a small deviation from the vertical direction is observed for the stationary rivulets at $b/L_0 = 12$ and the same Ca and β, see Figure 3(a). The wetted area fraction of the stationary liquid distribution is also computed for both $b/L_0 = 12$ and $b/L_0 = 8$, given that decreasing the corrugation length lowers the wetted area of about 12% and, thus, worsens the mass transfer. On the other hand, such a configuration diminishes the risk of flooding, since interference or coalescence between the rivulets flowing down adjacent, corrugated sheets is avoided.

5 Conclusion

The evolution of a liquid layer down a corrugated sheet, which is the characteristic configuration of structured packing, was investigated via the lubrication theory. The reduction to a 2D mathematical model allowed a sensible reduction of the computational cost respecting a fully 3D numerical approach. Compared to previous works, the simulation of a corrugated sheet was possible via the additional implementation of: a non-uniform gravity field in the lubrication framework; an additional force induced by the substrate curvature. The full curvature formulation, together with the disjoining pressure model, allowed to investigate surface wettabilities in the range of applications characterizing absorption through structured packing. Results show that the corrugation geometry affects the liquid flow pattern, which prediction is crucial to estimate the effective mass transfer rate in absorption applications. In particular, it was observed that low corrugation angles incentivate the formation of a continuous wetting layer, but augment the risk of flooding, that dramatically reduces the mass transfer performances. Furthermore, different possible patterns were observed for the rivulet regime, with the liquid moving downward along the vertical direction or following the corrugations, that further decreases the wetted area but avoids the occurrence of flooding.

The proposed modelization represents a novel approach and an effective tool for the design of the structured packing, since real configurations can be accurately investigated with low computational costs. The non-dimensional approach allows to extend results from a single simulation to a family of physical problems (for example, the capillary number is correlated to the liquid flux, given the liquid solvents properties). Parametric computations at different corrugation geometries and flow conditions provide useful statistical information, such as the interface area between solvent and solute, to be used for mapping the optimal geometry as a function of the required mass transfer rate. As a future work, the effect of the shear stress applied by the gas solute, which flows countercurrent to the liquid solvent, should be included, coupling the film solver with the CFD analysis of the core flow. Furthermore, this would allow to accurately predict flooding occurrence, which is as important as the liquid flow pattern over the corrugated sheet.

References

[1] Croce G, De Candido E, Habashi W G, Munzar J, Aubé M S, Baruzzi G S and Aliaga C 2010 *J. Aircraft* **47** 1283–1289
[2] Suzzi N and Croce G 2024 *J. Phys. Conf. Ser.* **2685** 012019
[3] Hoffmann A, Ausner I, Repke J U and Wozny G 2005 *Comput. Chem. Eng.* **29** 1433–1437
[4] Subramanian K and Wozny G 2012 *Int. J. Chem. Eng.* **2012** 838965
[5] Singh R K, Galvin J E and Sun X 2016 *Chem. Eng. Sci.* **142** 244–257
[6] Rocha J A, Bravo J L and Fair J R 1993 *Ind. Eng. Chem. Res.* **32** 641–651
[7] Rocha J A, Bravo J L and Fair J R 1996 *Ind. Eng. Chem. Res.* **35** 1660–1667
[8] Podgorski T, Flesselles J M and Limat L 1999 *Phys. Fluids* **11** 845
[9] Limat L R 2006 *Phys. Fluids* **18** 032102
[10] Johnson M F G, Schluter R A, Miksis M J and Bankoff S G 1999 *J. Fluid Mech.* **394** 339–354
[11] Oron A, Devis S H and Bankoff S G 1997 *Rev. Mod. Phys.* **69** 931–980
[12] Kataoka D E and Troian S M 1997 *J. Colloid Interf. Sci.* **192** 350–362
[13] Beltrame P, Knobloch E, Hänggi P and Thiele U 2011 *Phys. Rev. E* **83** 016305
[14] Ma C and Liu J 2021 *Phys. Fluids* **33** 022101
[15] Schwartz L W and Eley R R 1998 *J. Colloid Interf. Sci.* **202** 173–188
[16] Zhao Y and Marshall J S 2006 *J. Fluid Mech.* **559** 355–378
[17] Sellier M 2015 *Int. J. Multiphas. Flow* **71** 66–73
[18] Blake T D 2006 *J. Colloid Interf. Sci.* **299** 1–13
[19] Bonn D, Eggers J, Indekeu J, Meunier J and Rolley E 2009 *Rev. Mod. Phys.* **81** 739–805
[20] Perazzo C A and Gratton J 2004 *J. Fluid Mech.* **507** 367–379
[21] Suzzi N and Croce G 2019 *Phys. Fluids* **31** 122106
[22] Suzzi N and Croce G 2021 *J. Phys. Conf. Ser.* **1868** 012010
[23] Suzzi N and Croce G 2017 *J. Phys. Conf. Ser.* **923** 012020
[24] Suzzi N and Croce G 2021 *Fluids* **6** 405
[25] Diez J A and Kondic L 2002 *J. Comput. Phys.* **183** 274–306
[26] Witelski T P and Bowen M 2003 *Appl. Numer. Math.* **45** 331–351

Acknowledgments

We wish to acknowledge NextGenerationEU for supporting the present work, performed within the project Interconnected Nord-Est Innovation Ecosystem (iNEST), placed within the context of National Recovery and Resilience Plan (PNRR).

Parametric analysis of delivered power during laser ablation for prostate cancer.

A. Andreozzi[1], G. Napoli[2*] and G.P. Vanoli[2]

[1] Università degli Studi di Napoli Federico II, Dipartimento di Ingegneria Industriale, Piazzale Tecchio 80, 80125 Napoli, Italy
[2] Università degli Studi del Molise, Dipartimento di Medicina e Scienze della Salute "Vincenzo Tiberio", Via Francesco de Sanctis 1, 86100 Campobasso, Italy
* Corresponding author: giovanni.napoli@unimol.it

Abstract. Prostate cancer is one of the most recurrent forms of cancer in men, occurring in peripheral zones. Laser ablation is an emerging non-invasive protocol for this disease, offering the possibility to preserve the prostate proper functioning. However, due to shortage of in-vivo experiments, for ethical and practical reasons, it is difficult to properly design the treatment, leading to incomplete tumor destruction and metastasis development, owed to inappropriate exposure time or laser intensity. Consequently, it is of primary importance to develop accurate models to aid and provide guidelines to surgeons, towards a treatment optimization. For these reasons, in this paper an accurate model has been developed and solved through the finite elements commercial software Comsol Multiphysics. A 2D domain, representative of the tumor inside the prostate, has been investigated, using the porous media approach. According to the Local Thermal Non-Equilibrium (LTNE) assumption, the tissue and the blood are treated as two distinct entities having different thermal properties and behavior. The laser source has been described by means of the Beer-Lambert's law, assuming a Gaussian laser distribution. To find the optimal laser setting to achieve the maximum tumor destruction, the effect of different laser intensity, and bare fiber diameter on temperature field and thermal damage is investigated.

Keywords: Laser ablation, bioheat transfer, prostate cancer, LTNE, porous media

Nomenclature

a	Volumetric heat transfer area $[m^{-1}]$	β	Velocity correction factor [-]
A	Frequency factor $[s^{-1}]$	ΔE	Activation energy $[J\ mol^{-1}]$
c	Specific heat $[J\ kg^{-1}\ K^{-1}]$	ΔH	Latent heat of vaporization $[kJ \cdot kg^{-1} \cdot K^{-1}]$
d	Blood vessel diameter $[m]$	ρ	Density $[kg \cdot m^{-3}]$
g	Anisotropy factor	ε	Porosity [-]
h_c	Convective heat transfer coefficient $[W\ m^{-2}\ K^{-1}]$	μ_{eff}	Effective absorption coefficient $[m^{-1}]$
$I(r,z)$	Laser intensity $[W\ m^{-2}]$	σ	Gaussian beam standard deviation
I_0	Laser intensity at tissue surface $[W\ m^{-2}]$	φ	Angle in the direction perpendicular to the laser surface [°]
k	Thermal conductivity $[W\ m^{-1}\ K^{-1}]$	Ω	Thermal damage [-]
P	Laser power $[W]$	ω_b	Damaged tissue blood perfusion rate $[s^{-1}]$
$\dot{Q}$	External heat generation $[W\ m^{-3}]$	ω_0	Healthy tissue blood perfusion rate $[s^{-1}]$
R	Universal gas constant $[J\ mol^{-1}\ K^{-1}]$	**Subscript**	
r_{fiber}	Laser diode radius $[m]$	b	Blood
$r_{prostate}$	Prostate radius $[m]$	$coag$	Coagulated tissue
r_{tumor}	Tumor radius $[m]$	g	Gas

r, z	Spatial coordinates $[m]$	l	Liquid
t	Time $[s]$	$laser$	Referred to the laser
T	Temperature [K]	max	Maximum
u	Blood velocity $[m\,s^{-1}]$	met	Metabolism
W	Water content $[kg\,m^{-3}]$	min	Minimum
Greek symbols		nat	Native tissue
α	Absorption coefficient $[m^{-1}]$	t	Tissue
α_s	Scattering coefficient $[m^{-1}]$	var	Variable

1. Introduction

Prostate cancer is a recurrent disease and one of the most frequent forms of cancer diagnosed in men. However, considering that it is more likely to appear in the prostate's peripheral zone [1] and the rate of mortality is remarkably decreased in the last decades [2], invasive techniques, such as surgery, can be avoided and replaced by minimally invasive therapies, like ablative procedures [3], [4] or heat-triggered drug delivery [5]. Ablation methods offer the advantage of preserving patient proper prostate function. Among different hyperthermia protocols, laser induced hyperthermia results the most suitable to treat prostate diseases, due to the small area interested. The effectiveness of transperineal laser ablation for low and intermediate risk prostate cancer was demonstrated in a clinical trial carried out by Manenti et al. [6], finding a disease-free rate of 80% over a 3-year period.

During ablative procedures thermal energy is delivered to the target body, namely the tumoral tissue, in order to cause a temperature increase and consequently blood coagulation and tissue necrosis. However, the main goal of this process is to achieve necrosis in the tumor avoiding overheating and damaging the surrounding healthy tissue, since an overexposure to the thermal source could cause damages to the prostate and to the rectal wall, determining also erectile dysfunction [7]. However, being less widespread, there is still the need for a recognized, reliable and approved guidelines for laser ablation. Nevertheless, in-vivo data are still scarce, due to practical and ethical reasons, thus accurate computational models, describing the heat transfer in biological tissues are needed to develop personalized and minimally invasive treatment, aiming to optimize the procedure outcomes maximizing the complete tumor necrosis, preventing tumor recurrence and avoiding healthy tissue damaging.

Marqa et al. [8] developed a FEM model, based on Pennes equation [9] to compare the numerical results of focal laser ablation for prostate cancer with in-vivo experiments on rats. A good agreement between numerical results and the experiments was found, although they did not model the thermal source, assumed to be constant. Then, other works also included the effect of the laser source in the model, such as in Zhang et al. [10] and Namakshenas et al. [11]. The possibility to successfully improve the treatment outcomes by means two laser fibers was investigated by Wu et al. [12].

As pointed out, the proper modelling and then the correct power to deliver during laser ablation is still on ongoing issue, thus research is still needed to improve the computational methods and aid the surgeon to design properly the ablation procedure. For this reason, among the different approaches to address heat transfer in biological tissue, the one developed by Xuan and Roetzel [13], based on the porous media theory, with the local thermal non equilibrium assumption, has been implemented in this work. According to this approach, the biological domain is split into two different matrixes, a solid one and a liquid one, coexisting simultaneously and having different thermal behavior. For the different biological entities, namely blood, tissue and tumoral mass, specific thermophysical properties are employed, also considering their variation with temperature and thermal damage development. The external heat source is modelled with Beer-Lambert's law. The bioheat transfer problem is solved by means a FEM commercial code. The results show the effect of different laser diode and laser power.

2. Model description

The prostate is assumed to be a sphere having a diameter of 30 mm, with an inner concentrical tumor with a radius of 10 mm. For the sake of simplicity and to reduce computational times, due to axisymmetric of the geometry chosen, the 3D domain is reduced to a 2D axial-symmetrical one, significantly lowering computational time without affecting results accuracy. To execute the thermal

Journal of Physics: Conference Series **2940** (2025) 012011 doi:10.1088/1742-6596/2940/1/012011

therapy a bare-fiber, 27 *mm* in length, is inserted within the tissue to provide the thermal energy. The investigated domain is shown in Figure 1.

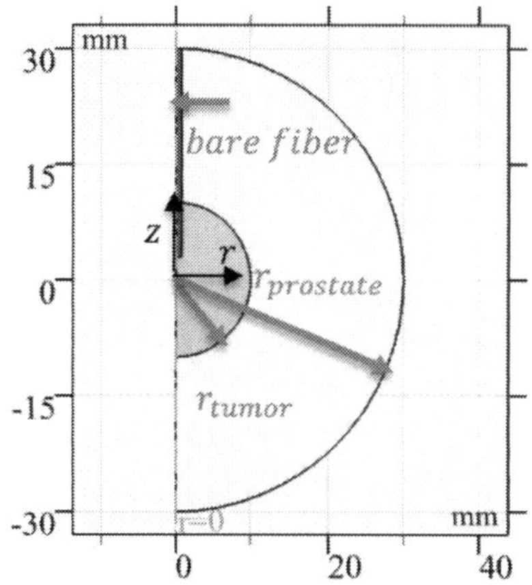

Figure 1. Investigated 2D domain. Spherical prostate in cyan, spherical tumor in yellow and bare fiber in violet.

To solve the problem boundary and initial conditions need to be fixed. Particularly, given the axisymmetric assumption, the geometrical symmetry axis is assumed to be also a thermal symmetry axis, thus a Neumann boundary conditions is set: $k\frac{T_t}{\partial t}(r = 0, t) = 0$. Then, since the investigated domain is sufficiently large to neglect the effect of the thermal therapy on the boundary of it, a Dirichlet boundary condition is applied on the prostate external surface, $T_t(r = r_{prostate}, t)$=37 °C

As initial condition, a temperature equal to 37 °C is assumed for both the investigated domains.

Different models have been proposed to account the heat transfer in biological tissue during hyperthermia, however, in this work the porous media approach has been followed, separating the biological tissue in two different regions, the fluid one (made up by vascular structures) assumed to be uniformly distributed into the computational domain, and the solid (tissue). Concerning the thermal response to the hyperthermia treatment, according to the approach developed by Xuan and Roetzel [13], a local thermal non-equilibrium (LTNE) assumption has been made and two different thermal behaviors are considered for the two phases, thus two energy equations are employed to address separately the solid and the liquid phases. Here, to take into account for blood perfusion, the modified version of the porous media model, formulated by [14], has been considered:

$$(1-\varepsilon)\cdot\rho_t\cdot c_t\cdot\frac{\partial T_t}{\partial t} = \nabla\cdot\left((1-\varepsilon)\cdot k_t\cdot\nabla T_t\right) - h_c\cdot a\cdot(T_t - T_b) - \rho_b\cdot c_b\cdot\omega_b\cdot\left((T_t - T_b)\right) + (1-\varepsilon)\cdot\dot{Q}_{laser} + (1-\varepsilon)\cdot\dot{Q}_{met} \quad (1)$$

$$\varepsilon\cdot\rho_b\cdot c_b\cdot\left(\frac{\partial T_b}{\partial t} + \beta\cdot u\cdot\nabla T_b\right) = \nabla\cdot(\varepsilon\cdot k_b\cdot\nabla T_b) + h_c\cdot a\cdot(T_t - T_b) + \rho_b\cdot c_b\cdot\omega_b\cdot\left((T_t - T_b)\right) + \varepsilon\cdot\dot{Q}_{laser} \quad (2)$$

In equations (3-4) b and t refer, respectively, to blood and tissue. ε is the tissue porosity. ρ, c and k are, the density, specific heat and the thermal conductivity. T and t are, respectively, the temperature and the time. Then, h_c and a are the convective heat transfer coefficient and the volumetric heat transfer area between blood and tissue, and ω_b is the perfusion coefficient. u is the blood velocity modulus and β is its corrective factor considering the thermal damage induced by temperature increase and thus blood coagulation. Finally, $\dot{Q}_{laser}$ and $\dot{Q}_{met}$ represent the external heat generation supplied by means the laser and the metabolic heat generation (assumed to be equal to 33.8 W/m^3), that during hyperthermia processes is significantly lower compared to external heat source.

Due to a lower vascularization inside the tumor, as reported in [15], the perfusion decreases, consequently a variable porosity is considered in the tumor, increasing with a quadratic law from the inner to the boundary. Thus, as done in [16] and [5] the porosity in the tumor is expressed as follows:

$$\varepsilon_{var} = \left(\frac{\varepsilon_{max} - \varepsilon_{min}}{r_{tumor}^2}\right) \cdot r^2 + \left(\frac{\varepsilon_{max} - \varepsilon_{min}}{r_{tumor}^2}\right) \cdot z^2 + \varepsilon_{min} \tag{3}$$

being ε_{min}=0.07 and ε_{max}=0.2, the minimum and maximum porosity, respectively, in the tumor centre and on its boundary (as well as in healthy tissue), and r_{tumor} is the tumor radius.

Due to the temperature increase in the tissue, that could be even higher than 100 °C, water vaporization could occur, determining water content variation. As done in [12] the water content in the prostate is:

$$W(T) = 0.778 \cdot \begin{cases} 1 - e^{\frac{T-106}{3.42}} & T \leq 103\,°C \\ 0.0289 \cdot T^3 - 8.924 \cdot T^2 + 916.6 \cdot T - 31573 & 103\,°C < T \leq 104\,°C \\ e^{\frac{-T-80}{34.37}} & T > 104\,°C \end{cases} \tag{4}$$

Thus, the thermal properties variation result affected by water content variation, as reported in [10], thus the density and specific heat as a function of water content are expressed as follows:

$$\rho(T) = 1000 \cdot W(T) + 0.222 \cdot 1300 \tag{5}$$

$$c(T) = \begin{cases} c_l - \frac{\Delta H}{\rho} \cdot \frac{\partial W(T)}{\partial T} & T \leq 100\,°C \\ c_g & T > 100\,°C \end{cases} \tag{6}$$

where c_l and c_{p_g} are the specific heat of, respectively, the liquid and the gas phases and ΔH is the water latent heat of vaporization at 100 °C. Since the model is developed under the LTNE assumption, equations (5-6) are computed separately for the two-phases, however here for brevity only an equation for each property is reported, where the temperature T refers to the phase considered.

As the temperature increases a thermal damage is induced inside the tissue, and it is evaluated with Arrhenius model, as proposed by [17], as a function of temperature and exposure time:

$$\Omega = \int_0^t A \cdot e^{\frac{-\Delta E}{R \cdot T_t(t)}} dt \tag{7}$$

where A is the frequency factor, ΔE the activation energy for the irreversible damage reaction and R is the universal gas constant.

The temperature increase and the consequent thermal damage propagation inside the tissue, significantly affect blood velocity and perfusion. Particularly, concerning the blood velocity, the coefficient β is introduced to account for a linear variation of the blood velocity with thermal damage:

$$\beta = 1 - \Omega(t) \tag{8}$$

Regarding the velocity vector, it is assumed to have two components along axial (r) and longitudinal (z) directions, evaluated as follows:

$$u_r = -|u| \cdot cos\varphi \tag{9}$$

$$u_z = -|u| \cdot sen\varphi \tag{10}$$

The negative sign in the previous is adopted because the flow is assumed to enter from the boundary to the inner of the computational domain, while the angle φ is measured from the direction perpendicular to the tip of the laser fiber.

As well as the blood velocity, also the perfusion term results affected by ongoing thermal damage:

$$\omega_b = \begin{cases} (1 + 25 \cdot \Omega(t) - 260 \cdot \Omega(t)^2) \cdot \omega_0 & \Omega(t) \leq 0.1 \\ (1 - \Omega(t)) \cdot \omega_0 & 0.1 < \Omega(t) < 1 \\ 0 & \Omega(t) = 1 \end{cases} \tag{11}$$

where ω_0 is the healthy tissue blood perfusion coefficient. The equation above indicates that when the thermal damage is lower than 0.1 the temperature increase causes a vasodilation involving a blood flow increase and thus the perfusion coefficient enhances. As the thermal treatment proceeds and the temperature increases, the blood clotting determines a strong decrease in blood perfusion, which ceases when the maximum thermal damage is reached, due to tissue necrosis and blood coagulation.

Regarding the convective heat transfer between tissue and blood, the convective heat transfer coefficient is evaluated considering a Nusselt number of 4.93 [18]:

$$h_c = Nu \cdot \left(\frac{k_b}{d}\right) \tag{12}$$

The volumetric heat transfer area between tissue and blood, evaluated with the hydraulic diameter is defined for porous media as:

$$a = \frac{4 \cdot \varepsilon}{d} \tag{13}$$

where d is the blood vessel diameter.

To evaluate the heat absorbed by the tissue, the Beer-Lambert's law has been employed to evaluate the heat provided by laser:

$$\dot{Q}_{laser}(r,z) = \mu_{eff} \cdot I(r,z) \cdot e^{(-\mu_{eff} \cdot z)} \tag{14}$$

where μ_{eff} is the tissue effective absorption coefficient, calculated considering the linear absorption coefficient α, the scattering coefficient, α_s and the anisotropy factor g:

$$\mu_{eff} = \sqrt{3 \cdot \alpha(\Omega) \cdot \left(\alpha(\Omega) \cdot +\alpha_s(\Omega) \cdot (1-g)\right)} \tag{15}$$

However, during the ablation process and the consequent thermal damage propagation the optical properties vary substantially, thus, as reported in [19], a linear variation, dependent on the thermal damage fraction, is considered between the values for the healthy tissue and those for the necrotic one:

$$\alpha(\Omega) = \alpha_{nat} \cdot (1-\Omega) + \alpha_{coag} \cdot \Omega \tag{16}$$

$$\alpha_s(\Omega) = \alpha_{s_{nat}} \cdot (1-\Omega) + \alpha_{s_{coag}} \cdot \Omega \tag{17}$$

The laser irradiance is described by a two-dimensional Gaussian profile:

$$I(r,z) = I_0 \cdot e^{-\left(\frac{r^2+z^2}{2\cdot\sigma^2}\right)} \tag{18}$$

where r is the radial distance from the source, σ Gaussian distribution standard deviation, assumed to be $\frac{r_{fiber}}{3}$ and I_0 is the initial intensity:

$$I_0 = \frac{P}{2 \cdot \pi \cdot \sigma^2} \tag{19}$$

The biological properties considered within this work are shown in Table 1.

The equations are numerically solved by means a finite-element software, COMSOL Multiphysics. The PARDISO solver is employed, and the transient study is carried out with a Backward Euler scheme, with a maximum time step of 0.001 s and an absolute tolerance of 0.0001. A triangular mesh is adopted and the results independence of the mesh elements has been verified. Particularly, the results are not affected for a number of elements higher of 47089, thus, to reduce computational cost, this value has been selected.

Table 1. Biological tissue properties

		Prostate	Blood	Tumor
$\rho\,[kg\,m^{-3}]$ [10]	Eq.7			
$c_l\,[J\,kg^{-1}\,K^{-1}]$ [20]	T<100°C	4000	4180	4000
$c_g\,[J\,kg^{-1}\,K^{-1}]$ [21]	T>100°C	2156	2156	2156
$k\,[\mathrm{W\,m}\,K^{-1}]$ [11]		0.52	0.51	0.6
$\omega_0\,[s^{-1}]$ [11]		$5.6\cdot10^{-3}$	-	$5.6\cdot10^{-3}$

$\alpha\ [m^{-1}]$ [12]	α_{nat}	30
	α_{coag}	40
$\alpha_s\ [m^{-1}]$ [12]	$\alpha_{s,nat}$	8000
	$\alpha_{s,coag}$	18000
g [-] [12]	0.95	
$\Delta E\ [J\ mol^{-1}]$	$2.768{\cdot}10^5$	
$A\ [s^{-1}]$	$5.51{\cdot}10^{41}$	

3. Results

The FEM simulations results are here shown in terms of temperature field and necrotic surface in both, the tumor and the healthy tissues. A treatment time of 500 seconds is assumed, and the effect of different delivered powers (from 1 to 7 *W*) and bare fiber diameters (from 0.1 to 1 *mm*) is investigated.

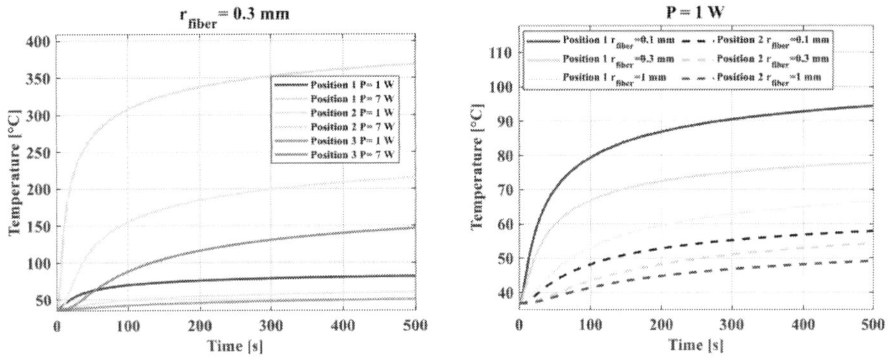

Figure 2. a) Effect of the laser power on different positions for a bare fiber diameter of 0.3 mm. b) Effect of bare fiber radius for a laser power of 1 W.

In Figure 2 the temperature profile during the whole treatment is shown for different points chosen in the computational domain. In Figure 2a the temperature evolution is shown in position 1, 2 and 3, respectively, 2, 4 and 6 *mm* distant from the laser fiber in radial direction and 3 *mm* in axial direction. Particularly, it appears evident, that, increasing the laser power the tissue temperature remarkably increases. Moreover, greater the distance from the laser source, smaller the temperature difference with the laser power rise. As a matter of fact, after 500 *s*, in position 1, with power increase, temperature augments from 81.7 *°C* to 368.3 *°C*, from 59.9 *°C* to 214.8 *°C* in position 2 and from 50.4 *°C* to 146.3 *°C* in position 3. Also, for a given point in the computational domain, the temperature transient lasts more as the laser power increases. Concerning the effect of bare fiber diameter, depicted in Figure 2b, it can be seen that as the laser fiber decrease, for a fixed power, the maximum temperature achieved augments. This is due to a higher intensity incoming on the tissue, as evident from Eqs. (17-18), since decreasing the bare fiber radius leads to a lower gaussian distribution standard deviation and thus a more focused light beam. This difference is more evident in position 1, while becomes weaker moving away from the bare fiber.

In Figure 3 the temperature fields at the end of the protocol are shown for a laser power of 3 *W* and a bare fiber radius of, respectively 1 *mm*, Figure 3a, and 0.15 *mm*, Figure 3b. Moreover, also three different temperatures contour, at 60, 55 and 50 *°C*, are shown in white. As previously pointed out, it is evident that lower the laser fiber diameter, higher the temperature in the tissue and consequently its spatial spreading in the tissue. Moreover, in the second case the temperature level corresponding to 60 *°C* is almost overlapped with the tumor contour, indicating that the thermal damage spreads outside its boundary with probably undesired overexposure to the treatment and healthy tissue damaging.

Journal of Physics: Conference Series **2940** (2025) 012011 doi:10.1088/1742-6596/2940/1/012011

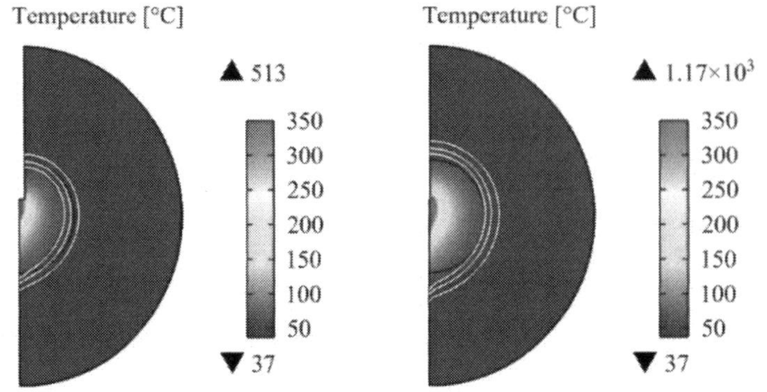

Figure 3. Temperature fields in the tumor and in the healthy tissue and temperature contours at 60, 55 and 50 *°C* for: a) bare fiber radius of 1 *mm* P = 3 *W*. b) bare fiber radius of 0.15 *mm* and P = 3 *W*.

This is also confirmed in Figure 4, where the thermal damage field is shown for a bare fiber radius of 0.5 *mm* and different laser power. Particularly, in Figure 4a, due to the high power, 7 *W*, the temperature contours at 60, 55 and 50 *°C* fall outside the tumor boundary, and coherently the thermal damage propagates outside it, clear sign of overtreatments and undesired over damage to the surrounding healthy tissue. On the other hand, in Figure 4b a laser power of 1 *W* determines an incomplete treatment, since only a fraction of the tumoral tissue is affected by it and necrotized. This could lead to tumor regrowth and metastasis formation.

These outcomes underline that it is necessary to optimize the design protocol power and time to avoid adverse effects. Consequently, it appears evident that the incidence of laser properties, in terms of dimensions, namely bare fiber diameter, and delivered power has a not negligible effect on protocol output.

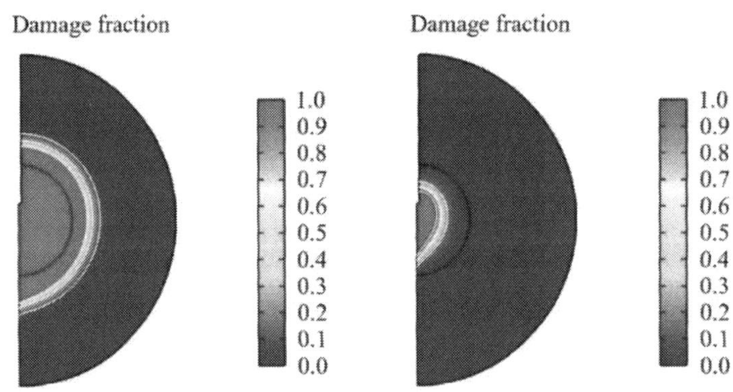

Figure 4. Damage fields in the tumor and in the healthy tissue and temperature contours at 60, 55 and 50 *°C* for: a) bare fiber radius of 0.5 *mm* and P = 7 *W*. b) bare fiber radius of 0.5 *mm* and P = 1 *W*.

4. Conclusions

In the present manuscript a numerical analysis of laser ablation of prostate cancer is carried out, in order to deepen the bare fiber radius and laser power on the protocol output. Particularly, the tumor and the surrounding healthy tissue are both considered as two spheres, simplified as a 2D axisymmetric domain to reduce computational time. The porous media approach is employed to address the biological tissue and the heat transfer is accounted with the local thermal non equilibrium approach, with a variable porosity inside the tumor. Also, the thermophysical properties variation with the temperature and the ongoing thermal damage, evaluated with the Arrhenius approach, are taken into account. The model is implemented in the commercial finite elements software Comsol Multiphysics.

The results highlight that increasing the laser power a remarkable rise in the thermal power delivered to the tissue is achieved and thus a faster and higher temperature increase is observed. Particularly, closer to the laser beam, higher the temperature increase. Moreover, it is also demonstrated that this effect is more evident at lower bare fiber diameter, when a more concentrated light beam causes a higher light intensity and thus a larger temperature increase in the tumor. Coherently, as the temperature field augment, also the thermal damage propagates more deeply in the tissue.

Finally, it is worth mentioning that the laser properties and treatment time must be carefully designed to avoid undertreatment of the target and undesired damage to the healthy tissue.

Acknowledgments
This work is partially founded by National Plan for NRRP Complementary Investments D^3 4 Health: Digital Driven Diagnostics, prognostics and therapeutics for sustainable Health care.

5. References

[1] J. G. Mygatt *et al.*, *Cancer Med*, vol. 9, no. 6, pp. 2235–2242, Mar. 2020, doi: 10.1002/cam4.2864.
[2] "American Cancer Society. Cancer statistics center, https://cancerstatisticscenter. cancer.org; 2021."
[3] A. Andreozzi, L. Brunese, M. Iasiello, C. Tucci, and G. P. Vanoli, Mar. 15, 2019, *Springer New York LLC*. doi: 10.1007/s10439-018-02177-x.
[4] A. Andreozzi, M. Iasiello, G. Napoli, and G. P. Vanoli, in *International Symposium on Advances in Computational Heat Transfer*, 2024, pp. 151–162.
[5] G. Adabbo, A. Andreozzi, M. Iasiello, P. A. Netti, and G. P. Vanoli, *International Journal of Thermal Sciences*, vol. 193, p. 108528, Nov. 2023, doi: 10.1016/J.IJTHERMALSCI.2023.108528.
[6] G. Manenti *et al.*, *Cancers (Basel)*, vol. 16, no. 7, Apr. 2024, doi: 10.3390/cancers16071404.
[7] M. Ahdoot, A. H. Lebastchi, B. Turkbey, B. Wood, and P. A. Pinto, May 01, 2019, *Lippincott Williams and Wilkins*. doi: 10.1097/CCO.0000000000000515.
[8] M. F. Marqa, P. Colin, P. Nevoux, S. R. Mordon, and N. Betrouni, *Biomed Eng Online*, vol. 10, Jun. 2011, doi: 10.1186/1475-925X-10-45.
[9] H. H. Pennes, J. Appl. Physiol. 1 (2) (1948) 93–122.," 1948.
[10] S. Zhang *et al.*, *Comput Methods Programs Biomed*, vol. 217, Apr. 2022, doi: 10.1016/j.cmpb.2022.106697.
[11] P. Namakshenas *et al.*, *Sci Rep*, vol. 13, no. 1, Dec. 2023, doi: 10.1038/s41598-023-37859-7.
[12] X. Wu *et al.*, *PLoS One*, vol. 13, no. 10, Oct. 2018, doi: 10.1371/journal.pone.0206065.
[13] Y. Xuan and W. Roetzel, *Chem. Eng. Technol*, vol. 20, pp. 268–276, 1997, doi: https://doi.org/10.1002/ceat.270200407.
[14] A. Nakayama and F. Kuwahara, *Int J Heat Mass Transf*, vol. 51, no. 11–12, pp. 3190–3199, Jun. 2008, doi: 10.1016/j.ijheatmasstransfer.2007.05.030.
[15] P. A. Netti *et al.*, 1999. [Online]. Available: www.pnas.org.
[16] A. Andreozzi, L. Brunese, M. Iasiello, C. Tucci, and G. P. Vanoli, *Computation*, vol. 9, no. 5, May 2021, doi: 10.3390/computation9050050.
[17] I. A. Chang, *Open Biomed Eng J*, vol. 4, pp. 3–12, Feb. 2010, doi: 10.2174/1874120701004020003.
[18] P. Yuan, *Med Eng Phys*, vol. 30, no. 2, pp. 135–143, Mar. 2008, doi: 10.1016/J.MEDENGPHY.2007.03.006.
[19] P. Namakshenas and A. Mojra, *Physica Medica*, vol. 89, pp. 104–113, Sep. 2021, doi: 10.1016/j.ejmp.2021.07.034.
[20] R. Mastrullo, A. W. Mauro, G. Napoli, and L. Viscito, *Case Studies in Thermal Engineering*, vol. 54, 2024, doi: 10.1016/j.csite.2024.104081.
[21] A. Cafarchio, M. Iasiello, G. P. Vanoli, and A. Andreozzi, *Comput Biol Med*, vol. 167, p. 107669, Dec. 2023, doi: 10.1016/J.COMPBIOMED.2023.107669.

Journal of Physics: Conference Series **2940** (2025) 012012 doi:10.1088/1742-6596/2940/1/012012

Thermal conductivity enhancement in carbon black composite PCM for thermal energy storage

Amit Kumar Mishra[1,*], Alessandro Ribezzo[1], Matteo Morciano[1], Luca Bergamasco[1], Elena Campagnoli[1], Valter Giaretto[1], Matteo Fasano[1], Eliodoro Chiavazzo[1,2,*]

[1] Department of Energy, Politecnico di Torino, Corso Duca degli Abruzzi 24, 10129 Torino, Italy
[2] Istituto Nazionale di Ricerca Metrologica, Strada Delle Cacce 91, 10135 Torino, Italy

* Corresponding author: amit.mishra@polito.it, eliodoro.chiavazzo@polito.it

Abstract. Phase Change Materials (PCMs) hold significant potential for developing next-generation energy storage technologies due to their cability to reversibly store and release thermal energy with energy density through isothermal phase transition. This work explores improving the heat conduction and thermophysical properties of PCMs, tackling issues such as low thermal conductivity, which can result in longer charge and discharge times. We demonstrated that the thermal conductivity of lauric acid-based PCM (LA-PCM) loaded with carbon black nanoparticles (CBNP) can be significantly enhanced. The PCM loading with 5 wt.% CBNP nanoparticles had an approximately 50% enhancement in thermal conductivity. Additionally, 25 wt.% of turmeric powder (Curcuma) significantly improved shape stability and stopped material leakage across the solid-liquid phase transition of composite PCM. The enhanced thermal conductivity of the composite PCM can be attributed to the development of interconnected percolation networks of CBNP particles with low thermal resistance during the freezing (solidification) of the PCM that provides a more effective phonon transport path. Sustainable energy materials (composite PCMs) with better thermal properties, shape stability, and low cast would be beneficial for practical application suitability and effective energy storage applications.

1. Introduction

Harnessing clean energy from sunlight and waste heat is a promising solution to address environmental challenges and energy shortages[1]. Latent heat-based TES using PCMs is an attractive choice for TES due to its high storage density (2–3 times compared to sensible heat storage systems) and ability to store/release latent energy at an almost constant temperature, corresponding to PCM's phase transition temperature[2]. PCMs have been extensively employed in various fields, like solar heat storage, waste heat recycling, and energy-saving buildings[3]. However, the practical use of organic PCMs may be greatly restricted due to their low thermal conductivity (TC) and poor heat storage capability, leading to slow charging/discharging transients during phase change processes[4]. In the past, various techniques were used to boost the heat conduction (thermal conductivity) of PCM[5],[6],[7]. However, leakage during melting (solid-liquid) phase transition makes conventional PCMs inappropriate for long-term implementation. Hence, shape-stable PCMs are being designed to cope with PCM leakage issues. In such context, selecting low-cost and eco-friendly alternative additives (fillers) and supporting materials to achieve superior thermophysical properties and energy storage efficiency of PCMs is particularly important.

2. Experimental Methods

2.1. Materials & Sample preparation

Lauric acid (LA) was purchased from Sigma-Aldrich (99% purity), and the carbon black nanoparticles (CBNP) were purchased from PowderNano (size ~20 nm) for TC enhancement of PCM (i.e., LA). In the study the materials purchased from the suppliers were used without any additional purification. Four different concentrations (2.0, 4.0, 5.0 and 6 wt.%) of CBNP were dispersed directly in the liquid PCM (sample temperature exceeds the respective PCM melting temperatures: and thereafter sample was undergone to water bath sonication for 1 hour at 50 °C. The resulting mixture was cooled at room temperature and the composite PCM prepared for thermal characterization (dim. 5.5 cm, height 1 cm).

2.2. Thermal Characterization of Composite PCMs

The latent heat values and phase transition temperatures of the PCM and CBNP-loaded composite PCM were investigated by the DSC (differential scanning calorimetry) technique under a nitrogen atmosphere with a 3 °C/min heating rate, using DSC 214 Polyma. The thermal conductivity/diffusivity of the pure PCMs (LA) and composite PCM were determined using a Thermal Analyzer (Hot Disk, TPS 500) with a nominal accuracy of ±5% and ±10%, respectively. The Hot Disk sensor is a double-spiral electrically conductive pattern etched from a thin nickel foil and its working principle and thermal properties measurement details are available at [8].

Figure 1 depicts heat stored/released characteristics during the melting and cooling of PCM across its phase transition potentially having superior TES behavior compared to sensible storage medium (pure solid/liquid) cases, owing to its high latent heat capacity.

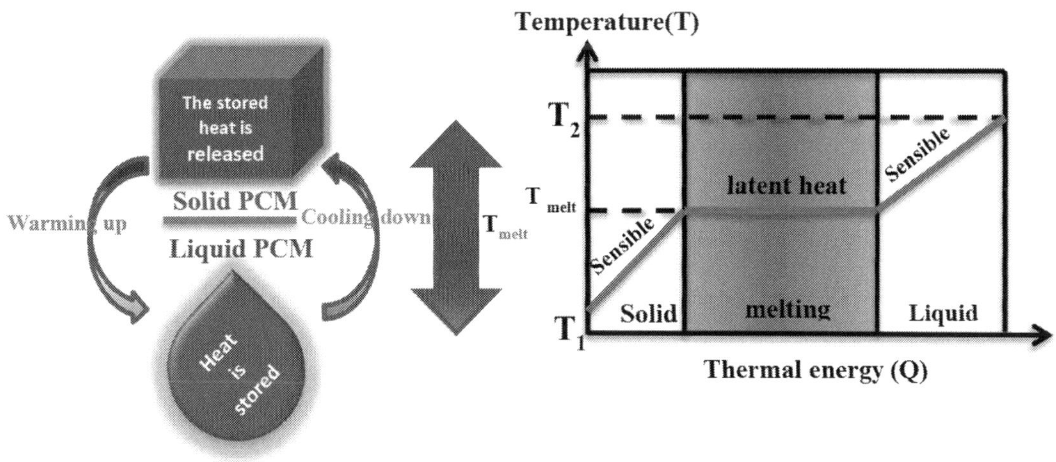

Figure 1 schematics of working principle of typical PCM across phase transition temperature.

Journal of Physics: Conference Series **2940** (2025) 012012 doi:10.1088/1742-6596/2940/1/012012

3. Results and discussion

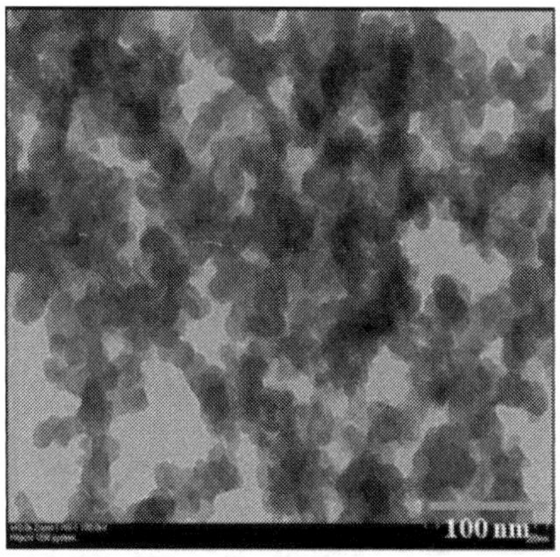

Figure 2 TEM image of CBNP nanoparticles, provided by (PowderNano) supplier.

Figure 2 shows the transmission electron microscopy (TEM) image of the CBNP nanoparticles, where the spherical morphology of the CBNP nodules (particles size in the order of 20 nm) was identifiable. It has been studied that CBNP nanoparticles form aciniform aggregates [2]. By DSC measurement the melting and solidification temperatures obtained are equal to 45.4 °C and 39.6 °C, respectively, and corresponding values of latent heat (calculated through the area under the curves) of pristine PCM amounted to 172.5 J g^{-1} and 171.2 J g^{-1}.

Table 1 illustrates the thermal conductivity and thermal diffusivity values of the PCM as the CBNP concentration changes. In detail, the thermal conductivity and diffusivity of PCM effectively increased from 0.219-0.333 W/(m.K) (i.e., ~50%) and from 0.102-0.276 mm^2/s, respectively, as the concentration of CBNP increased to about 5%. However, with further high loading concentration (i.e., 6.0 wt. %) the TC enhancement significantly decreased, which can be possibly attributed to the formation of larger particle clusters. The TC of composite PCMs was enhanced more significantly with the loading amount of CBNP nanoparticles because internal stress fields (during solidification: microstructure change) generated by CBNP particles are able to form a quasi-2D percolating network structure and to lower interface thermal resistance[9],[10], providing higher phonon heat conduction efficiency.

Table 1. Thermal properties of composite PCMs.

Samples	Thermal conductivity W/(m.K) (±5%)	Thermal diffusivity (mm^2/s) (±10%)	Thermal conductivity (% Enhancement)
LA (PCM)	~ 0.219	~ 0.102	0
2 wt.% CBNP	~ 0.241	~ 0.135	~10
4 wt.% CBNP	~ 0.327	~ 0.222	~ 49
5 wt.% CBNP	~ 0.333	~ 0.276	~ 50
6 wt.% CBNP	~ 0.275	~ 0.235	~ 25

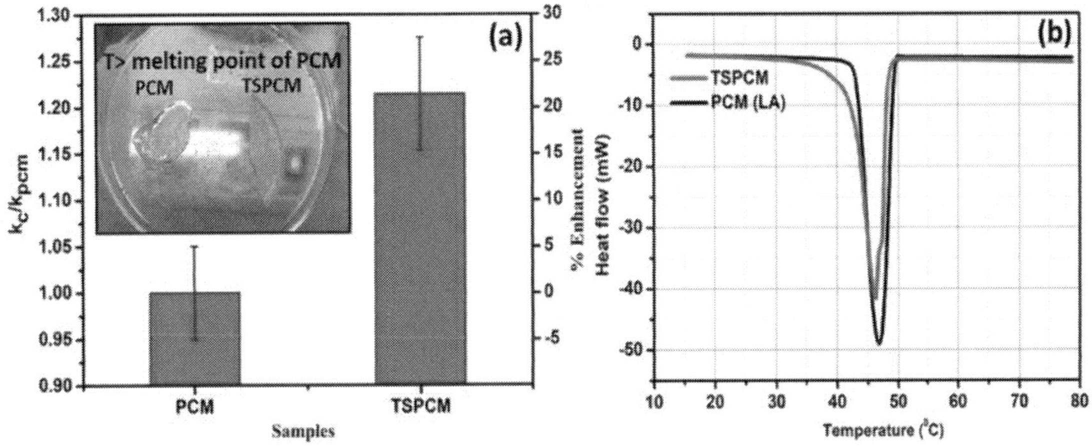

Figure 3 (a) Thermal conductivity enhancement of shape-stable PCM (TSPCM) vs conventional PCM. (b) DSC curves of the PCM & TSPCM, k_{pcm} and k_c refer to TC of pristine PCM and shape-stable composite PCM, respectively.

Figure 3 (a) shows thermal conductivity enhancement of shape-stable PCM (TSPCM) vs conventional PCM. As to prevent PCM leakage, during the solid-liquid phase transition, 25 wt. % turmeric powder was incorporated into PCM to fabricate shape-stable PCM (TSPCM). For leakage test (shape stability), the conventional and shape-stable PCMs were placed on hot plate (at ~50 ^{0}C) above to its phase transition temperature. The results of figure (3a) show that the stable-shape composite PCM based on turmeric powder has high thermal conductivity (increased by about 22%), and the inserted image (3a) shows that it effectively prevents the leakage of PCM during melting. It can be noticed from (3b) that the melting temperature of TSPCM , while not being significtly different as compared to pristine PCM, it still has sufficient high latent heat (~ 135 J/g). The present findings reveal that the designed composite PCM has high potential for use in industrial applications. The investigated thermal properties of pure PCM and shape-stable composite PCM are listed in Table 2, indicating the phase change characteristics and latent heat of shape-stable composite PCM are suitable for TES applications.

Table 2. Thermal properties of shape-stable composite PCMs.

Samples	Thermal conductivity W/(m.K)	Melting temperature (°C)	Latent heat (J/g)
LA (PCM)	~ 0.219	45.4	172.5
TSPCM (shape stable composite PCM)	~ 0.266	45.1	134.1

4. Conclusions

We studied the enhancement in heat conduction characteristics of lauric acid-based PCM at various concentrations of CBNP nanoparticles. We achieved ~50% thermal conductivity improvement with 5 wt.% CBNP loaded in the host matrices (PCM). The presence of turmeric powder improved the shape stability of PCM by arresting leakage across the melting temperature and obtained a high latent heat value of ~135 J/g. The designed low-cost and biodegradable (CBNP & turmeric powder) based composite PCM is a new result that contains desired thermophysical properties (enhanced heat conduction and shape stability) and can play an important role in combination with new emerging concepts in latent thermal energy storage (e.g. Dynamic PCM[11]), thus possibly boosting the thermal energy storage performance.

5. References

[1] Lin W, Lai J, Xie K, Liu D, Wu K and Fu Q 2022 *ACS Appl. Mater. Interfaces* **14** 38981–9
[2] Mishra A K, Lahiri B B and Philip J 2020 *Energy* **191** 116572
[3] Ribezzo A, Falciani G, Bergamasco L, Fasano M and Chiavazzo E 2022 *J. Energy Storage* **53** 105140
[4] Zhang Y, Wu K and Fu Q 2022 *Adv. Funct. Mater.* **32** 2109255
[5] Yuan K, Shi J, Aftab W, Qin M, Usman A, Zhou F, Lv Y, Gao S and Zou R 2020 *Adv. Funct. Mater.* **30** 1904228
[6] Ribezzo A, Bergamasco L, Morciano M, Fasano M, Mongibello L and Chiavazzo E 2023 *Appl. Therm. Eng.* **231** 120907
[7] Morciano M, Alberghini M, Fasano M, Almiento M, Calignano F, Manfredi D, Asinari P and Chiavazzo E 2023 *J. Energy Storage* **65** 107350
[8] Palacios A, Cong L, Navarro M E, Ding Y and Barreneche C 2019 *Renew. Sustain. Energy Rev.* **108** 32–52
[9] Mishra A K, Lahiri B B and Philip J 2019 *J. Mol. Liq.* **285** 640–57
[10] Harish S, Orejon D, Takata Y and Kohno M 2015 *Appl. Therm. Eng.* **80** 205–11
[11] Fu W, Yan X, Gurumukhi Y, Garimella V S, King W P and Miljkovic N 2022 *Nat. Energy* **7** 270–80

Acknowledgments

We acknowledge the financial support of the research contract PTR 2022/24 - ENEA (*Studio di sistemi di accumulo termico innovativi basati sull'utilizzo di nano-PCM dinamici: Modellazione e analisi sperimentale delle proprietà termofisiche, realizzazione e testing di un prototipo, e analisi delle possibili applicazioni di tale tecnologia nell'ambito dell'accumulo termico*) funded by Ministero dell'Ambiente e della Sicurezza Energetica (MASE).

Journal of Physics: Conference Series **2940** (2025) 012013 doi:10.1088/1742-6596/2940/1/012013

Infrared thermography applied to the study of the local heat transfer enhancement in a wall corrugated tubular heat exchanger for food applications

M Malavasi[1,*], F Bozzoli[2,3], L Cattani[2,3], L Pagliarini[1], M W Azam[1], S Rainieri[1,3]

[1] Department of Engineering and Architecture, University of Parma, Parco Area delle Scienze 181/A, I-43124 Parma, Italy
[2] Department of Engineering for Industrial Systems and Technologies - DISTI, University of Parma, Parco Area delle Scienze 181/A, I-43124 Parma, Italy;
[3] Interdepartmental Centre on Safety, Technologies and Agri-food Innovation SITEIA, University of Parma, Parco Area delle Scienze 181/A, I-43124 Parma, Italy

[*] e-mail: matteo.malavasi@unipr.it

Abstract. The combination of global and local performance analysis represents a powerful tool for the optimal design and sizing of heat exchangers, especially when wall corrugations are employed to enhance inner convection. In this paper, both approaches are applied to a transversely corrugated pipe to propose a thermal correlation and investigate the effects of corrugation on the local heat transfer performance. For the local analysis, infrared thermographic measurements, combined with the solution of the inverse heat conduction problem in the solid wall domain, were used to evaluate the internal convective heat transfer coefficient. The combined approach offers a valuable tool for the optimal design of heat exchangers and related thermal processes, particularly in applications where knowledge or prediction of the local temperature-time history of fluid particles is critical. This is especially relevant in the food industry, where such passive heat transfer enhancement solutions are widely used.

1. Introduction

Wall corrugated tubes can be surely considered the most frequently adopted passive heat transfer enhancement solution in thermal processes within the food industry [1]. The application of this strategy in shell and tube heat exchangers is very common, as it is easy and low cost to be implemented on the wall of the pipes of heat transfer equipment. In literature [2–4] it has been shown that this kind of technique promotes a substantial increase of the thermal performance in comparison to devices using smooth pipes and several correlations have been proposed for the different flow regimes. The ability of corrugated pipes to induce transition to a turbulent flow regime, thereby enhancing the heat transfer performance of heat exchangers, has made them highly valued in the food industry, particularly in applications involving medium/high-viscosity fluids. It is clear that, for this industrial sector, the study of the effect of corrugation on the heat transfer in pipes has a key role in the design of energy-efficient heat exchangers for unit operations on foodstuffs, such as in pasteurization and sterilization. In this perspective, descriptors such as heat transfer correlations predicting the overall enhancement effect, can represent a powerful tool to facilitate this goal. However, in the authors' opinion, it is important understanding also the mechanism at the base of the heat transfer enhancement effect promoted by

Journal of Physics: Conference Series **2940** (2025) 012013 doi:10.1088/1742-6596/2940/1/012013

corrugations, using alternative approaches, such as a local analysis. The local analysis of the convective heat transfer coefficient is of great interest for some specific industrial processes, such as the case of sterilization and pasteurization treatments of food, in which the occurrence of an irregular distribution of inner corrugation wall temperature, could compromise their effectiveness.
In this work, transverse corrugation has been investigated. This kind of geometry has been experimentally studied both considering the average heat transfer performance, suggesting a correlation for the overall thermal behavior, and investigating the local heat transfer enhancement as well. The experimental approach was based on the resolution of the Inverse Heat Conduction Problem (IHCP) within the tube's wall domain for estimating the convective heat transfer coefficient distribution at the fluid-internal wall interface by adopting the outer temperature distribution as input parameter. To acquire the temperature of the external side of the wall, a cooled infrared camera was used, capable of obtaining highly spatially resolved temperature map for a local monitoring of the effect of the surface morphology on the thermal field.

2. Experimental setup and data processing for global performances analysis

In this study, a 3-meters-long pipe made of AISI 304 was experimentally tested. The thickness was equal to 1 mm, while the outer diameter was equal to 16 mm. The pipe featured a transverse corrugation profile having 32 mm pitch and 1 mm depth; Figure 1 provides a detailed image of the corrugation.

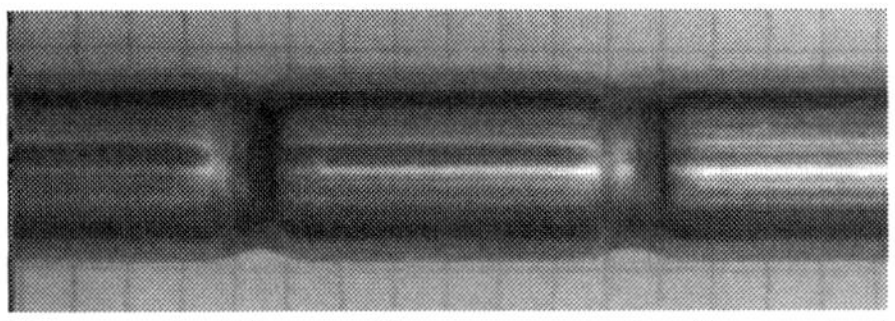

Figure 1: The wall corrugated tube tested.

The average performances of the tested pipe were evaluated by inserting the corrugated pipe horizontally in a specially designed setup, schematically reported in Figure 2. The test rig is composed of a volumetric pump that conveys the working fluid from the holding reservoir to the test section and then to an auxiliary heat exchanger fed with cold water. Heat generation by Joule effect in the tube wall within the test section is obtained through a power supply (HP 6671A, 0-8 V, 0-220 A). The generated heat flux is assumed to be uniformly distributed along the section of interest. The system is thermally insulated towards the environment with expanded polyurethane. The working fluid returns to the holding reservoir after leaving the corrugated pipe. Inlet/outlet and wall temperature were monitored by means of forty calibrated T-type thermocouples, placed at various axial and circumferential points. Two diametrically opposite thermocouples, one on the upper surface and one on the lower surface, were positioned for every axial distance taken into consideration. The steady-state heat conduction problem under uniform heat generation conditions in the tube wall was resolved to determine the internal wall temperature. For determining the inlet and outlet temperatures, the probes were placed on the tube's wall at the two ends of the heated section. The bulk temperature was calculated along the heated section by imposing energy balance. The volumetric flow rates were determined using the gravimetric method, by measuring the mass of working fluid exiting the test section within a known time interval. The differential pressure drop under isothermal conditions was monitored by a Rosemont® 3051S transducer. Temperatures were acquired for 5 min at 0.33 Hz once steady state was reached, for Reynolds number varying in the range $4 \cdot 10^3$- $1.6 \cdot 10^4$.

The experimental approach allowed the estimation of the Nusselt number axial distribution along with its average value over the whole heated section as follows:

$$Nu_z = \frac{h_z \cdot D_{int}}{\lambda} \quad (1) \qquad Nu = \frac{1}{L_H}\int_0^{L_H} Nu_z dz \quad (2)$$

where z is the axial coordinate, L_H is the test section length, and λ is the working fluid's thermal conductivity. Equation (3) determines the local convective heat transfer coefficient, h_z. Here, q is the heat exchanged per unit area, T_y. is the circumferentially averaged wall temperature, and T_b is the bulk fluid temperature. The fluid properties were evaluated by adopting the computed values of bulk temperature. The Nusselt number was evaluated as a function of the Reynolds number, which is defined in equation (4), being v the mean fluid velocity, ρ the fluid density, D_{int} the pipe's envelope internal diameter, and μ the fluid dynamic viscosity.

$$h_z = \frac{q}{(T_w - T_b)} \quad (3) \qquad Re = \frac{\rho \cdot v \cdot D_{int}}{\mu} \quad (4)$$

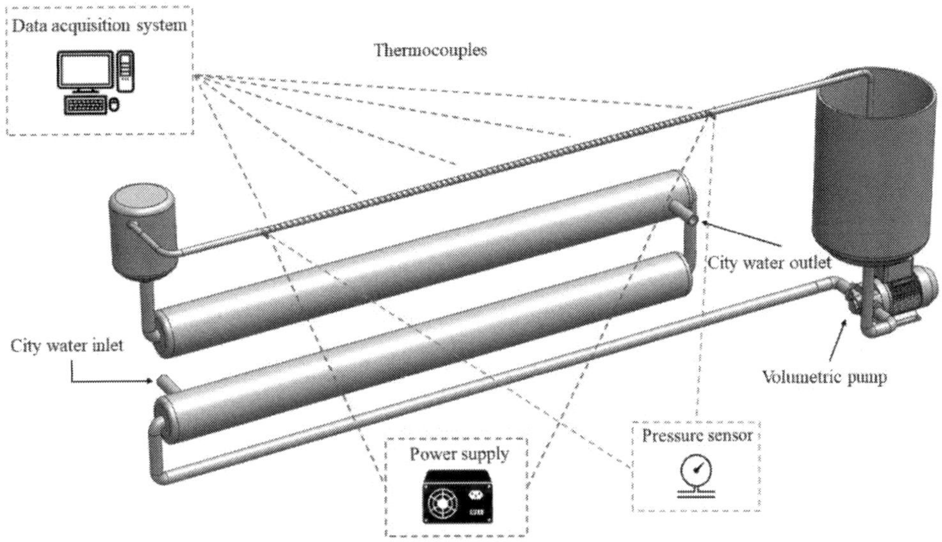

Figure 2: Schematic of the test rig.

3. Experimental setup and data processing for local performances analysis

h_z was estimated by InfraRed (IR) thermography on the outer tube's wall within the test section, achieved by removing a small portion of the thermal insulation layer. The length of such an uninsulated section was specifically chosen to capture a complete rotational period of the corrugation. This wall portion was coated with a thin film of high-emissive paint having known emissivity to perform IR acquisitions by means of an IR camera FLIR A6701 (640 x 512 pixels sensor, sensitivity = 20 mK, accuracy = ± 1 K). The advanced cooled sensor IR camera was made available within the METROFOOD-IT, the Research Infrastructure for promoting Metrology in Food and Nutrition [5]. The acquisition procedure consisted in rotating the IR camera around the tube's axis while acquiring six images. Thermal images were acquired at three Reynolds numbers, namely $Re = 4 \cdot 10^3$, 10^4 and $1.6 \cdot 10^4$ once the thermal steady state condition was reached. To reduce acquisition noise, the acquisition time of each image was set to 3s using a frame rate of 25 Hz, so that each image obtained is the result of averaging 75 frames. The six images were acquired using a support that allowed to move the camera around the pipe keeping the optic in normal position with respect to the tube axis and minimizing perspective artifacts and involuntary movements during the acquisition. In the present experimental set-up, the viewing angle lower than ± 30° allowed to consider the surface of the pipe as a diffuse grey emitter [6]. The emissivity of the applied paint was determined by capturing images of the target at various known temperatures, resulting in a value of 0.95. With the procedure of image processing validated by Bozzoli et al. [7], starting from the six separated pictures acquired around the pipe, it was possible to obtain the entire external temperature distribution on the pipe's wall. Starting from the obtained temperature distribution, to evaluate the value of the local convective heat transfer coefficient, the IHCP in the wall domain was solved.

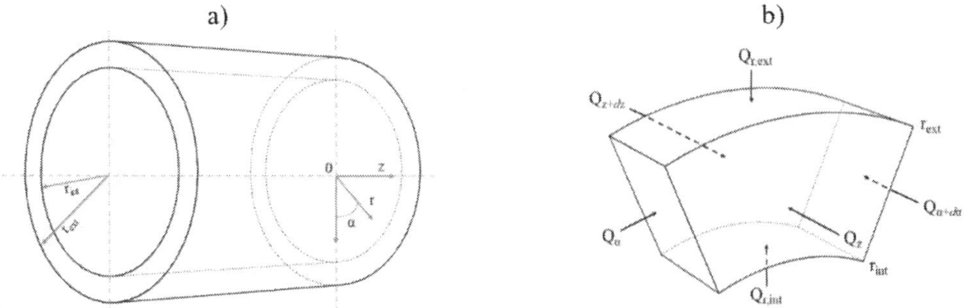

Figure 3: a) Geometrical domain and coordinate system; b) a portion of solution domain.

The physical problem to consider is the heat conduction in the wall of a cylindrical pipe (Figure 3), defined in its steady-state energy balance formulation by:

$$k\nabla^2 T + q_g = 0 \tag{5}$$

where q_g is the heat generated per unit volume, and k the wall thermal conductivity. Equation (6) and equation (7), which express the boundary conditions for the external and internal surface of the pipe, respectively, complete the description of the problem:

$$-k\frac{\partial T}{\partial r} = \frac{(T - T_{env})}{R_{env}}, \qquad r = r_{ext} \tag{6}$$

$$-k\frac{\partial T}{\partial r} = h_{int}(T - T_b), \qquad r = r_{int} \tag{7}$$

being T_{env} the ambient temperature, R_{env} the outer heat transfer resistance, and T_b is the fluid bulk local temperature.

By considering the thin-wall approximation, it results:

$$T(\alpha, r, z) \cong T(\alpha, r_{int}, z) \cong T\,(\alpha, r_{ext}, z) \tag{8}$$

The steady state local energy balance equation with an internal source term Q_g for the tube's section of Figure 3b reads as follows:

$$Q_{\alpha+d\alpha} + Q_\alpha + Q_{r,int} + Q_{r,ext} + Q_{z+dz} + Q_z + Q_g = 0 \tag{9}$$

By rearranging Equation (9) as specified in [8], the local convective coefficient at the wall-working fluid interface is expressed as follows:

$$h_{int} = \frac{k \cdot \ln\left(\frac{r_{ext}}{r_{int}}\right) \cdot \frac{\partial^2 T}{\partial \alpha^2} + \frac{k}{2}\left(r_{ext}^2 - r_{int}^2\right) \cdot \frac{\partial^2 T}{\partial z^2} - \frac{r_{ext}}{R_{env}}(T - T_{env}) + \frac{q_g}{2}\left(r_{ext}^2 - r_{int}^2\right)}{r_{int} \cdot (T - T_b)} \tag{10}$$

Equation (10) is ill-conditioned due to the existence of noise in the experimental data, detrimentally amplified by the second order derivative [8] Eliminating unnecessary noise from the raw data allows the Laplacian to be calculated directly. Many scholars have shown how effective the Gaussian filter is in this kind of approach [9]. The transfer function of this type of filter in the 2-D frequency domain assumes the following form:

$$H(u,v) = e^{-(u^2+v^2)/2u_c^2} \tag{11}$$

where u_c is the cut-off frequency, assumed equal for both axial and circumferential coordinates, chosen by means of the criterion provided by the discrepancy principle originally formulated by Morozov [8].

4. Results

In Figure 4 the behaviour of average Nusselt number, normalized as $Nu/Pr^{1/3}$, in function of Reynolds number is shown. The data show that the corrugation promotes an enhancement of the thermal performances up to 140% with respect to the smooth surface, estimated by the Dittus-Boelter correlation [10] on the entire range of Reynolds number tested. This behaviour confirms the results obtained for the laminar flow condition [11].

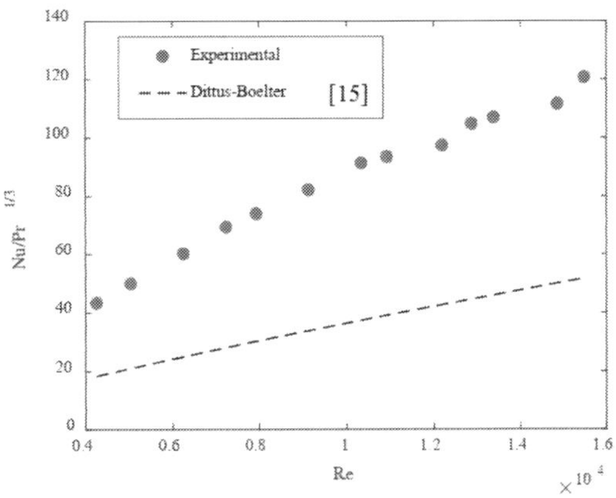

Figure 4: Experimental value of the average Nusselt number for the corrugated pipe compared to the smooth wall behaviour.

The experimental data obtained were used for suggesting a proper heat transfer correlation for this type of wall corrugation. At this purpose, the model of the Dittus-Boelter correlation wad adopted as follows:

$$Nu = C\,Re^{\alpha} Pr^{1/3} \tag{12}$$

where C and α express the effect of the tested geometry. The values of the Nusselt number estimated using the equation (12), namely Nu_{est}, have been easily calculated by imposing trial values for C and α and by forcing, through a least-square approach, the matching with the values of the Nusselt number experimentally obtained (namely Nu_{exp}), by tuning the two coefficients themselves. The C and α coefficients are therefore estimated by minimizing the following expression:

$$S(\boldsymbol{P}) = \sum_{n=1}^{N} \left[\boldsymbol{Nu}_{exp,n} - \boldsymbol{Nu}_{est,n}\right]^2 \tag{13}$$

where $\boldsymbol{P}$ is the vector of parameters to be estimated, $\boldsymbol{Nu}_{exp}$ and $\mathbf{Nu}_{est}$ are the experimental and predicted variables, respectively, and N is the total number of experimental samples. The results of the minimization are reported for every parameter in Table 1. It can be observed that the confidence interval (95% C.I.) and the coefficient of variation (CV) for the Reynolds parameter α are very small, whereas for the C parameter, these values are higher but still within acceptable limits. This can be confirmed by the observation, in Figure 5, of the distribution of the values of the experimental Nusselt number Nu_{exp} and the values of estimated ones, Nu_{est}: it is possible to state that the correlation is able to describe very well the behaviour of the experimental data, also confirming the high value of R^2 achieved.

Table 1. Results of the estimation procedure.

Coefficient	Estimated value	C.I. $^{95\%}$	*CV (%)*	R^2
C	0.092	(0.045, 0.139)	46.810	0.992
α	0.744	(0.689, 0.798)	5.799	

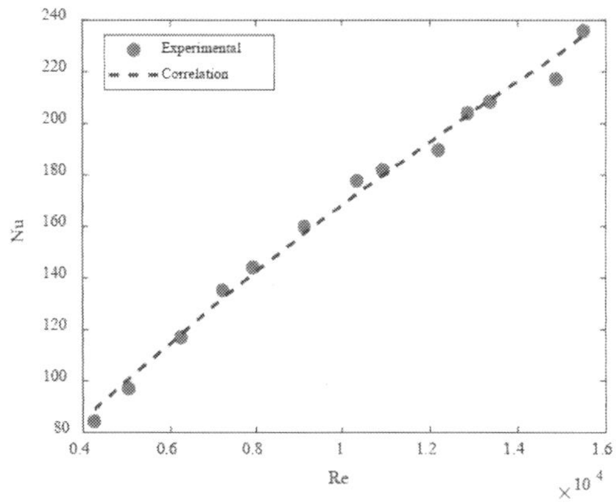

Figure 5: Experimental average Nusselt number vs estimated one by using the optimal correlation.

To deeply analyze the impact of the corrugation on the heat transfer performance enhancement, a local investigation has been carried out as well. The solution of the IHCP described by Equations from (3) to (9) has been obtained by means of outer wall temperature distributions, and the local Nusselt number map at the internal side of the pipe wall has been obtained. In Figure 6 the local Nusselt number distribution is reported for the case of Re=4000, (fluid flowing from left to right). Considering the position of the corrugations on the wall, identified within the area marked by the red dashed lines, it can be observed that in the immediate region downstream the indentation, there is a significant decrease in the heat transfer enhancement effect compared to the rest of the distribution, while further downstream a significant increase is observed. This behavior can be explained by the fluid recirculation generated near the indentation, which causes a significant local reduction in heat transfer performance. However, further downstream, increased turbulence triggers intensive fluid mixing, enhancing heat transfer. This phenomenon persists until the next corrugation is reached. However, it is important to note that in the Nusselt number map shown in Figure 6, the areas where the corrugation has a beneficial effect on heat transfer are clearly more prevalent than those where the effect is detrimental. As a result, the overall thermal performance is significantly improved compared to a smooth wall, as highlighted by the data in Figure 4.

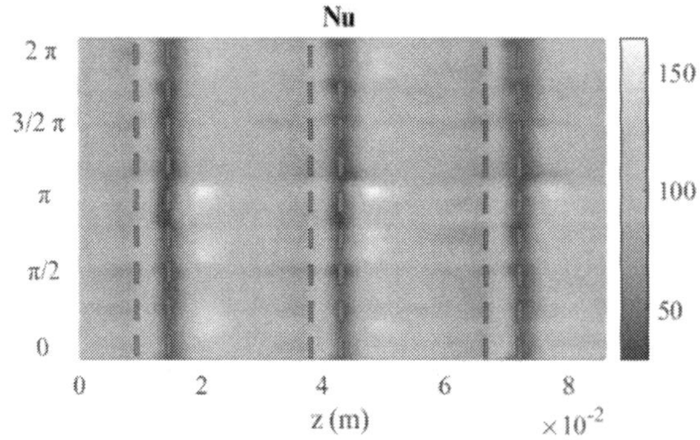

Figure 6: Local Nusselt number distribution and, in dashed red lines, the position of the three corrugations.

This behavior is observable also in Figure 7 where, for all the three Reynold numbers considered, the Nusselt number normalized as: $Nu/(Re^{\alpha}Pr^{1/3})$ is averaged along the circumferential coordinate and reported in function of z^*, defined as:

$$z^* = \frac{z}{D_i} \quad (14)$$

where, D_i is the internal diameter of the pipe and z the axial coordinate.

In Figure 7, additionally, it is possible to note that the behavior generated by the corrugation is proportionally the same, as scaled with $Re^{\alpha} Pr^{1/3}$, both in terms of the intensity of the convective heat transfer enhancement effect and of its spatial distribution along the longitudinal axial coordinate.

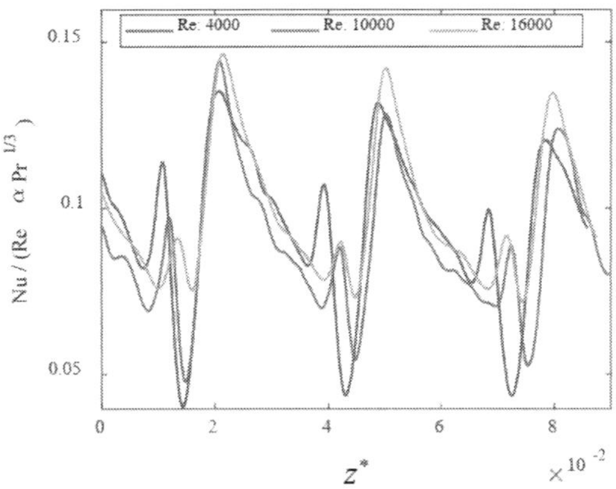

Figure 7: $Nu/(Re^{\alpha}Pr^{1/3})$ averaged along the circumferential coordinates and reported in function of z^*.

5. Conclusions

In this paper, global and local thermal analysis on a pipe with transversal corrugation was performed. The geometry tested belongs to a class of passive heat transfer enhancement solutions widely adopted in the food industry. Global analysis allowed to obtain a first overall description of the positive effect of the corrugated solution on the overall heat transfer, defining an increase in efficiency of about 140% when compared to a smooth tube. Experimental results were adopted for the estimation of a robust correlation describing the Nusselt number versus the Reynolds number, by providing a powerful tool for the design and sizing of heat transfer devices in which these solutions are implemented. Local analysis, at the same time, while confirming the results observed in global analysis, allowed to investigate the cause effect relationship between surface morphology and heat transfer augmentation mechanism. The combined approach (global and local) offers a valuable tool for the optimal dimensioning of heat exchangers and related thermal processes, particularly in applications where knowledge or prediction of the local temperature-time history of fluid particles is critical. This is especially relevant in the food industry, where such passive heat transfer enhancement solutions are widely used.

6. References

[1] M.A. Alfellag, H.E. Ahmed, M.G. Jehad, A.A. Farhan, The hydrothermal performance enhancement techniques of corrugated channels: a review, J Therm Anal Calorim 147 (2022) 10177–10206. https://doi.org/10.1007/s10973-022-11247-1.

[2] S. Rainieri, G. Pagliarini, Convective heat transfer to orange juice in smooth and corrugated tubes, Int. J. Heat Technol 15 (1997) 69–75.

[3] G. Wang, S.P. Vanka, Convective heat transfer in periodic wavy passages, Int J Heat Mass Transf 38 (1995) 3219–3230. https://doi.org/10.1016/0017-9310(95)00051-A.

[4] Yu.Ya. Trifonov, Calculation of the linear stability of liquid flow in a flat channel with streamwise wavy walls, Journal of Applied Mechanics and Technical Physics 64 (2023) 1000–1010. https://doi.org/10.1134/S0021894423060093.
[5] https://www.metrofood.it, METROFOOD-IT, (2022).
[6] T. Astarita, G.M. Carlomagno, Infrared thermography for thermo-fluid-dynamics, Springer Science & Business Media, 2012.
[7] L. Cattani, F. Bozzoli, S. Rainieri, Experimental study of the transitional flow regime in coiled tubes by the estimation of local convective heat transfer coefficient, Int J Heat Mass Transf 112 (2017) 825–836. https://doi.org/10.1016/j.ijheatmasstransfer.2017.05.003.
[8] V.A. Morozov, Methods for solving incorrectly posed problems, Springer Science \& Business Media, 2012.
[9] D.A. Murio, The mollification method and the numerical solution of ill-posed problems, John Wiley & Sons, 1993.
[10] TheodoremL. Bergman Adrienne S. Lavine, Fundamentals of heat and mass transfer, 2015.
[11] F. Bozzoli, L. Cattani, S. Rainieri, Cross-helix corrugation: The optimal geometry for effective food thermal processing, Int J Heat Mass Transf 147 (2020) 118874. https://doi.org/10.1016/j.ijheatmasstransfer.2019.118874.

Acknowledgments
The work was funded under the National Recovery and Resilience Plan (NRRP), Mission 4 Component 2 Investment 3.1 - Call for tender No. 3264 of 28/12/2021 of Italian Ministry of University and Research funded by the European Union – NextGenerationEU. Award Number: Project code IR0000033, Concession Decree No. 120 of 21/06/2022 adopted by the Italian Ministry of University and Research, CUP I83C22001040006, Project title "Strengthening of the Italian Research Infrastructure for Metrology and Open Access Data in support to the Agrifood" (METROFOOD-IT).

Journal of Physics: Conference Series **2940** (2025) 012014 doi:10.1088/1742-6596/2940/1/012014

Measurement of thermal diffusivity with a modified Angstrom method

S Piva[1*]

[1] Department of Engineering, Università di Ferrara, via Saragat 1, 44124 Ferrara, I

*E-mail: pvs@unife.it

Abstract. A methodology has been developed to measure the thermal diffusivity of high thermal conductivity solid materials. The basic idea traces back to the Ångström method, but the implementation differs in some points. Originally the specimen was free to the environment, here the specimen is thermally insulated to reduce, and control, the effects of natural convection. An appropriate analytical solution is used to calculate temperature distribution. The original two-point approach is replaced by a parameter estimation technique to consider multiple measurement points and reduce the final uncertainty. The methodology is used for a pure copper bar. The results are compared with the data available in the literature. The measured thermal diffusivity is $1.16{\pm}0.07{\cdot}10^{-4}$ m^2/s while the most reliable data available in the literature is $1.17{\pm}0.04{\cdot}10^{-4}$ m^2/s. The agreement of the present value with those available in the literature is excellent. This makes the use of this simple analytical approach attractive.

1. Introduction

Knowledge, within acceptable levels of uncertainty, of thermal diffusivity is currently of relevance, considering the potential for predicting the thermal behaviour of complex systems offered by numerical calculation procedures and the availability of new materials.

Electronics cooling, for example, shows an evolution that directly involves the knowledge of the thermophysical properties of materials, and in particular of thermal diffusivity. First, there is a widespread awareness that in many electronics thermal management applications involving time-varying workloads, ignoring the non-steady state behaviour of the systems certainly leads to oversized cooling [1] and that the driver of this time behaviour is thermal diffusivity [2]. Furthermore, new alternative materials to Al are proposed for the fabrication of finned heat sinks and other heat transfer components (alumina Al2O3, aluminium nitride AlN, beryllium oxide BeO etc.). With these new materials, in addition to the ever-present temperature dependence, small differences in alloy composition are sufficient to have large variations in thermophysical properties. For example, for Al2O3 (alumina) the thermal conductivity strongly depends on the Al concentration and can vary from 20 to 30 W·m^{-1}·K^{-1}. For AlN (aluminium nitride) the thermal conductivity ranges from 170 to 220 W·m^{-1}·K^{-1}. For BeO (beryllium oxide) it varies from 190 to 250 W·m^{-1}·K^{-1}, especially with temperature. Furthermore, while the thermal properties of pure metallic elements are well known [3], e.g. aluminium, those of alloys (e.g. Al 6060, 6061, 6063 used for heat sinks in electronics) can vary over a wide range, depending on the composition of the alloy. Measuring Al's thermal properties is further made troublesome by their variability with temperature.

A thermal property is generally determined by experimentally reproducing a situation, if possible a simple one, for which it is known how the property itself is related to some directly measurable quantities: temperature, thermal flows, etc. Thermal diffusivity must necessarily be determined in a non-stationary thermal condition since it characterizes this regime. These methods are based on measuring the temperature response to a time-varying heat source. In [4] an extensive discussion on the different related measurement methods and applications is proposed. Their classification distinguishes between transient and periodic temperature methods. In transient methods, thermal diffusivity is estimated from the temperature response of a sample to a sudden change in input heat. An example is the Flash method. In periodic temperature methods, thermal diffusivity is estimated from the response of a sample to a heat input that varies periodically with time. Examples include the 3ω-method and the Ångström method.

Conversely, methods that can be used to determine thermal conductivity involve measurements under steady-state conditions. However, these schemes are conveniently used only for materials with small thermal conductivity (less than 1 $W \cdot m^{-1} \cdot K^{-1}$), such as non-metallic materials. In the case of metallic materials, steady-state measurement methods, although conceptually very simple, can nevertheless be affected by uncertainties, even high ones. Therefore, in this case it is usual preferable to measure the thermal diffusivity in a non-steady state and possibly, through knowledge of the value of thermal capacity per unit volume, to trace back to that of the thermal conductivity. According to Abeles et al [5] the well-known calorimetric methods allow to measure the thermal capacity per unit of volume with an accuracy better than 1%, even at high temperatures (typically density and specific heat are, respectively, measured by the immersion method and by differential scanning calorimetry).

An extensive review of the main transient methods used to measure thermal diffusivity is reported by Danielson and Sidles [6]. Among these, processes in which a periodic thermal regime is reproduced deserve special attention, in particular those based on the Ångström method [7].

In this paper, a methodology for measuring the thermal diffusivity (and other thermal properties) of high conductivity solid materials is discussed. The basic idea goes back to the Ångström method [8], but the implementation differs in some points (see the section "Experimental data" for a discussion of the differences). The methodology is used for a highly pure copper bar. The results are compared with the data available in the literature and with the results obtained with a methodology [9] based on the use of a specifically developed FEM code [10]. In [9] an analytical solution was also used, but in that case the coupling with the environment was not considered. Necessarily, a simplified analytical approach gives results that differ from the most reliable data available in the literature, as we were able to measure in [9]. However, the complexity of the required FEM code [10] makes the use a relatively simple analytical approach attractive.

2. Experimental data

The basic idea of the present methodology for measuring the thermal diffusivity of high conductivity solid materials dates back to the Ångström method [8], but the implementation differs in some points.

In the Ångström technique [8], a "periodic square" thermal load is applied to one end of the specimen, while its entire length is exposed to the surrounding environment. A temperature wave propagates along the rod. In this wave propagation the phase shifts and the amplitude decreases. With measurements at two points spaced along the rod, the thermal diffusivity is determined.

Here, as in the case of the Ångström method [8], the data reduction is based on an analytical solution to the thermal problem. However, the square heating, which characterizes the Ångström method, is replaced by a monochromatic thermal wave produced by Joule heating. This simplifies the comparison between analytical solution and experiment.

A further specificity of Ångström's original method is the absence of thermal insulation on the specimen. This is because the tests are intended for a range of high temperatures, where the thermal radiation at the surface is high enough to contrast the characteristic fluctuations of natural convection. However, at relatively low temperature (less than 100°C), typical when using metals in finned heat sinks, the role played by thermal radiation is not sufficient to contrast the effect of natural convection, which may increase experimental uncertainty. To underline the order of magnitude of the situation, for a specimen temperature of about 450°C, the heat transfer coefficients are 13.4 $W \cdot m^{-2} \cdot K^{-1}$ for convection and 31.5 $W \cdot m^{-2} \cdot K^{-1}$ for radiation, respectively. For a specimen temperature of about 35°C, the heat transfer coefficients are 6.8 $W \cdot m^{-2} \cdot K^{-1}$ for convection and 5.8 $W \cdot m^{-2} \cdot K^{-1}$ by for radiation, respectively. Since here the focus was on low-temperature applications, the specimen was thermally insulated to reduce and control the effects of natural convection.

The role played by an external material in thermal contact with the specimen, such as a layer of expanded rubber, has been discussed by Eichhorn [11]. A parameter is proposed to evaluate whether this material, although having a non-negligible thermal capacity, does not participate and can be neglected in the mathematical model of the specimen. This parameter should be lower than 1. Here it is 0.24. For this reason, the presence of the insulating material has been considered in the boundary term and neglected in the unsteady term of the model.

An analytical solution is presented able to cover the data trend. The two-points approach of the original Ångström method is replaced by a more recent parameter estimation technique, using the analytical solution. In this way, a large number of measurement points can be considered, and the final uncertainty is reduced.

For an application of this modified Ångström method to the measurement of thermal diffusivity in high thermal conductivity materials, the data collected by Pagliarini and Piva in [9] are used here. The required data are essentially temperatures measured with thermocouples along a specimen. For a better understanding of the following data reduction procedure, a brief description of the system is provided. For further details please refer to the original article [9].

A cylindrical sample (12.8 mm diameter) of pure copper (99.9% purity) was examined (figure 1). One end of the sample is heated with a sinusoidal power by dissipating electrical energy in a constantan wire spirally wound on the surface; the other end is kept at a constant temperature by immersing the specimen in a thermostatic bath (free length between heater and bath 206.1 mm). The specimen is covered with an insulating layer of expanded rubber (8.9 mm thick) to reduce heat transfer to the environment. The temperature of the specimen is measured with T thermocouples glued to its surface.

The thermal power, variable over time with a sinusoidal law (period 50 s) around an average value:

$$Q(t') = Q_m\left[1 + \sin\left(\omega\, t' - \frac{\pi}{2}\right)\right] \tag{1}$$

is produced by dissipating the pulsed voltage:

$$V(t') = V_m\left[\sin\left(\frac{\omega}{2} t'\right)\right] \tag{2}$$

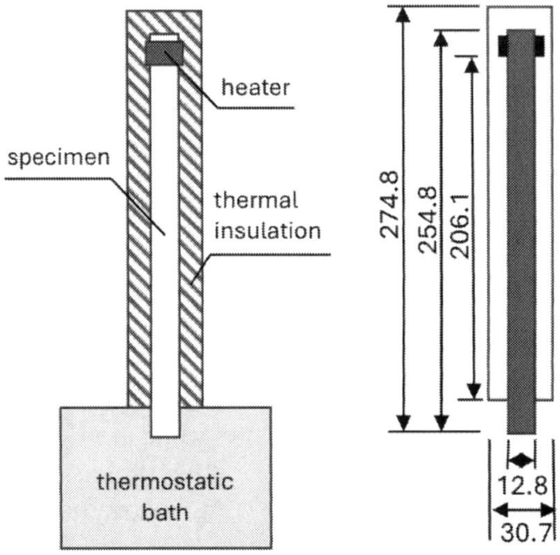

Figure 1. System and specimen schematics.

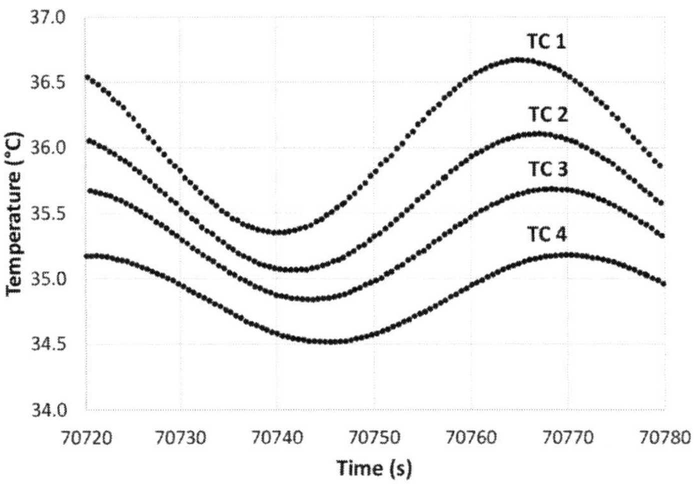

Figure 2. Time distribution of temperature in thermocouples 1-4.

In [9] the temperature data was used directly as a voltage signal, without any conversion. Here the same data is converted into temperature with a thermocouple calibration curve; then these temperatures are processed.

After a sufficiently long transient heating (never less than 10 hours), for each thermocouple a temperature composed of an average and an oscillating part is experienced, given as:

$$T'(x_i',t') = T'_{m,i}(x_i') + A_i(x_i')\sin\big(\omega\, t' - \varphi_i(x_i')\big) \tag{3}$$

In figure 2 the time distribution of temperature data is shown for the first four thermocouples. The amplitude reduction and phase delay superimposed on a decreasing mean temperature are clearly readable.

In [12] a different methodology is proposed in which the heater with steady positive contribution, equation (1), is replaced by a source based on the Peltier effect, capable to oscillate around zero power. However, as will be shown in the "Result" paragraph, the stationary gradient of temperature along the specimen due to the steady power contribution did not introduce difficulties in data reduction and calculations.

3. Data reduction

For each thermocouple, in the steady-periodic regime the temperature can be decomposed into an average part and an oscillating part, equation (3). To determine its unknown parameters T'_m, A and φ, a numerical code has been developed that allows to sinusoidally interpolate the temperature values collected for each thermocouple. Specifically, we search for the values of the parameters that minimize the sum of the squares of the deviations of the experimental values from the corresponding theoretical ones. In more detail, given the interpolation function T', where T'_m, A and φ are the parameters sought, once the experimental temperature values Y_i in a number of n successive instants of time are known, the function to be minimized is:

$$S = \sum_{i=1}^{n} \frac{1}{\sigma_i^2} (Y_i - T_i')^2 \tag{4}$$

where σ_i^2 is the variance associated with each experimental observation.

The solution of this research was obtained using the Gaussian linearization method described in [13]. The solution starts with the choice of trial values for the unknown parameters and then proceeds with the search for successive approximations of the corrections that lead to the minimization of equation (4). The search is stopped when for each of the parameters Π_i:

$$\frac{\Pi_i^{k+1} - \Pi_i^k}{\Pi_i^k} \leq 10^{-5} \tag{5}$$

In table 1 data is shown for the first eight thermocouples. In figure 3 amplitude, A_i, and phase delay, φ_i, are shown. The figure covers the free length of the rod (206.1 mm); the amplitude decreases exponentially and at the end of the rod it is reasonably equal to zero. The phase delay increases linearly.

In figure 4 the average temperature is shown. This parameter decreases linearly along the rod. This linear trend means that the coupling with the environment is barely appreciable, thus confirming the assumption of a limited participation of the thermal insulation in the oscillating heating.

The results of the data reduction for the first eight thermocouples are reported in table 1. Although there are more than eight thermocouples glued on the wall of the rod, only the first eight are reported because for the following ones the amplitude becomes less than 0.1 °C, clearly without statistical significance.

4. Theorctical analysis

The problem under consideration is schematized as a steady-periodic diffusion in a 1D semi-infinite body (fin approximation for negligible transverse temperature gradients).

To theoretically analyse the experimental problem of heat transfer in periodic regime, for the specimen under examination the energy balance equation for the cylinder thin fin must be written:

Journal of Physics: Conference Series **2940** (2025) 012014 doi:10.1088/1742-6596/2940/1/012014

Table 1. Temperature data reduction based on equation (3).

TC n.	1	2	3	4	5	6	7	8
x'_i (mm)	6.2	16.5	26.1	35.5	46.0	55. 7	66.2	85.3
T'_{mi} (°C)	36.03	35.60	35.27	34.86	34.46	34.12	33.70	32.94
A_i (°C)	0.66	0.52	0.42	0.33	0.26	0.21	0.16	0.10
φ_i (s)	15.11	16.82	18.39	20.10	22.25	23.86	25.83	29.67

$$\frac{\partial T'}{\partial t'} = \frac{k}{\rho c}\frac{\partial^2 T'}{\partial x'^2} - \frac{4K}{\rho c D}(T' - T'_a) \tag{6}$$

where for the surface facing the environment the heat transfer coefficient K includes both conduction in the layer of thermal insulation and convection coupled to radiation from the surface towards the environment.

The domain of interest is between the end of the heater and the beginning of the immersion zone in the thermostatic bath. Here the boundary conditions are, respectively:

$$x'=0 \qquad T' = T'_1 + \Delta T'_2 \cos(\omega t') \tag{7}$$

$$x' \rightarrow +\infty \qquad T' = T'_a \tag{8}$$

The problem is then decomposed into its stationary and periodic components:

$$T'(x', t') = T'_m(x') + T'_p(x', t') \tag{9}$$

In this way, the energy balance equation (equation (6)) can be separated into a stationary part and a periodic part. Regarding the periodic part we obtain:

$$\frac{\partial T'_p}{\partial t'} = \frac{k}{\rho c}\frac{\partial^2 T'_p}{\partial x'^2} - \frac{4K}{\rho c D}\left(T'_p - T'_a\right) \tag{10}$$

with the boundary conditions:

$$x'=0 \qquad T'_p(0, t') = \Delta T'_2 \cos(\omega t') \tag{11}$$

$$x' \rightarrow +\infty \qquad T'_p(x' \rightarrow +\infty, t') = 0 \tag{12}$$

Equation (10) is rewritten as a function of the following dimensionless variables:

$$T_p = \frac{T'_p - T'_a}{\Delta T'_2} \qquad t = \frac{t'}{P} \qquad x = \frac{x'}{D}$$

turning in:

$$\frac{\partial T_p}{\partial t} = Fo\frac{\partial^2 T_p}{\partial x^2} - 4\,Bi\,Fo\,T_p \tag{13}$$

with:

$$Fo = \frac{\alpha P}{D^2} \qquad Bi = \frac{KD}{k}$$

and the boundary conditions:

$$x=0 \qquad T_p(0,t) = \cos(2\pi\, t) \tag{14}$$

$$x \to +\infty \qquad T_p(x \to +\infty, t) = 0 \tag{15}$$

The solution to this problem is obtained analytically using the complex temperature method, as proposed by Arpaci [14]. The time-dependent real temperature is replaced by the corresponding complex temperature:

$$T_p(x,t) = F(x)\exp(i\, 2\pi\, t) \tag{16}$$

where the variable F belongs to the complex field:

$$T_p(x,t) = \left(F_{Re}(x) + i\, F_{Im}(x)\right)\exp(i\, 2\pi t) \tag{17}$$

Substituting equation (16) in equation (13) we get:

$$i\, 2\pi\, F = Fo \frac{d^2F}{\partial x^2} - 4\, Bi\, Fo\, F \tag{18}$$

$$x=0 \qquad F(0) = 1 \tag{19}$$

$$x \to +\infty \qquad F(x \to +\infty) = 0 \tag{20}$$

The solution of equation (18) is given by:

$$F(x) = \exp(-(\beta_{Re} + i\, \beta_{Im})x) \tag{21}$$

where:

$$(\beta_{Re} + i\, \beta_{Im})^2 = +\left(4\, Bi + i\, \frac{2\pi}{Fo}\right) \tag{22}$$

Equation (22) gives:

$$\beta_{Re}\beta_{Im} = \frac{\pi}{Fo} \tag{23}$$

$$\beta_{Re}^2 - \beta_{Im}^2 = 4\, Bi \tag{24}$$

The solution of the periodic component of the temperature distribution is finally:

$$T_p(x,t) = \exp(-\beta_{Re}\, x)\exp\left(i\, 2\pi\left(t - \frac{\beta_{Im}}{2\pi}\, x\right)\right) \tag{25}$$

5. Results

To validate the measurement procedure, data already collected for a 99.99% pure copper specimen [9] were used. There the data collected as voltage were processed directly, assuming that for a narrow temperature range the voltage-temperature correlation of the thermocouples T

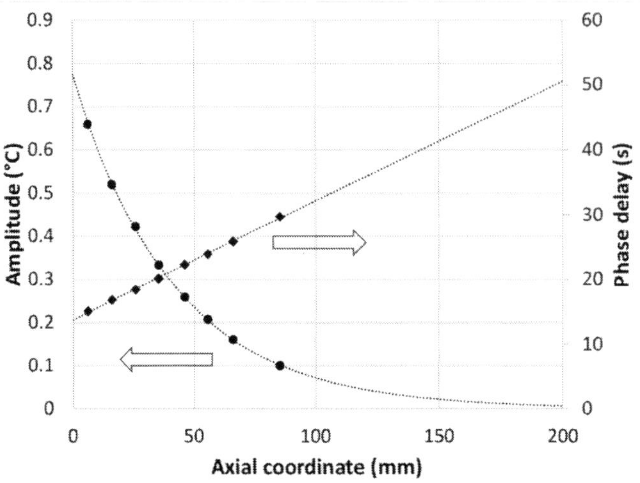

Figure 3. Amplitude (●) and phase delay (♦) distribution along the specimen.

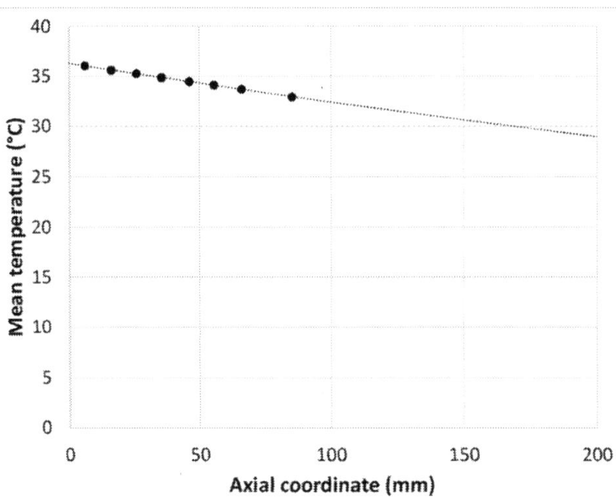

Figure 4. Average temperature distribution along the specimen.

was linear and no conversion into temperature was necessary. Here after calibrating the thermocouples, it was found that in the temperature range under examination a proper voltage-temperature correlation requires a parabolic, and therefore non-linear, correlation. For this reason, the original data were converted into temperature and reprocessed with the parameter estimation technique described in section 3. Results of these calculations are of values amplitude and phase delay of the temperature waves for the different thermocouples.

The parameter estimation technique discussed in section 3 was also used to calculate, from the amplitude and phase delay data, the parameters β_{Re} and β_{Im} in equation (21). The thermal diffusivity α of the specimen is then evaluated with equation (23) of the analytical model proposed in section 4, different from the one used in [9], which did not consider the thermal coupling with the environment.

The results of thermal diffusivity obtained by applying the parameter estimation method described in section 3 are compared in table 2 with those obtained in [9] and with others available

Table 2. Thermal diffusivity values for pure copper.

	α 10^4 m^2/s	ε_a 10^4 m^2/s
Present solution	1.16	0.07
Toulukian [3]	1.17	0.05
Gandia and Lopez-Baeza [15]	1.20	0.13
Leden et al [16]	1.17	0.04
Pagliarini and Piva [9]: numerical	1.17	0.04
Pagliarini and Piva [9]: analytical	1.14	0.07

in the literature. The agreement of the present value with those available in the literature is excellent. Therefore, the methodology based on the amplitude and phase delay of the periodic signal allows to model the experimental details with adequate precision.

The expanded uncertainty of the measured parameters is estimated at the 95% confidence level, following the procedure proposed in [17]. In the estimation of thermal diffusivity with equation (23), the parameters involved are: diameter, D, period, P, real and imaginary part of the complex temperature, β_{Re} and β_{Im}. The corresponding standard uncertainties are: ε_D = 0.2 mm, ε_P = 0.055 s, $\varepsilon_{\beta Re}$ = 0.001 and $\varepsilon_{\beta RIm}$ = 0.001. The last two come from interpolating the amplitude and phase delay of the oscillatory part of the temperature along the rod using the parameter estimation technique. In this application a standard uncertainty of 0.03°C is assumed for temperature measurements (see [9] for details). The combined standard uncertainty assigned to the calculated parameter Π is then estimated using the root-sum-square propagation rule.

6. Concluding remarks

A variant of the Ångström method is used to measure the thermal diffusivity of a specimen with high thermal conductivity. The first variant consists of having the specimen thermally insulated to reduce the thermal coupling with the environment. Second, the two-point measurement method has been replaced by a parameter estimation procedure, which allows the result to be based on many data, with lower uncertainties. Third, a monochromatic thermal wave produced by Joule heating is used. A simplified mathematical model has been solved analytically.

The measurement procedure was validated on a high purity copper sample, in order to refer to literature data for comparison. The results allowed to consider the procedure reliable.

References

[1] Mathew J and Krishnan S A, 2022 *J. Electronic Packaging* **144** / 010801-1/20
[2] Meysenc L, Jylhäkallio and Barbosa P, 2005 *IEEE Trans Power Electronics* **20** 687-693
[3] Toulukian Y S, Powell R W, Ho C Y and Nicolau M C 1973 *Thermal Diffusivity* (New York: Plenum)
[4] Jannot Y, Degiovanni A 2018 *Thermal Properties Measurement of Materials*, (Hoboken, NJ: Wiley),
[5] Abeles B, Cody G D and Beers D S, 1960 *J. Applied Physics* **31** 1585-1592
[6] D'Alessandro G, Potenza M, Corasaniti S, Sfarra S, Coppa P, Bovesecchi G, de Monte F 2022 *Energies* **15**, 8807
[7] Danielson G C and Sidles P H 1969 Thermal Diffusivity and Other Non-steady-state Methods, in *Thermal Conductivity vol 3* ed R P Tye (London: Academic Press) chapter 3 149-201
[8] Angstrom A J 1861 *Annalen der Physik und Chemie* **114** 513-530

[9] Pagliarini G and Piva S 1993 *CMEM VI* eds Brebbia C A and Carlomagno G M (Southampton: CMP) 429-438
[10] Pagliarini G 1991 *Numerical Heat Transfer, Part A: Applications* **20** 533-544
[11] Eichhorn R 1964 *Int. J. Heat Mass Transfer* **7** 675-679
[12] Muscio A 2008 *QIRT2008* July 2-5, Krakow – Poland
[13] Beck J V and Blackwell B 1988 Inverse Problems in *Handbook of Numerical Heat Transfer* eds W J Minkowycz et al. (New York: Wiley) chapter 19 pp 787-834
[14] Arpaci V 1966 *Conduction Heat Transfer* (Reading: Addison Wesley)
[15] Gandia V and E Lopez-Baeza E 1988 *J. Phys. E: Sci. Instrum.* **21** 757-759
[16] Leden B Hamza M H and Sheirah M A 1976 *Automatica* **12** 445-456
[17] ISO IEC 2008 *Uncertainty of measurement Part 3: Guide to the expression of uncertainty in measurement*

7. Nomenclature

Symbols

A	K	Amplitude
Bi	-	Biot number (K·D)/k
c	$J \cdot kg^{-1} \cdot K^{-1}$	Specific heat
D	m	Diameter
F	-	Complex dimensionless temperature
Fo	-	Fourier number $(\alpha \cdot P)/L^2$
i	-	Imaginary number
k	$W \cdot m^{-1} \cdot K^{-1}$	Thermal conductivity
K	$W \cdot m^{-2} \cdot K^{-1}$	Heat transfer coefficient
P	s	Period
Q	W	Thermal power
S	-	Objective function
t'	s	Time
t	-	Dimensionless time
T'	K	Temperature
T	-	Dimensionless temperature
V	V	Voltage
x'	m	Axial coordinate
x	-	Dimensionless axial coordinate
Y	K	Measured temperature

Greek symbols

α	m^2/s	Thermal diffusivity
β	-	Complex number
ε	-	Uncertainty
ρ	kg/m^3	Density
σ^2	K^2	Temperature variance
ω	rad/s	Angular frequency $2\pi/P$
Π	-	Generic parameter
φ	rad	Phase delay

Subscripts

a	Ambient
Im	Imaginary part
m	Mean
p	Periodic
Re	Real part

Acknowledgments

The author wishes to acknowledge Giorgio Pagliarini for the long and always fruitful cooperation.

An analytical model of evaporator performance under frosting conditions for lumped-parameter simulations of heat pumps

R Sedoni[1,*], P E Santangelo[1,2], D Angeli[1,3], M Romani[4], L Fioravanti[4]

[1] DISMI - Dipartimento di Scienze e Metodi dell'Ingegneria, Università degli Studi di Modena e Reggio Emilia, Reggio Emilia (Italy)
[2] Centro interdipartimentale per la ricerca InterMech – MO.RE., Modena (Italy)
[3] Centro Interdipartimentale En&Tech, Reggio Emilia (Italy)
[4] Zehnder Group – Competence Center Campogalliano, Campogalliano (Italy)

* Corresponding author: roberto.sedoni@unimore.it

Abstract. Enhancing the energy efficiency of HVAC systems is crucial for meeting the nZEB standards. This involves developing analytical models for simulating building-system dynamics over time with low computational resources. However, existing models struggle with accuracy in heating mode due to evaporator frosting, a factor not usually accounted for commercial software. To address this, a new heat exchanger block that accurately models frosting and its impact on system performance is developed using the Simscape™ language in Simulink® environment. This innovative block enhances the predictive capability of heat-pump simulations that use a lumped-parameter approach, ensuring compatibility and easy integration with existing toolboxes. The improved model overcomes conventional limitations, offering reliable simulations in heating mode and providing insights into energy transport and efficiency. Incorporating accurate frosting representation aids in developing more effective defrost strategies, minimizing energy losses and improving compliance with energy efficiency standards.

1. Introduction

The issue of evaporator frosting in air-to-air heat pumps arises due to the formation of frost on the evaporator coils during the heating mode operation. As the heat pump extracts heat from the outside air, moisture in the air can desublimate on the evaporator coils, leading to a frost buildup. This frost accumulation insulates the coil, reducing its ability to effectively absorb heat from the air and also increases the pressure drop by restricting the available flow cross-section. Consequently, the heat pump performance decreases, resulting in reduced heating efficiency and potential system malfunction or shutdown. In the past years these phenomena were experimentally studied and empirically modeled. Hayashi et al. [1, 2] analyzed frost formation by photographic observation, and frost properties, such as density and effective thermal conductivity, were evaluated in connection with the classified frosting growth types. O'Neal et al. [3] characterized the effect of Reynolds number, air humidity, air temperature and plate temperature on frost growth and density on a vertical plate. A more theoretical modeling approach for this phenomena, was establish by Dietenberger [4], who laid the foundation for the more recent frost formation models. Le Gall et al. [5] re-derived a one-dimensional transient formulation of the same model based on a local volume averaging technique to predict frost growth and densification on a cold wall subjected to a moist air flow. Padhmanabhan et al. [6] successfully coupled that model in a segment-by-segment heat exchanger (HX) model and introduced an air redistribution algorithm to improve predictability. Quiao et al. [7] integrated the same frost formation model into a distributed-parameter dynamic heat exchanger model embedded in a two-stage vapor injection heat pump model.

As a question still open in the literature, the main objective of this work is to propose an elementary exchange block developed in Simulink® environment, that features low computational cost and considers the possible frost formation on the heat transfer surface, according to the most recent models available.

2. Frost growth model

2.1. Frost initiation

The frost on the evaporator surface may start to form when the frost point at the heat exchanger (HX) wall is reached. The DP calculation in the model is based on a formula presented by Górnicki et al. [8]:

$$DP = \frac{Alog_{10}(RH) + \frac{BT_a}{A+T_a}}{B - log_{10}(RH) - \frac{BT_a}{A+T_a}} \tag{1}$$

where DP is the dew point in °C, RH is the relative humidity in decimal, T_a is the air temperature in °C, $A = 237$ °C and $B = 7.5$.

2.2. Frost formation

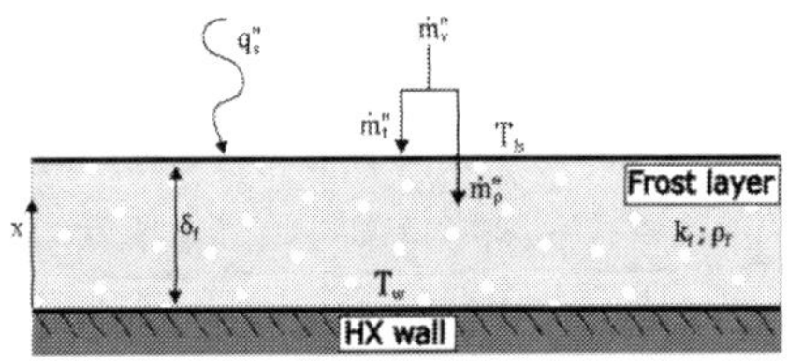

Figure 1. Energy and mass transport within the frost control volume

2.2.1. Frost surface balances To simplify the modeling approach, thermal capacity of the frost layer is neglected in this work. Therefore, the developed model is quasi steady-state. Padhmanabhan [6] demonstrated the validity of the quasi steady-state hypothesis in his work. If we consider a uniform frost layer (fig. 1), at the frost surface the steady-state energy balance can be expressed as:

$$k_f \frac{(T_{fs} - T_w)}{\delta_f} = h_{MA}(T_a - T_{fs}) + (\dot{m}"_v - \dot{m}"_\rho)\Delta h_{sub} \tag{2}$$

where $\dot{m}"_v$ is the total water mass flux, $\dot{m}"_\rho$ is the diffusive mass flux, Δh_{sub} is the latent heat of sublimation, δ_f is the frost thickness, T_w is the the air side tube wall temperature and h_{MA} is the heat transfer coefficient between frost (or coil wall) and air. Equation (2) is resolved every time step as the frost grows in time, to calculate frost surface temperature (T_{fs}).

2.2.2. Frost growing and densification Following Qiao et al. [7], the total water vapor mass flux ($\dot{m}"_v$) and sensible heat flux ($q"_s$) transferred from the air stream to the frost surface can be evaluated as:

$$\dot{m}"_v = \frac{\dot{m}_{a,d}(\omega_a - \omega_{fs})\left\{1 - exp\left[-\frac{h_m(A_{t,MA}+\eta_{fin}A_{f,MA})}{\dot{m}_a}\right]\right\}}{A_{f,MA} + A_{t,MA}} \tag{3}$$

$$q"_s = \frac{\dot{m}_a c_{p,a}(T_a - T_{fs})\left\{1 - exp\left[-\frac{h_{MA}(A_{t,MA}+\eta_{fin}A_{f,MA})}{\dot{m}_a c_{p,a}}\right]\right\}}{A_{f,MA} + A_{t,MA}} \tag{4}$$

The total water vapor mass flux can be divided over two contributions:

$$\dot{m}"_v = \frac{d}{dt}(\overline{\rho}_f\delta_f) = \dot{m}"_t + \dot{m}"_\rho = \overline{\rho}_f\frac{d\delta_f}{dt} + \delta_f\frac{d\overline{\rho}_f}{dt} \tag{5}$$

where $\dot{m}_{a,d}$ is the dry air mass flow rate, $\dot{m}_a$ is the moist air mass flow rate, ω_a is the humidity ratio of the air side inflow, w_{fs} is the humidity ratio at the frost surface (or at the wall if frost is not present), h_m is the mass transfer coefficient, $A_{t,MA}$ is the external tube surface area, $A_{f,MA}$ is the fin area, $c_{p,a}$ is the air specific heat, $\dot{m}"_t$ is the water mass flux responsible for frost thickening, $\dot{m}"_\rho$ is the water mass flux responsible for frost densification and $\overline{\rho}_f = \frac{1}{\delta_f}\int_0^{\delta_f}\rho_f dx$ is the average density of the frost layer. So, if $\dot{m}"_v$ and $\dot{m}"_\rho$ are known $\dot{m}"_t$ can be determined from equation (5).

Molecular diffusion of water vapor into the frost layer is mainly due to Fickian diffusion caused by the vapor partial pressure gradient [5]:

$$\dot{m}"_\rho = D_{eff}\left(\frac{M_v}{RT}\right)\left(\frac{p_t}{p_t - p_v}\right)\frac{dp_v}{dx} \tag{6}$$

where D_{eff} is the effective diffusion coefficient, M_v is the molecular weight of water, R is the universal gas constant, T is the frost temperature, p_t is the total pressure and p_v is the partial pressure of water vapor. According to the local thermodynamic equilibrium hypothesis ($p_v = p_{vs}$), and in combination with the perfect gas law and Clausius-Clapeyron equation [9], $\dot{m}"_\rho$ can be also expressed in terms of a temperature gradient. At the frost surface:

$$\dot{m}"_\rho = D_{eff}\left(\frac{M_v^2 L_s}{R^2 T_{fs}^3}\right)\left[\frac{p_t p_{vs}}{p_t - p_{vs}}\right]\frac{dT}{dx}\bigg|_{fs} \tag{7}$$

with $p_{vs} = 0.61094\text{exp}\left(17.625\frac{T}{T+243.04}\right)$, where p_{vs} is the saturation pressure of water vapor expressed using the August-Roche-Magnus formula [10] with pressure in kPa and temperature (T) in °C. The temperature gradient at the frost surface is given by the Fourier's law:

$$\frac{dT}{dx}\bigg|_{fs} = \frac{q"_t}{k_f} \tag{8}$$

with $q"_t = q"_s + q"_l = q"_s + (\dot{m}"_v - \dot{m}"_\rho)\Delta h_{sub}$, where $q"_t$ is the total heat flux sum of the sensible and latent heat fluxes at the frost surface, k_f is frost thermal conductivity, that is determined by the Sanders's correlation [11] based on frost average density ($\overline{\rho}_f$): $k_f = 0.001202\overline{\rho}_f^{0.963}$

Finally, from equation (5) the time evolution of the average density and thickness of the frost layer are:

$$\delta_f(t) = \int_0^t \frac{\dot{m}"_t}{\overline{\rho}_f}dt \tag{9}$$

$$\overline{\rho}_f(t) = \int_0^t \frac{\dot{m}"_\rho}{\delta_f} dt \tag{10}$$

The initial conditions for the frost formation model are: $\delta_{f,0} = 1 \cdot 10^{-5}$ m for frost thickness, and $\rho_{f,0} = 25$ kg m^{-3} for frost density.

2.2.3. Diffusion coefficient The diffusion path of water vapor follows the pore volume in the porous frost layer, so the actual length of the diffusion path is longer than that of the path in free space for the same displacement, since the pore region is generally tortuous in the porous medium. Hence, the effective diffusion coefficient is redefined [7]:

$$D_{eff} = D_{AB}\frac{\epsilon}{\tau} \tag{11}$$

with $\epsilon = 1 - \frac{\rho_f}{\rho_{ice}}$, $\tau = \frac{\epsilon}{1-(1-\epsilon)^{0.5}}$, $D_{AB} = 2.6 \cdot 10^{-5}\left(\frac{T_{film}}{T_0}\right)^{1.75}$, where ϵ and τ are the porosity and the tortuosity of the porous medium, ρ_{ice} is the ice density and T_{film} is $\frac{T_a+T_{fs}}{2}$ with $T_0 = 298$ K. An empirical equation for the binary diffusion coefficient (D_{AB}) [12] for water vapor diffusion in the frosting process is adopted.

2.3. Fin efficiency

For a dry HX surface with anular circular fins, fin efficiency can be computed as [13]:

$$\eta_{f,MA} = \frac{\tanh(m\phi L)}{m\phi L}\cos(m\phi Ls) \tag{12}$$

with $m = \sqrt{\frac{h_{MA}}{0.5k_{fin}t_{fin}}}$, $L = R_e - R_i$, where $s = 0$, $\phi = 1$ (Schmidt efficiency), R_i is the fin internal radius (which is equal to the tube outer diameter of the coil), R_e is the fin external radius, k_{fin} is the fin thermal conductivity and t_{fin} is the fin thickness. These relationship can be applied to determine the fin efficiency for a dry finned-tube heat exchanger using a 1D-SERF model [14].

2.3.1. Frost integration Fin efficiency strongly decreases as frost grows on the evaporator surface [15]. This effect can be considered by modelling the frost layer as a low conductivity coating layer [16]:

$$\eta_{fin,f} = \left[cos(m\delta_f) - \frac{mk_{fin}t_{fin}}{2k_f}sin(m\delta_f)\right]\eta_{fin} \tag{13}$$

where $m = \sqrt{\frac{h_{MA}k_f}{\delta_f k_f^2 + 0.5\delta_f h_{MA} k_{fin} t_{fin} + 0.5 k_{fin} k_f t_{fin}}}$

2.4. Heat and mass transfer

The Nusselt number (Nu) computation is based on a standard Colburn relationship:

$$\mathrm{Nu} = j_a \mathrm{Re} \mathrm{Pr}^{1/3} \tag{14}$$

with the Reynolds number (Re) based on the hydraulic diameter of the minimum frosted flow area (Dh_f) of the HX, Pr the Prandtl number, j_a the Colburn j-factor. The correlations by Wang et al. are utilized to estimate j_a for both plain fins [17] and louvered fins [18] types.

The mass transfer coefficient (h_m) is determined based on the Chilton-Colburn analogy from the heat transfer coefficient (h_{MA}) [19]. As suggested by Lee et al. [20] the Lewis number is evaluated at the current time step based on its definition. All the geometrical parameters of the HX, as well as those in the heat transfer coefficient correlations, are corrected (i.e., by adding the δ_f term) to account for frost formation. This allows the heat transfer and, consequently, the mass transfer coefficients to vary during frost formation.

2.5. Air side pressure drop

The air side pressure drop across each segment of the HX is calculated using the equation proposed by Kays and London [21]:

$$\Delta p = \frac{G_{max}^2}{2\rho_1}\left[f\frac{A_0}{A_c}\frac{\rho_1}{\rho_m} + (1+\sigma^2)\left(\frac{\rho_1}{\rho_2} - 1\right)\right] \tag{15}$$

where G_{max} is the maximum mass flux ($G_{max} = \dot{m}_a/A_c$), ρ_1 is inflow density, ρ_2 is outflow density, ρ_m is their arithmetic mean, σ is the ratio of the minimum free flow area (A_c) to the HX frontal-area and A_0 is the total surface area of the HX. The fanning friction factor (f) is obtained from the correlations of Wang et al. for both plain fins [17] and louvered fins [18].

All the geometrical parameters of the HX as well as those in the fanning friction factor correlations, are corrected (i.e., by adding δ_f term) to account for frost formation. This allows the pressure drop to increase during frost formation.

2.6. Heat transfer and pressure drop refrigerant side

The heat transfer coefficient and the pressure drop characteristics, refrigerant side, for each segments are computed using a built-in 'Pipe (2P)' block [22]. This block calculates an average heat transfer coefficient and an average friction factor (for the head losses) with respect to the temperature and state of the fluid volume in each segment.

2.7. Overall heat transfer

The sensible heat flow rate ($\dot{Q}_{sens}$), from the moist air domain to the two-phase fluid domain is calculated for each segment using a modified ϵ-NTU method to consider the effect of frost layer on the heat transfer surface:

$$\dot{Q}_{sens} = \epsilon C_{min}(T_{in,MA} - T_{in,2P}) \tag{16}$$

where C_{min} is the minimum heat capacity, $T_{in,MA}$ is the moist air inlet temperature and $T_{in,2P}$ is the two-phase fluid inlet temperature.

The effectiveness (ϵ) is calculated according to the flow arrangement, which in this case is cross-flow with both fluids unmixed:

$$\epsilon = \begin{cases} 1 - \exp\left\{\dfrac{\exp(-C_r NTU^{0.78}) - 1}{C_r NTU^{-0.22}}\right\} & \text{if single-phase} \\ 1 - \exp(-NTU) \quad \text{if two-phase} \end{cases} \tag{17}$$

with $C_r = \frac{C_{min}}{C_{max}}$ where C_{max} is the maximum heat capacity and $NTU = \frac{1}{R_{tot}C_{min}}$. The modified relation is the total thermal resistance between the two flows (R_{tot}):

$$R_{tot} = R_{c,2P} + R_{wall} + R_{frost} + R_{c,MA} \tag{18}$$

Thermal resistances are defined as: $R_{c,2P} = \frac{1}{h_{2P}(A_{t,2P}+\eta_{f,2P}A_{f,2P})}$, $R_{wall} = \frac{\ln(\frac{OD}{ID})}{2\pi k_t L_t}$, $R_{frost} = \frac{\ln\left(\frac{OD+2\delta_f}{OD}\right)}{2\pi k_f L_t}$, $R_{c,MA} = \frac{1}{h_{MA}(A_{t,MA}+\eta_{f,MA}A_{f,MA})}$, for internal convective resistance, wall resistance, frost resistance and external convective resistance, respectively, where OD is the tube outer diameter, ID is the tube inner diameter, $A_{t,2P}$ is the tube area refrigerant side, L_t is the tube length and k_t is the tube thermal conductivity. A constant fin efficiency ($\eta_{f,2P}$) and an additional fin area at the refrigerant side ($A_{f,2P}$) are included in the model to consider the effect of possible inner micro-fins in the tube.

To close the energy balance the overall heat transfer rate between the two streams (from moist air to two-phase flow) is:

$$\dot{Q}_{tot} = \dot{Q}_{sens} + \dot{Q}_{lat} = \dot{Q}_{sens} + \dot{m}_{w,cond}\Delta h_{sub,vap} \tag{19}$$

where $\dot{Q}_{lat}$ is the latent heat flow rate due to desublimation (or condensation) that occurs on the moist air side, $\Delta h_{sub,vap}$ is the latent heat of sublimation (or vaporization) evaluated at the average frost surface (or wall) temperature.

3. Model results and validation under standard conditions

Results from the frost-free (i.e., clean) heat exchanger model are compared with the standard Evaporator-Condenser (2P-MA) block in the Simscape toolbox, which cannot capture frosting phenomena. Three inlet refrigerant conditions are tested: an evaporation temperature of -15 °C with a vapor quality of 0.2 (fig. 2a), an evaporation temperature of -5 °C with 5 °C of subcooling (fig. 2b), and a condensation temperature of 45 °C with 5 °C of superheating (fig. 2c). The refrigerant inlet mass flow rate is 40 kg·h^{-1}, while air-side conditions are 20 °C and 50% RH with an inlet mass flow rate of 360 kg·h^{-1}.

Since the developed model is a segment by segment model, the comparison is carried out by varying the number of segments.

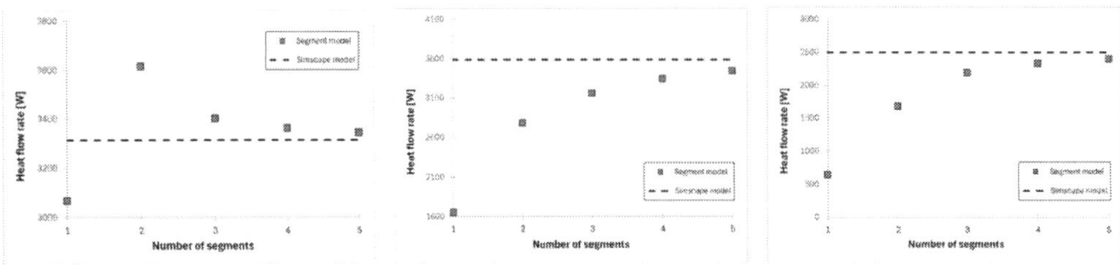

(a) Evaporator with two-phase flow inlet (b) Evaporator with single-phase flow inlet (c) Condenser with single-phase flow inlet

Figure 2. Segments sensitivity study

As shown in figure 2, the accuracy of the solution depends on the number of segments, as it approaches the Simscape™ model solution. In fact the implemented approach requires a sufficiently accurate representation of the fluid state throughout the entire circuit to properly calculate the maximum transferable power and effectiveness per segment. Moreover, due to the same underlying principle, figure 2 illustrates how the minimum number of segments for achieving a sufficiently consistent solution depends on the initial state of the inlet flow. Specifically, it correlates with the number of phase changes the fluid undergoes between the inlet and outlet. For instance, as depicted in figure 2a, when the inlet flow is in a two-phase state and transitions to a single-phase vapor state at the outlet, a consistent solution can be obtained with merely three segments. Conversely, as shown in figure 2b, if the inlet state is liquid and transitions to vapor at the outlet, a minimum of four segments is necessary to achieve an acceptable accuracy. The same principle applies to figure 2c, where the inlet state is vapor and the outlet is liquid; in this case the HX is acting as a condenser.

4. Air flow redistribution algorithm

The non-uniform frost growth on the HX surface, caused by the subdivision into segments, leads to various degrees of blockage across the different segments. On the air side, the HX can be viewed as a parallel arrangement of heat exchangers; consequently, the total air mass flow ($\dot{m}$) must be divided according to the classical parallel arrangement so that the pressure drop (Δp) across each segment remains the same, as described by the system of equation (20).

$$\begin{cases} \dot{m}_i^2 R_i = \Delta p \\ \dot{m} = \sum_{i=1}^{n} \dot{m}_i \\ R_i = \dfrac{1}{2\rho_1 A_{c,i}} \left[f \dfrac{A_{0,i}}{A_{c,i}} \dfrac{\rho_1}{\rho_m} + (1+\sigma_i^2)\left(\dfrac{\rho_1}{\rho_2} - 1\right) \right] \end{cases} \tag{20}$$

The specific flow resistance (R_i) is defined for each segment by equation (15).

5. Modeling results under frosting condition

To demonstrate predictability under frosting conditions, the selected inlet conditions are tested: a refrigerant evaporation temperature of -15 °C with a vapor quality of 0.2 and $\dot{m} = 40$ kg·h^{-1}, and air at $T_a = 8$ °C, $RH = 80\%$, and $\dot{m} = 375$ kg h^{-1}.

In this scenario, a single-circuit heat exchanger subdivided into four segments is analyzed (fig. 3a). The results of the frost thickness are presented in a dimensionless form ($\delta_f^* = \frac{\delta_f}{FP/2}$), with half of the fin spacing (FP) as the parameter for normalization. Indeed, if δ_f^* is one, the air-side segment is fully obstructed, on the other hand, if δ_f^* is zero, the air-side segment is free of frost.

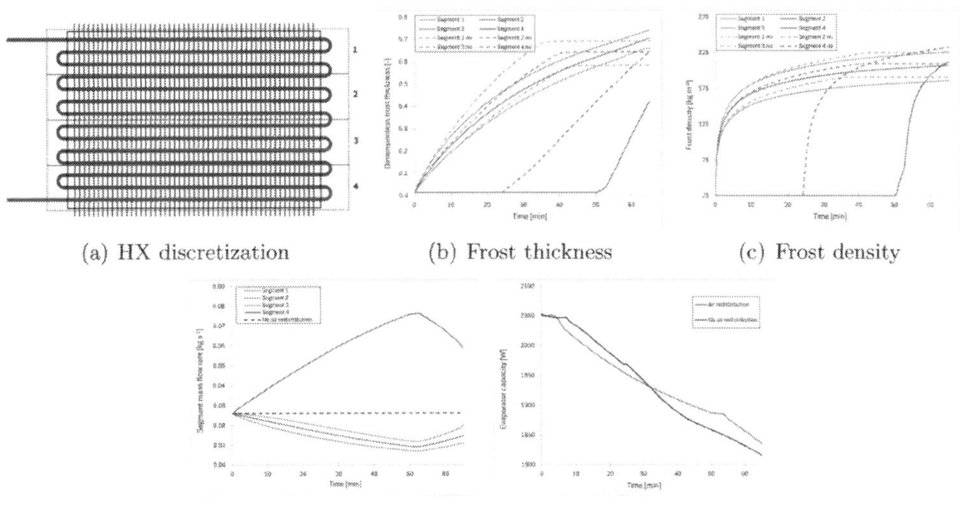

(a) HX discretization (b) Frost thickness (c) Frost density

(d) Segment mass flow rate (e) Evaporator capacity

Figure 3. Model results under frosting conditions

The results of the simulations in figures 3(b-e) show the frosting dynamics at the evaporator, and how it significantly affects the heat transfer throughout a simulation period of 65 minutes. In particular, the frost growth dynamics with the activation of the air redistribution algorithm is also shown and compared to an unrealistic simulation without air redistribution. In figure 3b, it is highlighted how flow redistribution significantly impacts on frost growth across different

Journal of Physics: Conference Series **2940** (2025) 012015 doi:10.1088/1742-6596/2940/1/012015

segments. Specifically, in the simulation without redistribution, the surface temperature of the frost reaches the frost point in just 30 min, halting any further growth. However, flow redistribution effectively reduces the airflow to these segments, consequently decreasing surface temperature of the segment and allowing for much more significant growth over time. Figure 3e shows that for this particular case the evaporator capacity does not change drastically with or without the flow redistribution algorithm; only a small change in the overall system behavior is predicted. However, in general, this difference could be much more significant.

6. Conclusions and future developments

A lumped-parameter model of an elementary block (i.e. segment) of a heat exchanger, including frosting, was developed in this work. In particular, a sensitivity study against the number of segments was conducted for the case of a clean heat exchanger (i.e. no frost), comparing the results with those of a commercial heat exchanger block, already implemented by MathWorks®, which does not include the possibility to model frosting. It is demonstrated that with the proposed modeling technique, it is possible to represent an evaporator with sufficient accuracy using only three segments, and a condenser with four segments. To further demonstrate the effect of frosting on the evaporator performance, a typical case study was simulated, focusing on the actual frosting dynamics. The model can ultimately be considered validated from the perspective of a clean heat exchanger, and the results under frosting conditions also appear realistic. Experimental validation of the frosted heat exchanger model represents the next challenge of this study, together with the integration of a defrost model.

7. References

[1] Hayashi Y, Aoki A, Adachi S and Hori K 1977 *J. Heat Transfer.*
[2] Hayashi Y, Aoki K and Yuhara H 1977 *Heat Transf. J. Res.* **6** 79–94,
[3] O'Neal D and Tree D 1985 *ASHRAE Transactions* **91** 267–281
[4] Dietenberger M A 1982 A frost formation model and its validation under various experimental conditions Tech. rep. NASA
[5] Le Gall R, Grillot J and Jallut C 1997 *Int. J. Heat Mass Transf.* **40** 3177–3187
[6] Padhmanabhan S 2011 *Study of frost growth on heat exchangers used as outdoor coils in air source heat pump systems* Ph.D. thesis Oklahoma State University
[7] Qiao H, Aute V and Radermacher R 2017 *Int. J. Refrig.* **84** 181–197
[8] Górnicki K, Winiczenko R, Kaleta A and Choińska A 2017 *Tech. Sci.* **20** 241–257
[9] Wark K 1988 *Thermodynamics* (McGraw-Hill Science, Engineering & Mathematics)
[10] Alduchov O A and Eskridge R E 1996 *J. Appl. Meteorol. Climatol.* **35** 601–609
[11] Sanders C T 1974 *The influence of frost formation and defrosting on the performance of air coolers* Ph.D. thesis Technical University of Delft, Netherlands
[12] Fossa M and Tanda G 2005 Modelling of frost growth in simple geometries and passages *Modelling and Simulation: 16 th IASTED International Conference Proceedings*
[13] Nemati H and Samivand S 2014 *Arab. J. Sci. Eng.* **39** 1–6
[14] Suárez F, Keegan S, Mariani J and Barreto G 2018 *Appl. Therm. Eng.* **149**
[15] Chen H, Thomas L and Besant R W 2003 *Int. J. Refrig.* **26** 140–149
[16] Tu P, Inaba H, Horibe A, Li Z and Haruki N 2006 *J. Heat Transfer.* **128**
[17] Wang C C, Chi K Y and Chang C J 2000 *Int. J. Heat Mass Transf.* **43** 2693–2700
[18] Wang C C, Lee C J, Chang C T and Lin S P 1999 *Int. J. Heat Mass Transf.* **42** 1945–1956
[19] Mitchell J W and Braun J E 2012 *Principles of heating, ventilation, and air conditioning in buildings*
[20] Lee K S, Kim W S and Lee T H 1997 *Int. J. Heat Mass Transf.* **40** 4359–4365
[21] Kays W M and London A L 1984 *Compact heat exchangers* (McGraw-Hill, New York, NY)
[22] MathWorks URL `https://it.mathworks.com/help/simscape/physical-modeling.html`

8. Acknowledgement

We express our gratitude to the Zehnder Group - Competence Center Campogalliano (MO) for their assistance and the allocation of necessary resources throughout this research endeavor.

Energy transition audits in industries located in Valencian region

Mar Cañada-Soriano[1*], Elisa Peñalvo-López[2], Carla Montagud-Montalvá[1] and Iván Valencia-Salazar[2]

[1] Applied Thermodynamics Department, Universitat Politècnica de València, València, Spain
[2] Electrical Engineering Department, Universitat Politècnica de València, València, Spain

*E-mail: macaaso@upv.edu.es

Abstract. The industrial sector is a major global consumer of energy, and meeting Europe's climate neutrality targets will require a complete transformation of the global energy industry. To this end, Energy Transition Audits (ETAs) provide industries with a comprehensive and tailored action plans to reduce their energy consumption while mitigating the impact of CO2 emissions. In this context, this work presents the first results of an Energy Transition Audit conducted in a Valencian food processing facility within the framework of the European project EnTRAINER- Energy Transition Audits towards Decarbonisation. Starting from the industry baseline in terms of energy consumption and costs and production through a one-year analysis, five improvement actions are proposed and evaluated using the following performance indicators: energy and cost savings along with emissions reduction. Finally, the improvement measures are prioritized based on the company's emissions reduction targets. Therefore, this innovative approach empowers the industries to monitor their progress towards achieving their decarbonization targets.

1. Introduction

The industrial sector plays a key role in the transition towards a climate-neutral economy since it is one of the largest consumers of energy globally, accounting for a significant portion of total energy consumption in numerous countries. The level of energy consumption in the industrial sector varies depending on factors including the types of industries present, the scale of operations, the efficiency of equipment and processes, and the availability and cost of energy sources.

In the case of Spain, as in many other European countries, the industry sector is a major energy consumer. It represents the second largest consumer of final energy in the Spanish economy, with 25.6 % of the total in 2021, behind transport (38.5%) and ahead of households (18.8%). Energy consumption in Spanish industry has experienced moderate growth in recent decades. Between 1990 and 2021, final energy consumption in Spanish industry increased by 24.4%. Fossil fuels continue to be the main source of energy for Spanish industry, accounting for 74.9% of total consumption in 2021. Regarding electricity, it is the most widely used fuel (43.4%), followed by natural gas (21.2%) and petroleum products (10.3%). It is notable that the share of renewable energy in the final consumption of Spanish industry has been increasing in recent

Journal of Physics: Conference Series **2940** (2025) 012016 doi:10.1088/1742-6596/2940/1/012016

years. In 2021, it reached a record 25.1%, which includes both direct consumption of renewables to produce heat (9.9%) and electricity generated from renewable sources (15.2%) [1]. The industrial sector in the Valencian Community follows similar patterns to those observed in Spain, although with some specific nuances. The Valencian Community is one of the most industrialised regions in Spain, with a variety of sectors contributing significantly to energy consumption including the ceramic, agrifood, textile, automotive, textiles, chemicals, ceramics, and food processing sectors.

The European Parliament's Economic, Scientific and Quality of Life Policy Department underlines this important role of industry in the transformation towards a carbon neutral economy, as it not only has to work to reduce its own carbon footprint, but also accelerate the transition by providing affordable and clean technological solutions as well as developing new business models [2-4]. The successful transition to a low-carbon energy system will require a complete transformation of the global energy industry, with a shift away from fossil fuels towards carbon-free alternatives by the second half of this century. This transformation will aim to significantly reduce CO2 emissions from energy consumption, thereby mitigating the impact of climate change and keeping the global temperature rise within 1.5 degree Celsius above pre-industrial levels. The European Union (EU) is a world leader in promoting energy transition through adopting ambitious measures to achieve decarbonization goals [5-7]. In this sense, conventional energy audits in industry, as foreseen by current legislation, have to be improved in their focus and scope towards the Energy Transition Audits (ETAs) [8]. Energy Transition Audits goes beyond conventional energy audits, and therefore, complies with the minimum criteria for energy audits, including those carried out as part of energy management systems as described in the 2012/27/EU directive [9]. The procedures and methodologies for conventional energy audits are standardized by UNE-EN 16247:2023, thus ensuring uniformity across European countries [10]. In this context, the scope of ETA is to focus not only on the conventional energy efficiency action plan and the evaluation of emissions profiles as equivalent CO2 emissions, but also on developing Decarbonization Action Plans (DAPs) that include a complete decarbonization strategy for companies. Therefore, the implementation of an ETA will result in a dedicated decarbonization and energy transition roadmap with scheduled actions and funding opportunities, tailored to the needs of each of the companies to reduce its environmental impact and to increase its sustainability and energy independence.

The purpose of this paper is to present the first results of the European project EnTRAINER - Energy Transition Audits towards Decarbonization, funded by the Environment and Climate Action (LIFE) programme. This work presents some of the steps included in the development of advanced energy transition audit of an industry of the Valencian region in order to achieve an innovative and fundamental shift from traditional energy audits to a holistic and comprehensive methodology known as Energy Transition Audits (ETA). The main objective of this new approach is to integrate a decarbonisation actions and strategies in the energy audits to facilitate industries the transition to a more sustainable energy development.

2. Methodology

2.1 The Energy Transition Audit (ETA)

As can be observed in Figure 1a, an Energy Transition Audit (ETA) consists of the implementation of different main steps with a particular emphasis on the in-depth analysis phase. Starting from the initial data (initial situation) of the company regarding the energy carriers, annual consumption, and costs, along with CO2 emissions and production figures among others, an energy scan is performed. To this end, a preliminary contact must be established in conjunction with the collection of preliminary data. This information is then used to assess the distribution of energy consumption, establish an emissions profile, identify improvement opportunities and setting goals seeking the increase of energy efficiency and the reduction of CO2 emissions. Once the energy scan has been completed and discussed with the company, the fieldwork is carried out in order to gain an understanding of the operating routines, processes being held and user among others.

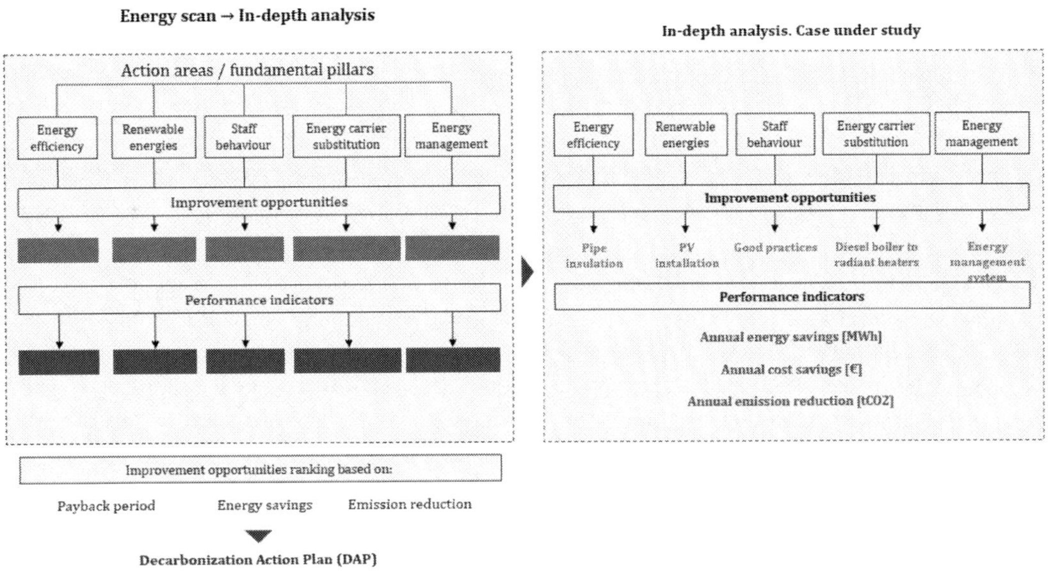

Figure 1. Energy Transition Audit workflow including the in-depth analysis specifically for the case under study.

The in-depth analysis entails the central core of the ETA. From an analysis of the existing energy performance situation of the company, five fundamental pillars can be identified on which to implement opportunities to achieve industrial decarbonisation. These include energy efficiency measures, such as improving system efficiency, optimising process performance as well as thermal energy recovery. The integration of renewable energy sources is a fundamental action area since makes a major contribution to the decarbonisation effort. Another essential part deals with raising awareness among the staff to change or adapt their actions towards a more sustainable and energy efficient behaviour. The substitution of carbon fuels with alternative energy sources, such as hydrogen, biofuels, or the "greening of the grid," when grid-purchased electricity is utilized, can result in a further reduction of combustion-associated greenhouse gas (GHG) emissions for industrial processes. The implementation of energy management systems to reduce the overall consumption would also entail another fundamental pillar of the in-depth analysis. Once each area of action has been subjected to a comprehensive analysis, improvement opportunities and associated performance indicators should be proposed for each pillar. After the proposal and evaluation of the improvement measures, they should be ranked based on relevant criteria for the company such as scopes targeted, investment costs, payback period, energy savings, and/or CO2 emission reduction.

The next step would include the proposal of the decarbonization action plan in which the proposed actions and implementation plan and schedule would be carry out. In this sense, the auditor(s) will propose and agree with the company on the ETA workflow, time schedule, and data that will be collected for the ETA implementation.

2.1.1 Case under analysis.

Based on the above-mentioned steps included in an ETA, the case study focuses primarily on the in-depth analysis step as depicted in Figure 1b. For the each of the five action areas, the following improvement measures are proposed:

Pipe insulation (Energy Efficiency). During the visit to the facilities, several uninsulated pipes were identified by means of infrared thermography. The lack of insulation in these pipes results in significant heat losses due to convection to the environment and to radiation since the fluid that circulates inside them is at high temperatures (approximately 100°C). To minimise these losses, the installation of commercial insulation on the external surface of these pipes was considered.

Photovoltaics installation (Renewable energies). Given that the orientation of the roof presents optimal solar radiation incidence, the installation of photovoltaic modules on the available roof area is evaluated.

Good practices (Staff behaviour). When the facilities were inspected, several counterproductive actions or bad operator habits, such as doors left open in air-conditioned rooms, or lights on in unoccupied chambers were identified. In this sense, the commitment by the staff along with behavioural changes are needed, and to achieve this, awareness-raising workshops are recommended.

Energy carrier substitution. Diesel-fired boilers are used as heating system in the industrial buildings where most of the workers spend their working day. Thus, the replacement of diesel boilers with the installation of electric radiant heaters is evaluated. In this regard, the heating mechanisms of these two systems are fundamentally distinct. While the boiler relies on the combustion of diesel to heat the air, electric radiant heaters warm the air through radiation. Since the first heating system provides warm air, air stratification takes usually place when industrial buildings are concerned. As a result, the warm air rises to the top of the building, where it is lost through the roof. Moreover, since the gates on the buildings are usually open, a significant proportion of the previously heated air is dissipated outside. Radiant heat emitters, in turn, transmit heat by radiation in the form of infrared waves that directly heat objects, floors, walls and people. Therefore, in warehouses where there is a high degree of air exchange with the external environment, or where gates or windows are frequently opened, radiant heat represents an effective solution, as it does not directly heat the air.

Implementation of an energy management system. Through this improvement measure, record energy consumptions and any anomaly that may occur would be tracked. As a result, both energy expenses and emissions could be reduced.

Once the improvement opportunities regarding each action area have been proposed, the most interesting performance indicators was defined. Specifically, the annual energy and cost savings, along with the annual emissions reduction were selected in the case evaluated. Moreover, based on the different performance indicators obtained along with the financial output such as the payback period, the ranking of the different measures would be carried out.

3. Results

3.1 Case under study. Initial situation

A preliminary analysis of the company was performed to establish a baseline for the decarbonization process. The analysed industry belongs to the food sector, specifically it is dedicated to the processing of honey with an annual production of 4 million kg. Table 1 shows the most relevant information regarding the company assessed on the case under study.

Table 1. Most relevant data regarding the initial situation of the company under study.

	Energy carrier	
Electricity	**Natural gas**	**Diesel**
Machinery HVAC Computers Lights	Boiler (steam)	Boiler (heating)
	Annual consumption [MWh]	
460	830	197
	Annual cost [€]	
136640	132810	26280
	Emissions [tCO2]	
118	167	52

The two primary forms of energy consumption in this industry are the thermal and the electrical one. On the one hand, natural gas is the fuel used by the boilers to generate the steam needed for the process. On the other hand, the machines and motors used in the process, as well as the office air conditioning, computers, and lighting, are electrically powered. The electricity is purchased from the grid through a contract with a capacity of 180 kW and a tariff of 6.1 TD. In this company, diesel was also used to heat the warehouses by means of boilers. Using the company's baseline data and invoices for the year under review (2022), annual consumption and costs were calculated for each energy supply. The highest energy consumption corresponds to natural gas, as well as the emissions (obtained using recognised emission factors for each emission source). Conversely, the economic costs are higher for electricity.

3.2 Case under study. Performance indicators

The performance indicators obtained for each improvement opportunity implemented for the case under study are shown in Figure 2. Thus, from the estimation of the annual energy savings [MWh], the annual savings in both economic costs [€] and emissions [tCO2] could also be obtained.

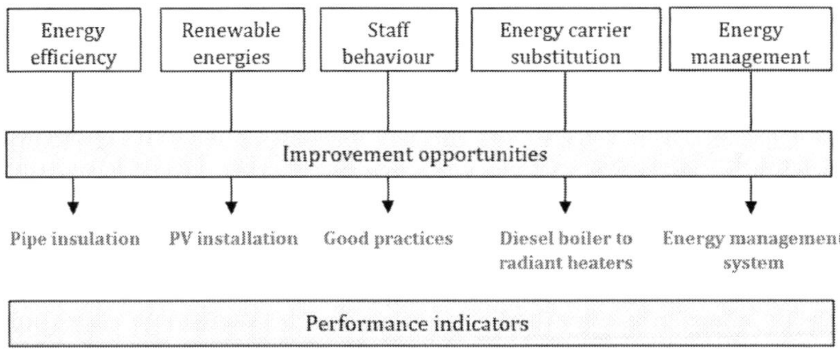

	Pipe insulation	PV installation	Good practices	Diesel boiler to radiant heaters	Energy management system
Annual energy savings	98 MWh	138 MWh	19 MWh	40 MWh	34 MWh
Annual cost savings	15663 €	40000€	4000€	-20000€	6700€
Annual emission reduction	20 tCO2	28.3 tCO2	4 tCO2	20.6 tCO2	7 tCO2
Payback	**0,3 years**	**2,9 years**	**0,12 years**	-	**5,2 years**
Ranking	2	3	1	4	5

Figure 2. Performance indicators and payback obtained for the case of study regarding each action area and proposed improvement measure.

Specifically, the results for each implemented measure are outlined below.

Pipe insulation (Energy Efficiency). From the thermographic images of the condensate return pipes obtained during the visit, it was possible to estimate the external surface temperatures, which in turn allowed the calculation of the heat losses by convection and radiation for the analysed pipe sections. The calculation was then repeated for the same pipe sections but taking into account the use of commercial insulation on the external surface of these pipes. The result of this improvement translates into an annual energy saving of 98 MWh of natural gas, an annual economic saving of 15663€ and an annual CO2 emission saving of 20t. Insulating the pipes would have a payback time of 3.5 months.

Photovoltaics installation (Renewable Energies). The PV installation consists of 390 photovoltaic modules with a power rating of 450 Wp, giving a total installed capacity of 176 kWp, and occupies approximately 852.57 m^2 of roof space. Based on the current electricity consumption data and the production rate of the PV installation (determined by the climate zone), it was estimated that the company would consume 30% less electricity from the grid, resulting in 138 MWh being removed from the grid. This would result in annual economic savings of 40000 € and a reduction of 35 tonnes of CO2 per year. This measure has a payback time of almost 3 years.

Good practices (Staff behaviour). The impact of this measure is not easy to quantify as it depends on the behaviour and attitudes of employees and the use of the equipment. This study estimates that the savings will be around 3% of the total, half of which will be in natural gas and half in electricity. Consequently, the annual savings in natural gas and electricity could be 12 MWh and 7 MWh respectively. The annual economic savings would be 2000€ for both gas and electricity, and 4t CO2 could be not emitted per year. As the inversion of this measure is quite small, this improvement would be paid off in one month.

Energy carrier substitution. As can be observed in Figure 2, the use of radiant heat emitters as the heating system for the industrial building has an impact on both the energy and emissions savings but this approach carries a higher economic cost. When evaluating this measure, a reduction of 20% in energy consumption was considered. Therefore, the annual electricity savings result of 40 MWh and annual CO2 emission savings results of 20t. Given that the cost of electricity is greater than that of diesel, this measure would result in an additional cost of 20400€.

Implementation of an energy management system. It is also challenging to quantify the potential energy savings that can be achieved by implementing an energy consumption management system. According to the Catalan Energy Agency, based on empirical evidence from the implementation of such systems in diverse sectors, savings were estimated to be between 2% and 7%. This study assumed energy savings of 3% in natural gas consumption and 2% in electricity consumption. This results in annual savings of 25MWh in natural gas and 9MWh in electricity. The annual economic are 3980€ and 2720€ for natural gas and electricity respectively, and 7tCO2 could be saved per year (5 tCO2 for NG and 2 tCO2 for electricity). This measure has a 5,2-year pay-back.

3.3 Case under study. Ranking of the proposed measures

The energy transition audit identifies proposed solutions and actions. These are then presented and discussed with the company to ensure they align with their overall strategy, including their climate goals and targets. The final step prioritizes these improvement measures based on the company's emission reduction ambitions. For this specific assessment, factors like annual energy savings, payback periods, and emission reductions were considered most relevant. The implementation calendar is then established based on the company's decarbonisation action plan. (Table 2).

Table 2. Decarbonisation action plan for the proposed measures.

Strategy category	Measure	Energy saving [MWh/year]	Year of implementation
Energy efficiency	Pipe insulation	97.9	2027
Renewable energy	Installing PV	138	2028
Staff behaviour	Good practices in electric and NG consumption	19.3	2025
Energy demand management	Integration of an energy monitoring system	34.1	2030
Energy carrier substitution	Substitution of diesel boilers with radiant heaters	39.4	2029

Journal of Physics: Conference Series **2940** (2025) 012016 doi:10.1088/1742-6596/2940/1/012016

Specifically, the measures with no or very little inversion cost were placed at first by management. In this sense, due to its low investment cost, raising awareness among employees would be the first measure to implement. The pipes that carry fluids at high temperatures should be insulated on second place due to the large energy savings achieved and the low payback. The PV modules would be following installed. Replacing diesel boilers with radiant heaters as the heating system does have an additional cost associated with the cost of electricity. Nevertheless, this measure aims to move away from fossil fuels in favour of electricity that could even be 100% green. Finally, the implementation of an energy management system would be considered in last place due to the low annual energy savings achieved.

4. Conclusions

EnTRAINER project transforms energy audits by introducing a novel, holistic methodology: the Energy Transition Audit (ETA). This innovative approach goes beyond traditional audits, delivering a comprehensive and customized action plan tailored to achieving complete decarbonization for the audited site, aligning perfectly with the company's emission reduction goals. We present the initial findings of the in-depth analysis phase of an ETA conducted at a Valencian food processing facility. Through a year-long analysis of the company's consumption patterns and associated costs, we established a detailed baseline. This data serves as the foundation for proposing a series of targeted improvement measures to reduce energy consumption and its corresponding emissions. The next project phase will evaluate the viability of sustainable energy solutions across short, medium, and long-term horizons. This will empower the company's top management to monitor their progress towards achieving their decarbonization roadmap and targets.

References

[1] International Energy Agency (IEA). (2022).

[2] Altenburg, T., & Rodrik, D. (2017). Green industrial policy: Accelerating structural change towards wealthy green economies. Green Industrial Policy.

[3] Rissman, J., Bataille, C., Masanet, E., Aden, N., Morrow III, W. R., Zhou, N., ... & Helseth, J. (2020). Technologies and policies to decarbonize global industry: Review and assessment of mitigation drivers through 2070. Applied energy, 266, 114848.

[4] Wesseling, J. H., Lechtenböhmer, S., Åhman, M., Nilsson, L. J., Worrell, E., & Coenen, L. (2017). The transition of energy intensive processing industries towards deep decarbonization: Characteristics and implications for future research. Renewable and Sustainable Energy Reviews, 79, 1303-1313.

[5] Tagliapietra, Simone, et al. "The European union energy transition: Key priorities for the next five years." Energy Policy 132 (2019): 950-954.

[6] Popescu, Catalin, et al. "Energy transition in European Union—Challenges and opportunities." Energy Transition: Economic, Social and Environmental Dimensions (2022): 289-312.

[7] Skjærseth, Jon Birger. "Towards a European Green Deal: The evolution of EU climate and energy policy mixes." International Environmental Agreements: Politics, Law and Economics 21.1 (2021): 25-41.

[8] Kandpal, Vinay, et al. "Challenges and Opportunities for Sustainable Energy Transition and Circular Economy." Sustainable Energy Transition: Circular Economy and Sustainable Financing for Environmental, Social and Governance (ESG) Practices (2024): 307-324.

[9] European Parliament and Council of the European Union. (2012). Directive 2012/27/EU of the European Parliament and of the Council of 25 October 2012 on energy efficiency, amending Directives 2009/125/EC and 2010/30/EU and repealing Directives 2004/8/EC and 2006/32/EC. https://eur-lex.europa.eu/legal-content/EN/TXT/?uri=celex%3A32012L0027.

[10] Asociación Española de Normalización y Certificación (Aenor). UNE-EN 16247:2027. Auditorías energéticas

This work has been carried out within the framework of the research project "Energy Transition Audits towards Decarbonization' — 'EnTRAINER" (Contract Ref: LIFE21-CET-AUDITS-EnTRAINER/101076424) of the European Union's LIFE Framework Program.

Journal of Physics: Conference Series **2940** (2025) 012017 doi:10.1088/1742-6596/2940/1/012017

Influence of Bed Medium on Bubbling Fluidized Bed Reactors for Biomass Gasification: Experimental Study

Lina Montuori[1,*], Carlos Vargas-Salgado[1], Manuel Alcazár-Ortega[1] and Emma Frosina[2]

[1] Institute for Energy Engineering, Universitat Politècnica de València, Camino de Vera s/n, 46022 Valencia, Spain

[2] Department of Engineering, Universitá degli Studi del Sannio, Piazza Roma, 27 – 821000 Benevento, Italy

* Corresponding author: lmontuori@iie.upv.es

Abstract. This research provides a contribution to biomass gasification technology development and to its establishment into the market as mature and functional technology, competitive for large and medium-scale applications. In this article, an experimental analysis of Bubbling Fluidized Bed Reactors (BFBR) for power generation with different sand mixtures is presented. This experimental study is based on a 50 kW and a 10 kW BFBR, respectively, with inner diameters of 10 cm and 22.2 cm. In the experiments, three different bed sands (molochite, silica and alumina) have been tested attending to their properties, such as operating temperatures, bulk density for fluidisation, homogeneous grain size and commercial availability. According to the implemented tests, the obtained results demonstrate that molochite (calcined kaolin) has best effectiveness as bed additive for optimizing the reactor pressure drop, operating temperature, mean particle size and fluidization conditions. On the other side, the most appropriate range for the relation between the sand bed height and the reactor inner diameter to reduce the energy consumption was determined. Finally, the distributor pressure drop and the range of pressured drop were experimentally evaluated.

1. Introduction

Biomass gasification is an emerging technology that is call to play a crucial role in modern renewable energy systems. Moreover, biomass can ensure energy flexibility and continuous production, providing a reliable green resource that can be easily stored [1]. In this way, it can help to overcome the intermittent and highly variable energy production by solar and wind resources.

The use of Bubbling Fluidized Bed Reactors (BFBR) began in the 20th century and, nowadays, its utilization is increasing due to such favourable characteristics as high combustion efficiency, flexibility and potential scalability [2], as well as high char circulation rate, which improves the overall gasification efficiency and reduces the emission of pollutants [3], [4]. However, in spite of these advantages, this technology is still struggling, with such challenges as significant tar and impurities obtaining or pressure drop instability, which can produce operational problems [5], [6].

The variable shape, sizes and densities of the biomass particles require the support of a fluidizing medium such as sands. This material act as a heat vector, ensuring the maintenance of isothermal

conditions for the bed, preventing the formation of hot spots and increasing the global efficiency of the process [7].
Fluidized bed reactors are widely used for biomass gasification because of the long residence time, uniform temperature distribution, and high mass and heat transfer rates that can be achieved. Since optimal fluidisation is required to enhance biomass gasification reactions with a low power consumption, the gasifying agent admission system is an essential component for a BFBR [8]. Thus, fluidization is well suited when the bed material is properly chosen and the admission system is well designed, obtaining the following advantages: high heat transfer, uniform and relatively controllable temperatures, favourable gas-solid contact and ability to handle a wide range of particulate materials [9]. Subsequently, in order to achieve a proper fluidization and a better efficiency in the BFBR, an appropriated combination between the bed material and the pressure drop in the admission system must be chosen. In consequence, keeping these goals in mind and considering the starring role of sand in the fluid bed gasification processes, the experimental study here presented compares three sand materials (molochite, silica, and alumina) for its utilization in this kind of reactor, identifying the most suitable operation conditions to optimize the gasification process.

2. Materials, methods and calculations

2.1. Experimental setup

The experimental setup, depicted in Figure 1, includes two transparent methacrylate reactors of diameters 10 cm and 22.2 cm, respectively, where a rectangular window has been placed to permit the visualization of the sand fluidization process. Both reactors share a vacuum pump to inject the necessary air (gasifying agent), an inverter to set up the motor speed to control the airflow, and a cleaning system to avoid sand going into the vacuum pump.

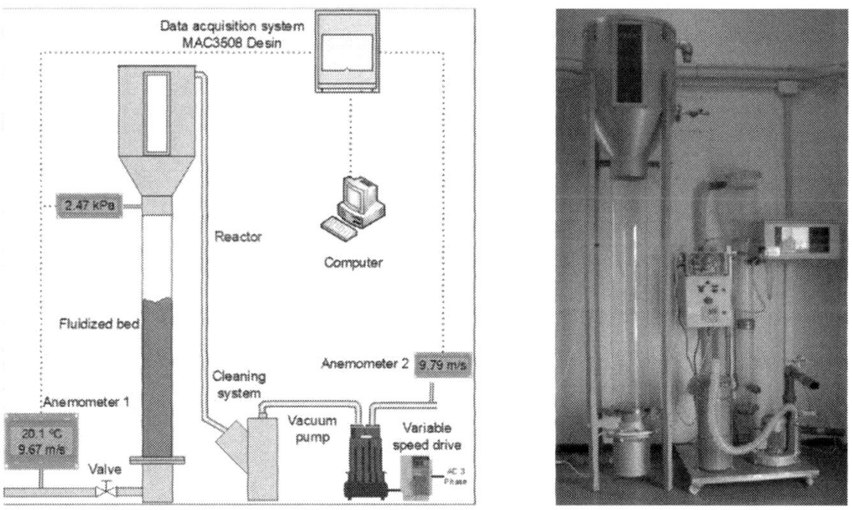

Figure 1. The schema of the pilot plant (diagram and picture)

A data acquisition system collects all the required information. Metering is done by means of the following devices: an anemometer (Anemometer 1), located in the reactor inlet pipe for measuring the air velocity (v_{air}), that will be used to calculate the input air flow ($\dot{Q}_{air}$.); an anemometer (Anemometer 2), located at the output of the vacuum pump, that is used to compare the inlet and outlet air flows, as well as to detect possible leaks; and a pressure transmitter, located at the top of the reactor column, to measure the total pressure drop (ΔP_{tot}).

2.2. *Characterization and comparative analysis of sands*

The experimental study tested three types of sand as material bed in the reactor: molochite, silica and alumina. These sands, whose properties are shown in Table 1 as provided by the manufacturer, were chosen due to their properties, such as high melting point, suitable bulk density for fluidization, uniform grain size and commercial availability on the market.

Table 1. Comparative table of characteristics of the selected sands

Characteristics	Molochite	Silica	Alumina
Density (kg/m^3)	2.70	2.20	3.72
Bulk density non-compressed (kg/m^3)	1.21	1.25	1.83
Bulk density compressed (kg/m^3)	1.25	1.37	2.06
Colour	Grey	Clear	White
Maximum working temperature (°C)	1700	1100	1700
Thermal conductivity (W/m·K)	11.54	1.38	25
Thermal expansion coefficient (10^6/°C)	4.4	0.55	174
Specific heat (J/kg·K)	800	740	880

Silica and alumina are fused in the manufacturing process, while molochite can be produced by the calcination of mined kaolin, being refined after to get a cheaper product [10]. Molochite is an aluminium silicate (42%-55%) which combines the high melting point of alumina and the low density of silica. The most common sand used in previous studies for BFBR was silica [3], [11], [12] while few applications of alumina or molochite can be found in literature [7].

Figure 2 shows a picture of one grain for each of the studied sands. In order to determine an average grain diameter for a specific type of sand, the size distribution for each one of them was calculated. For this purpose, six mesh sieves stacked of different sizes were used (550, 500, 300, 200, 100 and 50 μm), being measured the weight of the remaining amount of sand in the sieve for each of them.

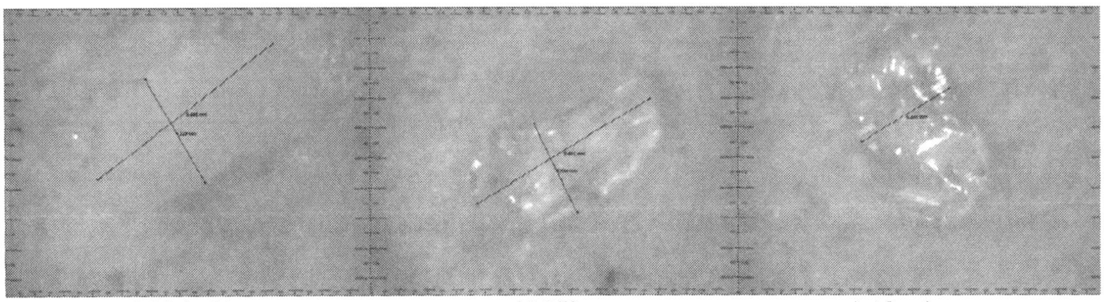

a) Molochite b) Silica c) Alumina

Figure 2. Detail of a grain for each of the considered sands

After that, for each type of sand, two tests were carried out in order to determine the average diameter (d_p), being calculated by equation (1).

$$d_p = \frac{1}{\sum_{i=1}^{5}\left(\frac{x_i}{d_{p,i}}\right)} \tag{1}$$

Where:

- d_p is the average sand diameter, in μm
- x_i is the weigh fraction of sand in the sieves size range, in per unit
- $d_{p,i}$ is the average mesh size of the sieves, in μm.

Size test results and the obtained average diameter for each sand are shown in Table 2.

Table 2. Size test results and determination of the average diameters d_p.

	Sieves size range	$d_{p,i}$	Molochite	Silica	Alumina
$m_{p,i}$ (kg)	550-500 μm	525 μm	0.0024	0.0000	0.0000
	500-300 μm	400 μm	0.208	0.4724	0.0043
	300-200 μm	250 μm	5.09	2.76	8.25
	200-100 μm	150 μm	0.18	1.61	1.74
	100-50 μm	75 μm	0.01	0.26	0.01
	Total weight (kg)		**5.49**	**5.11**	**10.01**
x_i (pu)	550-500 μm	525 μm	0.0004	0.0000	0.0000
	500-300 μm	400 μm	0.0379	0,0924	0.0004
	300-200 μm	250 μm	0.9271	0.5401	0.8242
	200-100 μm	150 μm	0.0328	0.3151	0.1738
	100-50 μm	75 μm	0.0018	0.0509	0.001
	d_p (μm)		***223***	***193***	***247***

According to the obtained values, the mean particle size for the three sands is bounded between 190 and 250 μm (in the case of molochite, 93% of particles are in this range). This size belongs to Class B according to the Geldart classification of particles [13], which are in the range of 150 to 1000 μm. Particles under this class present a good fluidization, appearing bubbles when the minimum fluidisation velocity is exceeded [14]. Sands with this kind of particles have been tested in BFBRs by different authors. Thus, Aznar et al. used silica with a size grain between 400 and 630 μm [15]. Amstrong et al. [16] used limestone with a mean particle size of 600 μm. Finally, Xue et al. [17] considered magnesite in the range from 355 to 600 μm, with mean particle size of 472 μm.

2.3. Configuration of the distributor plate and calculation of the hole size

In tests, the performance of the reactor for each of the considered sands is calculated by varying the section of the sieves' hole with different sections. The sand distributor consists of a 125 μm mesh located at the top and a holed plate below. The configuration of the distributor, as well as the dimensions of the holed plates used in the distributor, are shown in Figure 3. In the figure, the following variables are referred to the distribution plate: N_h is the number of holes and d_h is the hole diameter, while *Hole (%)* is the relation between the holed surface and the total section of the plate, according to its geometry.

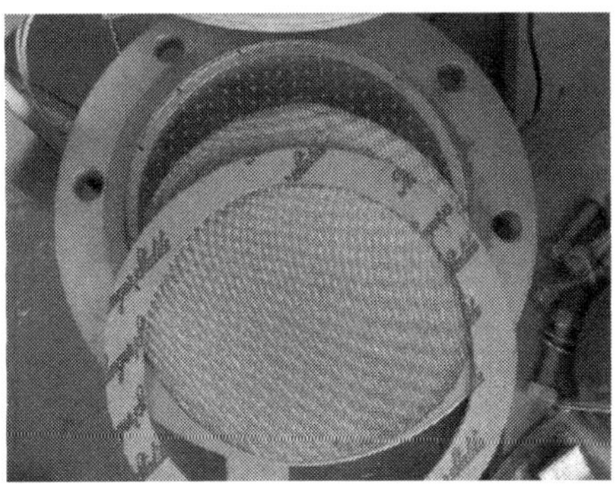

50 kW Reactor	**Hole (%)**
Plate 1 (N_h = 97, d_h =2,5 mm)	1.23%
Plate 2 (N_h =199, d_h =2 mm)	1.62%
Plate 3 (N_h =385, d_h =2 mm)	3.12%
Plate 4 (N_h =385, d_h =2.5 mm)	4.88%
Plate 5, conical holes (N_h =385, d_h =2.5 mm)	1.62%
10 kW Reactor	
Plate 1 (N_h =102, d_h =3mm)	9.2%
Plate 2 (N_h =5800, d_h =0,5 mm)	14.5%
Plate 3 (N_h =180, d_h = 1 mm)	1.8%
Plate 3 (N_h =182, d_h = 1,5 mm)	4.1%

Figure 3. Distributor configuration and dimensions of the holed plates

3. Experimental procedure

3.1. Sand mass to introduce in the reactor

The quantity of sand to be introduced in the reactor has been determined by means of the sand fixed height/reactor inner diameter ratio (H_{fix}/d_R), as itis shown in Table 3.

Table 3. Sand mass to introduce in the reactor (kg)

	d_R = 10 cm			**d_R = 22.2 cm**		
H_{fix}/d_R	Molochite	Silica	Alumina	Molochite	Silica	Alumina
0.25	0.24	0.25	0.36	2.60	2.69	3.94
0.5	0.48	0.49	0.72	5.20	5.37	7.88
0.75	0.71	0.74	1.08	7.80	8.06	11.82
1	0.95	0.98	1.44	10.40	10.74	15.76
1.15	1.09	1.13	1.66	11.96	12.35	18.12

3.2. Airflow evaluation

In order to evaluate the air flow that needs to be introduced in the reactor in order to evaluate the suitability of sands under study, the process explained in Figure 4 has been followed.

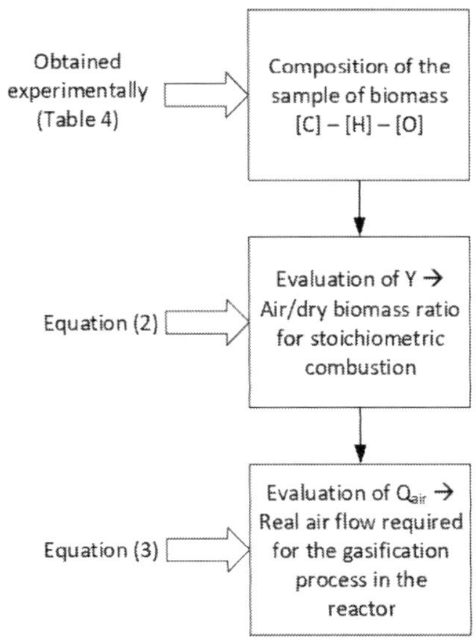

Figure 4. Process followed to calculate the air flow necessary for the gasification process in the reactor

First of all, the sample of biomass used for tests, consisting on woodchips, was characterized experimentally at the IIE-UPV Laboratory, obtaining the results shown in Table 4.

Table 4. Biomass characterisation

Length (mm)		10 - 40
	C	53%
Composition (%)	H	6%
	O	39%

Next, the air/dry biomass ratio for stoichiometric combustion Y (which means that all the fuel is completely burned with the considered air mass) is calculated according equation (2). The air quantity must be able to provide proper sand fluidisation.

$$Y = \frac{m_{air}}{m_{fuel}} = \frac{137.3}{\rho_{air}} \cdot \left(\frac{[C]}{12.011} + \frac{1}{4} \cdot \frac{[H]}{1.008} - \frac{[O]}{32.00} \right) \quad (2)$$

Considering the air density ρ_{air} equals to 1.19 kg/m^3 (STP conditions: 25 °C and 10^5 Pa abs), it was evaluated that the stoichiometric combustion requires 5.45 m^3 of air per kg of biomass. The quantity of air required for the gasification process in a BFBR is then calculated according to equation (3).

$$\dot{Q}_{air} = G \cdot A_R \cdot Y \cdot ER \quad (3)$$

where A_R is the inner reactor section; G is the biomass flow per area in the reactor; Y is the air/dry biomass ratio for stoichiometric combustion calculated before; and ER is the Equivalence Ratio, defined as the relation between the amount of air actually introduced into the reactor and the air that should be introduced for stoichiometric combustion. Applying equation (3), the airflow obtained is 16 m^3/h for the reactor of 10 cm, and 79 m^3/h for the reactor of 22.2 cm. The value used for ER was 0.3 [11], [18], while G took the value of 1.250 kg/h·m^2 [11], [19].

3.3. Pressure drops determination

Several ratios H_{fix}/d_R combined with the pressure drop at the holed plates have been tested to obtain a proper fluidisation. To avoid bed fluctuations, fluidisation has been considered adequate when, keeping the load constant, there are neither pressure nor flow oscillations before and after the sand bed. To do the test, the sand was introduced into the reactor and the pressure drop (ΔP_{tot}) was adjusted by means of a valve located in the inlet pipe. Then, the maximum and minimum ΔP_{tot} values at different H_{fix}/d_R ratios were obtained keeping $\dot{Q}_{air}$ constant by means of a vacuum pump (at 16 m^3/h for the 10 cm reactor and 79 m^3/h for 22.2 cm reactor) and obtaining values of ΔP_{tot} without pressure and velocity oscillations in both sides of the fluidized bed (inlet and outlet of the reactor). Later, the pressure drop at the distributor (ΔP_{dist}) was experimentally determined and, finally, the range of pressured drop at the sand bed (ΔP_{sand}) was calculated as the difference between ΔP_{tot} *and* ΔP_{dist}.

3.4. Experiments uncertainty: accuracy of metering equipment

The accuracy of the obtained results depends on the metering equipment used during the experiments. Accordingly, the main characteristics of the considered devices are:

- Anemometer (2 devices), model PCE-VA 11-ICA, measurement range 0,4 to 30 m/s, accuracy ±0,2 m/s.

- Differential pressure meter, model PCE-HVAC 10, measurement range -1 to 60 bar, accuracy ±0,03 bar
- Compound Optical Microscope 40X-5000X Magnification, accuracy ±6 μm

4. Results

4.1. Pressure drop range

Figure 5 shows the range of total pressure drop for a proper fluidisation at different values of ratio H_{fix}/d_R for the three types of sand tested. The pressure drop range for molochite is more extensive than for silica and alumina. Alumina requires the highest pressure drop, while silica is in between and molochite has the lowest one. Because alumina has a higher bulk density, the energy consumption is higher to reach a proper fluidisation. As shown in the figure, the results are similar in both tests (10 and 22.2 cm reactors).

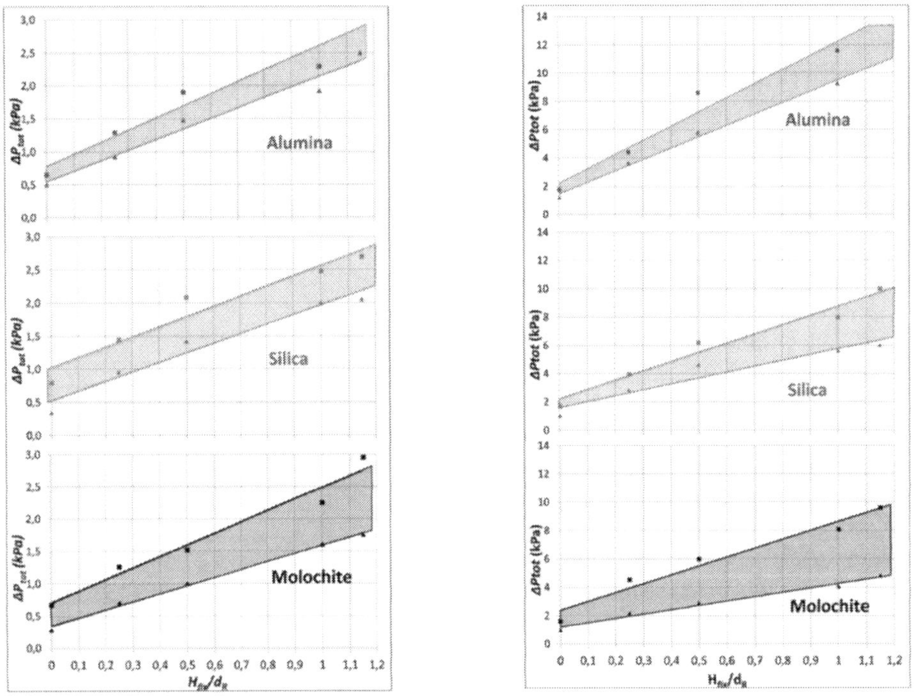

Figure 5. Total pressure drop in the 10 cm (left) and 22.2 cm (right) reactors for different sands. In both cases, the obtained values for molochite is more extensive considering the same relation H_{fix}/d_R, taking extreme values from 0.3 to 2.9 in tests for the 10 cm reactor, and varying from 1.0 to 9.8 in the reactor of 22.2 cm.

As mentioned above, in order to compare the different sands without taking into account the pressure drop of the distributor, ΔP_{sand} was calculated as the difference between ΔP_{tot} and $\Delta P_{dist.}$ The value of ΔP_{tot} is obtained as the average between the maximum and minimum values of the range for proper fluidisation. Several tests were carried out to estimate ΔP_{dist}. The test was based on injecting a predefined airflow into the empty reactor (without sand), using different holed plates and with the valve in the position of the average value between the maximum and minimum ΔP_{tot} range for proper fluidisation and at different H_{fix}/d_R ratios. The results are shown in Figure 6.

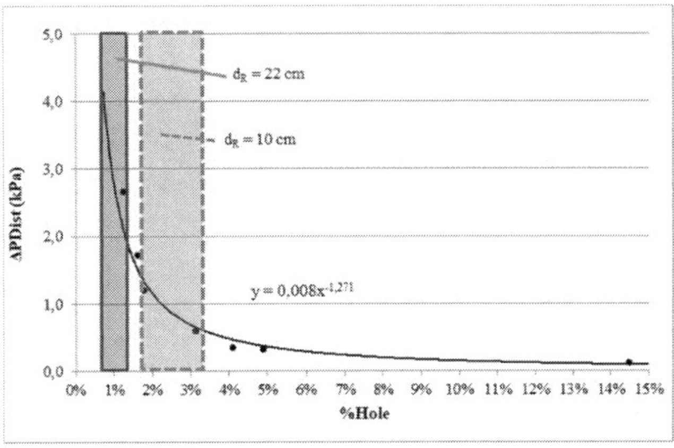

Figure 6. Pressure drop of the distributor in the 10 cm and 22.2 cm reactors for different sands. Several tests were done using different holed plates with different rate of holed section in order to determine the pressure drop of the distributor. % Hole represents the relation between the holed surface and the total section of the plate, according to its geometry. Plates between 1.8% and 3.2% of holed section are used in the 10 cm reactor, while the holed section of plates in the 22.2 cm reactor took values from 1.8% to 3.4%.

As deduced previously, ΔP_{sand} is higher for alumina, and it is similar for molochite and silica; nevertheless, molochite has a more extensive ΔP_{sand} range than silica, which means that molochite can work properly at lower ΔP_{sand}. (Figure 7).

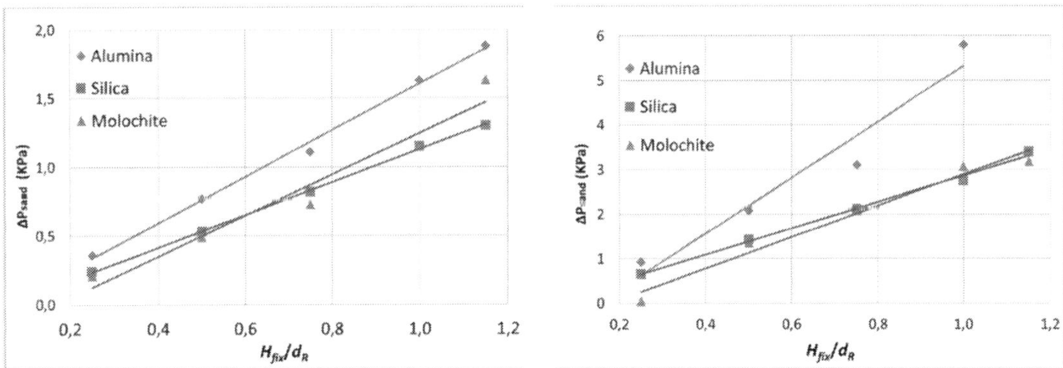

Figure 7. ΔP_{sand} as a function of the H_{fix}/d_R ratio for the 10 cm (left) and 22.2 cm (right) reactors. In all cases, the pressure drop for Alumina is higher than for the other sands. In both reactors the pressure drop for Molochite is lower for a ratio H_{fix}/d_R, overcoming the Silica value for higher ratios. However, the pressure drop range for Molochite is the most extensive, varying from 0.4 to 1.8 in the 10 cm reactor, and taking values from 0.1 to 3.2 for the 22.2 cm reactor.

5. Conclusions

In this experimental research, three types of sand (molochite, silica and alumina) were tested as bed material for a BFBR. They were initially chosen due to such properties as their ability to work at high temperatures, their suitable bulk density, their homogeneous grain size and their easy availability on the market. However, one of them has arisen as the most appropriate for this purpose compared to the others. Tests demonstrated that molochite has the best behaviour because of several factors. The first one is economic: due to the manufacturing process, silica and alumina are more expensive than molochite. Secondly, the pressure drop range for molochite is more extensive than for silica and alumina and according to the ratio H_{fix}/d_R obtained in tests, molochite presents the lowest values for this factor, which means that less energy is required to overcome the additional pressure drop. On the other hand, the maximum operating temperature of use is better in the case of molochite and alumina (1700 °C) than for silica (1100 °C), which reduces the problem of fused ashes. Finally, regarding the mean particle size, 93% of molochite particles is within the range from 200 to 300 μm, which means that this sand is more homogeneous and contains much less fine particles than silica (54%) or alumna (82%).

References

[1] Erdiwansyah, A. Gani, M. Zaki, R. Mamat, M. Nizar, S. Rosdi, S. Yana and R. Sarjono, "Analysis of technological developments and potential of biomass gasification as a viable industrial process: A review," *Case Studies in Chemical and Environmental Engineering,* vol. 8, no. 100439, 2023.

[2] B. K. M. Mahgoub, S. A. Sulaiman and S. Hassan, "Syngas Dual-Fuelling," in *Alternative Fuels for Compression Ignition Engines*, Singapore, Springer Briefs in Energy. ISBN 978-981-10-7753-1, 2018, pp. 37-52.

[3] J. Xie, W. Zhong, B. Jin, Y. Shao and H. Liu, "Simulation on gasification of forestry residues in fluidized beds by Eulerian-Lagrangian approach," *Bioresource Technology,* vol. 121, pp. 36-46, 2012.

[4] A. Suwatthikul, S. Limprachaya, P. Kittisupakorn and I. M. Mujtaba, "Simulation of steam gasification in a fluidized bed reactor with energy self-sufficient condition," *Energies,* vol. 10, no. 3, pp. 1-15, 2017.

[5] L. Montuori, C. Vargas-Salgado and M. Alcázar-Ortega, "Impact of the throat sizing on the operating parameters in an experimental fixed bed gasifier: Analysis, evaluation and testing," *Renewable Energy,* vol. 83, pp. 615-625, 2015.

[6] A. Akbarian, A. Andooz, E. Kowsari, S. Ramakrishna, S. Asgari and Z. A. Cheshmeh, "Challenges and opportunities of lignocellulosic biomass gasification in the path of circular bioeconomy," *Bioresource Technology,* vol. 362, no. 127774, 2022.

[7] H. Chi, M. A. Pans, C. Sun and H. Liu, "Effectiveness of bed additives in abating agglomeration during biomass air/oxy combustion in a fluidised bed combustor," *Renewable Energy,* vol. 185, pp. 945-958, 2022.

[8] C. Vargas-Salgado, L. Montuori y M. Alcazár-Ortega, «Experimental analysis of a bubbling fluidized bed gasification plant fed by biomass: Design, implementation and validation of the control system,» de *ICREN*, Rome, 2020.

[9] A. Suwatthikul, S. Limprachaya, P. Kittisupakorn and I. M. Mujtaba, "Simulation of steam gasification in a fluidized bed reactor with energy self-sufficient condition," *Energies,* vol. 10, no. 314, 2017.

[10] IMERYS, "Product ranges. Molochite," Imerys S.A., [Online]. Available: https://www.imerys.com/product-ranges/molochite. [Accessed 03 May 2024].

[11] I. Narváez, A. Orío, M. P. Aznar and J. Corella, "Biomass gasification with air in an atmospheric bubbling fluidized bed. Effect of six operational variables on the quality of the produced raw gas," *Industrial & Engineering Chemistry Research,* vol. 35, no. 7, pp. 2110-2120, 1996.

[12] S. Koppatz, C. Pfeifer and H. Hofbauer, "Comparison of the performance behaviour of silica sand and olivine in a dual fluidised bed reactor system for steam gasification of biomass at pilot plant scale," *Chemical Engineering Journal,* vol. 175, no. 1, pp. 468-483, 2011.

[13] D. Geldart, "Types of gas fluidization," *Powder Technology,* vol. 7, no. 5, pp. 285-292, 1973.

[14] R. Cocco, S. R. Karri and T. Knowlton, "Introduction to fluidization," American Institute of Chemical Engineers, New York, 2014.

Journal of Physics: Conference Series **2940** (2025) 012017 doi:10.1088/1742-6596/2940/1/012017

[15] M. P. Aznar, M. A. Caballero, J. A. Sancho and E. Francés, "Plastic waste elimination by co-gasification with coal and biomass in fluidized bed with air in pilot plant," *Fuel Process,* vol. 87, no. 5, pp. 409-420, 2006.

[16] L. M. Armstrong, S. Gu and K. H. Luo, "Parametric study of gasification processes in a BFB coal gasifier," *Industrial & Engineering Chemistry Research,* vol. 50, no. 10, pp. 5959-5974, 2011.

[17] G. Xue, M. Kwapinska, A. Horvat, W. Kwapinski, L. Rabou, S. Dooley, K. Czajka and J. Leahy, "Gasification of torrefied Miscanthus×giganteus in an air-blown bubbling fluidized bed gasifier," *Bioresource Technology,* vol. 159, pp. 397-403, 2014.

[18] S. Sarker, F. Bimbela, J. L. Sánchez and H. K. Nielsen, "Characterization and pilot scale fluidized bed gasification of herbaceous biomass: A case study on alfalfa pellets," *Energy Conversion and Management,* vol. 91, pp. 451-458, 2015.

[19] S. Kraft, F. Kirnbauer and H. Hofbauer, "CPFD simulations of an industrial-sized dual fluidized bed steam gasification system of biomass with 8 MW fuel input," *Applied Energy,* vol. 190, pp. 408-420, 2017.

Journal of Physics: Conference Series **2940** (2025) 012018 doi:10.1088/1742-6596/2940/1/012018

Investigating the Impact of Water-Glycol Mixture Ratios as Heat Transfer Fluid on the Performance of a Rotary Magnetic Refrigerator

J A Bocanegra*, **F Scarpa and L Tagliafico**

[1] Department of Mechanical, Energy, Management and Transportation Engineering, University of Genoa, Genoa, GE 16145, Italy

* Corresponding author: augusto.bocanegra@unige.it

Abstract. Magnetic refrigeration offers an eco-friendly and sustainable alternative for cooling applications. The selection of an appropriate working fluid is critical for heat removal and regeneration efficiency in Active Magnetic Regenerators (AMRs). This study evaluates various working fluids, focusing on water mixed with an antifreeze agent, glycol, to extend the temperature range and prevent corrosion. The thermophysical properties of such mixtures differ from pure water, influencing the refrigerator's performance. A parametric investigation using a numerical model of a rotative magnetic refrigerator was conducted to analyze the effects of varying the water-glycol mixture ratios. The study systematically varied operational parameters, such as rotation frequency and mass flowrate, to determine optimal performance conditions. Results indicate that the performance, measured by the Coefficient of Performance (COP), cooling load, and efficiency, declines with increasing glycol content, particularly at lower temperature spans. Increased viscosity in the mixture necessitates higher pumping power and torque for optimal operation, affecting the optimal rotational frequency. This study underscores the importance of balancing the mixture composition to enhance temperature range while minimizing negative impacts on performance, advocating for mixtures with lower glycol content.

1. Introduction

Common vapor-compression refrigerators rely on gas refrigerants with high Global Warming Potential (GWP). In contrast, magnetic refrigeration represents a viable alternative, utilizing solid-state materials instead of these environmentally detrimental gases. An early comprehensive overview of magnetic refrigeration technology was presented in 2007 [1], followed by a review of various prototypes developed before 2010 [2] and a subsequent state-of-the-art report in 2021 [3]. Magnetic refrigerators can be classified based on various factors, including the type of motion, which can be either linear reciprocating or rotary [4]. Magnetic refrigeration operates based on the Active Magnetic Regenerative (AMR) cycle, which employs magnetocaloric materials (MCMs) that exhibit the magnetocaloric effect (MCE). This effect involves a temperature increase when the MCM is subjected to a magnetic field and a corresponding temperature decrease when it exits the field. The magnitude of the MCM's temperature change depends on the magnetic field strength; commonly used materials exhibit a temperature change of a few degrees Celsius with a field strength of 1 Tesla. The regenerative cycle enhances the achievable temperature span by continuously moving the MCM in and out of the magnetic field. The AMR cycle employs a working fluid to transport heat throughout the cycle. The fluid must be precisely synchronized to transport heat in the correct direction, achieving the desired cooling load and facilitating heat rejection.

The AMR cycle consists of four cyclical steps:

i. **Adiabatic Magnetization:** During this step, the magnetocaloric material (MCM) is exposed to a magnetic field. As a result, the MCM undergoes adiabatic magnetization, leading to an increase in its temperature without exchanging heat with its surroundings.
ii. **Hot Blow:** In this phase, the heated working fluid flows from the MCM to the hot heat exchanger. The heat absorbed during magnetization is transferred to the fluid. The fluid then carries this heat to the hot heat exchanger, where it is expelled into the environment.
iii. **Adiabatic Demagnetization:** After the hot blow, the magnetic field is removed. The MCM undergoes adiabatic demagnetization, during which its temperature decreases. This step cools the MCM without any heat exchange with the surroundings.
iv. **Cold Blow:** In the final step, the working fluid flows back to the MCM. The cold fluid absorbs heat from the MCM. This cooled fluid then flows to the cold heat exchanger, where it absorbs heat from the space or substance being cooled.

In this way, the working fluid exchanges heat with the MCM matrix, and the energy is transported from the cold side heat exchanger (performing the cooling load) to the hot heat exchanger (performing the heat rejection).

Choosing a suitable working fluid for efficient heat removal and regeneration in AMRs is crucial for optimizing the performance of magnetic refrigerators. The working fluid must alternately and synchronously transport heat in coordination with the changes in the magnetic field. Various fluids have been studied, including common heat transfer fluids such as alcohol and water and advanced options like nanofluids [5] and liquid metals [6]. When selecting water as the working fluid, it is essential to consider mixing it with antifreeze agents, such as glycol, to create a water-glycol blend. This blend extends the operational temperature range by lowering the freezing point and mitigates corrosion effects. However, the thermophysical properties of the water-glycol mixture differ significantly from those of pure water, which impacts the overall performance of the magnetic refrigerator.

Numerical methods have been widely used to study AMR systems [7], typically employing one-dimensional (1D) or two-dimensional (2D) spatial representations of the regenerator. An example of a 1D model applied to near-room temperature refrigeration was presented by [8]. The present study conducts a parametric investigation into the performance of a rotary magnetic refrigerator by varying the ratio of the water-glycol mixture used as the working fluid. A well-established numerical model is utilized to simulate the performance of a rotary AMR with different working fluids. Systematic variations in operational parameters, such as rotation frequency and mass flowrate, are considered to ensure optimal refrigerator performance.

2. Methods

2.1. Numerical model

The numerical model employed in this research is a one-dimensional model that discretizes the solid and fluid regions. The energy balance is calculated for each interacting solid and fluid element. For the solid elements, the energy balance is represented by equation (1), and for the liquid domain, by equation (2). Quantities related to the solid and fluid domains are denoted by the subscripts s and f, respectively, with the fluid velocity denoted by u. Two geometrical ratios characterize the regenerator: the ratio of the heat transfer surface area to the total volume inside the regenerator ($\beta = A/V$) and the ratio of the fluid volume to the total volume of the channel ($\alpha = V_f/V$), where $V = V_f + V_s$. The solid and fluid thermophysical properties include density (ρ_s, ρ_f), specific heat (c_s, c_f), and thermal conductivity (k_s, k_f).

$$\rho_s c_s(T_s, B)\frac{\partial T_s}{\partial t} = k_s\frac{\partial^2 T_s}{\partial x^2} - h_{eff}\,\frac{\beta}{1-\alpha}\,(T_s - T_f) + \dot{Q}_{MCE} \tag{1}$$

$$\rho_f c_f\left(\frac{\partial T_f}{\partial t} + u\frac{\partial T_f}{\partial x}\right) = k_f\frac{\partial^2 T_f}{\partial x^2} + h_{eff}\,\frac{\beta}{\alpha}\,(T_s - T_f) - u\frac{dP}{dx} \tag{2}$$

The first equation includes the MCE experienced during the magnetization and demagnetization steps, denoted as $\dot{Q}_{MCE}$. The MCE term $\dot{Q}_{MCE}$ is calculated directly using the adiabatic temperature change ΔT_{span}, obtained from the change in the thermophysical properties by the action of the applied magnetic field, as described in detail in our previous publications [9]. The other two terms correspond to thermal conduction and the heat exchanged with the fluid domain. The second equation encompasses fluid conduction, heat exchange with the solid domain, and pressure losses due to friction. While this term has limited relevance when using water as the working fluid, it was included because its importance increases under certain conditions, such as when the fluid properties or operating conditions change, inducing an increased required pumping power, as shown in the results section. It was assumed that the regenerator is isolated from the environment and only connected to the thermal reservoirs. The boundary conditions for this study include imposed mass flow rate and fluid inlet temperatures, as illustrated in Fig. 1. Regarding the numerical modeling, an explicit finite difference scheme was applied to solve the discretized equations. Thermophysical properties are calculated at each solid and fluid node during each timestep. The thermophysical properties of the fluids used in this study have been evaluated using the SecCool software (www.ipu.dk/products/seccool/), which calculates properties based on data in Melinder's book [10].

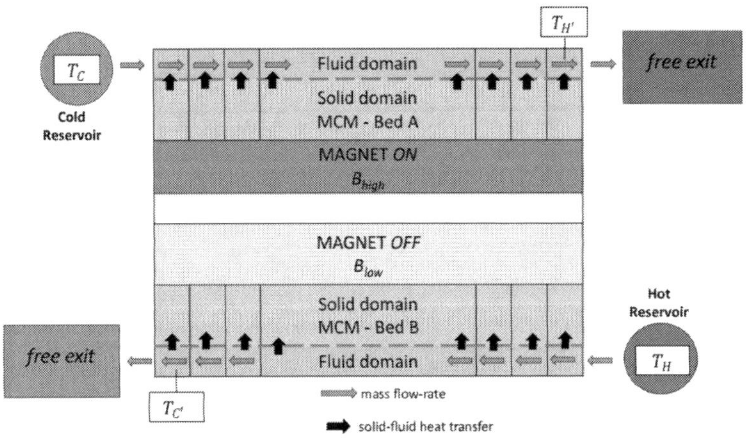

Figure 1. Sketch of the numerical model

Figure 1 presents a sketch of the numerical model, illustrating the solid and fluid domains. The two beds are exchanged in each half-cycle to complete the AMR cycle. It is important to note that an ideal thermal reservoir imposes the inlet temperature of the fluid for each bed. Table 1 lists the main parameters that were fixed during this study.

Table 1. Main geometrical parameters of the considered case study

Parameter	
α	1/3
β	2660 m^{-1}

Lx	0.17 m
Ly	0.08 m
B_H	0.945 T
B_L	0.056 T

2.2. *Performance indicators*
The performance of the device is measured by two key indicators: the coefficient of performance (COP) and the weighted cooling load (Q_{in}/M_s). The COP, given by equation (3), measures the efficiency of the refrigeration cycle. The weighted cooling load, given by equation (4), is the cooling load's ratio to the solid material's mass. In these equations, τ_c represents the cycle time, $\dot{m}_f$ denotes the fluid mass flowrate and $\dot{Q}_{out}$ represents the heat rejection.

$$COP = \frac{\dot{Q}_{in}}{\left|\dot{Q}_{out} + \dot{Q}_{in}\right|} \tag{3}$$

$$\dot{Q}_{in}/M_s = \frac{1}{M_s\tau_c}\int_0^{\tau_c/2} \dot{m}_f c_f\,(T_{C'} - T_C)dt \tag{4}$$

Additionally, equation (5) presents the exergetic efficiency (η_{II}), defined as the ratio of the COP to the coefficient of performance of a Carnot cycle (COP_{carnot}) operating between the same temperatures.

$$\eta_{II} = \frac{COP}{COP_{carnot}} = COP\frac{T_H - T_C}{T_C} \tag{5}$$

2.3. *Operative parameters*
The primary operative parameters include the rotation frequency f and the fluid mass flowrate of the working fluid, $\dot{m}_f$. In some studies, the utilization factor (Φ), a dimensionless parameter, is used instead of the fluid mass flowrate, as shown in equation (6):

$$\Phi = \frac{\dot{m}_f c_f}{M_s c_s}\tau_c \tag{6}$$

Operational optimization typically involves maximizing performance indicators, such as the cooling load. While machine learning techniques have been applied to optimize the performance of such devices [11], this study employs a different approach. Here, a parametric sweep is conducted, systematically varying the frequency and utilization factor to obtain optimal parameters that ensure the maximum cooling load for each temperature span.

3. Results
Simulations demonstrate a significant performance dependence on the mix ratio, as measured by the Coefficient of Performance (COP), cooling load, and efficiency. Particularly noteworthy is the inverse relationship observed, whereby performance deteriorates with increasing glycol content in the mixture. This effect is especially pronounced at low-temperature spans, as illustrated in Figure 2. At low ΔT_{span}, the observed increase in COP aligns with the theoretical trend of the Carnot COP; however, it is important to note that this study considers only the internal losses of the regenerator, while a complete system would also include additional losses across the entire circuit.

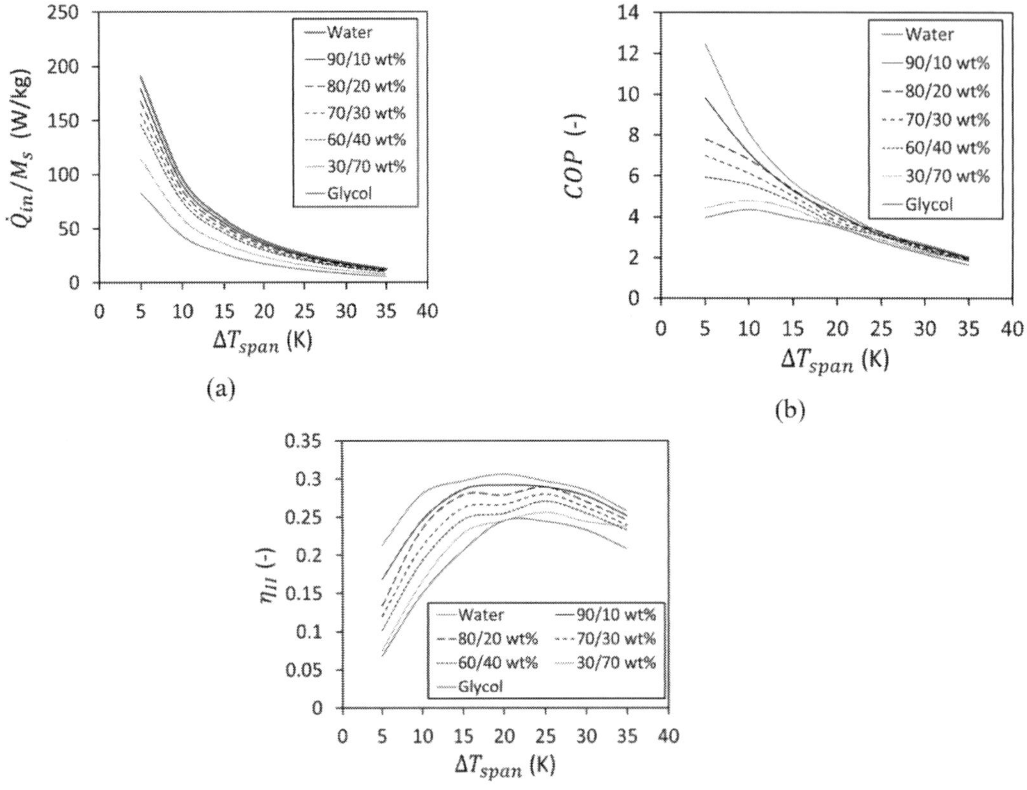

(c)

Figure 2. Effects of the mix ratio on the performance of the rotary AMR.

Due to changes in thermophysical properties, such as increased viscosity, optimal operation requires higher pumping power and torque (refer to Figure 3). The oscillations observed in the torque figure may correspond to numerical artifacts.

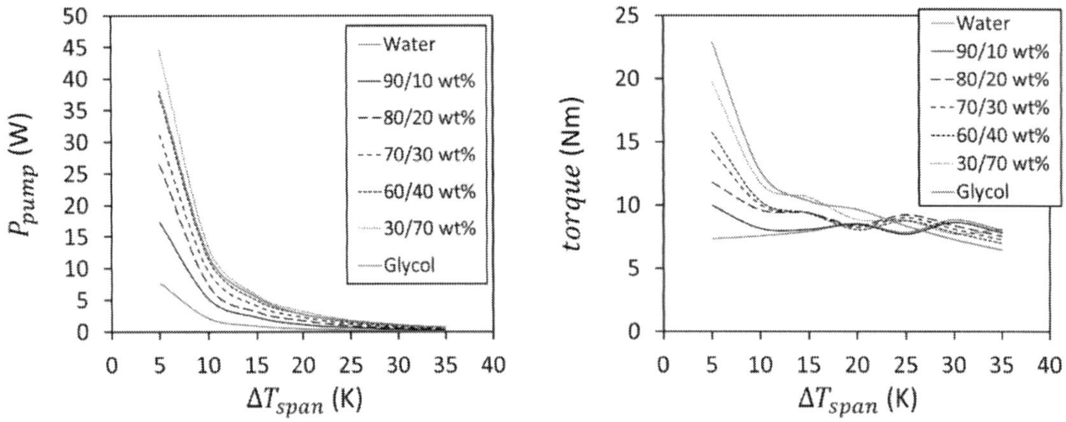

(a) (b)

Figure 3. Effects of the mix ratio (wt%) on the pump power (a) and torque (b)

Each mixing ratio modifies the optimal operational parameters, primarily decreasing the optimal rotational frequency (see Figure 4). Optimal control curves are presented only for pure water and a 70/30 wt% mixture to enhance clarity. Notably, the 70/30 wt% mixture is of practical interest, while ratios with higher glycol concentrations are included in the paper to illustrate their effects. The curves represent optimal values for different temperature spans; it is evident that to operate at a higher temperature span, both the frequency and mass flowrate must be decreased. Furthermore, although the required utilization factor remains practically unaltered when comparing the mixture with pure water, representing the mass flowrate instead of the utilization factor reveals a reduction in the optimal value.

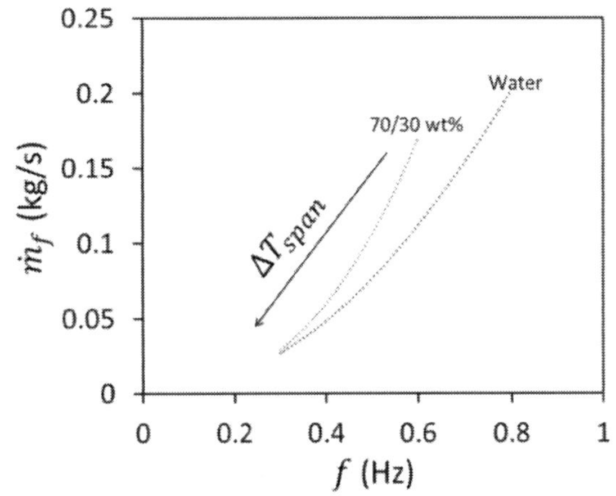

Figure 4. Optimal operative parameters (fluid mass flowrate and frequency) for different temperature spans. Comparison of water and 70/30 wt% water/glycol mixture.

4. Conclusions

This paper presents a numerical investigation of a rotary magnetic refrigerator, exploring the effects of varying the water-glycol mixture ratio as the working fluid. Performance evaluation was based on cooling power, coefficient of performance, and exergetic efficiency, while optimal control parameters, torque, and pumping power were also studied across different temperature spans.

Insights into the impact of working fluid composition on the regenerative thermodynamic cycle of rotary magnetic refrigeration were provided. A delicate balance is necessary when selecting a mixture, weighing the benefits of widening the operative temperature range (anti-freezing effect) against potential negative impacts on performance. Emphasis is placed on mixtures with low wt%.

While numerical results offer valuable insights, experimental validation is essential. Future studies will focus on refining the numerical method, including enhanced fluid properties modeling and considering alternative fluids.

While this technology shows promising potential, its current limitations must be acknowledged. One significant challenge is the high cost of both the magnetic materials and, in particular, the magnetocaloric materials (MCM), which represent a key barrier to widespread adoption. Additionally, the system's scalability remains an area of focus, as transitioning from small-scale prototypes to industrial applications requires addressing technical complexities and economic feasibility. However, with increasing interest in cleaner and more sustainable energy solutions, this technology holds great promise for integration into larger energy systems. Future advancements in material science and system design are expected to reduce costs and improve efficiency, further enhancing the appeal of this innovative approach.

5. References

[1] Engelbrecht, K., Nellis, G., Klein, S., and Zimm, C., 2007, "Review Article: Recent Developments in Room Temperature Active Magnetic Regenerative Refrigeration," HVAC&R Res., **13**(4), pp. 525–542. https://doi.org/10.1080/10789669.2007.10390970.

[2] Yu, B., Liu, M., Egolf, P. W., and Kitanovski, A., 2010, "A Review of Magnetic Refrigerator and Heat Pump Prototypes Built before the Year 2010," International Journal of Refrigeration, **33**(6), pp. 1029–1060. https://doi.org/10.1016/j.ijrefrig.2010.04.002.

[3] Alahmer, A., Al-Amayreh, M., Mostafa, A. O., Al-Dabbas, M., and Rezk, H., 2021, "Magnetic Refrigeration Design Technologies: State of the Art and General Perspectives," Energies, **14**(15), p. 4662. https://doi.org/10.3390/en14154662.

[4] Scarpa, F., Tagliafico, G., and Tagliafico, L. A., 2015, "A Classification Methodology Applied to Existing Room Temperature Magnetic Refrigerators up to the Year 2014," Renewable and Sustainable Energy Reviews, **50**, pp. 497–503. https://doi.org/10.1016/j.rser.2015.05.029.

[5] Scarpa, F., and Bianco, V., 2024, "Improving the Performance of Room Temperature Magnetic Regenerators Using Al2O3-Water Nanofluid," Applied Thermal Engineering, **236**, p. 121711. https://doi.org/10.1016/j.applthermaleng.2023.121711.

[6] Scarpa, F., and Slimani, S., 2023, "Galinstan Liquid Metal as the Heat Transfer Fluid in Magnetic Refrigeration," Applied Thermal Engineering, **232**, p. 120971. https://doi.org/10.1016/j.applthermaleng.2023.120971.

[7] Nielsen, K. K., Tusek, J., Engelbrecht, K., Schopfer, S., Kitanovski, A., Bahl, C. R. H., Smith, A., Pryds, N., and Poredos, A., 2011, "Review on Numerical Modeling of Active Magnetic Regenerators for Room Temperature Applications," International Journal of Refrigeration, **34**(3), pp. 603–616. https://doi.org/10.1016/j.ijrefrig.2010.12.026.

[8] Aprea, C., and Maiorino, A., 2010, "A Flexible Numerical Model to Study an Active Magnetic Refrigerator for near Room Temperature Applications," Applied Energy, **87**(8), pp. 2690–2698. https://doi.org/10.1016/j.apenergy.2010.01.009.

[9] Tagliafico, G., Scarpa, F., and Canepa, F., 2010, "A Dynamic 1-D Model for a Reciprocating Active Magnetic Regenerator; Influence of the Main Working Parameters," International Journal of Refrigeration, **33**(2), pp. 286–293. https://doi.org/10.1016/j.ijrefrig.2009.10.001.

[10] Melinder, A., 2010, *Properties of Secondary Working Fluids (Secondary Refrigerants or Coolants, Heat Transfer Fluids) for Indirect Systems*, IIF-IIR, France/France.

[11] Aprea, C., Greco, A., and Maiorino, A., 2017, "An Application of the Artificial Neural Network to Optimise the Energy Performances of a Magnetic Refrigerator," International Journal of Refrigeration, **82**, pp. 238–251. https://doi.org/10.1016/j.ijrefrig.2017.06.015.

Acknowledgments

The present work was supported by MUR (Italian Ministry of University and Research). Project FISR2019_04798 SUSSTAINEBLE A Solution Using Solid-STate-cooling: An INvestment Eco-compatiBLE).

Journal of Physics: Conference Series **2940** (2025) 012019 doi:10.1088/1742-6596/2940/1/012019

Numerical investigation on geometrical parameter effects of metal foams in liquid cooling efficiency of battery thermal management within phase change materials

A Arumugam[1], B Buonomo[1], S Nardini[1], R E Plomitallo[1] and O Manca[1,*]

[1] Department of Engineering, University of Campania "Luigi Vanvitelli", Via Roma 29, 81031, Aversa (CE), Italy

* Corresponding author: oronzio.manca@unicampania.it

Abstract. The objective of this numerical study is to evaluate the effects of geometrical parameters of a metal foam inserted along with a chosen PCM material on the efficiency of liquid cooling of lithium polymer-based batteries. A three-dimensional model of a lithium manganese dioxide battery module is chosen for the analysis purpose. The analysis is performed by placing the module inside a parallelopiped based arrangement at the centre position, which is cooled by liquid flowing through the tubes by means of convection phenomenon. The tubes placed in the arrangement are assumed at a constant temperature and the entire module is adiabatic in nature. The governing physical equations involved in the analysis are based on the Local Thermal Non-Equilibrium constrains. The numerical model is solved using the finite volume technique using the Ansys-Fluent software. The study involves the analysis of thermal management for different pore density values of metal foam such as 5, 10, 20 and 40 with different C-rates of 0.5 and 1C respectively. The change in the internal resistance of the battery is determined using the Generalized Reduced Gradient technique. The analysis is interpreted by the temperature and liquid fraction profiles of the battery module.

1. Introduction

Rechargeable batteries, which are essential for storing electrochemical energy by the process of reversible chemical reactions, play a vital role in enhancing the efficiency of electric vehicles. These vehicles contribute an estimated efficiency of around 70% as compared to the internal combustion engines performance. Since 2005, there has been a renewed emphasis on electric mobility. Unlike the previous focus centered on urban pollution, the current scenario motivates towards electrification which serves as recommended solution concerning about climate change and rising oil prices [1-2].

Within the transportation sector, the challenges of rising oil prices and increasing carbon emissions have propelled the popularity of electric vehicles. These vehicles, being free from oil dependency and minimal greenhouse gases have made them increasingly popular. These vehicles contribute to the environmental benefits, known for their reduced air pollution by less reduced greenhouse gases emissions. Since these vehicles are used as a substitute for fossil fuels by the utilization of electricity, they have significantly less operation costs. These vehicles are known for their simpler designs, greater durability and fewer moving parts completely distinguishing themselves from IC based engine vehicles. These vehicles are promoted by their potential use of renewable energy resources. These vehicles are

Journal of Physics: Conference Series **2940** (2025) 012019 doi:10.1088/1742-6596/2940/1/012019

facilitated by the availability of charging stations at various locations. These vehicles are also characterized by several mechanical and dynamic features which makes them quite popular [3].

Within the domain of rechargeable batteries, there exists a family whose characteristic trait is the movement of lithium ions from positive to negative electrodes during the process of discharge. Lithium-ion based batteries, known for their exceptional energy density, hold a significant advantage over various other battery chemistries. This phenomenon has driven an increase in their utilization and the gradual replacement of older battery technologies across various industries. Lithium being one of the lightest and most reactive metals with high electro-positivity nature emerges as a prime candidate for electrochemical cells in the periodic table. The evolution from conventional lithium-ion batteries leads to the emergence of lithium-polymer based battery cells which resemble their structure but with notable enhancements in performance [4]. A significant safety advantage of lithium polymer-based batteries is the elimination of the explosion risk upon full charge [5]. These batteries incorporate plastic-based anodes and solid electrolytes, prioritizing safety, and stability [6]. The on-going advancements in lithium-polymer based batteries technology introduce gel-based electrolytes and innovative separators, contributing to improved battery characteristics. Considering the improvisation involved in the technological process, there are several advantages of lithium-based batteries as compared with their counterparts. These batteries are characterized with spontaneous lower discharge and easy to maintain. These batteries are inbuilt with protection circuits to avoid the risks of explosion. These batteries are not required to be fully discharged and are popular due to their light weight. Their life span isn't strongly affected by the number of charges, and they operate better across a wide temperature range, withstanding up to 60°C without capacity loss. These batteries are known for their high energy density and highest electrochemical potential as compared to other chemistries. These batteries generally are notified for their absence in memory effect. These batteries can also be manufactured in delicate thin shapes and possess suitable voltage ranges.

The lithium-based batteries are also characterized by few limitations as compared to various other chemistries which are recommended to be taken into consideration during their application. There is considerable damage to the battery's health and capacity when the voltage fluctuates from its normal operating range. These batteries possess high internal resistance leading to potential overheating and explosions during short circuits. These batteries are usually expensive as compared to various other battery chemistries. These batteries are usually characterized by short service life in electric vehicle applications and are associated with high temperatures capable of melting metals during ignition and explosion. Additionally, overloading a fully charged lithium battery by increasing their voltage or current can lead to electrode corrosion and complete capacity destruction [4].

Enhancing performance across diverse operational conditions, particularly amidst frequent charging and discharging cycles, while implementing measures to prevent thermal runaway and deploying effective Battery Thermal Management Systems (BTMS) to maintain optimal temperature thresholds. For electric vehicles, a compact, cost-effective, reliable, and easily maintainable BTMS with minimal parasitic power consumption is essential [7].

Latent heat storage, employing phase change materials (PCMs), emerges as a beneficial approach for storing energy generated by renewable energy sources like solar and wind. PCMs, selected with care, offer advantages such as higher energy storage density, suitable phase transition temperature ranges, stability, and reduced supercooling. Effective PCMs for thermal management must exhibit essential characteristics like latent heat, appropriate thermal conductivity, and a stable chemical structure to fulfill both heat storage and transfer requirements. Despite their favorable heat storage properties, organic PCMs encounter challenges such as low thermal conductivity, liquid leakages, and diminished efficiency [8]. Utilizing metal foams presents a viable approach to enhance the thermal conductivity of PCMs. Salient features of metal foams include high porosity, favorable thermophysical properties, and robust mechanical strength [9].

The objective of this study is to analyze a numerical model for cooling of a lithium polymer-based battery module using a hybrid battery thermal management technique utilizing both active and passive mechanisms of cooling techniques. The passive mechanism involves a chosen PCM material which is embedded with metal foams. The role of the various geometrical parameters of the metal foams chosen to be inserted with the PCM material during the process of cooling is investigated in detail for a Local Thermal Non-Equilibrium (LTNE) battery system.

2. Materials and Methodology

Theoretical Framework

This study examines a standard lithium-polymer based battery which is cylindrical in nature designed for electric mobility applications. Thermal control management was assessed through numerical analysis using Ansys-Fluent software. The chosen cylindrical batteries, comprising of lithium-manganese oxide were housed within a parallelopiped shaped casing filled entirely with a chosen PCM material embedded with metal foams of four different pore densities namely 5,10, 20 and 40 respectively. Additionally, convective tubes adjacent to the batteries allowed fluid flow, with water chosen as a coolant, maintained at a constant temperature of 293.15K. This circulating coolant ensures uniform temperature regulation along the casing wall, enhancing overall thermal control of the battery pack system. In figure 1the cylindrical lithium manganese oxide batteries are denoted in red, while the liquid-conveying channels are depicted in blue, and the remaining grey area is indicated by the chosen PCM material embedded with metal foams of four different pore densities depending on the chosen case.

While figure 1 (a) showcases an entire battery pack, the study involves a single battery cell, considering the symmetrical arrangement of the battery cells in the pack. The schematic of a three-dimensional battery cell model considered in the analysis is shown in figure 1 (b) [10]. The elements of the battery cell are indicated using the arrows. The top portion of the cell is termed as the connector which is generally composed of steel. Followed by the connector, is the mandrel, which is a nylon isolator wounded around the cell. The pre-final element corresponds to the active material of the battery cell and the final element indicates the canister. The thickness of the battery canister is neglected due to its minimal impact on the temperature profile of the cells. Each component of the battery cell possesses distinct physical and chemical properties, owing to which the thermal conductivity of each element varies correspondingly [10]. The battery considered for the study was an Efest IMR 18650 Lithium manganese dioxide cell. It is 0.064 m in length and 0.184 m in diameter. The nominal voltage and capacity of the cell is 3.7V and 3000 mAh. The cut-off voltage of the cell was noted to be 2.5V. The maximum current of discharge is 20A [11,12]. Moreover, the thermophysical properties are density 2700 kg/m^3, specific heat 0.823 kJ/kg K and thermal conductivity 8.9 W/m K [11].

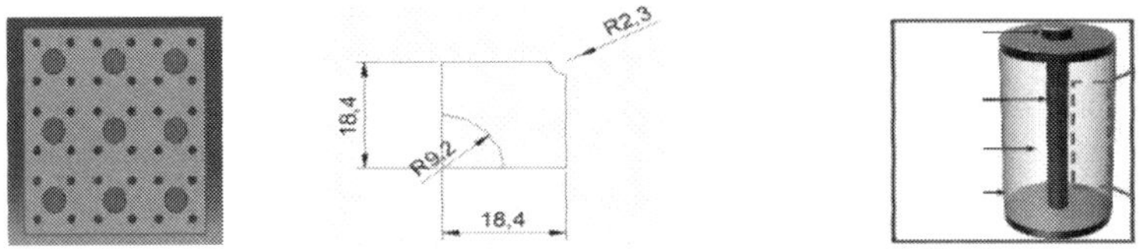

(a) (b) (c)

Figure 1. (a) Schematic representation of battery pack; (b) 3-D model of a cell with dimensions (c) Specifications of the battery cell with their individual element [10]

Specifications of various cooling elements of Battery pack

The various components involved in the process of cooling of the battery pack for the numerical investigation are namely the cooling fluid flowing through the convective tubes, the PCM material chosen for the purpose of cooling and the metal foams of different pore densities. The detailed specifications of these elements are listed out in the following table 1.

Table 1. Specifications of the cooling elements in battery pack

Specifications	PCM RT25HC	Specifications	Metal foam
Density (ρ) [kg/m^3]	767	Pore density (ω) [PPI]	5,10,20,40
Thermal conductivity [W/m K]	0.185	Porosity (ε)	0.946, 0.954, 0.949,0.952
Specific heat c_p [J/kg K]	2100	Fibre diameter [10^{-3}m]	0.47, 0.326, 0.325, 0.239
Viscosity [kg/m s]	0.0030	Pore diameter [10^{-2}m]	0.39, 0.27, 0.27, 0.198
Thermal expansion coefficient [1/K]	0.0006	Inertial drag coefficient (C_f)[m]	414.60, 600.34, 600.16, 819.08
Melting heat [J/kg]	232000	Permeability (K) [10^{-7}m^2]	2.234, 1.112, 1.085, 0.591
Solidus Temperature [K]	295.16		
Liquidus Temperature [K]	299.16		

Experimental analysis to determine the change in internal resistance of battery module

The charging and discharging of the chosen battery cell were performed using an ITECH IT6054B-1500-120 regenerative power system, which served as both a source and load, which in turn was connected to a computer using an IEEE 488 GPIB bus. The analysis was carried out for both 0.5C and 1C respectively. The output voltage and current were measured using SIGLENT SDM-3055SC mustimeters. The temperature of the battery cell was measured using a J-type thermocouple operating within -40°C to 750°C temperature range. Data acquisition was performed using NI-based DAQ system which was integrated with the experiment using the LabVIEW software.

The uncertainty of J-type thermocouples chosen was between ± 0.1°C. The voltmeter as well as the ammeters were initially adjusted for the zero and span correction. The programming accuracy of the voltmeters and ammeters used for the experiment is 0.05% + 10 mV and 0.2% + 10 mA. The reading accuracy of the meters were 0.05% + 5 mV and 0.15% + 5 mA respectively. The maximum power of the meters was estimated to be 700 VA with full load.

Further the variability in internal resistance was estimated using the nonlinear Generalized Reduced Gradient (GRG) algorithm. The internal resistance was estimated considering the discharge process of the cell for the mentioned C-rates. An exponential regression model proposed by [13,14] was adopted to fit the curve using the GRG algorithm, from which the internal resistance was estimated. The various equations involved in the formulation of the internal resistance is given by equation (1) to equation (3).

$$R(T) = R_0\ exp\ b_1T + b_2T^2 \pm \gamma \tag{1}$$

$$V_{oc}(s) = K_0 - \frac{K_1}{s} + K_2e^{-\alpha(1-s)} \tag{2}$$

$$V(t,T) = V_{oc}(s) - I(t) * R(T) \tag{3}$$

where, $R(T)$ represents the variable internal resistance in ohm, γ denotes the offset parameter in mΩ, K_0, K_1 and K_2 corresponds to the model of vector parameters and R_0 was assumed as 0.01Ω [15]. The regression analysis of the estimated internal resistance for the C-rate are given by equation (4) respectively.

$$R_{(1C)} = 6.717e^{-06t} + 0.0596 \tag{4}$$

Fundamental equations governing the process of thermal cooling

The physical model of thermal cooling process employs the extended Darcy-Forchheimer's model and the hypothesis of the Linear thermal Non-Equilibrium (LTNE) to depict the heat exchange process between chosen paraffin material and the aluminium foams of different geometrical parameters. The various fundamental equations which control the battery system are namely the equation of continuity and the equation of momentum which are expressed by equations (5) to (8).

$$\frac{\partial u_{pcm}}{\partial x} + \frac{\partial v_{pcm}}{\partial y} + \frac{\partial w_{pcm}}{\partial z} = 0 \tag{5}$$

$$\frac{\rho_{pcm}}{\varepsilon}\left(\frac{\partial u_{pcm}}{\partial t}+\frac{u_{pcm}}{\varepsilon}\frac{\partial u_{pcm}}{\partial x}+\frac{v_{pcm}}{\varepsilon}\frac{\partial u_{pcm}}{\partial y}+\frac{w_{pcm}}{\varepsilon}\frac{\partial u_{pcm}}{\partial z}\right)= -\frac{\partial p_{pcm}}{\partial x}+\frac{\mu_{pcm}}{\varepsilon}\left(\frac{\partial^2 u_{pcm}}{\partial x^2}+\frac{\partial^2 v_{pcm}}{\partial y^2}+\frac{\partial^2 w_{pcm}}{\partial z^2}\right)+S_x \quad (6)$$

$$\frac{\rho_{pcm}}{\varepsilon}\left(\frac{\partial v_{pcm}}{\partial t}+\frac{u_{pcm}}{\varepsilon}\frac{\partial v_{pcm}}{\partial x}+\frac{v_{pcm}}{\varepsilon}\frac{\partial v_{pcm}}{\partial y}+\frac{w_{pcm}}{\varepsilon}\frac{\partial v_{pcm}}{\partial z}\right)= -\frac{\partial p_{pcm}}{\partial y}+\frac{\mu_{pcm}}{\varepsilon}\left(\frac{\partial^2 v_{pcm}}{\partial x^2}+\frac{\partial^2 v_{pcm}}{\partial y^2}+\frac{\partial^2 v_{pcm}}{\partial z^2}\right)+S_y \quad (7)$$

$$\frac{\rho_{pcm}}{\varepsilon}\left(\frac{\partial w_{pcm}}{\partial t}+\frac{u_{pcm}}{\varepsilon}\frac{\partial w_{pcm}}{\partial x}+\frac{v_{pcm}}{\varepsilon}\frac{\partial w_{pcm}}{\partial y}+\frac{w_{pcm}}{\varepsilon}\frac{\partial w_{pcm}}{\partial z}\right)= -\frac{\partial p_{pcm}}{\partial z}+\frac{\mu_{pcm}}{\varepsilon}\left(\frac{\partial^2 w_{pcm}}{\partial x^2}+\frac{\partial^2 w_{pcm}}{\partial y^2}+\frac{\partial^2 w_{pcm}}{\partial z^2}\right)+S_z \quad (8)$$

On employing the LTNE hypothesis during the process of heat transfer in the porous medium of the metal foams, the PCMs and metal foams are characterized by distinct temperature fields. The source terms involved in the momentum equations are expressed by equations (9) to (11).

$$S_x=\left(\frac{(1-\beta^2)}{(\beta^3+0.001)^3}A_{mush}+\frac{\mu_{pcm}}{K}+\frac{C_F}{\sqrt{K}}\rho_{pcm}\left|\vec{V_{pcm}}\right|\right)v_{pcm} \quad (9)$$

$$S_y=\left(\frac{(1-\beta^2)}{(\beta^3+0.001)^3}A_{mush}+\frac{\mu_{pcm}}{K}+\frac{C_F}{\sqrt{K}}\rho_{pcm}\left|\vec{V_{pcm}}\right|\right)v_{pcm}+\rho_{pcm}g\gamma_{pcm}(T_{pcm}-T_0) \quad (10)$$

$$S_z=\left(\frac{(1-\beta^2)}{(\beta^3+0.001)^3}A_{mush}+\frac{\mu_{pcm}}{K}+\frac{C_F}{\sqrt{K}}\rho_{pcm}\left|\vec{V_{pcm}}\right|\right)v_{pcm} \quad (11)$$

Based on the hypothesis of LTNE, two energy equations are involved unlike the LTE principles, which are described by equations (12) and (13).

$$\frac{\partial}{\partial x}\left(K_s\frac{\partial T_s}{\partial x}\right)+\frac{\partial}{\partial y}\left(K_s\frac{\partial T_s}{\partial y}\right)+\frac{\partial}{\partial z}\left(K_s\frac{\partial T_s}{\partial z}\right)-h_{sf}a_{sf}(T_s-T_f)=(\rho c_p)_s(1-\varepsilon)\left(\frac{\partial T_s}{\partial t}\right) \quad (12)$$

$$\frac{\partial}{\partial x}\left((K_f+k_d)\frac{\partial T_{pcm}}{\partial x}\right)+\frac{\partial}{\partial y}\left((K_f+k_d)\frac{\partial T_{pcm}}{\partial y}\right)+\frac{\partial}{\partial z}\left(K_f\frac{\partial T_{pcm}}{\partial z}\right)+h_{sf}a_{sf}(T_s-T_{pcm})-\varepsilon\rho_{pcm}H_L\frac{\partial\beta}{\partial t}= (\rho c_p)_{pcm}\varepsilon\left(\frac{\partial T_{pcm}}{\partial t}\right)+(\rho c_p)_{pcm}\left(u_{pcm}\frac{\partial T_{pcm}}{\partial x}+v_{pcm}\frac{\partial T_{pcm}}{\partial y}+w_{pcm}\frac{\partial T_{pcm}}{\partial z}\right) \quad (13)$$

p denotes the relative pressure in N/m^2, A_{mush} refers to the mushy zone constant accounting for the velocity damping during solidification. These equations are correlated separately with the applications of PCM and metal foams during the cooling process, involving two effective thermal conductivities (K_s and K_f). The k_d involved in equation (13) is related to dispersion thermal conductivity, which is neglected in the designed model as the velocity is too slow in a natural convection process. The effective thermal conductivities of the solid and fluid medium are estimated using the empirical equations proposed by Calmidi and Mahajan [16,17] which are expressed by equations (14) and (15) respectively.

$$K_f{}^*=\varepsilon k_f \quad (14)$$

$$K_s{}^*=0.181(1-\varepsilon)^{0.763}k_s \quad (15)$$

The heat transfer coefficient parameter involved in equation (13) is associated with the Nusselt's number expressed by equation (16)

$$Nu=\frac{h_{sf}d_f}{k_f} \quad (16)$$

The Nusselt's number is estimated using the Churchill's correlation which is implemented using a UDF function in C++. The correlation is given by equation (17)

$$Nu=0.36+0.518\left(\frac{Re}{\left[1+(0.559/Pr)^{9/16}\right]^{16/9}}\right)^{1/4} \quad (17)$$

Similarly, the a_{sf} associated with the process refers to the interfacial surface area is given by the equation (18).

$$a_{sf} = \left(\frac{3\pi d_f}{(0.59 * d_p)^2}\right)\left(1 - e^{-(1-\varepsilon/0.04)}\right) \quad (18)$$

The metal foams embedded with the PCM adhere to the principle of Enthalpy-Porosity method. Based on the concept, a mixed solid-liquid area of pseudo-porous zone is established. Porosity equals the phenomenon of liquid fraction (β) in this zone fluctuating between 0 and 1 respectively, where at 0, the zone is completely solid and at 1, it is completely in the liquid form. Based on this principle, source terms are added to the momentum equations, corresponding to the solid portions present in the mixed zone. The phenomenon is established by the mathematical equation (19).

$$\beta = 0 \quad for\ T < T_{solidus}$$
$$\beta = \left(\frac{T - T_{solidus}}{T_{liquidus} - T_{solidus}}\right) \quad for\ T_{solidus} < T < T_{liquidus}$$
$$\beta = 1 \quad for\ T > T_{liquidus} \quad (19)$$

Modeling of the porous zone integrates the Darcy and Forchhheimer's laws. These laws address the pressure loss caused by the viscous and inertial effects due to the existence of PCM and metal foam. Based on this context, the permeability of the porous medium is denoted by K, while the inertial drag factor is represented by C_F. These two parameters are expressed by equation (20) and (21).

$$K = 0.00037(1-\varepsilon)^{-0.224}\left(\frac{d_f}{d_p}\right)^{-1.11} {d_p}^2 \quad (20)$$

$$C_F = 0.00212(1-\varepsilon)^{-0.132}\left(\frac{d_f}{d_p}\right)^{-1.63} \quad (21)$$

here,

$$\frac{d_f}{d_p} = 1.18\sqrt{\frac{1-\varepsilon}{3\pi}}\left(\frac{1}{1-e^{1(1-\varepsilon)/0.04}}\right) \quad (22)$$

$$d_p = \frac{0.0254}{\omega} \quad (23)$$

where d_p refers to diameter of pore in m and d_f corresponds to the ligament diameter of pore in m. The value of the mushy constant (A_{mush}) is 10^6 kg m^{-3} s^{-1} [18].

The presence of battery in the model is given by the heat conduction equation:

$$\frac{\partial}{\partial x}\left(K_b\frac{\partial T_b}{\partial x}\right) + \frac{\partial}{\partial y}\left(K_b\frac{\partial T_b}{\partial y}\right) + \frac{\partial}{\partial z}\left(K_b\frac{\partial T_b}{\partial z}\right) + \dot{u}_{gen} = \left(\rho c_p\right)_b\left(\frac{\partial T_b}{\partial t}\right) \quad (24)$$

with $\dot{u}_{gen}$ the ratio between V^2/R.

The initial temperature and velocity are assumed equal to the ambient temperature at 293.17 K and 0 m/s, respectively. The boundary conditions are as follows: the lateral surfaces of the module were symmetrical. The top portions of the battery cell module were adiabatic, the channel of liquid is at 293.15 K, no slip conditions for the liquid PCM.

Numerical modelling design of the chosen battery cell

The various assumptions based on which the model is designed are that the model follows the Boussinesq's approximation. The thermophysical properties of both PCM and metal foam are assumed to be homogeneous and constant. The initial condition is assumed to be ambient temperature and radiation losses are neglected. Based on the assumptions, the model was designed using the finite volume technique using the Ansys-Fluent software. The pressure-based calculations and velocity pressure coupling are solved using the PRESTO and SIMPLE algorithms. The time chosen for analysis is 0.2s. It was observed that the convergence errors in case of energy equation was 10^{-8} and in case of flow equations it was 10^{-3} respectively. Four meshes were chosen to maintain the optimal selection of trade-off between time and accuracy. Mesh A accounted for 400680 cells, whereas mesh B comprised of 218463 cells. Mesh C was designed to have 61380 cells and finally Mesh D resulted in 108108 cells. By the Grid independence analysis, the error percentage of the maximum temperature attained by the cell was estimated. Among the four meshes, Mesh D resulted in the best optimal conditions. The grid independence test based on the error analysis for a C-rate of 1C for the maximum temperatures acquired by the designed cell indicated that the chosen mesh error resulted in a value of less than 0.1% [19]. The present model was compared with a simplified model given in [18] without the generation term and an assigned temperature along the surface of battery equal to 305.15. The comparison in terms of average temperature was less than 0.2%.

3. Results and Discussion

This study aimed to estimate and analyse the average surface temperature and the maximum temperature of the battery during the discharge process at a specific C-rate of 1C. The maximum temperature of PCM along with the liquid fraction was also estimated for the thermal cooling process. Moreover, the PCM material used in the analyses, was embedded with an aluminium foam of different porosities of 5,10,20 and 40PPI respectively. The discharge process was experimentally tested for the C-rate of 1C. Additionally, a numerical model was created and simulated for the mentioned C-rate using Ansys-Fluent software. The experimental results allowed for the estimation of the terminal output voltage and the discharge time at various depth of discharge based on the C-rate. The GRG algorithm was utilized to fit a curve that estimated the variable thermal resistance and open circuit voltage. A linear regression function, dependent on these parameters and the depth of discharge, was obtained as described in equation (4) and this was further applied in the model. After selecting an appropriate mesh through a grid independence test, simulations were performed. Detailed discussion of the parameters for the chosen PCM material with different porosities at a C-rate of 1C is described as follows.

The maximum temperature of the battery along with the average surface temperature of the battery responses is displayed in the figure 2 (a) and (b) respectively. From the characteristic curves, it is observed that, the maximum temperature of the battery and the average surface temperature of the battery was at a maximum for porosity equalling 5PPI, and the least for 40PPI. However, the 10PPI and 20PPI, were higher displaying a higher temperature than 40PPI, but less than 5PPI respectively.

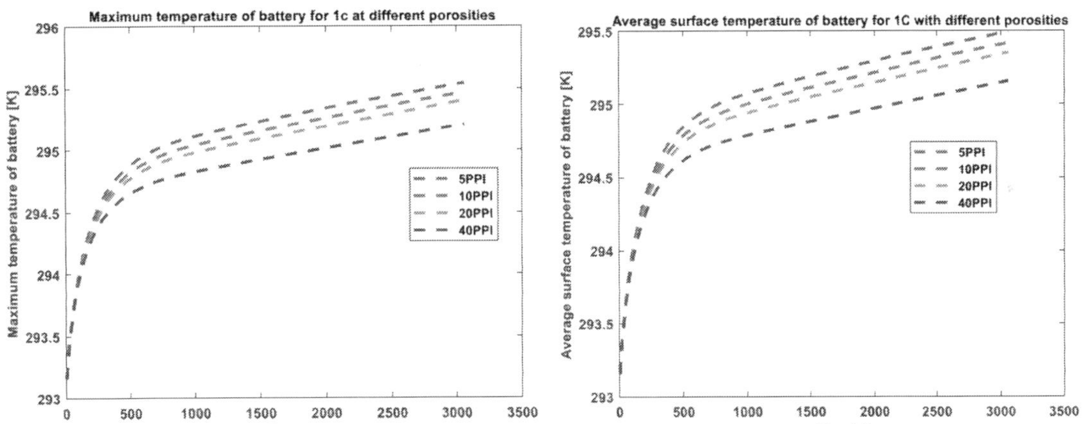

(a) (b)

Figure 2. (a) Characteristic response of maximum temperature of battery with different porosities. **(b)** Characteristic response of the average surface temperature of the battery with different porosities.

Similarly, in case of the PCM material used for the process of thermal cooling which was embedded with metal foams of different porosities, the maximum temperature was recorded by 5PPI, and the least was recorded by 40PPI respectively. It resembled the same profiles similar to the ones displayed by the battery's temperature in figure 2 respectively. The maximum temperature profile of the PCM material is displayed in figure 3a.

The liquid fraction of the PCM material used for the process of cooling is displayed in figure 3b. From the figure it is observed that, the metal foams due to their increased conductivity, tries to enhance the activity of the PCM. Due to the role of metal foams in the cooling process, the PCM phenomenon of liquid fraction was not observed much as in the case of only PCM conducted in the study of [18].

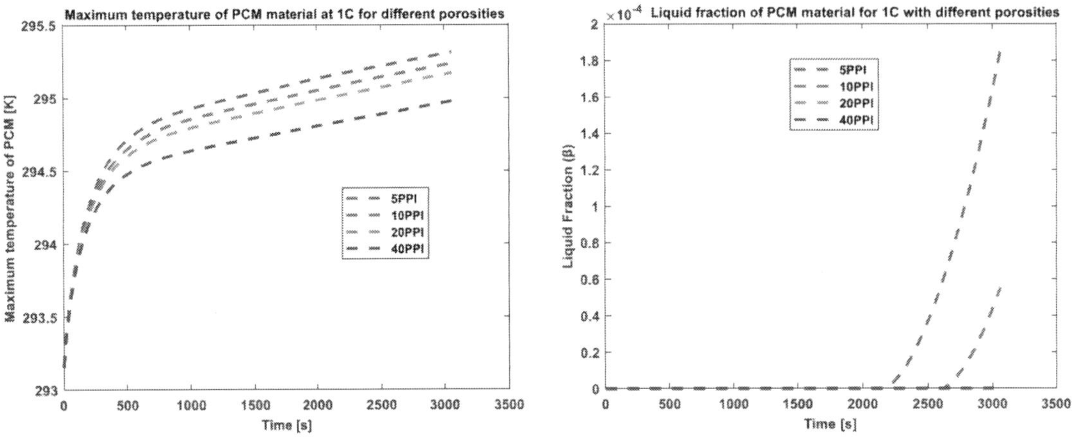

(a) (b)

Figure 3. Characteristic response of PCM with different porosities: (a) maximum temperature (b) liquid fraction

4. CONCLUSION

This study focused on the thermal control of battery using a PCM material with metal foams of different porosities using the principle of Local thermal non-equilibrium. The battery was discharged with a C-rate of 1C and the various parameters of maximum temperature, average surface temperature of the battery chosen along with PCM material's maximum temperature and liquid fraction was estimated. The results indicated a maximum value at 5PPI and a least value at 40 PPI respectively. It was observed that the metal foams with a greater value of PPI, indicated a better profile than the one with the lower PPI. It is suggested that various other forms of cooling hybrid mechanisms can be implemented to understand the thermal characteristics of the system.

5. References

[1] Dijk M, Orsato R.J., and Kemp R 2013 *Energy Policy*. **52**(2013).
[2] Arumugam A, Buonomo B, Luiso M and Manca O 2023 *Int.J.Energy Prod. Manag.8(4)*.
[3] Alanazi F 2023 *Applied Sciences*. **13** 6016.
[4] Hsu C, et.al. 2024 e-Prime- *Advances in Elect Engg and Elect and Energy* . **8**(2024).
[5] Esfahani N.N, Garmestani H, Rozati M and Smaisim G.F 2023 *J. Energy Storage*. **63** 107061.
[6] Farahani S.D, Mohammadi S, Farahani A.D., and Smaisim G.F 2023 *J. Energy Storage*. **60** 106628.
[7] Kim J, Oh J and Lee H 2019 *App .Ther. Engg*. **149** 192-212.
[8] Amba Prasad Rao G., and Shravan Kumar K 2024 e-Prime- *Advances in Elect Engg and Elect and Energy*. **8**(2024) 100526.
[9] Buonomo B, Ercole D, Manca O and Menale F 2018 *Energy Proced*. **148**(2018) 1175-1182.
[10] Du J, Tao H, Chen Y, Yuan X, Lian C and Liu H 2021 *Chin. Phys.Lett*. **38** (11)118201.
[11] Arumugam A, Cipolletta G, Delle Femine A, Gallo D, Landi C and Luiso M 2022 *AMPS* Conference, Italy.
[12] Efest IMR 18650 3000 mAh (Purple), Index of tested Li ion batteries, 2017 [online] at: https://lygte-info.dk/info/batteryIndex.html
[13] Neumann D.E. and Lichte S 2011 MSTV Conference, Michigan.
[14] Zhang C, Jiang J, Zhang S, Liu L, Wang P and Loh C 2016 *Energies* **9** (2016).
[15] Pillai P, Sundaresan S, Kumar P, Pattipati K.R, and Balasingam B 2022 *Energies* **15** (2022).
[16] Calmidi V.V and Mahajan R.L, 2000 *ASME Transact on J. Heat Trans*. **122** (2000).
[17] Bhattacharya A, Calmidi V.V, and R.L. Mahajan 2002 *Int. J Heat and Mass Trans*. **45** (2002).
[18] Arumugam A, Buonomo B, D'Arienzo C, Romano P and Manca O 2023 *Phys: Conf. Series* **2648** (2023).
[19] Ansys-Fluent, 2018 "Computational Fluid Dynamic Code Version 19.2 User Guide."

Journal of Physics: Conference Series **2940** (2025) 012020 doi:10.1088/1742-6596/2940/1/012020

Optimal design of thermal storage in hybrid energy systems: application to a laboratory-scale case study with experimental validation

N Bianco[1], M Di Somma[1, *], G Graditi[2], L Mongibello[2]

[1] Dipartimento di Ingegneria Industriale, Università degli Studi di Napoli Federico II, Napoli, Italy
[2] ENEA– Agenzia nazionale per le nuove tecnologie, l'energia e lo sviluppo economico sostenibile, Roma, Italy

[*] e-mail: marialaura.disomma@unina.it

Abstract. The benefits of thermal storage in hybrid energy systems (HESs) are multiple, being among others the enhancement of flexibility of the network and of the efficiency in the energy resources use, by also reducing the overall cost of the energy system. Choosing the capacity of TSS to minimize the operational cost of the HES is not a trivial task, depending on several variables existing at both supply and demand sides. The contribution of this paper is to present the experimental validation of an innovative optimization model aiming to identify the optimal capacity of TSS in HESs that maximizes the economic performances. The optimization model is based on a mixed-integer linear programming approach and allows determining the optimal hourly operating strategies of the system components as well as the optimal capacity of the TSS, while minimizing the total daily energy costs. The optimized hourly operation schedule of the HES was simulated experimentally by using one of the systems from the TSS laboratory of ENEA Portici Research Center. The experimental results demonstrated the practical applicability of the proposed optimization model. In fact, a good overlap was found between the results obtained from the implementation of the optimization model with those obtained from the experimental simulation, both considering energy and economic data.

1. Introduction

The benefits of thermal storage in hybrid energy systems (HESs) are multiple, being among others the enhancement of flexibility of the network and of the efficiency in the energy resources use, by also reducing the overall cost of the energy system [1-3]. In fact, thermal storage systems (TSSs) can store the excess thermal energy produced when the heat generation equipment operates in optimal conditions, while releasing it when the demand exceeds the capacity of the heating equipment [4-6]. Despite these benefits, choosing the capacity of TSS to minimize the operational cost of the HES is not a trivial task, depending on a number of variables at both supply (e.g., energy prices) and demand (e.g., time-varying energy demand) sides [7,8]. Based on the literature review, it emerges that most tools developed for the optimal sizing of TSS have been tested through numerical case studies, and there is a gap in the experimental implementation and/or validation of these tools in real-world contexts, including lab-scale applications.

The contribution of this paper is to present the experimental validation of an optimization model aiming to identify the optimal capacity of TSS in HESs that maximizes the economic performances. The optimization model is based on a mixed-integer linear programming (MILP) approach and allows determining the optimal hourly operating strategies of the system components as well as the optimal capacity of the TSS, while minimizing the total daily energy costs. The HES under study consists of a combined heat and power (CHP) system, an auxiliary boiler and a TSS, and one of the systems from the TSS laboratory of ENEA Portici Research Center was used for the experimental tests and validation. The considered user is in the Italian E climatic zone, of domestic type and represented by a single-family house with a useful surface area of 200 m^2, a shape factor of 0.9 m^{-1}, and was selected to experimentally simulate real thermal loads. The MILP model has been implemented with IBM ILOG CPLEX Optimization Studio Version 12.6 optimization software and solved by using branch-and-cut. The optimized hourly operation schedule of the CHP was simulated experimentally, and the experimental results demonstrated the practical applicability of the proposed optimization tool. In fact, a good overlap was found between the results obtained from the implementation of the optimization model with those obtained from the experimental simulation, both considering energy and economic data. Finally, a comparison was made between the experimental results obtained with the optimized capacity of the TSS, and those obtained from the experimental validation of the optimization model applied to the HES with the original TSS. From this comparison, the efficacy of the optimization model was demonstrated for further reduction of the total net daily cost by identifying the optimal capacity of the TSS.

2. Description of the ENEA experimental laboratory

Figure 1a shows a photo of the HES used for the experimental validation of the optimization tool presented in this study, whereas Figure 1b shows the TSS system of the experimental plant, which consists of commercial cylindrical thermally insulated tank 1.27 m high, with an internal diameter of 0.65 m, and a total capacity of approximately 420 liters, including the volume occupied by the two 1" coil heat exchangers with which it is equipped, each with a heat exchange area of approximately 2 m^2. The current TSS has a thermal capacity of 9.5 kWh. The lower coil of the storage tank is connected to the thermal circuit of the heat generator, while the upper coil is connected to the user circuit, i.e., the circuit in which there is a heat exchanger that allows simulating the user thermal loads for space heating (SH). For both circuits, water is the heat transfer fluid (HTF). Furthermore, the tank is equipped with nine type T class 1 thermocouples (accuracy: ±0.5°C) with the hot junctions uniformly distributed along the axis, and the mass flow rate measurements of HTF in the coils circuits is done with magnetic flow meters (accuracy: 0.2%, min flow rate: 0.05 liters/s).

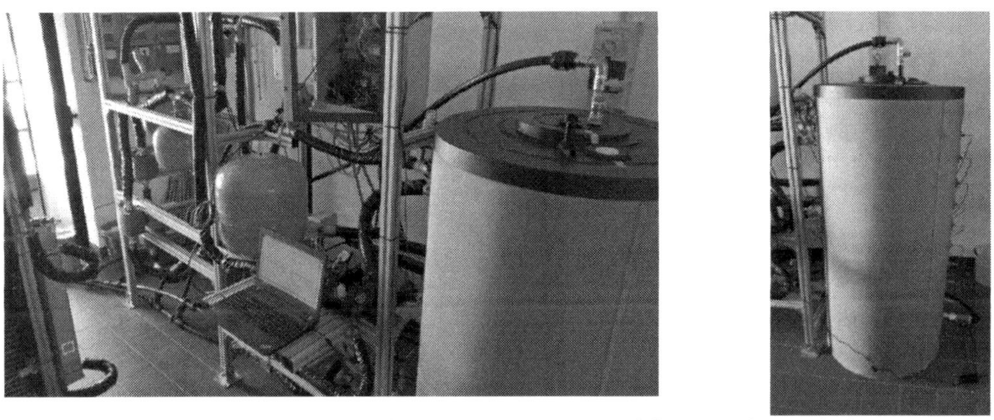

Figure 1. a) Experimental HES; **b)** TSS of the experimental plant.

For the experimental tests, the thermal production of the CHP and the thermal loads were simulated experimentally, while the thermal production of the auxiliary boiler, the electrical production of the CHP, and the electrical loads were simulated in a software environment using the data reported in next sections. As concerns the prime mover of the CHP, in both the experimental tests carried out, the ON/OFF operation at full load is implemented, without the possibility of partial load, on an hourly basis.

3. Optimization model formulation

The HES under study, represented in Figure 4, consists of a CHP system, an auxiliary gas-fired boiler and a TSS aiming at satisfying the multi-energy demand (electricity, DHW and SH) of a single-family house. As shown in Figure 2, the CHP and the power grid are used to satisfy the user's electrical demand, considering that the electricity supplied by the CHP can be both used for self-consumption and fed into the grid. To satisfy the thermal demand for DHW and SH, the CHP, the auxiliary boiler, and the TSS can be used.

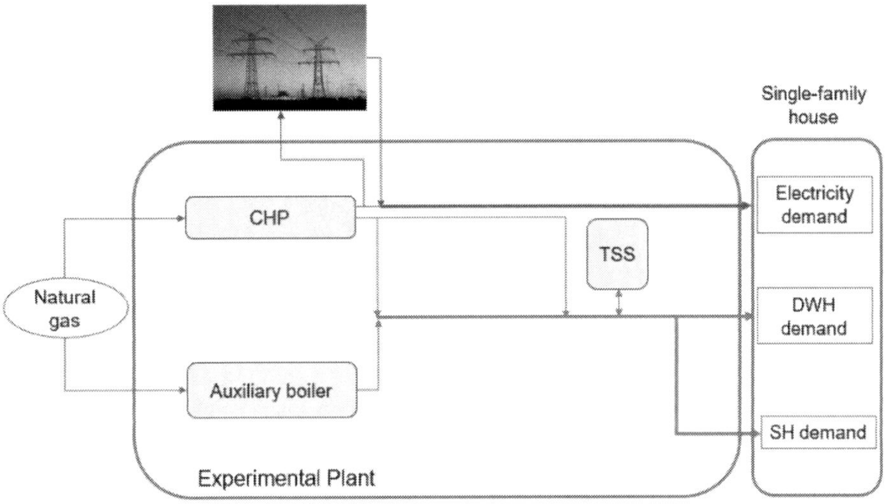

Figure 2. Scheme of the HES under study for the optimization model formulation.

The decision variables of the optimization model include the on/off status of each technology, the electrical and thermal power supplied by each technology, the electricity purchased and fed into the grid, and the capacity of the TSS with charging and discharging heat rates.

3.1. *Objective function*

The total daily net energy cost to minimize represents the objective function of the optimization problem, taking into account the costs incurred for the purchase of electricity from the grid and natural gas to power the CHP and the boiler, the daily fraction of the annualized investment cost of the TSS, and the revenues obtained from the sale of the electricity produced by the CHP. This objective function is formulated as:

$$Net\ daily\ cost = \sum_t \left(E_t^{Buy} \Pi_t^{Grid} + G_{i,t}^{Buy} \Pi^{Gas} + C^{Inv,TSS} - E_{CHP,t}^{Sell} \Pi_t^{DA}\right) Dt,\ i \in \{CHP, Boil\} \quad (1)$$

where E_t^{Buy} is the share of electricity purchased by the grid at time *t* at the price Π_t^{Grid}; $G_{i,t}^{Buy}$ is the total quantity of natural gas purchased at time *t* at the price Π^{Gas}; $E_{CHP,t}^{Sell}$ represents the share of electricity produced by the CHP fed into the grid at time *t* at the day-ahead market price Π_t^{DA}; and *Dt* is the hourly time-step. The annualized investment cost of the TSS is formulated as:

$$C^{Inv,TSS} = \frac{CRF_{TSS}(C_{c,TSS}Cap_{TSS})}{N_{workingdays}}, \text{ with } CRF_{TSS} = r(1+r)^{N_{TSS}}/[(1+r)^{N_{TSS}}-1] \quad (2)$$

where CRF_{TSS} is the capital recovery factor of the TSS, $C_{c,TSS}$ is the specific investment cost, Cap_{TSS} is the optimal capacity of the TSS to determine, and $N_{workingdays}$ is the number of working days in one year, and allows considering the daily fraction of this annualized investment cost to compute in the objective function.

The optimization of the operating schedule is on a daily basis and has been conducted on several consecutive days, characterized by the same input data, in order to find a periodic solution to be validated experimentally. Therefore, the economic objective function considered for the experimental validation has been formulated as the sum of the daily objective functions.

3.2. Model constraints

The constraints of the optimization model consist of operating constraints of the technologies in the HES and energy balance constraints for satisfying the hourly electrical and thermal loads of the user. With reference to the operating constraints of the technologies, in order to maintain the linearity of the optimization model, the hypothesis of constant efficiency has been made, therefore assuming that it does not vary with the load. The common constraint for all technologies is the capacity constraint, formulated below for the CHP:

$$E_{CHP}^{min} x_{CHP,t} \leq E_{CHP,t} \leq E_{CHP}^{max} x_{CHP,t}, \forall t \quad (3)$$

The electrical power delivered is therefore limited by the minimum load and the maximum power, if the technology is in use, namely if the binary decision variable, $x_{CHP,t}$ is equal to 1. The ramp constraint, which limits the fluctuation of the total electrical power delivered between two following time-steps, within the corresponding ramp-down and ramp-up is also established for the CHP. The overall amount of electrical power supplied is equal to the sum of the power fed into the grid and the power used for personal consumption.

The quantity of natural gas for the CHP is formulated as:

$$G_{CHP,t}^{Buy} = E_{CHP,t}/(\eta_{CHP,e} LHV_{gas}), \forall t \quad (4)$$

whereas the recovered heat rate is formulated as:

$$H_{CHP,t} = E_{CHP,t}\eta_{CHP,th}/\eta_{CHP,e}, \forall t \quad (5)$$

As for the auxiliary boiler, the quantity of gas consumed is formulated similar to Eq. (4), by considering the conversion efficiency of the boiler.

The state dynamic of the TSS is formulated as:

$$H_{TES,t} = H_{TES,t-1}\left(1 - \varphi_{TES}(Dt)\right) + (H_{TES,t}^{Ch} - H_{TES,t}^{Disch})Dt, \forall t \quad (6)$$

Energy balance constraints are necessary to ensure that assigned user's loads are met. With reference to the electricity energy balance, the electricity load must be satisfied by the electricity supplied by the CHP and the electricity from the electrical grid:

$$E_t^{Dem} = E_{CHP,t}^{Self} + E_t^{Buy}, \forall t \quad (7)$$

With reference to the thermal energy balance for DHW, the load must be satisfied by the CHP, the boiler and the storage, i.e.:

$$H_t^{DHW} = H_{CHP,t}^{DHW} + H_{Boil,t}^{DHW} + H_{TES,t}^{DHW,Disch} - H_{TES,t}^{DHW,Ch}, \forall t \quad (8)$$

A similar formulation can be used for the thermal energy balance for SH demand.

The branch-and-cut approach, which is especially effective for MILP-type models, is used to solve the linear optimization problem, which involves both binary and continuous variables.

4. Technical characteristics of the user and of technologies in the HES

The user considered for the experimental validation of the tool has been selected such that to experimentally simulate the real thermal loads. Therefore, a single-family house has been considered, characterized by a useful surface area of 200 m^2, a shape factor of 0.9 m^{-1}, and located in the Italian E climatic zone. The hourly profiles of the thermal and electrical loads used for the simulations are related to a typical day in the month of March and have been calculated using the approach presented in [9]. In details, the hourly load profiles have been obtained considering an annual thermal demand for SH equal to 68 kWh/m^2/year, an annual thermal demand for DHW of 15 kWh/m^2/year, and an annual electrical demand, relating to domestic electricity consumption excluding that for air conditioning in summer, equal to 18 kWh/m^2/year. The simulated CHP system consists of an internal combustion engine as prime mover. Table 1 shows the technical characteristics of the technologies in the HES.

As regards the economic data, the price of natural gas has been set equal to 0.85 €/Nm^3, or 0.77 €/Nm^3 if consumed by the CHP system due to the discount on the excise duty, and the price of purchasing electricity from the grid has been set equal to 0.17 €/kWh. The economic value of the electricity produced and fed into the grid has been considered varying with time. The specific investment cost of the TSS is set equal to 100 €/kWh.

Table 1. Characteristics of the simulated technologies in the HES.

Technology	Size	Efficiency	
		Electric	Thermal
CHP	5.4 kW_{th}; 2.275 kW_{el}	0.286	0.679
Auxiliary boiler	5.4 kW		0.80

5. Experimental results

The MILP model has been implemented with IBM ILOG CPLEX Optimization Studio Version 12.6 optimization software. The hourly schedule resulting from the application of the optimization model to the case study prescribes that the CHP, in the simulated standard day, operates from hour 8 to 12, at hours 14 and 16, from hour 18 to 21, and at hour 23. The optimized capacity of the TSS is found equal to 5.6 kWh, that is much lower that the capacity of the current TSS (9.5 kWh). The optimized schedule was applied to the experimental system using the following configuration of the experimental flow rates: the mass flow rate of the HTF in the generator circuit is equal to 0.4 kg/s; the mass flow rate of the HTF in the user circuit is equal to 0.15 kg/s. The consumption of DHW was simulated by oversizing the load for SH, i.e., by adding, hour by hour, the thermal load relative to the consumption of DHW to that of SH. During the hours in which the CHP is scheduled to operate, the electric heater is turned off if the temperature at the outlet of the electric heater exceeds 95°C, or if the maximum temperature in the storage tank exceeds 85°C. If one of the aforementioned conditions occurs, the electric heater is subsequently reactivated when the average temperature inside the tank drops below 70°C, provided that at the moment in which this occurs the optimized schedule prescribes that the CHP is ON, otherwise the electric heater remains off. The same schedule and the same loads were simulated for several consecutive days to obtain periodic results day after day. The results reported below refer to periodic operating conditions, obtained starting from the third day of experimental simulation.

Figure 3 shows the thermal power transferred from the electric heater to the HTF (water), that is in turn dissipated by the heat exchanger which simulates the load for SH, that also includes the DHW load, while Figure 4 shows a comparison between the thermal energy transferred from the electric heater to the HTF, and that generated by the CHP resulting from the optimization tool. The thermal power transferred from the electric heater to the HTF is evaluated through an enthalpy balance at the electric heater, considering the temperatures measured at the inlet and outlet, while the thermal power dissipated by the heat exchanger was calculated through an enthalpy balance at the exchanger using the measured input and output temperatures. Figure 4 shows that in the hours 8, 14, 16, 18 and 23 the experimental heat production does not correspond to that resulting from the optimization model. This result is due to the fact that, at the beginning of the hours 8, 14, 16, 18 and 23, or, equivalently, at the transition from the OFF state to the ON state of the heater, the latter takes some time to bring the output temperature to the set-point value corresponding to the theoretical power, and to the fact that, during hour 23, the maximum temperature measured in the tank reaches the maximum allowable value (85°C), as can be seen from Figure 5 which shows the trend of the average temperature in the storage tank over the course of the day, i.e., the storage tank reaches its maximum capacity, whereby the electric heater is deactivated. With reference to Figure 5, in this case the difference between the maximum and minimum values of the average temperature is equal to 17°C, corresponding to a storage capacity of 7.8 kWh. The difference between the theoretical capacity of the TSS (5.6 kWh) and that resulting from the experimental test is due to the uncertainty on the thermal losses relating to the circuit and the TSS, assumed equal to 12 % of thermal energy produced in the optimization tool.

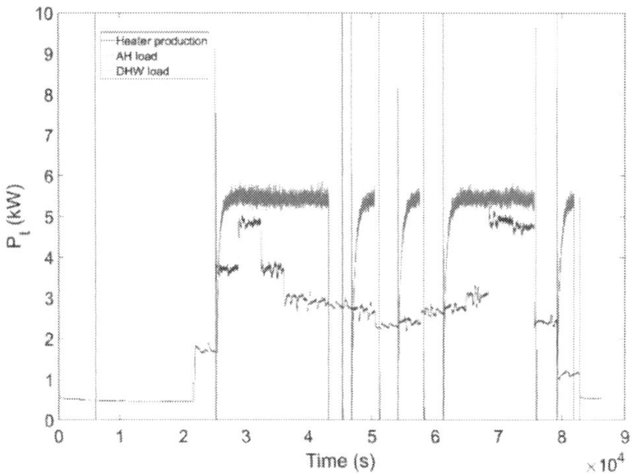

Figure 3. Thermal power generation and experimental thermal loads.

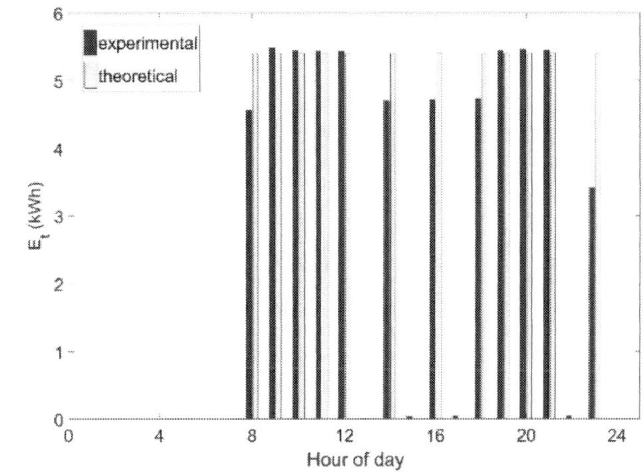

Figure 4. Comparison between the thermal energy produced by the electric heater and that resulting from the optimization tool.

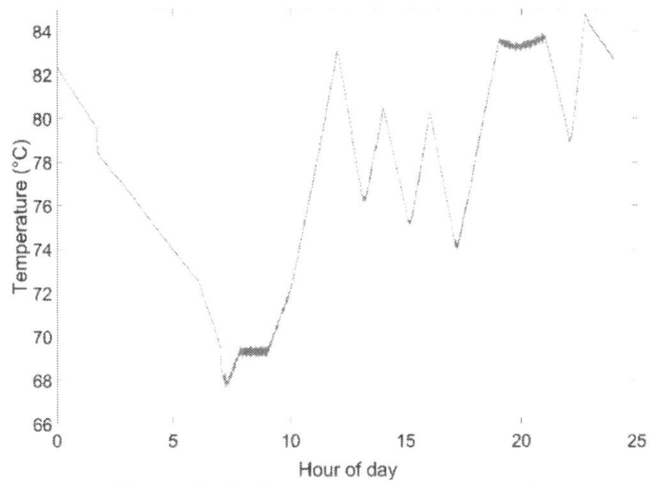

Figure 5. Tank water average temperature.

Figure 6 compares the experimental data and results from the implementation of the optimization tool, relative to the thermal load for SH satisfied by the CHP. It can be noted that the experimental data are almost coincident with the assigned theoretical loads, which implies that the gas consumption of the auxiliary boiler resulting from the optimization model is practically zero at all hours of the day.

IOP Publishing
doi:10.1088/1742-6596/2940/1/012020

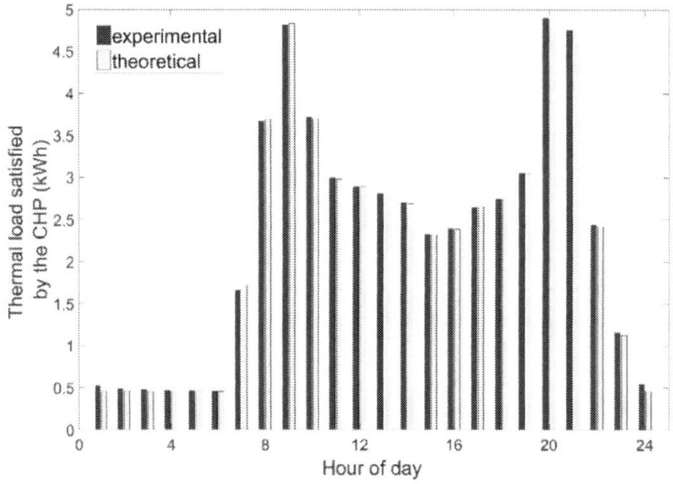

Figure 6. Comparison between the experimental thermal load satisfied by the CHP and the theoretical one.

Figure 7 shows the values of the electrical energy produced by the CHP resulting from the experimental simulation and the optimization tool. As stated before, in the experimental case, the electrical production was simulated in a software environment, using the experimentally measured thermal production of the CHP and the thermal and electrical efficiency of the CHP.

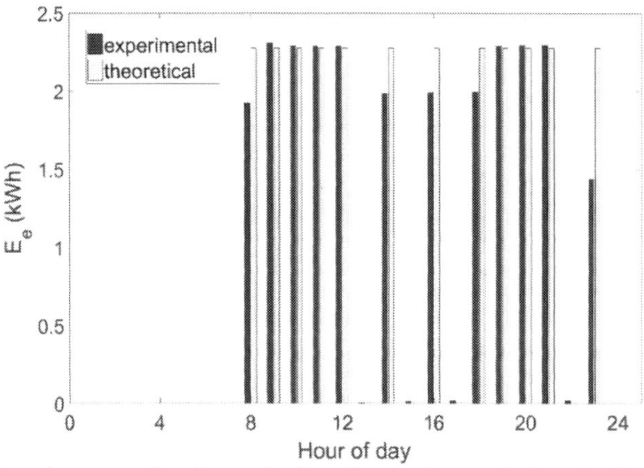

Figure 7. Comparison between the theoretical and experimental electrical production of the CHP.

Table 2 compares the consumption of gas and electricity and the amount of electricity injected into the grid obtained with the optimized capacity of the TSS, and those obtained from the experimental validation of the optimization model applied to the HES with the original TSS, while Table 3 compares the costs and revenues resulting from the two aforementioned experimental simulations. The results in Table 2 indicate that in the two cases, there are very similar values of both gas and electricity consumption and electricity fed into the grid. Consequently, the costs of the energy carriers and the economic value of the electricity fed into the grid are also very similar in the two cases, as is evident from the results in Table 3. The substantial difference in economic terms concerns the reduced investment cost of the TSS (daily fraction), which is significantly lower in the case relating to the presented optimization model that also optimizes the size of the TSS.

Table 2. Gas and electricity consumption, and electricity fed into the grid for one day.

	CHP gas consumption (Nm^3)	Boiler gas consumption (Nm^3)	Electricity from the grid (kWh)	Electricity injected into the grid (kWh)
Optimized operation and TSS size	9.24	0	4.15	15.54
Optimized operation (current TSS)	9.24	0.02	4.11	15.67

Table 3. Costs and revenues (€) for one day.

	Cost gas for CHP	Cost gas for boiler	Cost electricity from the grid	Revenue for selling electricity	Daily fraction of annualized TSS investment cost	Daily economic balance
Optimized operation and TSS size	-7.11	0	-0.71	+1.30	-0.37	-6.89
Optimized operation (current TSS)	-7.11	-0.02	-0.70	+1.32	-0.60	-7.11

From this comparison, the effectiveness of the optimization model is demonstrated for further reduction of the total net daily cost by identifying the optimal capacity of the TSS.

Additionally, in order to illustrate the efficacy of the optimization model from an environmental standpoint, the total daily CO_2 emissions have been computed for both of the cases under analysis, assuming that the power grid carbon intensity is 0.354 $kgCO_2/kWh$ [10] and that natural gas carbon intensity is 0.202 $kgCO_2/kWh$ (i.e., 1.93 $kgCO_2/Nm^3$) [11].

Results show that through the optimal capacity of the TSS, it is possible to reduce daily CO_2 emissions by 0.10%. It must be highlighted that to attain a higher reduction of CO_2 emissions, a multi-objective approach is needed to be implemented in the optimization model.

6. Conclusions

This works presents the experimental validation of an optimization model aiming to identify the optimal capacity of TSS in HESs that maximizes the economic performances. The HES under study consists of a CHP system, an auxiliary boiler and a TSS, and one of the systems from the TSS laboratory of ENEA Portici Research Center was used for the experimental tests and validation. The experimental results demonstrated the practical applicability of the proposed optimization tool, as a good overlap was found between the results obtained from the implementation of the optimization model with those obtained from the experimental simulation, both considering energy and economic data. From the comparison between the experimental results obtained with the optimized capacity of the TSS, and those obtained from the experimental validation of the optimization model applied to the HES with the original TSS, the effectiveness of the optimization model was demonstrated for further reduction of the total net daily cost by identifying the optimal capacity of the TSS. Future work will implement a multi-objective approach to take into account also environmental concerns in the optimal design problems for TSS.

7. References

[1] Roberto R, De Iulio R, Di Somma M *et al.* 2019 A multi-objective optimization analysis to assess the potential economic and environmental benefits of distributed storage in district heating networks: A case study *International Journal of Sustainable Energy Planning and Management*, **20**.

[2] Ahmadisedigh H and Gosselin L. 2022 How can combined heating and cooling networks benefit from thermal energy storage? Minimizing lifetime cost for different scenarios *Energy* **243**,

Journal of Physics: Conference Series **2940** (2025) 012020 doi:10.1088/1742-6596/2940/1/012020

123112.

[3] Caliano M, Delfino F, Di Somma M, Ferro G, Graditi G *et al.* 2022. An Energy Management System for microgrids including costs, exergy, and stress indexes. *Sustainable Energy, Grids and Networks* **32** 100915.

[4] Mongibello L, Bianco N, Di Somma M, and Graditi G 2017 Experimental validation of a tool for the numerical simulation of a commercial hot water storage tank *Energy Procedia* **105** 4266-4273.

[5] Chicco G, Di Somma M, Graditi G 2021 Overview of distributed energy resources in the context of local integrated energy systems. *Distributed Energy Resources in Local Integrated Energy Systems* 1-29, Elsevier.

[6] Mongibello L, Bianco N, Di Somma M, Graditi G, Naso V. 2014. Numerical simulation of a solar domestic hot water system. In *Journal of Physics: Conference Series* **547**. IOP Publishing.

[7] Di Somma M, Graditi G, Mongibello L *et al.* 2018, June. Trade-Off Solutions between Economy and CO 2 Emissions for the Daily Operation of a Distributed Energy System: A Real Case Study in Italy. In *EEEIC/I&CPS Europe* (pp. 1-6). IEEE.

[8] Yan B, Di Somma M and Graditi G. 2021 Multiobjective operation optimization of DER for short-and long-run sustainability of local integrated energy systems. In *Distributed energy resources in local integrated energy systems* (pp. 89-123) Elsevier.

[9] Mongibello L, Bianco N, Caliano M, and Graditi G 2016 Comparison between two different operation strategies for a heat-driven residential natural gas-fired CHP system: Heat dumping vs. load partialization *Applied Energy* **184** p 55.

[10] IPCC, Guidelines for National Greenhouse Gas Inventories E CO2 Emissions from Fuel Combustion, Highlights, 2018 ed., IEA, 2018.

[11] Educogen, The European Educational Tool on Cogeneration, second ed., 2001.

Acknowledgments

This research was funded by the Italian Ministry of Economic Development, within the research project "RdS-PAR 2019–2021".

Journal of Physics: Conference Series **2940** (2025) 012021 doi:10.1088/1742-6596/2940/1/012021

Preliminary assessment of a two-phase direct cooling of Lithium-Ion battery pack for e-bike mobility

L Giammichele[1*], D Colarossi[2], M Falone[2] and F Fioretti[2]

[1] Department of Life and Environmental Sciences (DiSVA), Università Politecnica delle Marche, 60131 Ancona, Italy
[2] Department of Industrial Engineering and Mathematical Sciences (DIISM), Università Politecnica delle Marche, 60131 Ancona, Italy

*E-mail: l.giammichele@staff.univpm.it

Abstract. Electric mobility is playing an increasingly central role in emission reduction policies to mitigate climate change effect. During the operation of electric vehicles, the batteries may be subject to high variation of the required current, which can lead to a sudden increase in the cell temperature. If this condition occurs repeatedly, there would be a reduction in battery capacity and useful life, and autonomy reduction of the electric vehicle. In the worst case, this problem can lead to the thermal runaway. Therefore, cooling of electric vehicle propulsion systems is a fundamental issue for the electric mobility development. In this article we propose an innovative cooling system using a dielectric low boiling fluid in which the batteries are directly immersed. The system was tested on an electric bicycle under real operative conditions. A special test bench was realized, consisting of a real electric bicycle, a training roller to simulate the load due to road slope and an external electric motor to simulate the pedaling of the cyclist. The results show a substantial decrease in the temperature of the cells with the proposed cooling system and there was a marked improvement in the temperature uniformity of the cells inside the battery pack.

1. Introduction

Electric vehicles (EVs) are keeping increasing attention by the transport manufacturer because of the central role they play in the reduction of fossil fuel consumption. This trend is even more emphasized by the common interest in the global warming and environmental pollution. Over the last years there was a wide diffusion of electric vehicles not only in the automotive sector, but also other applications, such as electric bike, scooter, and marine transport [1].

Lithium batteries (Li-ion) is the most common power supply option for EVs. They have sevral advantages: lower weight, higher specific power and energy density, longer life cycle, and lower self-discharge rates [2, 3]. However, they could be subjected to high temperature increase during their functioning due to electrochemical heat generation. The consequence is a reduction in electrical performance and lifetime. Furthermore, a very dangerous consequence is the thermal runaway that could occur if the batteries work at extreme temperatures, or if subjected to electrical or mechanical abuse [4].

In this context, the batteries thermal management is a fundamental topic in the development of EVs and to overcame safety problem of EVs. Several studies addressed this problem focusing on the design of an effective battery thermal management system (BTMS).

Due to their simplicity and cost-effectiveness, air cooling systems are often adopted, both in natural and forced convection, in low power applications. They are usually combined with system to enhance heat transfer such as heat sinks, pin-fins, pulsating heat pipe (PHP) [5-7]. However, air systems have a lower heat transfer coefficient and heat capacity. Using a liquid could be a solution to this problem since the higher thermal capacity. Indirect liquid cooling, single-phase or two-phase, were deeply studied [8, 9]. The problem of this solution is the limited surface contact area, and the thermal resistance between the surfaces in contact.

Direct contact liquid cooling is a promising solution since the batteries are directly immersed in a dielectric fluid. In this case the advantages are: (i) thermal power generated is directly transferred to the coolant avoiding the thermal contact resistance problem, (ii) larger heat transfer surface area [10]. Wu et al. [11], used a silicon oil and obtained a temperature 20% lower than the indirect liquid cooling. Patil et al. [12], study the effect of static and flowing condition on a large-format pouch cell. Wu et al. [13], noted a substantial reduction of maximum temperature and a higher temperature uniformity by submerging a pouch cell into a low boiling dielectric liquid. Jithin and Rajesh [14], numerically show the effect of three different dielectric coolants on a 1P4S battery pack. Williams et al. [15] experimentally investigated a direct liquid cooling on a small battery pack under several discharge rates and both in single and double phase natural convection.

The system proposed in this paper is based on three fundamental points: (i) the batteries are directly submerged in a dielectric fluid ensuring the maximum heat transfer surface; (ii) the fluid has a low boiling point in order to create a thermal buffer for the heat power peaks generated by high electric power requirement; (iii) the system is totally passive, avoiding any other electric consumption to move the fluid inside the battery pack. A passive cooling system is particularly attractive for electric vehicle. Indeed, the energy requirements of the battery cooling system weighs on the battery itself, resulting in a higher discharge rate and a lower range for the EVs. Furthermore, the battery undergoes a higher number of charge-discharge cycle reducing the operative life and thus must be disposed and replaced earlier, with clear environmental consequences. In previous works the system was tested on a single cylindrical cell [16] and on a small battery pack [17] under several rates of discharge using an electronic load. However, the discharge current was always set to a specific and constant value. Instead, in this paper, our aim is to test the system in real operating conditions for the first time. For this purpose, a special test bench was realized, consisting of a real electric bicycle (e-bike), a training roller to simulate the load due to road slope and an external electric motor to simulate the pedaling of the cyclist. A 5P10S battery pack was realized with similar dimensions and electrical characteristics to a real e-bike battery and it was tested with the dielectric low boiling fluid inside the container. The results were compared with the case without dielectric fluid.

2. Experimental setup

Four macro elements composed the test bench (Fig. 1): the e-bike, the battery pack, the road slope simulator, and the simulator of the cyclist pedaling.

The e-bike is the Trek model Powerfly 4 Gen 3 equipped with a Bosch power train. The electric motor (Bosch Performance CX Line) has a maximum torque of 85 Nm, a maximum continuous power of 250 W, and a power peak of 600 W. It is supplied by the Bosch PowerTube

Figure 1. Experimental setup.

625 with a nominal voltage of 36 V and a capacity of 16.7 Ah. The on-board computer (Bosch Nyon) allows to monitor the performance of the cyclist in terms of cadence, mechanical power to the pedal, and travel speed. These data are acquired by the on-board computer every second and transferred to the software eBike Connect by bluetooth connection. Furthermore, it is possible to set four different levels of electric support.

The battery pack was composed by 50 cells arranged in a 5P10S configuration in order to obtain the same voltage and similar capacity of the original battery of the bicycle without overcame the volume occupied by the original battery. The cell is a cylindrical NMC (Lithium nickel cobalt manganese oxide) battery model 18650 (height 65 mm and diameter 18 mm) with a weight of 46 g. The nominal cell voltage, maximum voltage, and nominal capacity were 3.7 V, 4.2 V, and 2.85 Ah, respectively. The battery pack is composed of 10 cells in series and 5 cells in parallel resulting in a nominal voltage of 37 V and a capacity of 14.25 Ah. The battery pack was placed inside a hermetic container. The gap between the battery pack and the internal wall of the container is 10 mm on each side, while the ratio between the external surface of the container and the battery pack is equal to 0.6.

The road slope was simulated by an Elite training roller model Direto XR-T. It allows to replicate a maximum road slope of 24% that can be set to a fixed value, according to a variable cycle or following the real conditions of a route. The system has an integrated power meter and cadence sensor. The power meter allows to measure the total mechanical power of the cyclist and the e-bike motor. The roller can be managed by the relative software (My E-Training) which allows to set the road slope and acquires the sensor data (mechanical power, cadence, and travel speed) once per second.

The cyclist's pedaling simulator is composed of an external DC brushless electric motor (Bafang model BBS02B) and an electronic controller. The motor has a nominal power of 500 W and a maximum torque equal to 110 Nm. It is supplied by a lithium battery pack of nominal

IOP Publishing
doi:10.1088/1742-6596/2940/1/012021

Figure 2. (a) Cyclist's pedaling simulator system; (b) motor shaft connection to the e-bike.

voltage 36 V and capacity 15.6 Ah. The motor is controlled by an ESC (Electronic Stability Control) FlyFun model HV130A OPTO V5 able to vary the power supply to the motor according to a PWM (Pulse Width Modulation) signal. The PWM signal was generated by the NI PXI-6289 device and allows to change the power supplied by the external electric motor. The motor shaft was finally connected to the crankset of the e-bike by means of an elastic joint. The system used to simulate the cyclist's pedaling is shown in figure 2.

Two current transducers were used to measure: the operating current supplied by the e-bike battery and that sunk by the external electric motor. The first has an uncertainty is ±0.02 A in the range ±85 A; the second has an uncertainty of ±2% of full scale in the range ±100 A. The Turnigy Reaktor charger manages the charge of the battery, balancing the charge of the cells and applying a typical constant current constant voltage charge (CCCV). Ambient temperature was measured with an uncertainty of ±0.6 °C. The voltages of the batteries of the e-bike and external motor were measured by two digital displays equipped with a high voltage module with an uncertainty of ±0.1% of the reading. The temperature of the cells surface was measured by 15 T-type thermocouples bonded in the middle height of the cells with a high conductive adhesive. Other 6 thermocouples were placed in the fluid between the cells and near the wall of the container at the

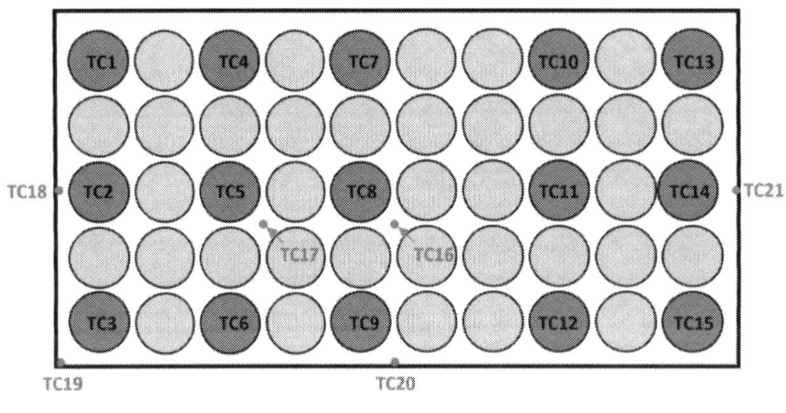

Figure 3. Thermocouples position inside the battery pack.

same height. The position of the thermocouples is showed in figure 3. Finally, one thermocouple was placed in the centre of the container in the vapour phase. An ice point reference was adopted for the thermocouples before the data logger acquisition. All the thermocouples were calibrated using a high precision bath, obtaining an uncertainty of ±0.06 °C.

The on-board computer allows to measure the mechanical power of the cyclist using a torque sensor integrated in the e-bike motor with an uncertainty of ±1.5% of the reading. The data logger is a NI PXI-1050 equipped with a data acquisition device for voltage signal (NI PXI-6289) and a device for thermocouples (NI SCXI-1303). The sensor signals of the current, battery voltage, and ambient temperature were acquired by the NI PXI-6289 device. The device has a resolution of 0.076 mV and the uncertainty is ±0.25 mV. The cadence is measured by a magnetic sensor placed

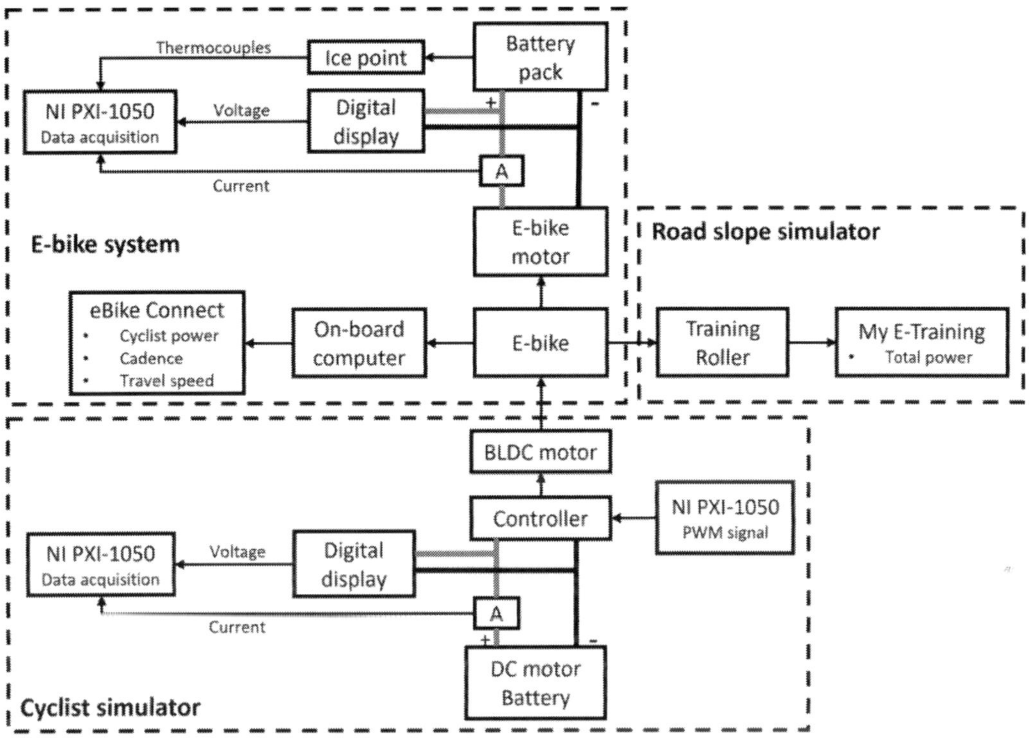

Figure 4. Schematic diagram of the experimental setup.

on the back wheel and this data is used by the on-board computer to calculate the travel speed. The integrated power sensor of the training roller allows to measure the total mechanical power (cyclist and e-bike motor) with un uncertainty of ±1.5% of measured value. Figure 4 shows a schematic diagram of the experimental setup.

The data measured by the on-board computer and the training roller were acquired at 1 Hz. Other parameters (i.e. thermocouples, ambient temperature, voltages and currents) were acquired at 25 Hz; then the mean values at 1Hz were calculated.

3. Results and discussion

The tests were performed with the following conditions: constant road slope equal to 12%, total weight (e-bike and cyclist) of 90 kg, and time duration of 1 h. The external electric motor was set to a fixed power equal to about 70 W through the PWM signal controller, while the e-bike motor was set to the maximum level of support. The mean resulting cadence was 75 rpm and the mean C-rate of the battery discharge was 0.65C. The same ambient temperature (25 °C) was adopted for all the tests.

The battery pack was placed inside a hermetic container. The test described above was firstly performed on the battery pack with only air in natural convection. Then, the container was filled with the dielectric fluid (Novec 7000) to completely cover the cells, and the test was repeated with the same conditions. Between two tests the battery was charged with a low current CCCV cycle, and enough time was waited for the cells to return to ambient temperature.

Table 1 shows the mean mechanical power of the e-bike motor ($P_{m,b}$), the mean mechanical power of the cyclist ($P_{m,c}$) measured by the on board computer, the mean total mechanical power ($P_{m,t}$) measured by the training roller, and the mean electrical power ($P_{e,b}$) supplied by the battery for the two configurations. It is worth nothing that the e-bike motor mechanical power was calculated as $P_{m,b} = \eta\, P_{e,b}$, where the electrical power was calculated as the product between voltage and current supplied by the e-bike battery pack, and η is the overall efficiency defined as

Table 1. Mean powers measured during the tests.

	Air	Novec 7000
$P_{m,c}$	70.6 ± 1.1	70 ± 1.1
$P_{m,t}$	347.9 ± 5.3	346.2 ± 5.2
$P_{e,b}$	331.8 ± 3.4	330.9 ± 3.7
$P_{m,b}$	277.4 ± 5.4	276.2 ± 5.3

$\eta = (P_{m,t} - P_{m,c})/P_{e,b}$. The mean powers measured for the two tests (with and without dielectric fluid) are summarized in table 1. The power data highlight that the same load conditions were respected for the test with air and with dielectric fluid since the measured powers were approximately the same.

The temperature plot of the cells for the two cases under investigation is showed in figure 5. The temperature of the cells during the test was subtracted to their initial value, in order to avoid the influence of ambient temperature and highlights the temperature rise. Only 6 thermocouples were taken as reference and showed in the graphs. These thermocouples were chosen because the system has two axes of symmetry, and the behaviour of the symmetrical thermocouples was found to be the same. The maximum uncertainty obtained for the thermocouples was equal to ±0.07 °C. This value includes both the instrument uncertainty, and the random uncertainty based on the standard deviation of the measurements.

The figures put in evidence a better temperature uniformity distribution achieved with the dielectric fluid. In this case all the temperatures have similar value with a very small difference by each other. The maximum difference obtained with dielectric fluid was 0.29 °C. Otherwise, in the case without dielectric fluid there are higher differences between the cells. Furthermore, the difference of temperature between the cells measured in the case without dielectric fluid is consistent with the position of the cells: the highest temperature was measured for TC5 and TC8

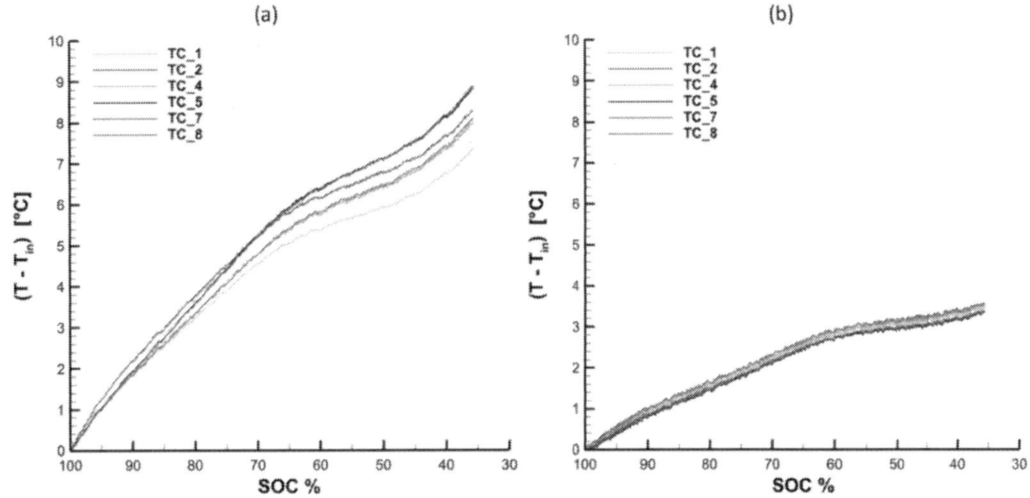

Figure 5. Temperature difference of 6 cells with air (a) and with dielectric fluid (b).

that are in the middle of battery pack and surrounded by other cells on each side; intermediate values was found for TC2, TC4, and TC7 since they are on the side of the battery pack and surrounded on three sides; the lowest temperature is for TC1 which is placed on the corner. Indeed, if the cell is surrounded by other cells there is a mutual heat transfer, and the temperature rise more rapidly. Furthermore, the convective heat transfer is limited by the small space between the cells which are placed in the middle of the battery pack compared with the external cells. Finally, it could be notice that the temperature of TC2 initially rises at a higher rate. This could be explained since the positive terminal of the battery pack is solder near this cell. In both cases the temperature rise more rapidly in the first part of the test; then at about the 60% of SOC there is a change of slope in the curves; finally, the temperature rise again over the 40% of SOC. This

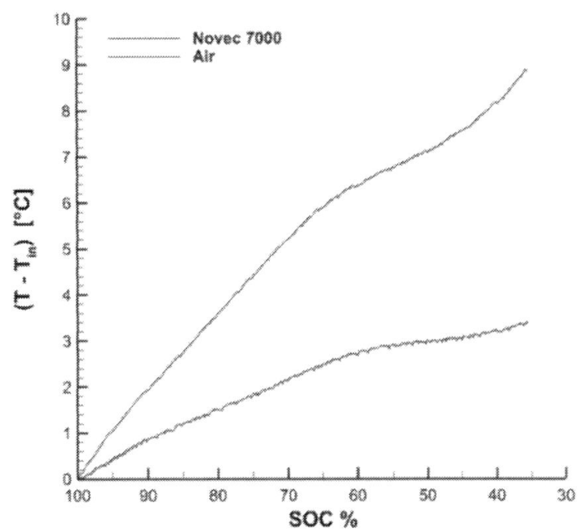

Figure 6. Comparison of TC8 temperature difference with air and with dielectric fluid.

behaviour is typical of Li-Ion battery discharge phase and is due to the battery internal reaction [16].

Figure 6 shows the comparison of the temperature measured by TC8 with and without dielectric fluid. It is very clear from the figure the positive effect of the cooling strategy since a lower temperature was reached by the case with dielectric fluid. In this case the maximum temperature difference was 8.5 °C; while, with the dielectric fluid, the maximum temperature difference lightly overcame 3 °C, obtaining a temperature increase reduction of 62% at the end of the test. The lower temperature at the end of the test obtained with the dielectric fluid can be explained by the higher heat capacity of the fluid and the higher convective heat transfer coefficient.

4. Conclusions

In this paper a preliminary experimental assessment of a novel cooling method of Li-Ion battery packs for e-bike under real operative conditions was presented. The batteries are directly submerged in a low boiling dielectric fluid. The proposed system is completely passive since there is no flowing fluid inside the battery pack, avoiding any additional power consumption. Furthermore, the low boiling point creates a thermal buffer in case of high electrical power requirements.

A special test bench was realized, consisting of an e-bike, a road slope simulator and a cyclist simulator. The original battery pack of the e-bike was replaced with a customized battery in configuration 5P10S with volume and electrical characteristics similar to the original battery. The pack was instrumented with several thermocouples placed on cells surface and in the fluid. Other relevant parameters were also acquired to obtain the mechanical power supplied by the electric motor and the cyclist.

Compared with the case of air in natural convection, a significant decrease of the temperature increase during battery discharge was observed. The temperature rise reduction was equal to 62% with the dielectric fluid. Another important result regards the temperature uniformity of the cells inside the battery pack. Indeed, the temperature was more uniform in the case with dielectric fluid. In this case, all the thermocouples measure very close values. Otherwise, in the case without dielectric fluid, there is a higher difference between the cells, according to their position inside the battery pack. Despite the excellent thermal control, the system introduces an additional weight to the e-bike of approximately 1 kg due to the dielectric fluid. However, it is important to remark that the percentage impact on the whole e-bike is only 5%.

The proposed system is capable to effectively control the temperature of the cells in their optimal range. The system could be also applied in other types of electric vehicles such as automotive or lightweight aircraft. In order to face off the higher thermal power to dissipate, in these applications the system could be improved by properly optimize the geometry of the external surface of the container, for example by adding some fins to enhance heat transfer area and facilitating the condensation of the dielectric fluid or by integrating other passive system such as heat pipe or pulsating heat pipe.

References

[1] Mali V, Saxena R, Kumar K, Kalam A and Tripathi B 2021 *Renew. Sustain. Energy Rev.* **151** 111611.

[2] Lowe M, Tokuoka S, Trigg T and Gereffi G 2010 Lithium-ion Batteries for Electric Vehicles: the U.S. Value Chain. Technical Report Duke University Center on Globalization, Governance and Competitiveness.

[3] Speirs J, Contestabile M, Houari Y, Gross R 2014 *Renew. Sustain. Energy Rev.* **35** 183-193.

[4] Feng X, Ouyang M, Liu X, Lu L, Xia Y and He X 2018 *Energy Storage Mater.* **10** 246-267.
[5] Mohammadian S K and Zhang Y 2017 *Appl. Therm. Eng.* **118** 375-384.
[6] Qin P, Liao M, Mei W, Sun J and Wang Q 2021 *Appl. Therm. Eng.* **195** 117212.
[7] Cattani L, Malavasi M, Bozzoli F, D'Alessandro V and Giammichele L 2024 *Energies* **17**(13) 3236.
[8] Wang H, Tao T, Xu J, Mei X, Liu X and Gou P 2020 *Appl. Therm. Eng.* **178** 115591.
[9] Fang Y, Ye F, Zhu Y, Li K, Shen J and Su L 2020 *Energy Rep.* **6** 238-247.
[10] Song Y F, Liu Z G, Li S W and Jin Q Y 2021 *Int. J. Heat Technol.* **39** 1923-1929.
[11] Wu S, Lao L, Wu L, Liu L, Lin C and Zhang Q 2022 *Appl. Therm. Eng.* **201**(A) 117788.
[12] Patil M S, Seo J H and Lee M Y 2021 *Energy Convers. Manag.* **229** 113715.
[13] Wu N, Ye X, Yao J, Zhang X, Zhou X and Yu B 2021 *Int. J. Heat Mass Transf.* **170** 121018.
[14] Jithin K V and Rajesh P K 2022 *Int. J. Heat Mass Transf.* **188** 122608.
[15] Williams N P, Trimble D and O'Shaughnessy S M 2023 *J. Energy Storage* **86** 111289.
[16] Giammichele L, D'Alessandro V, Falone M and Ricci R 2022 *Int. J. Heat Technol.* **40**(1) 1-8.
[17] Giammichele L, D'Alessandro V, Falone M and Ricci R 2022 *IOP Conf. Ser.: Earth Environ. Sci.* **1106** 012017.

Preliminary experimental studies on coalescer efficiency for liquid-liquid separation

I M Carraretto[1,*], D Scapinello[1], R Bellini[1], L P M Colombo[1] , L Molinaroli[1] and G Manzolini[1]

[1] Dipartimento di Energia, Politecnico di Milano, via Lambruschini 4a, 20156 Milano (MI), Italy

* Corresponding author: igormatteo.carraretto@polimi.it

Abstract. The international organizations have identified both water demand and accessibility as critical challenges with both current and future implications for human well-being. In reason of that due to the growing demand, it is essential to advance the technologies used in freshwater production, focusing on sustainability while maintaining operational efficiency. Within the HORIZON 2020 program, it was thought to integrate the desalination processes with Renewable Energy Sources (RES). The desalination process is based on Forward Osmosis (FO), which uses a membrane to split pure water from seawater by exploiting the osmotic pressure difference between the feed solution (i.e., seawater) and the draw solution (i.e., permeate side). For the process to be effective, a low-cost method for regenerating the diluted draw solution is necessary. Hence, a viable alternative is to deploy a thermo-responsive polymeric draw agent that can be regenerated by the heat power rejected from a CO_2 power cycle integrated with a Concentrated Solar Power. To perform the separation, a liquid-liquid separator (i.e., coalescer), mostly deployed in the oil and gas sector, is proposed to split the fresh-water from the polymeric draw solution. Hence, the aim of this work is to preliminarily assess the performances of the coalescer, aiming at the formulation of an experimental efficiency expression. Expression that is function of: (1) temperature at which the regeneration is performed, (2) residence time, and (3) draw concentration in the initial solution. The draw agent used in the experimental campaign is the PAGB2000, a thermo-responsive co-polymer that has a Lower Critical Solution Temperature (LCST) above which the fluid splits in two distinct phases: a polymer-rich (more concentrated), and a polymer-poor (more diluted). From the analysis it results that the coalescer performances strongly increase with both temperature and residence time, indeed efficiency achieves values near to 100%. On the other hand, the performances slightly worsen with increasing the initial draw concentration. Eventually, having an efficiency formulation can both help the design process of the desalination plant and provide an understanding on how much this technology can be applied in context far from the ones it was developed for.

1. Introduction

Water demand and accessibility have been recognized as two of the most critical tasks for human well-being in the near future [1]. Hence, because of the increasing need in fresh-water production through desalination processes, there is a strong interest in improving the technologies employed without neglecting both sustainability and efficiency.

Journal of Physics: Conference Series **2940** (2025) 012022 doi:10.1088/1742-6596/2940/1/012022

Specifically, the desalination of seawater is a well-known technology to provide communities with freshwater where there is limited access to conventional resources. Historically, most desalination plants have relied on either Multi-Effect Distillation or Multi-Stage Flash technologies, often in conjunction with fossil fuel power stations [2]. Nowadays membrane based processes have gained space in the market [2]. The process can be made more environmentally sustainable by having a coupling with renewable energy sources. Perhaps, Reverse Osmosis membranes can be coupled with Photovoltaic systems [3] or, as in the case here proposed, Forward Osmosis membranes can be coupled with Concentrated Solar Power plants exploiting the waste heat that otherwise would be dissipated.

The FO process, similarly, to the RO process, is a membrane-based technology, however it deploys a different governing mechanism which is the osmotic pressure difference between the feed and the draw sides. It is indeed the above-mentioned pressure difference that promotes the flow of freshwater across the membrane. On the other hand, the solute or salt molecules present in the seawater are selectively prevented from passing [4]. The now-diluted draw requires additional treatment to split the fresh-water from the re-concentrated draw, enabling its future reuse. In particular, the regeneration process depends on the selected draw. Hence, to guarantee a low-cost regeneration a draw agent [5] that shows a thermo-physical response was selected, with the intention to regenerate it with the heat rejected from a power cycle as presented in Figure 1.

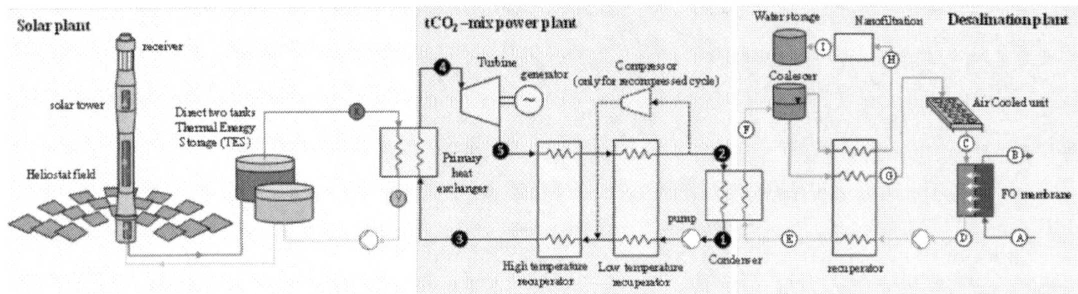

Figure 1. Layout of the integrated CSP+D system for a current central tower CSP plant.

This concept underpins the European Project DESOLINATION, which utilizes an advanced power cycle based on CO_2 blends to release heat at temperatures exceeding 75 °C, facilitating the regeneration step of the forward osmosis process [6].

Based on the analysis conducted by Inada et al. [7], [8], a LCST polymer can be the right draw agent, providing both high osmotic pressure and an easily-accomplishable regeneration process. However, the simplicity, feasibility, and effectiveness of the polymer regeneration process are critical factors that must be examined to ensure proper plant design. As noted by Colciaghi et al. [9], a coalescer is selected to perform the regeneration step.

The coalescer is a static separator, mainly used in the petrochemical sector, able to split two distinct immiscible liquids from a stable emulsion (liquid-liquid) or purifying a gas from impurities to achieve high-quality gas (gas-liquid).

The idea of employing a coalescer gravity separator was for the first time presented by TREVI System and then subsequently investigated by Ahmed et al. [10]. Their work evaluates the technical feasibility of using a thermo-responsive polyelectrolyte draw solution in a forward osmosis desalination pilot-scale system with a capacity of 10 m^3/day. In particular, they consider both DS and feed flowrates effect on the net water recovery and product water flowrates of the system. The selected draw solution demonstrated its potential for use in commercial-scale forward osmosis desalination plants due to its low viscosity and ease of phase separation, utilizing a coalescer regeneration system operating at a temperature of 85 °C.

Given the limited studies available in the open literature on both experimental and modeling results, a preliminary experimental campaign was conducted to evaluate the performance of the coalescer in regenerating a thermo-responsive draw agent at varying temperatures, flow rates, and initial solution concentrations. The aim was to gather data to develop an empirical expression for its efficiency.

2. Materials and Methods

2.1. Description of the test facility

The experimental setup (Figure 2) is created to reproduce the regeneration step of the diluted draw agent deployed in the FO plant (Figure 1 blue box), considering a draw agent composition as it is at the outlet of the FO membrane.

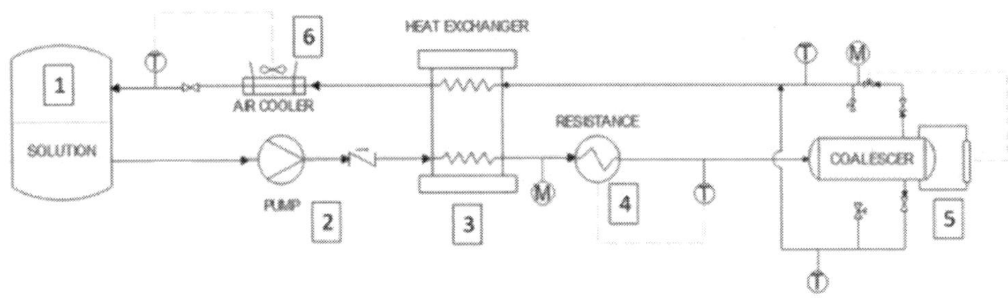

Figure 2. Experimental facility layout.

A tank (1), positioned 1 meter above the centrifugal pump (2) to avoid cavitation during operation, holds the draw solution at atmospheric pressure and at a temperature lower than the LCST. A centrifugal pump is used to set the fluid flowrate and to force it to pass via first the pre-heating section (3) and then via the heating one (4). Electric resistances are adopted to heat the draw solution from the inlet temperature up to the regeneration one (T_{reg}). The fluid then flows into the coalescer (5), where the split between the two phases, poor and rich respectively, occurs. A metal mesh is inserted within the coalescer to facilitate the coalescing process. The flow is subsequently divided into two streams: (i) the polymer-rich phase, which exits the coalescer from the bottom, and (ii) the water-rich phase, which exits from the top. Each section has its own tap for sampling. The two streams are subsequently recombined and directed into the hot side of the heat exchanger (3). Finally, at the heat exchanger outlet, the fluid is cooled by an air cooler (6) before returning to the storage tank (1).

Within the coalescer (D = 0.20 m, L = 1.60 m) a metal mesh (STYLE42C, Costacurta S.p.A.-VICO, Milano, Italy) is tested, the mesh density (ρ_M) is of 262 kg/m^3.

The selected polymer is PAGB2000 whose thermo-physical properties are presented in [11]. The polymer has already been recognized in studies published in the open literature as a promising option in Forward Osmosis processes [9]. The product is available on the marked with the commercial name of UNILUBE 50MB-26, and it is produced by NOF Corporation® (Tokyo, Japan) and deployed without further purification.

2.2. Measurement and instrumentation

In the experimental setup, four thermo-resistances (RTD Pt100 IEC751 cl.A, Smeri, Assago (MI), Italy) are adopted to measure the temperatures and are positioned at: (1) the inlet of the coalescer measuring T_{reg}, (2) the outlet of the coalescer on both branches measuring either T_{rich} or T_{poor}, (3) the outlet of the air cooler measuring T_{cooler}. Two flow-meters are used to measure the volumetric flowrate and are placed at: (1) the outlet of the heat exchanger measuring the total flowrate (F) (PromagP300, 0 - 100 l/min, ±0.5% r.v., Endress Hauser, Reinach, Switzerland), (2) at the outlet of polymer-poor (water-rich) branch (HTLD-MAG, 0 - 100 l/min, ±0.5% f.s., Smeri, Assago (MI), Italy) measuring the

Journal of Physics: Conference Series **2940** (2025) 012022 doi:10.1088/1742-6596/2940/1/012022

water produced (F_{poor}). The concentration of the polymer solution is measured using a portable refractometer (MA871, 0 - 85% brix, ±0.2%, Milwaukee, Rocky Mount (NC), United States), which measures the refractive index (n [brix]). Eventually, to get the weight concentration the calibration curve equation is applied. Table 1 reports in a synthetic form the instrumentation deployed.

Table 1 Instrumentation deployed.

Instrument	#	Name, range, accuracy	Producer
Thermo-resistance	4	RTD Pt100 IEC751 cl.A	Smeri
Flowmeter	1	PromagP300, 0 - 100 l/min, ±0.5% r.v.	Endress Hauser
Flowmeter	1	HTLD-MAG, 0 - 100 l/min, ±0.5% f.s.	Smeri
Refractometer	1	MA871, 0 - 85% brix, ±0.2%	Milwaukee

2.3. Test protocol and operating conditions

The experimental conditions analyzed in this work are reported in Table 2. Specifically, 4 initial concentrations, 5 flowrates and 6 regeneration temperatures are assessed.

The procedure implemented, starting from only-water tank, is here reported:

1. Add the required mass of polymer to obtain the chosen concentration (wt./wt.).
2. Obtain a homogeneous concentration by re-circulating the solution.
3. Check the concentration by measuring the refractive index, 3 samples are extracted:
    a. storage tank
    b. water-rich branch
    c. polymer-rich branch
4. Once the correct concentration is achieved, adjust the flow rate to the desired value.
5. Activate the modular resistance and adjust it to achieve the precise temperature setting.
    a. Check that T_{reg}, T_{rich} and T_{poor} are equal.
6. Adjust the flow rate on the polymer-poor side (F_{poor}) using the valve located at the coalescer's output. F_{poor} should match, within the margin of error, a target value calculated using the lever rule, based on T_{reg}, c_{in} and the polymer's LCST curve.
7. Collect three samples each of the rich and poor solutions, allowing a 15-minute interval between each sampling. Perform three concentration measurements on each sample.
8. Measure the refractive index three times once the samples are at 20 °C.
9. Modify either flowrate or temperature and repeat (5 – 7), ensuring that T_{cooler} remains below 40 °C (i.e., lower than LCST) to ensure a single-phase fluid at the start of the process for each initial concentration tested.

Table 2. Operating conditions.

c_{in} [–]	F [l/min]	T_{reg} [°C]
0.30	4	55
0.40	6	60
0.50	8	65
0.60	10	70
	12	75
		80

Due to technological limitations, some combinations of experimental conditions are not tested. Specifically, at low flow rate and high temperature there is a very high risk of overheating the system. Whereas at high temperature and high flow rate the experimental apparatus cannot ensure steady-state conditions.

As already stated in Section 1, the main goal of the experimental campaign is to identify the influence of: (1) regeneration temperature (T_{reg}), (2) initial fluid concentration (c_{in}), (3) residence time

(τ), on the efficiency of separation for a coalescer. The residence time is defined as the ratio of the inner coalescer volume to the total flow rate.

2.4. Data processing

The coalescer separation efficiency is defined as:

$$\varepsilon = \left(c_{in} - c_{exp,poor}\right)/\left(c_{in} - c_{th,poor}\right) \tag{1}$$

indicating the coalescer's capacity to achieve the theoretically obtainable concentration (as determined by the phase diagram) in comparison to the concentration obtained experimentally.

The modeling performance is evaluated against the experimental values using the Mean Relative Deviation (*MRD*) and Mean Absolute Relative Deviation (*MARD*), defined as follows:

$$MRD = \frac{1}{N}\sum_{i=1}^{N}\frac{z_{i,exp} - z_{i,th}}{z_{i,exp}} \text{ (i) and } MARD = \frac{1}{N}\sum_{i=1}^{N}\frac{\left|z_{i,exp} - z_{i,th}\right|}{z_{i,exp}} \text{ (ii)} \tag{2}$$

3. Experimental results and discussion

This section provides a summary of the results from the experimental campaign and characterizes the performance of the coalescer as a function of the operating conditions.

3.1. Concentration measurements

Considering an initial concentration of 0.40, the concentrations for the water-rich side (c_{poor}), results are presented as a function of the total flow rate and regeneration temperature in Figure 3 (a).

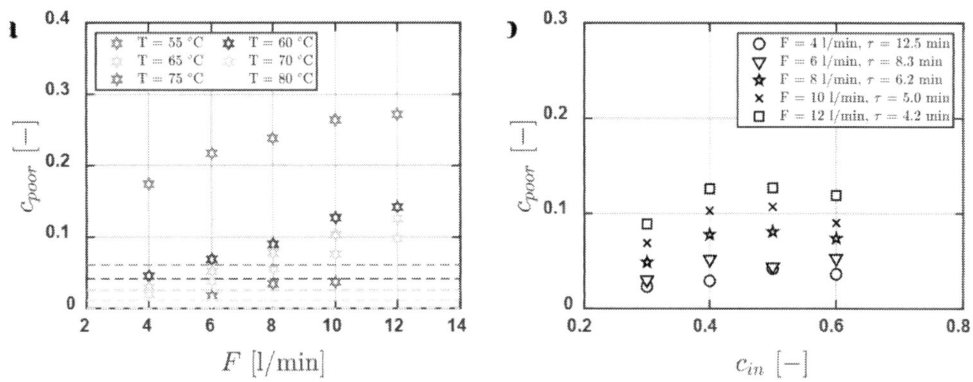

Figure 3. (a) c_{poor} having c_{in} = 0.40 and (b) c_{poor} vs. c_{in} at T_{reg} = 65 °C.

The results indicate that the regeneration temperature is likely the most significant parameter for achieving high-quality separated phases; at elevated temperatures, the concentrations rapidly approach the theoretical values. (dashed lines in Figure 3 (a)).

Finally, the initial polymer concentration appears to have a minimal impact on the separation process across the four tested conditions (c_{in}) at fixed flowrate (Figure 3 (b)).

Figure 4 graphically summarizes the results of the experimental campaigns, presenting the concentrations of the water-rich phase (left side of the charts) and polymer-rich (right side of the charts) measured relative to the LCST [11]. The charts also include error bars indicating measurement uncertainty, consistently ranging from ±0.002 to ±0.02. Having the concentration a limited impact on the process performance, only one condition is plotted (i.e., c_{in} = 0.40).

Ultimately, it is observed that phase separation closely aligns with the LCST curve for the given case when a high regeneration temperature (above 65 °C) is selected. These results demonstrate the excellent performance of the separation process.

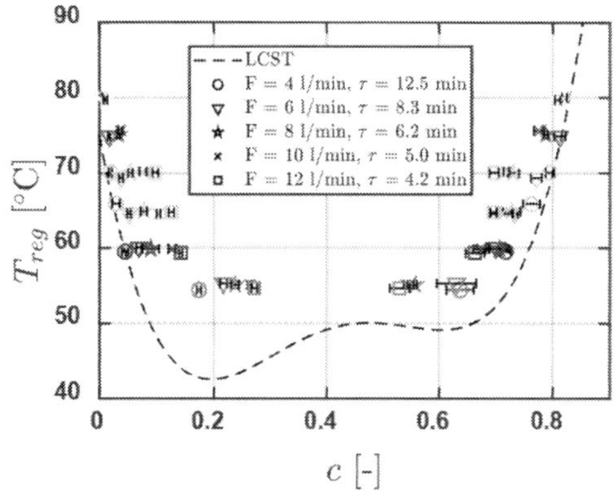

Figure 4. c_{poor} and c_{rich} measured vs. PAGB2000 LCST [11], at different T_{reg}, τ and c_{in} = 0.40.

3.2. Coalescer efficiency

Collecting the experimental concentrations outcomes and applying Equation (1), it is possible to determine the efficiency values (Table 3), which are then analyzed as function of both regeneration temperature and residence time at fixed initial concentration (see Figure 5). The values indicate that the coalescer operating temperature should exceed 70 °C, which is suitable for the intended application (i.e., in advanced power cycles, heat is typically rejected at around 80 °C). Furthermore, at temperatures above 70 °C and flow rates below 10 l/min, efficiency can exceed 95%.

Table 3. Efficiency values (ε) according to Equation (1).

T_{reg}	F	c_{in} [-]				T_{reg}	F	c_{in} [-]			
[°C]	[l/min]	0.30	0.40	0.50	0.60	[°C]	[l/min]	0.30	0.40	0.50	0.60
	4	51.8%	67.5%	75.0%	78.1%		4	99.5%	99.9%	98.3%	99.3%
	6	38.4%	54.0%	64.5%	73.6%		6	91.2%	92.6%	94.0%	95.0%
55	8	36.1%	48.0%	57.2%	69.8%	60	8	83.1%	87.0%	87.1%	88.9%
	10	26.5%	40.3%	51.8%	65.1%		10	59.4%	76.6%	80.1%	84.5%
	12	17.5%	38.1%	48.2%	59.4%		12	56.0%	72.7%	78.6%	82.7%
	4	100.0%	98.7%	96.8%	98.5%		4	99.1%	98.1%	98.7%	99.6%
	6	98.7%	93.5%	96.3%	95.2%		6	97.4%	93.9%	93.5%	93.2%
65	8	91.4%	86.4%	88.1%	91.7%	70	8	94.1%	89.2%	89.8%	93.6%
	10	84.4%	79.7%	83.4%	88.9%		10	89.6%	83.4%	85.2%	86.5%
	12	76.8%	73.6%	78.8%	83.6%		12	86.4%	77.9%	81.8%	86.6%
	6	96.4%	95.7%	91.9%	95.5%	80	6	97.6%	98.1%	98.0%	97.9%
75	8	95.5%	91.6%	94.0%	94.0%		8	97.1%	97.4%	98.0%	97.6%
	10	92.0%	90.8%	91.7%	94.1%						

Once the efficiency is determined, a mathematical expression to calculate the coalescer performance as function of the above-mentioned quantities is proposed. Specifically, the expression is an exponential function developed starting from mass-exchangers efficiency, reported in [12]:

$$\varepsilon = 1 - C_1(c_{in}) \exp\left\{\frac{C_2(c_{in})}{\tau} - C_3(c_{in})T^*\right\} \tag{3}$$

where τ is the residence time and T^* is the dimensionless temperature defined as:

$$T^* = \frac{T_{reg} - \mathrm{LCST}(c_{in})}{T_{max} - \mathrm{LCST}(c_{in})} \tag{4}$$

having fixed T_{max} = 85 °C to account for all the regeneration temperatures measured.

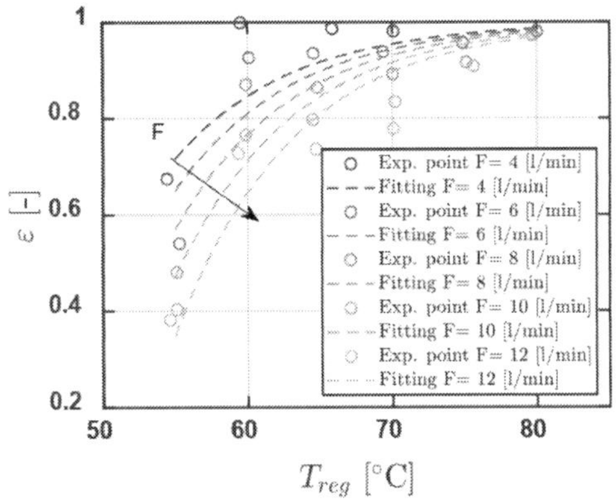

Figure 5. Efficiency vs. Regeneration Temperature at different flowrates for c_{in} = 0.40.

This expression incorporates the key control variables of the coalescence process: regeneration temperature, residence time, and initial concentration. Ultimately, the proposed equation aligns with the physics of the phenomenon and exhibits a monotonic trend from a mathematical perspective, approaching a value of 1 within the experimental domain considered.

A code developed with Matlab® was used to analyze the results and calculate the three fitting constants, that are function of the initial solution concentration.

The presented fitting model demonstrates reliability, with all cases showing $R^2 > 0.80$ and both *MRD* and *MARD* falling within the ±10% range.

Eventually, it results that the three fitting constants are independent of the initial concentration, hence the median value is chosen for each constant. Heading to, an efficiency expression equal to:

$$\varepsilon = 1 - 0.32 \exp\left\{\frac{5.05}{\tau} - 4.29T^*\right\} \tag{5}$$

4. Conclusions

- The study aimed to assess the effectiveness of a coalescer for liquid-liquid separation in a forward osmosis (FO) plant.
- The coalescer reduces energy consumption during the separation of a water-rich stream from the draw agent, particularly with a thermo-responsive material.
- The thermo-responsive polymer PAGB2000 was selected, based on previous research showing its ability to produce pure water with low energy requirements.
- Experiments evaluated the coalescer under varying conditions: initial solution concentrations, regeneration temperatures, and flow rates.
- Results showed separation efficiencies above 90% in most cases, regardless of initial concentration.
- Efficiency was significantly affected by regeneration temperature and flow rate/residence time.
- An expression for mass exchanger efficiency was derived from the findings.
- Future research will test performance with alternative thermo-responsive draw agents and metal meshes.

5. References

[1] Unesco, "United Nations World Water Development Report." https://www.unesco.org/en/wwap

[2] J. Eke, A. Yusuf, A. Giwa, and A. Sodiq, "The global status of desalination: An assessment of current desalination technologies, plants and capacity," *Desalination*, vol. 495, p. 114633, 2020, doi:10.1016/j.desal.2020.114633.

[3] N. Ghaffour, J. Bundschuh, H. Mahmoudi, and M. F. A. Goosen, "Renewable energy-driven desalination technologies: A comprehensive review on challenges and potential applications of integrated systems," *Desalination*, vol. 356, pp. 94–114, 2015, doi:10.1016/j.desal.2014.10.024.

[4] M. Mohammadifakhr, J. de Grooth, H. D. W. Roesink, and A. J. B. Kemperman, "Forward osmosis: A critical review," *Processes*, vol. 8, no. 4, p. 404, 2020, doi:10.3390/pr8040404.

[5] D. J. Johnson, W. A. Suwaileh, A. W. Mohammed, and N. Hilal, "Osmotic's potential: An overview of draw solutes for forward osmosis," *Desalination*, vol. 434, no. September 2017, pp. 100–120, 2018, doi:10.1016/j.desal.2017.09.017.

[6] E. Morosini, A. Ayub, G. di Marcoberardino, C. M. Invernizzi, P. Iora, and G. Manzolini, "Adoption of the CO2+ SO2 mixture as working fluid for transcritical cycles: A thermodynamic assessment with optimized equation of state," *Energy Convers. Manag.*, vol. 255, p. 115263, 2022, doi:10.1016/j.enconman.2022.115263.

[7] A. Inada, K. Yumiya, T. Takahashi, K. Kumagai, Y. Hashizume, and H. Matsuyama, "Development of thermoresponsive star oligomers with a glycerol backbone as the draw solute in forward osmosis process," *J. Memb. Sci.*, vol. 574, no. September 2018, pp. 147–153, 2019, doi:10.1016/j.memsci.2018.12.067.

[8] A. Inada, K. Kumagai, and H. Matsuyama, "Effect of the molecular weights of thermoresponsive polyalkylene glycol draw solutes on forward osmosis performance," *Sep. Purif. Technol.*, vol. 252, p. 117462, 2020.

[9] R. Colciaghi, R. Simonetti, L. Molinaroli, M. Binotti, and G. Manzolini, "Potentialities of thermal responsive polymer in forward osmosis (FO) process for water desalination," *Desalination*, vol. 519, no. September 2021, p. 115311, 2022, doi:10.1016/j.desal.2021.115311.

[10] M. Ahmed, R. Kumar, B. Garudachari, and J. P. Thomas, "Performance evaluation of a thermoresponsive polyelectrolyte draw solution in a pilot scale forward osmosis seawater desalination system," *Desalination*, vol. 452, no. November 2018, pp. 132–140, 2019, doi:10.1016/j.desal.2018.11.013.

[11] I. M. Carraretto *et al.*, "Characterization of the physical properties of the thermoresponsive block-copolymer PAGB2000 and numerical assessment of its potentialities in Forward Osmosis desalination," *Polym. Test.*, vol. 128, no. October, p. 108238, 2023, doi:10.1016/j.polymertesting.2023.108238.

[12] R. E. Treybal, *Mass transfer operations*, 3rd Editio., vol. 466. New York: McGraw-Hill, 1980.

Nomenclature

c	Concentration	[kg/kg]	reg	Regeneration	
F	Flowrate	[l/min]	th	Theoretical	
n	Refractive index	[Brix]	rich	Polymer-rich/ Water-poor	
T	Temperature	[°C]	ε	Efficiency	[-]
cooler	Air cooler		τ	Residence time	[min]
in	Inlet/Tank		ρ	Density	[kg/m^3]
exp	Experimental		MRD	Mean Relative Deviation	
poor	Polymer-poor/ Water-rich		$MARD$	Mean Absolute Relative Deviation	

Acknowledgments

The project has been partly funded by the European Union's Horizon 2020 research and innovation program under grant agreement No. 101022686 (DESOLINATION).

The study has been carried out within the NEST - Network 4 Energy Sustainable Transition (D.D. 1243 02/08/2022, PE00000021) and partly funded under the National Recovery and Resilience Plan (NRRP), Mission 4 Component 2 Investment 1.3, funded from the European Union - NextGenerationEU. This manuscript reflects only the authors' views and opinions, neither the European Union nor the European Commission can be considered responsible for them.

Testing of a prototype substation for a 5th generation district heating and cooling network

G Martinazzoli[1,2]*, D Pasinelli[2], A M Lezzi[1] and M Pilotelli[1]

[1] Dipartimento di Ingegneria Meccanica e Industriale, Università degli Studi di Brescia, Brescia 25123
[2] A2A Calore & Servizi S.r.l., Brescia 25124

*E-mail: g.martinazzoli002@unibs.it

Abstract. 5th generation district heating (5GDH) networks can be an effective way to increase the penetration of renewable energy and achieve decarbonisation in the heating and cooling sector. This article describes 5GDH networks by focusing on network substations, the complexity of which increases compared to a conventional district heating substation due to the presence of a decentralised heat pump and a bidirectional pumping system. The purpose of the article is to show the components and layout of a prototype substation and a pilot network, realised to recover low-temperature waste heat and cover the heating and cooling demand of a building. The results present the performance and energy consumption of the prototype during the first heating season in which the prototype is in operation. The prototype was tested through two kinds of tests: tests under design and off-design conditions. In tests under design conditions, the average COP of the heat pump is between 3.3 and 4.2, while the values drop to 2.6 – 3.1 considering the overall COP. The off-design tests were performed to define the performance of the heat pump for a wider temperature range. For these tests, the COP is between 2.9 and 6.1.

1. Introduction

As in the electricity sector, where the transition to more sustainable energy sources is taking place, various efforts are also being made in the heating and cooling sector to replace fossil fuels as primary energy sources. The role of the heating and cooling sector within the energy field is crucial, as it is responsible for over 50 % of global final energy consumption and one third of carbon emission [1]. In Europe, the 37 % of greenhouse gasses emissions comes from activities related to heating and cooling [2]. This sector is characterised by a slower decarbonization process, especially if compared to the electricity sector. However, innovative solutions are also being developed in the heating and cooling sector to achieve a transition to sustainable energy sources, and one of the main drivers in this direction is the district heating (DH) [3].

In conventional DH systems, heat is generated in thermal plants, such as cogeneration, waste-to-energy and renewable energy plants, in order to have the advantages of centralised heat generation, as well as wasted heat can be also recovered from industrial processes. Thus, plant efficiency in energy conversion is higher, whereas pollutant, CO2 and consumption of primary energy sources are lower, in line with the European strategy for efficient energy use for heating and cooling [4]. The generated heat is distributed at a temperature high enough to connect the DH network directly to the end users through heat exchangers.

Along with the development of conventional DH, which has evolved to the 4th generation [5], an innovative generation of DH is being developed to overcome some of the limitations of conventional networks such as the requirement of high-density population areas, heat losses that count for about 20 % in conventional DH [6], separation from the power grid, as well as the limitated application of renewable energy. In this kind of DH networks, called 5th generation DH (5GDH), heat is shared at a temperature close to ground temperature, typically between 10 – 30 °C [7]. The low temperature of the heat transfer fluid in the network allows the use of naturally available low-temperature heat sources, such as shallow geothermal energy [8], or the recovery of low-temperature heat, wasted from industrial processes [9]. The low temperature of the heat in the network does not enable a direct connection between the 5GDH network and the users via heat exchangers as in conventional DH, but requires the installation of decentralised heat pumps in the network substations. Decentralised heat pumps require electricity to generate heat or cold for the building, allowing the connection of the power grid to the DH network, as depicted in Figure 1. Figure 1 shows the differences between a conventional DH network and a 5GDH network, in which the low-temperature of the heat shared with the network and the electricity used by heat pumps facilitate the penetration of renewable energy. Most of the energy is used to supply the grid substations, which are strategic network hub. Another advantage is that the same infrastructure can be used to cover the heating and cooling demand of the buildings connected to the network, creating a district heating and cooling system.

On the other hand, in the 5GDH the complexity of the network substation increases due to the presence of a heat pump and the possibility of having a heating and cooling configuration that leads to considering a user as a heat consumer in the heating configuration, but also as a heat producer, in cooling configuration. Another complication is the decentralised pumping system: a pump is installed in each substation to pump the flow to the heat pump evaporator to generate the required heat to the condenser, in heating configuration. In addition, pumping costs are higher than in the conventional DH network, where the pumping system is centralised, due to the low temperature difference between the supply and return line of the network.

As this is a new and potentially revolutionary topic, some papers in literature refer to this kind of networks with techno-economic evaluations [6], others propose different kinds of low-temperature heat sources [9], a few focus on the use of heat pumps in 5GDH, explaining the main parameters that should be optimised [10] or even regarding the developments that heat pump manufacturers should implement to improve performance in 5GDH [7]. However, authors do not know any article in which the layout of the substation for 5GDH network, including the pumping system, as well as the related energy consumption and performance analysis for a real case are deeply described. Therefore, this paper aims to describe the 5GDH substation layout adopted for a prototype of a real case study and a pilot 5GDH network. The prototype substation serves a real building with a heating and cooling demand and it is equipped with a measurement system to acquire data to analyse the performance and energy consumption of the substation. In this way, it is possible to evaluate and verify the proper operation of the 5GDH substation and to collect data to evaluate the possibility of deploying this kind of network on a large scale.

The paper is structured as follows: after a brief introduction in Section 1, Section 2 describes the configuration of the prototype substation and how it will be tested. Section 3 shows the

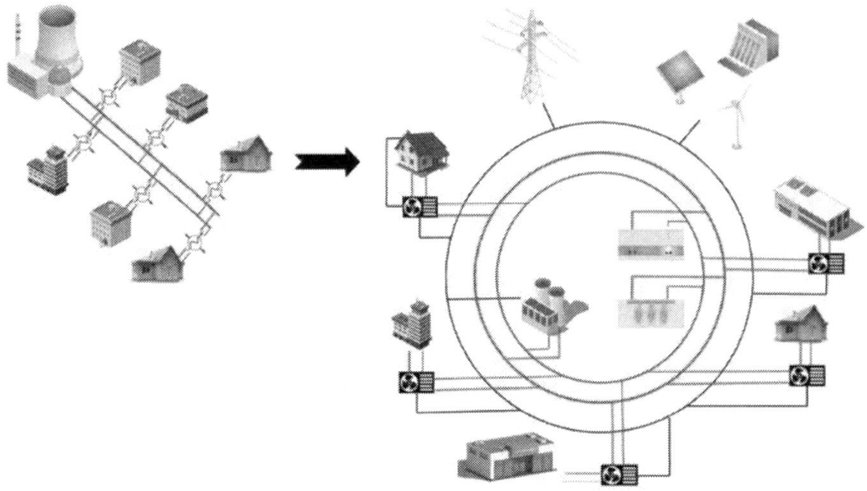

Figure 1. Difference between conventional DH (on the left) and 5GDH (on the right) network.

substation energy consumption results for the first heating season and an analysis of the performance of the heat pump installed in the substation. Section 4 presents the conclusions and a description of the future works on the prototype substation.

2. Materials and methods

The project concerning the development of a prototype substation for a 5GDH network has been carried out in cooperation between the University of Brescia and the company A2A Calore & Servizi. The city of Brescia has an extensive and capillary conventional DH network of more than 670 km of double pipes. This DH network is managed by A2A Calore & Servizi, which is interested in the development of new solutions to decarbonize the conventional DH networks and in the development of innovative DH such as the 5GDH.

The prototype has been installed inside the "Centrale Nord" in Brescia, a thermal plant of A2A Calore & Servizi that also houses chillers that feed a district cooling network and dissipate the heat produced through evaporative towers. Therefore, this site was selected due to the presence of low-temperature waste heat from the chillers and the presence of a building with heating and cooling demand, where the offices and control room of the plant are located. The prototype substation recovers the low-temperature heat wasted by the chillers through a 5GDH pilot network, supplying the heating and cooling demand of the building inside the Centrale Nord. Figure 2 shows the components of the prototype substation, which are:

- Gas-side reversible heat pump (HP)
- Bidirectional pumping system on 5GDH network (P-1; M3V-01; M3V-02)
- User pumping system (P2; P3)
- Inertial thermal energy storage (TES)
- Heat exchanger (HE).

A gas-side reversible heat pump has been selected to cover the heating and cooling demand of the building. The heat pump consists of two separated refrigeration circuits with a fixed-rpm scroll compressor for each one. The refrigerant R134a was selected as the best choice for this specific case study, as explained in [11] where the design phase of this project is well described.

The bidirectional pumping system is realised with a couple of motorised three-way valves and a pump (P-1), which is not equipped with an inverter, so that the flow rate in the pipes connecting the heat pump to the network is regulated by the FCV-01 motorised valve. In the heating configuration, the two three-way valves are opened in the main direction (ab-a) to pump water from the highest temperature line of the 5GDH network to the evaporator of the heat pump and then pumped it into the lowest temperature line of the grid. Conversely, in the cooling configuration, the three-way valves vary the set-up (ab-b) to pump water from the lowest temperature line and pump it, after passing through the chiller condenser, to the highest temperature line of the grid. In this way, the substation operates as a thermal prosumer, taking heat from the network in heating configuration and releasing heat to the grid in cooling configuration.
The inertial thermal energy storage has been adopted to decouple the heating and cooling demand from the heat pump and to enable an increase in heat pump performance.
Instead, the heat exchanger has been only installed in the substation prototype and will not be replicated when the prototype will be scaled up. The heat exchanger is being used to increase the temperature range on 5GDH side during the testing phase. Through the heat exchanger, the temperature of the 5GDH network, mostly between 18 and 20 °C during the winter, is increased up to 40 °C at the inlet of the heat pump evaporator, using heat from the thermo-hydraulic circuit itself. The reason for monitoring heat pump performance over a wider temperature range is to test the possibility of suppling the 5GDH network with different kind of low-temperature heat sources, available at different temperatures. The upper water inlet limit temperature has been set at 40 °C as a compatible limit with the 5GDH networks described in the literature and the sources available in Brescia.
The main role of the two motorised three-way valves is to allow the change of the flow direction in the network without reversing the direction of the flow in the evaporator or condenser. Moreover, the three-way valves (M3V-01, M3V02) can be used to test the heat pump at lower temperatures compared to those of the evaporative towers. Indeed, with an intermediate degree of opening, they can mix the flow entering the heat pump evaporator with the flow leaving the evaporator, resulting in temperature at the inlet of the evaporator in the range of 10-20 °C.
The nominal thermal capacity of the heat pump is 100 kW. The pipes of the 5GDH network have a diameter of 300 mm and are made by uninsulated carbon steel, as well as the pipes connecting the grid to the substation, which have a diameter of 80 mm. While the pipes of the intermediate circuit, between the heat pump condenser and the inertial storage tank, are made of insulated carbon steel, again with a diameter of 80 mm, as well as the pipes of the circuit connecting the inertial storage tank to the building's thermo-hydraulic circuit manifolds. The inertial storage tank, with a capacity of about 1000 l, is also insulated. The insulation is made with an extruded and vulcanized synthetic rubber (elastomer) foam suitable for insulating both heat and cold.

Figure 2 also shows the measurement system installed in the substation, consisting of flow meters (FT), temperature probes (TE) electricity and thermal energy counters (JQI). Specifically, four wires Pt500 temperature probes in the supply and return lines with magnetic-inductive (MID) flow meters and thermal energy counters are installed on the pipes of the network connection circuit (1), the intermediate circuit (2) and the building circuit (3). In addition, there are two electricity energy counters, one dedicated to the two heat pump compressors and the

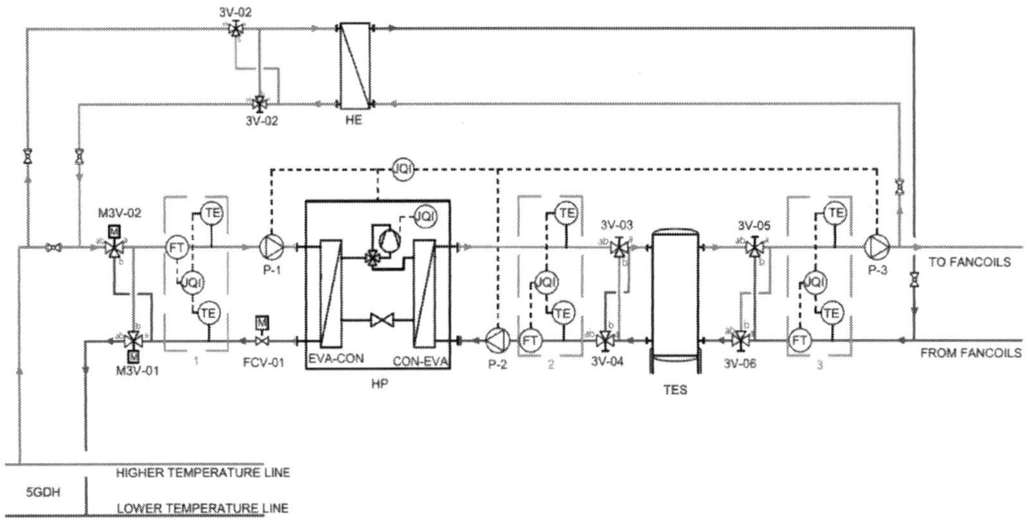

Figure 2. Process flow diagram of the prototype substation.

second one measuring the total amount of energy consumed by the substation prototype, including pumps and auxiliaries.

The COP is calculated as the ratio between the thermal power measured in the intermediate circuit (2) and the electricity absorbed by the heat pump compressors, while the global COP (GCOP) as the ratio between the heat supply to the building and the total electricity absorbed by the prototype substation. The former parameter does not take into account heat losses related to the building's thermo-hydraulic circuit and the energy consumption of pumps and other auxiliaries, but only defines the performance of the heat pump.

Two different kinds of tests were developed to analyse the performance of the 5DGH substation: tests under design conditions and under off-design conditions. The tests under design conditions are used to evaluate the energy consumption by the substation during the first heating season 2023-2024, taking into account the energy absorbed by the heat pump and the network pump (P1) under designed conditions, i.e. using the waste heat available from the centralised chillers at evaporator and condenser temperatures is in line with the temperatures required by the building's fancoil system. The tests under off-design condition were performed to evaluate the performance of the heat pump at different temperatures at the heat pump evaporator and condenser through the heat exchanger, the three-way valves and the variation of the condenser setpoint temperature. In this way, different kinds of waste heat (evaporator) and different kinds of thermo-hydraulic system (condenser) can be simulated. The data acquisition system has been developed to historise measurements with a sampling rate of one minute.

3. Results and Discussion

For the design tests, energy consumption has been evaluated during 24 hours for 64 days from December 21st to March 31st, and off-design tests have not been performed on these days.

The results of these tests for different temperatures at the heat pump condenser in the range between 49 and 60 °C are shown in Table 1, where the average values and standard deviation of the inlet evaporator temperature *(T_{eva})*, the outlet condenser temperature *(T_{con})*, the outdoor room temperature *(T_{out})*, the heat output to the condenser *($\dot{Q}_{con}$)*, the power input to the heat pump compressors *($\dot{W}_{hp}$)*, the power input to the network pump *($\dot{W}_{np}$)* and the *COP* are represented. Each row in Table 1 relates to a single day and has been obtained by considering only those instants of time when the heat pump was in operation during 24 hours, for a total of six days chosen as indicative among the 64 of the heating season due to the different temperatures at the condenser outlet and outside. In Table 2 are reported the 24-hour consumption for the six days of Table 1, in the same order. Specifically, Table 2 shows: the heat supplied to the building *(Q_{bui})*, the heat supplied to the heat pump condenser *(Q_{con})*, the electricity absorbed by the heat pump *(W_{hp})*, the electricity absorbed by the network pump (W_{np}), the total amount of electricity absorbed by the substation *(W_{tot})* and the overall *COP (GCOP)*.

Table 1. Average values and standard deviation of substation performance in steady-state conditions.

T_{eva} (°C)	T_{con} (°C)	T_{out} (°C)	$\dot{Q}_{con}$ (kW)	$\dot{W}_{hp}$ (kW)	$\dot{W}_{np}$(kW)	*COP*
18.7 ± 0.4	49.0 ± 0.3	8.7 ± 3.5	52.4 ± 2.5	12.5 ± 0.6	2.2 ± 0.6	4.2 ± 0.1
18.8 ± 0.4	51.0 ± 0.3	8.3 ± 1.2	52.1 ± 1.0	13.0 ± 0.1	2.3 ± 0.4	4.0 ± 0.1
18.6 ± 0.2	53.6 ± 2.4	8.3 ± 2.1	51.3 ± 4.2	13.5 ± 1.3	2.4 ± 0.4	3.8 ± 0.2
18.6 ± 0.1	55.7 ± 0.5	7.6 ± 0.3	51.1 ± 3.4	14.1 ± 0.9	2.5 ± 0.6	3.6 ± 0.1
18.4 ± 0.2	57.7 ± 0.5	7.6 ± 0.5	50.8 ± 1.0	14.8 ± 0.2	2.1 ± 0.3	3.4 ± 0.1
18.0 ± 0.3	59.4 ± 0.6	2.5 ± 3.6	50.5 ± 2.1	15.3 ± 0.6	3.0 ± 0.2	3.3 ± 0.1

Table 2. Energy consumption of the prototype substation for the 6 days of Table 1.

Day	Q_{bui} (kWh)	Q_{con} (kWh)	W_{hp} (kWh)	W_{np}(kWh)	W_{tot}(kWh)	*GCOP*
February 15th	450.6	474.9	109.9	23.9	145.7	3.1
March 28th	367.3	396.5	96.2	19.8	128.0	2.9
February 24th	454.3	472.4	121.7	24.3	158.0	2.9
December 31th	633.2	664.5	180.5	37.4	229.9	2.8
January 7th	619.2	648.0	183.8	38.0	233.9	2.7
January 13th	850.0	881.0	263.7	55.8	331.5	2.6

Under design operating conditions, only one compressor of the two heat pump compressors operates, because the average building demand is between 20-40 kW and the heat pump condenser can supply more than 50 kW, depending on the temperature setting at the condenser outlet. The heat supply to the condenser cannot be modulate due to the use of compressors without inverters for the heat pump, so the heat output of the heat pump can only be modulated by switching the compressors on or off and using the inertial storage.

Each bar in Figure 3 shows the sum of the daily amount of electrical energy absorbed by the heat pump (in blue), by the network pump (in orange) and by the auxiliaries (in yellow) to supply

IOP Publishing
doi:10.1088/1742-6596/2940/1/012023

heat to the building, represented on the x-axis, for all 64 days of the heating season. The continue lines represent the outdoor temperature and the dashed lines the waste-heat temperature. The daily consumptions are divided into four plots with different temperature ranges at the outlet of the heat pump condenser (Tc). From Figure 3, it can be seen as the electricity absorbed by the heat pump can vary between 72.6 and 80.7 %, while the network pump absorbs ca. 15.8 % of the substation's total electricity consumption. Auxiliaries account for 7.4 %.

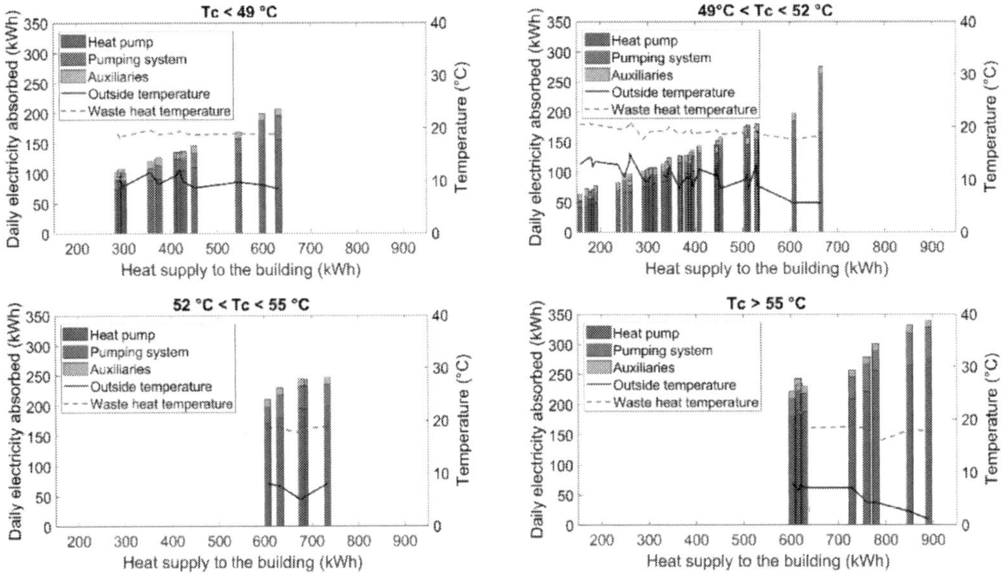

Figure 3. Energy consumption of the heat pump (blue), network pump (orange) and auxiliaries (yellow), with outside and waste heat temperatures. The four charts differ for the temperatures at the heat pump condenser.

The off-design tests have been carried out with temperatures at the evaporator inlet between 12 and 30 °C with increments of 2 °C, with the intention of increasing this range up to 40 °C in the coming months. On the other hand, the temperature range at the condenser is from 40 to 60 °C in steps of 5 °C.

These tests have been performed by keeping the heat pump under the test conditions for three hours. The data acquired during the three hours were filtered to retain only those values in which the heat pump was operating under steady-state conditions. The filter consists of disregarding *COP* values that deviate more than 3 % from the five minutes moving average, in order to disregard transient values due to the compressors switching on or off.

The results are presented through Figure 4, which describes the performance of the heat pump using the *COP* as key parameter index. The intention is to represent the *COP* as a function of the condenser temperature and the difference between condenser and evaporator temperature (ΔT_{lift}). Figure 4 shows the *COP* for different temperatures at the condenser outlet in different shades of colours. The COP value increases as the ΔT_{lift} decreases with a trend described by the centre of the crosses corresponding to the average *COP* value of the individual test. The vertical lines represent the standard deviation of the *COP*, the horizontal lines the standard deviation of ΔT_{lift}. The measurements that deviate from the main trend and the greatest value of the standard deviation of these are found when both heat pump compressors are running. When the two compressors are in operation, the increase in heat absorbed by the evaporator is achieved by decreasing the outlet temperature to the evaporator, while maintaining the same temperature at the inlet. In this way, the evaporation temperature decreases. The *COP* also increases by decreasing the condenser temperature, indeed, the measurement with the same ΔT_{lift} resulting in higher *COP* value with a lower condenser temperature.

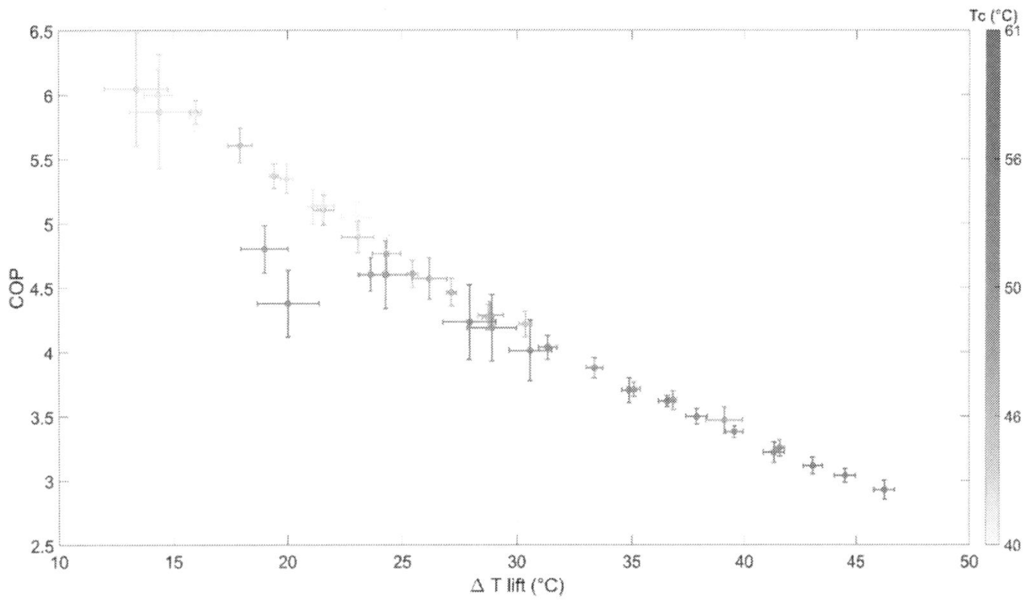

Figure 4. Performance of the heat pump for different temperatures at the condenser and evaporator.

The intention to study different temperature at the heat pump evaporator is due not only to the possibility of having different kinds of low-temperature heat sources, but also to the fact that temperatures within a 5GDH network may change over time depending on the heating or cooling demand of the end users connected to the grid. Therefore, it is useful to know the behaviour of the heat pump under different operating conditions. Lindhe et al. [7] explain that the heat pump for 5GDH network substation should have high performance for both heating and cooling configuration and for different operating points, not only under design conditions.

4. Conclusions

The aim of this work is to test a prototype substation for the 5GDH networks in a real case study. The substation covers the heating and cooling demand of a real building using low-temperature waste heat. This article describes the components, the substation layout and the 5GDH network that enables the prototype to recover the waste heat. The results show the energy consumption of the substation in heating configuration during the 2023-2024 heating season and the performance of the heat pump under design conditions. Moreover, the heat pump was tested under off-design conditions, which are obtained when the heat pump operates in a wider temperature range at the evaporator and condenser than in design condition.

The average value of the COP for the design tests is between the 3.3 and 4.2 maintaining 18 °C at the evaporator inlet, while the condenser temperature is between 49 and 59 °C. The GCOP has values between 2.6 and 3.1 considering the heat supply to the building and the overall energy consumption of the heat pump, pumps and auxiliaries.

The off-design tests show an average COP of 2.9 to 6.1 for different values of temperature difference between condenser and evaporator.

Future tests will be developed to analyse the performance of the substation in the cooling configuration in the summer of 2024. These results are useful for the techno-economic evaluation of the development of 5GDH networks. The data obtained from this work can therefore be used as input parameters in models studying the feasibility of this kind of networks.

References

[1] Eisentraut, Anselm, and Adam Brown. Heating Without Global Warming: Market Development and Policy Considerations for Renewable Heat: Featured Insight. OECD/IEA, 2014.

[2] Sadeghi, H., Jalali, R., & Singh, R. M. (2024). A review of borehole thermal energy storage and its integration into district heating systems. Renewable and Sustainable Energy Reviews, 192, 114236. https://doi.org/10.1016/J.RSER.2023.114236

[3] Lund, H.; Moller, B.; Mathiesen, B.; Dyrelund, A. The role of district heating in future renewable energy systems. Energy 2010, 35, 1381–1390.

[4] European Commission. An EU strategy on heating and cooling. J Chem Inf Model 2016;53(9):1689-99.https://energy.ec.europa.eu/topics/energy-efficiency/heating-and-cooling_en

[5] Lund, H., Werner, S., Wiltshire, R., Svendsen, S., Thorsen, J. E., Hvelplund, F., & Mathiesen, B. V. (2014). 4th Generation District Heating (4GDH). Integrating smart thermal grids into future sustainable energy systems. In Energy (Vol. 68, pp. 1–11). Elsevier Ltd. https://doi.org/10.1016/j.energy.2014.02.089

[6] Saini, P., Huang, P., fiedler, F., Volkova, A., & Zhang, X. (2023). Techno-economic analysis of a 5th generation district heating system using thermo-hydraulic model: A multi-objective analysis for a case study in heating dominated climate. Energy and Buildings, 296, 113347. https://doi.org/10.1016/J.ENBUILD.2023.113347

[7] Lindhe, J., Larsson, M., Willis, M., Tiljander, P., & Johansson, D. (2024). Challenges and potentials of using a local heat pump in a 5 GDHC solution: Results from field and laboratory evaluations. Energy, 289, 129807. https://doi.org/10.1016/J.ENERGY.2023.129807

[8] García-Céspedes, J., Herms, I., Arnó, G., & de Felipe, J. J. (2023). Fifth-Generation District Heating and Cooling Networks Based on Shallow Geothermal Energy: A review and Possible Solutions for Mediterranean Europe. In Energies (Vol. 16, Issue 1). MDPI. https://doi.org/10.3390/en16010147

[9] Buffa, S., Cozzini, M., D'Antoni, M., Baratieri, M., & Fedrizzi, R. (2019). 5th generation district heating and cooling systems: A review of existing cases in Europe. In Renewable and Sustainable Energy Reviews (Vol. 104, pp. 504–522). Elsevier Ltd. https://doi.org/10.1016/j.rser.2018.12.059

[10] Reiners, T., Gross, M., Altieri, L., Wagner, H. J., & Bertsch, V. (2021). Heat pump efficiency in fifth generation ultra-low temperature district heating networks using a wastewater heat source. Energy, 236. https://doi.org/10.1016/j.energy.2021.121318.

[11] Martinazzoli, G.; Pasinelli, D.; Lezzi, A.M.; Pilotelli, M. Design of a 5th Generation District Heating Substation Prototype for a Real Case Study. Sustainability 2023, 15, 2972. https://doi.org/10.3390/su15042972

Acknowledgments

The authors thank the other people who are working on the development of the substation prototype.

Funding

This research was funded by MUR, fund PON "Ricerca e Innovazione" 2014–2020, Azione IV.4 "Dottorati e contratti di ricerca su tematiche dell'innovazione" and Azione IV.5 "Dottorati su tematiche green".

Journal of Physics: Conference Series **2940** (2025) 012024 doi:10.1088/1742-6596/2940/1/012024

Urban Building Energy Modeling for Self-Sufficiency Scenarios with solar technologies in Turin, Italy

Yasemin Usta[1], Ahad Montazeri[1] and Guglielmina Mutani[1*]

[1] Department of Energy, Politecnico di Torino, 10129 Torino, IT

*E-mail: guglielmina.mutani@polito.it

Abstract. Understanding the complexities of building energy dynamics is crucial as it offers a pathway to reducing energy consumption and associated emissions. By using Urban Building Energy Modeling (UBEM), this work aims to drive sustainable urban development, by integrating renewable energy sources in district heating networks. This study focuses on UBEM as an effective scale for analyzing buildings within their specific characteristics that influence their consumptions and limit their production from renewables. The analysis focuses on utilizing solar thermal collectors (STC) for space heating consumption in residential buildings connected to the district heating network. The aim is to allow the integration of renewables into district heating network by testing lower supply temperatures. The validated model is used to analyze the potential self-sufficiency achieved through STC production for all residential buildings in three analyzed zones. The results indicate the potential of achieving self-sufficiency using STCs with different inlet and outlet temperatures; a self-sufficiency of 24% with 90°C and 21% with 105°C supply temperatures in March while for October it could only reach 9% and 7% respectively. The results of this work are crucial for optimizing solar energy production and to explore a distributed production system for individual and collective self-consumption strategies, offering insights into sustainable energy scenarios and policies.

1. Introduction

District heating networks (DHN) play a crucial role in urban energy systems, providing efficient heating and hot water to densely populated areas through centralized heating plants. These networks have predominantly relied on conventional heating plants, mainly fossil fuel based. However, with the growing need to mitigate GHG emissions and transition towards a sustainable energy future, there is a pressing need to integrate renewable energy sources into energy infrastructures. The Solar Heat Worldwide report analyze the global solar thermal market, revealing its steady growth and significant potential [1]. Notably, Italy experienced remarkable expansion, indicating a promising step for solar thermal energy [2]. While the dominance of fossil fuels persists in meeting global heating needs, there is growing recognition of the potential of solar thermal energy, particularly in urban district heating systems, to contribute to sustainability goals. The urgent need for emissions reduction in district heating, highlighted by significant CO_2 emissions particularly in regions like China, underscores the necessity to integrate renewables, such as solar thermal energy, into existing systems [3]. Recent studies have shown promising results regarding the integration of solar thermal collectors into district heating networks,

Journal of Physics: Conference Series **2940** (2025) 012024 doi:10.1088/1742-6596/2940/1/012024

indicating substantial contributions to heating demands [4]. Another comprehensive analysis on solar technologies concluded that the optimal choice depends on the specific requirements of the application. If maximizing electricity output is the primary goal, the PV/T (photovoltaic/thermal) system is a better option, especially for rural house applications, while solar thermal systems are preferable when the thermal energy is more critical especially in cold climates [5]. However, challenges persist, including cost, technical barriers, and policy uncertainties, hindering widespread integration [6].

To effectively implement the targets of the European Union, it is essential to comprehensively analyze current technologies and assess the potential for integrating renewable energy sources into DHN. This necessitates not only evaluating the technical feasibility but also considering practical aspects such as available roof areas for solar installations. Following that, this work aims to provide a flexible methodology that uses a place-based approach for further advancement in sustainable solutions to integrate more renewable energy sources in energy infrastructure.

1.1 Objective of the work

The objective of this study is to evaluate the feasibility of integrating solar thermal collectors (STC) into DHNs with a focus on utilizing lower supply temperatures to support the integration of renewable energy sources (RES) into DHN. The aim is to assess whether this integration can effectively meet the space heating demand of residential buildings. By analyzing the energetic indexes of the buildings connected to the DHN with installed STC, the study seeks to determine the extent to which STC can contribute to reducing reliance on conventional heating sources within the DHN.

Moreover, the analysis will highlight both the contribution of using solar technologies and the effect of building and urban characteristics in the final energy consumption and production. This will be realized by using Urban Building Energy Modeling (UBEM) [7] for:

- preparing the geo-database with building geometry, typological characteristics and urban parameters to be used in a bottom-up process-driven model.
- a detailed urban analysis for the received solar irradiation on the rooftops, excluding technical and regulatory constraints, to be used for calculating the energy production with different solar collector technologies, i.e., flat plate and vacuum tube panels, with different inlet and outlet temperatures.

The methodology presented in this work can be used for different urban contexts, since this place-based and bottom-up approach allows to evaluate the behaviour of each single building according to its characteristics, but also the spatial distribution of energy consumption and production in different points of the DHN. Furthermore, this energy modeling is hourly and therefore there will be more favourable and less favourable periods during the year but overall, an annual share of renewables can be evaluated. The proposed space-time modeling allows to develop different scenarios and identify the most suitable solution for the specific urban context and climate. In this way there will be areas of the city with higher shares of renewable energy, capable of compensating other areas which, due to various constraints, are unable to reach high values. In this way, a high share of renewables could be achieved globally with the most suitable solutions.

2. Materials and methods

The general workflow of the presented methodology is provided in Figure 1. Urban building energy modeling (UBEM) is used as the bottom-up and place-based tool for calculating buildings' hourly energy consumption and production. The input data are collected and pre-processed in QGIS (Quantum Geographic Information System, version 3.34.6) to create a complete geo-database for both the hourly process-driven model and for the simulation of solar irradiation on building rooftops and energy production with roof-integrated solar collectors.

These input data include building scale data (e.g., building footprint area and height), urban scale data (e.g., digital surface model DSM), and population data (e.g., number of inhabitants).

Finally, the energetic indexes are calculated for the residential buildings connected to the existing DHN using hourly consumption and production modeling.

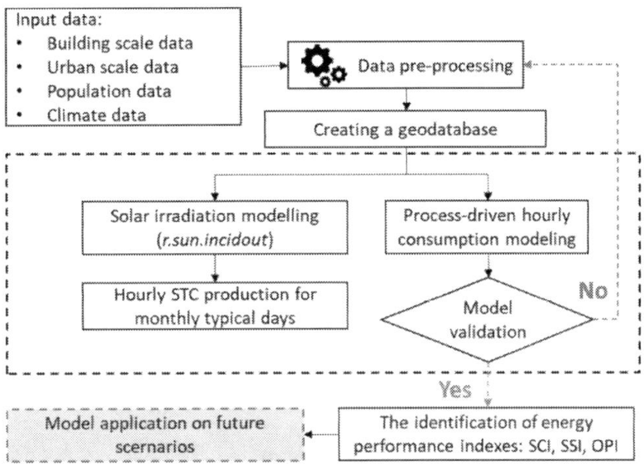

Figure 1. The general workflow of the presented methodology

2.1 Bottom-up engineering model for space heating consumption calculation

The energy consumption of residential buildings for space-heating is calculated with a bottom-up process-driven model using three energy balance equations of the three thermodynamic systems (TSs) for each building that exchange heat with the external environment. The equations of the three TSs are represented with [7 for more details]: the opaque envelope (E), the transparent envelope (G), and the internal part of a building (B). Equation 1 provides the hourly heat balance of the thermodynamic systems during the heating season:

$$C_{TS}\frac{dT_{TS}}{dt} = \emptyset_{sol} + \emptyset_I + \emptyset_H - (\emptyset_T + \emptyset_V) \quad (1)$$

where: C is heat capacity (JK^{-1}), T is the temperature (K), t is the time (s), $\emptyset_{sol}$ is the heat flow rate from solar gains (W), $\emptyset_I$ is the heat flow rate from internal gains (W), $\emptyset_H$ is the heat flow rate released from the heating system (W), $\emptyset_T$ is the heat flow rate lost by transmission (W), $\emptyset_V$ is the heat flow rate lost by ventilation (W).

This modeling uses, for each building within a city, its geometric data and typological characteristics according to the Standards: ISO 52016-1:2017, ISO 52017-1:2017, ISO 13786:2018, UNI/TR 11552:2014 and UNI/TS 11300. In this work, the hourly consumptions were calculated using climate data related to the heating season 2022-2023 (starting from

Journal of Physics: Conference Series **2940** (2025) 012024 doi:10.1088/1742-6596/2940/1/012024

November 3rd until April 7th). The results of the hourly consumption modeling were calibrated with the available measured data for a sample of 110 residential buildings from various periods of construction and building typologies (e.g., little and big condominiums).

2.2 Production from solar technologies

The energy production from solar collectors Qu (Wh) was calculated with equations 2 and 3 using the hourly solar irradiation simulated in QGIS with *r.sun.incidout* tool for monthly typical days:

$$Q_u = A_c \times H_{sol} \times \eta_{coll} \tag{2}$$

$$\eta_{coll} = \eta_0 - a_1 \times \frac{\Delta T_m}{I} - a_2 \times I \times \left(\frac{\Delta T_m}{I}\right)^2 \tag{3}$$

where: A_c is the active surface of solar collectors (m^2), H_{sol} is the solar irradiation (Wh/m^2), η_{coll} is the efficiency of the collectors, η_0 is the optical efficiency, a_1 and a_2 are the heat loss coefficients, ΔT_m is the difference between water inlet and outlet temperatures, and I is the solar irradiance (W/m^2); the monthly typical days for solar analyses were selected based on the literature [8].

In this work, only roof-integrated solar collectors were considered.

To describe the 3D urban environment a Digital Surface Model (DSM) of 1 meter precision was used. The DSM allows a detailed spatial analysis of the received solar irradiation on the buildings' roofs with different orientations and slopes, excluding the technical, historical, cultural and regulatory constraints. For this analysis the roofs used for the installation of STC were the ones with a good solar exposition (East -90° – West +90°) excluding obstacles and constraints due to cultural heritage (Italian Decree 42/2004) and historical restrictions from the Municipal Master Plan.

Solar irradiation results were used to calculate the production from STC with different solar collectors' technologies and water inlet and outlet temperatures (in Table 1).

Table 1. Technical characteristics of STC and the used water supply temperatures

		η_0		a_1 (W/m^2/K)		a_2 (W/m^2/K^2)
Beretta (vacuum tube)	Technical characteristics	0.70		1.15		0.011
Vaillant (flat plate)		0.855		2.41		0.039
Inlet temperature (°C)	Operating conditions of STC	60	50		60	15.6
Outlet temperature (°C)		105	90		70	80

Considering having to reduce the operating temperatures of the DHN, two preliminary scenarios were hypothesized for the integration of solar collectors for space-heating, one at high and one at low temperatures, with inlet and outlet temperatures of: 60-105 °C and 50-90 °C respectively. The third scenario with 60-70 °C considers also the implementation of a heat pump. The last scenario can be used to make comparisons considering the buildings that are not connected to the DHN with a supply temperature of 80 °C and inlet temperature of 15.6 °C (the average annual outdoor air temperature in Turin).

3. The case study

Turin has the largest DHN in Italy with 726 km of double pipeline connecting 73.2 million m^3 of buildings and about 650,000 inhabitants. Almost 90% of the energy fed into the grid is produced through cogeneration systems, using heat accumulators installed in the power plants and along the network [9]. The design to reduce operating temperatures determines a challenge for decarbonization with the use of STC systems distributed along the network. The aim is to test if lower supply temperatures, achievable through STC, can effectively met the energy demand for space heating in residential buildings.

Turin is in a temperate climate with cold winters and hot humid summers and, for this work, the heating season of 2022-23 was analyzed with an average outdoor air temperature of 15.6 °C. For energy consumption modeling, the hourly data of 110 buildings were analyzed considering an inside constant air temperature of 19 °C during the heating season (according to Italian Decree 383 of 6/10/2022). Besides, two more hypotheses were added in energy modeling: the nighttime interruption of the heating system from 10 pm to 5 am and two more interruptions mainly (for about the 80% of the buildings) during the day at about 10 am and 2 pm.

The hourly engineering model was applied to the district heating area and the results will be shown for three different zones in Turin. Figure 2 provides the annual solar irradiation map of Turin with the selected zones and the general characteristics of the buildings. Each zone has a different period of construction, but they are all built before the first energy saving law (Law 373/1976) so the thermal transmittance values (*U*) are high. The available "roof area-to-volume ratio" is very low in San Salvario which can be explained by its dense urban morphology (BD) with large blocks of buildings and bigger condominiums compared to the other zones.

	Cenisia	Crocetta	San Salvario
N. connected buildings	24	33	8
Main period of construction	1961-70	1919-45	before 1918
Avg. U_{wall} (W/m²/K)	1.11	1.26	1.27
Avg. U_{window} (W/m²/K)	4.03	3.88	3.74
S/V(m²/m³)	0.36	0.35	0.30
BD (m³/m²)	4.99	4.70	7.72
Available roof area (m²)	2,454	4,604	1,016
Available roof area/ building volume (m⁻¹)	0.03	0.02	0.003

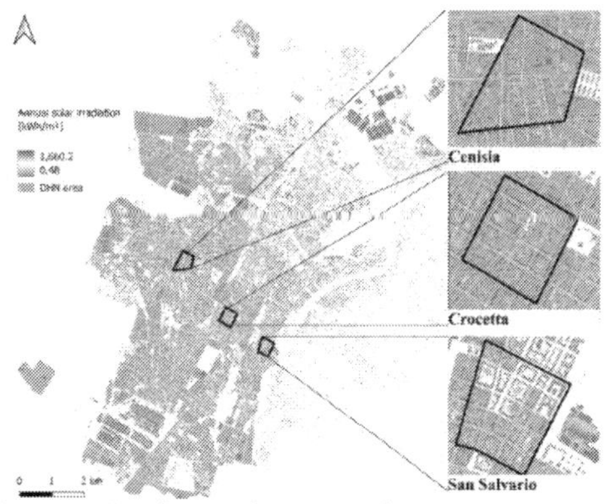

Figure 2. Solar irradiation map of Turin with the selected zones for the application of energy modelling.

4. Results and discussion

This section presents the results of:

- a statistical analysis on the sample of 110 residential buildings and blocks of buildings for validating the use of process-driven modeling in Turin,
- the hourly profiles of energy consumption modeling for space heating,
- the hourly profiles of energy production with STC using solar irradiation simulated in QGIS,
- and finally, the self-sufficiency scenarios for the selected three zones in Turin.

The energy production from STCs is analyzed comparing the flat plate and vacuum tube panels with the following inlet-outlet temperatures: 60-105 °C, 50-90 °C, 60-70 °C and 15.6-80 °C.

4.1 Statistical analysis of the 110 buildings for energy modeling

The methodology is tested on a sample of 110 buildings with different characteristics: surface-to-volume ratio (S/V), building density (BD) and building volume (V). Figure 3 illustrates the statistical distribution of these buildings in comparison to the total number of connected buildings and those within the DHN area. This comparison reveals that the sample of 110 buildings for training and testing the model fall within the average range for S/V, BD and building' volume. This analysis is crucial for verifying the model's applicability in contexts with different buildings and urban characteristics.

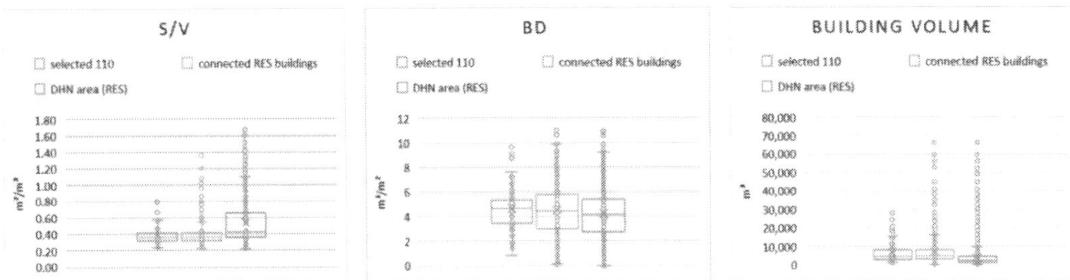

(a) (b) (c)

Figure 3. The distribution of the sample of buildings among other datasets considering: (a) surface-to-volume ratio S/V; (b) buildings density BD; (c) building volume V.

4.2 Energy consumption modeling

The results of the hourly consumption using the energy balance modeling is illustrated in Figure 4 for an old and a new building (with similar volumes) in a cold day in December and a warm day in March during the heating season 2022-23. The two sample buildings show that the modeling relative error (RE) is lower for old buildings (Figure 4, a) and cold climate, which is concluded in Table 2 for all the analyzed 110 buildings. Table 2 provides the monthly mean absolute percentage errors for buildings with different periods of construction, focusing on the months during the heating season 2022-2023 (from November 3rd to April 7th).

The errors about the period of construction 2006-2012 are in grey because we have only one building built in this period, thus this error is not meaningful to evaluate. The overall results show that the modeling performs better for older buildings (before 1960), newer buildings (after 2001) and in the colder months of December and January. Considering the period of construction, the variation in the error between buildings can be attributed to the reliability of thermal transmittance data, which are generally more predictable for older and newer buildings.

The maximum error excluding the single building of 2006-2012 is 24.58% in November, while considering only December, January and February the maximum error reached is 15.14%.

The higher errors occur in buildings built from 1960 to 2000, which is certainly related to greater variability in the use of energy and building technologies, and in warmer months, in which the model is underestimating the consumption. This is related to the fact that the modeling relies on the climatic data and building physics for predicting the energy consumption, while the user behaviour cannot be reflected precisely. For this reason, the set point temperature in November, March and April is increased by 2 °C, with respect to 19 °C that is defined in the Italian Decree 383 of 6/10/2022. With this increase, the modeling showed a better correlation with the measured data which can be explained by the user behaviour that is not strictly following the limits in air temperature according to regulations.

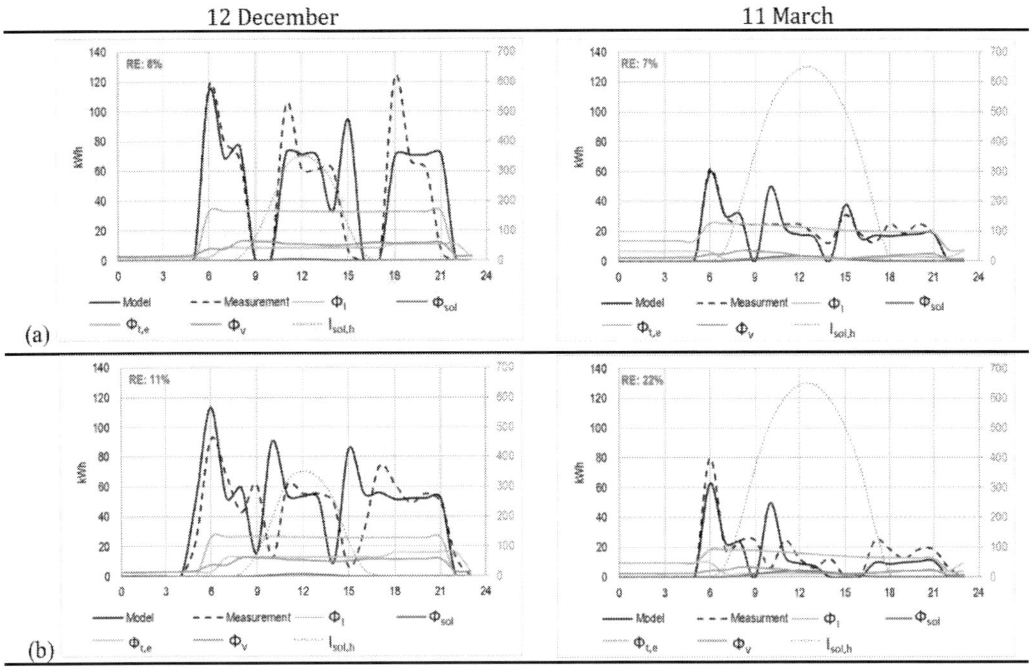

Figure 4. Energy consumption profiles in December and March for two buildings with different periods of construction: (a) 1946-1960 (V_{gross} = 5691 m^3) and (b) 1991-2000 (V_{gross} = 5565 m^3)

Table 2. The mean absolute percentage error of monthly consumptions for residential buildings

Period of construction	Nov	Dec	Jan	Feb	Mar	Apr
Before 1918	9.57%	11.34%	6.51%	8.51%	11.57%	16.78%
1919-1945	16.68%	14.20%	4.20%	9.60%	12.66%	22.38%
1946-1960	24.58%	13.56%	9.50%	11.12%	14.40%	22.17%
1961-1970	16.31%	11.00%	7.48%	8.74%	11.67%	16.61%
1971-180	22.28%	12.16%	9.95%	11.58%	14.00%	21.02%
1981-1990	15.40%	11.56%	9.39%	12.25%	12.98%	19.29%
1991-2000	11.42%	6.32%	7.12%	15.14%	12.86%	21.81%
2001-2005	11.55%	7.56%	4.11%	13.56%	10.86%	18.64%
2006-2012	17.40%	18.00%	26.00%	34.00%	33.90%	34.64%
2012-2015	12.42%	7.50%	6.00%	12.00%	18.63%	23.97%

4.3 Solar irradiation results and the production from STC

The simulated solar irradiation with QGIS is compared with the hourly irradiation data obtained from PVGIS. The data about roof azimuth and slope were calculated with QGIS for the 110 sample of buildings and the DH area and used in PVGIS for the comparison.

Figure 5 illustrates the solar irradiance on different roof orientations for a block of building; it is possible to observe the higher similarities during summer months when solar height is higher, compared to winter months. This discrepancy can be attributed to the fact that solar irradiance in QGIS is simulated with 1 meter precision considering all the surrounding built context, while the PVGIS results only consider the shadows from the topography of the territory (with a precision of 90 meters).

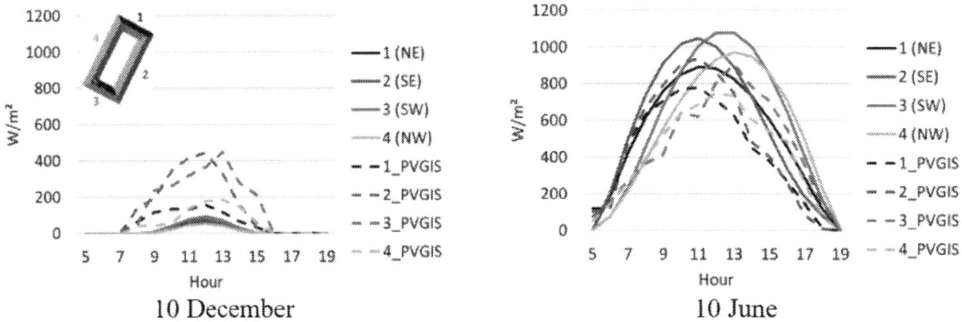

Figure 5. Hourly solar irradiance for typical days in winter (10 December) and summer (10 June)

The calculation of the energy production from solar thermal collectors considers two different solar technologies: flat plate and vacuum tube collectors. Figure 6 illustrates the hourly panel efficiency using different inlet and outlet temperatures, with both panels: the flat plate (Vaillant VFK) in blue and the vacuum tube (Beretta) in orange. It can be observed that, during the heating season, the vacuum tube works better mainly because of the lower heat loss correction values a_1 and a_2. Moreover, the efficiency of the panels decreases with higher operating temperatures; this gives important insights on the effect of decreasing the supply temperatures, allowing a higher integration of STC into DHN.

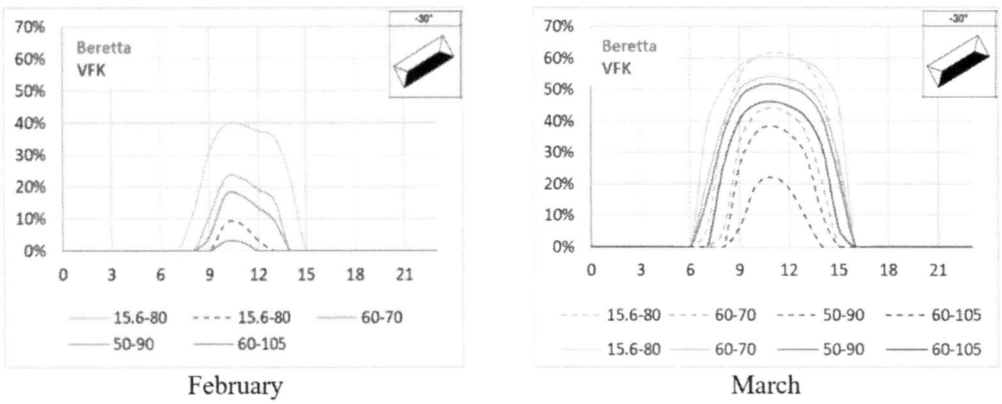

Figure 6. Hourly efficiency for South-East roofs with different inlet/outlet temperatures (°C)

Figure 7 provides hourly consumption and production data for the representative buildings illustrated previously in Figure 4. The results of STC production are provided using 50-90 °C inlet-outlet temperatures with the vacuum tube panel Beretta in March. This example shows the modeling performance by comparing the calculated and measured hourly consumption and providing the relative error (RE), the self-consumption SC, and self-sufficiency SS. During the morning hours with the peak demand, all the energy produced by STC is consumed, resulting in SC 100%. However, the SS reaches 100% when there is an over-production which results in decreasing the SC. The daily self-sufficiency index (SSI_d), provided in green in the bottom right, reaches an average of 37% for the illustrated two buildings.

4.4 Hourly SCI and SSI analysis

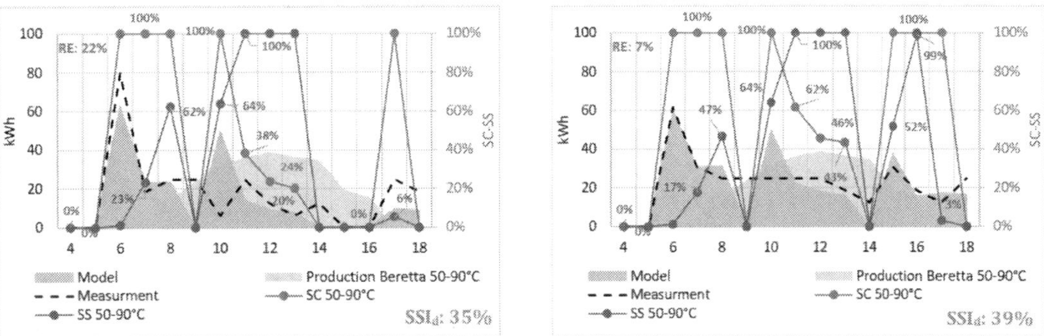

1991-2000 (Gross volume: 5565 m³) 1945-1960 (Gross volume: 5691 m³)

Figure 7. Hourly consumption and production modeling in March with the energetic indexes.

Table 3. The monthly SCI and SSI for the three analyzed zones

		Cenisia				Crocetta				San Salvario			
		Oct	Nov	Feb	Mar	Oct	Nov	Feb	Mar	Oct	Nov	Feb	Mar
60 - 105 °C	production [kWh]	4,313	0	1,238	92,490	11,575	0	11,726	211,059	2,530	0	2,423	46,747
	SCI_h	32%	0%	66%	33%	37%	0%	74%	39%	53%	0%	79%	55%
	SSI_h	5%	0%	0%	17%	7%	0%	1.0%	20%	6%	0%	1%	21%
	SSI_d	16%	0%	0%	51%	20%	0%	1.4%	51%	11%	0%	1%	38%
50 - 90 °C	production [kWh]	5,702	686	8,117	111,581	14,634	5,277	31,388	253,037	3,203	1,058	6,806	56,092
	SCI_h	31%	61%	70%	31%	36%	78%	74%	37%	52%	88%	79%	53%
	SSI_h	6%	0.1%	2%	19%	9%	0.6%	3%	22%	7%	0.5%	2.4%	24%
	SSI_d	21%	0.2%	2%	62%	25%	0.8%	4%	61%	13%	0.6%	3%	46%
15.6 - 80 °C	production [kWh]	8,471	10,778	28,065	144,478	20,602	35,301	72,285	318,240	4,540	7,800	15,924	70,520
	SCI_h	30%	55%	68%	27%	35%	72%	72%	33%	49%	86%	78%	50%
	SSI_h	9%	2%	5%	22%	12%	4%	6%	25%	9%	4%	5%	29%
	SSI_d	31%	4%	7%	80%	35%	6%	9%	77%	19%	4%	7%	57%
roof area [m²]		2,454				4,604				1,016			

The results of self-sufficiency index (SSI) and self-consumption index (SCI) for the analysed three zones in Turin (described in Figure 2) are represented in Table 3. The effect of using lower supply temperatures results in higher self-sufficiency values, especially in March, while in November and February the self-sufficiency is lower due to lower solar irradiation. The daily self-sufficiency index (SSI_d) indicates the presence of a storage system (with the capacity of daily storage), and in Table 3 it is possible to analyze the effect of having a storage system by comparing the results with the hourly self-sufficiency index (SSI_h). Thus, Table 3 summarized the results of using different operating temperatures and the presence of a storage system. It is also possible to analyze the total monthly production and compared to the available roof area, excluding any local constraints. However, December and January are not reported as the energy production by STC is zero.

5. Conclusions

This study examines the feasibility of integrating solar thermal collectors (STC) in district heating networks (DHNs) by exploring the utilization of lower supply temperatures to facilitate RES integration. The analysis highlights the importance of understanding building energy dynamics employing Urban Building Energy Modeling (UBEM) as a tool to consider buildings' specific characteristics that influence the energy consumption and renewable energy production. The model validation with measured data revealed that this approach works better with older and newer buildings in the cold months, with a maximum error of 15.14 %. The results of the analyzed three zones in Turin showed a potential self-sufficiency SSI level of up to 24% with 50-90 °C and 21% with 60-105 °C inlet-outlet temperatures in March without a storage system. However, self-sufficiency levels decreased to 9% and 7% in October, highlighting the seasonal variability and the importance of the operating temperatures.
The analysis revealed the impact of lower operating temperatures on panel efficiency, emphasizing the need for temperature optimization together with an appropriate selection of solar technology to enhance renewable energy utilization. The scale of this analysis and its applicability for different urban contexts can help in forcing the decarbonization of energy infrastructures in dense urban environments (where solar is mainly the only renewable resource), by considering the specific characteristics and constraints of the buildings' heritage. Future developments on exergy analyses are ongoing to provide more accurate energetic indexes.

References

[1] IEA SHC | Solar Heat Worldwide. [accessed on 2024 Feb 19]. Available from: https://www.iea-shc.org/solar-heat-worldwide
[2] IEA - Renewables Report. IEA. 2021 [accessed on 2024 Feb 19]. Available from: https://www.iea.org/reports/renewables-2021
[3] Net Zero Emissions by 2050 Scenario (NZE) – Global Energy and Climate Model – Analysis. IEA. [accessed on 2024 Feb 19]. Available from: https://www.iea.org/reports/global-energy-and-climate-model/net-zero-emissions-by-2050-scenario-nze
[4] Bilardo M, Kaempf JH, Fabrizio E. Energy assessment of a district by integrating solar thermal in district heating network: a dynamic analysis approach. J Phys: Conf Ser 2023; 2600 052005. https://doi.org/10.1088/1742-6596/2600/5/052005
[5] Huide F, Xuxin Z, Lei M, Tao Z, Qixing W, Hongyuan S. A comparative study on three types of solar utilization technologies for buildings: Photovoltaic, solar thermal and hybrid photovoltaic/thermal systems. Energy Conversion and Management. 2017; 140:1–13. https://doi.org/10.1016/j.enconman.2017.02.059
[6] Selvakkumaran S, Axelsson L, Svensson I-L. Drivers and barriers for prosumer integration in the Swedish district heating sector. Energy Reports. 2021; 7:193–202. https://doi.org/10.1016/j.egyr.2021.08.155
[7] Mutani G, Todeschi V, Santantonio S. Urban-Scale Energy Models: the relationship between cooling energy demand and urban form. J Phys: Conf Ser 2022; 2177 012016. https://doi.org/10.1088/1742-6596/2177/1/012016
[8] Solar Engineering of Thermal Processes | Wiley Online Books. [accessed on 2024 May 15]. Available from: https://onlinelibrary.wiley.com/doi/book/10.1002/9781118671603
[9] Gruppo Iren: la rete di teleriscaldamento. [accessed on 2024 Mar 5]. Available from: https://www.gruppoiren.it/it/i-nostri-servizi/teleriscaldamento/la-nostra-rete.html.

HPC modelling for nuclear safety assessment

S A Cancemi[1,*], R Lo Frano[1], M Angelucci[1], A Pucciarelli[1], F Subba[2] and H Wu[2]

[1] DICI, University of Pisa, lg. L. Lazzarino n.1 Pisa, IT
[2] DENERG, Politecnico di Torino, Corso Duca degli Abruzzi n. 24 Torino, IT

* Corresponding author: salvatore.cancemi@phd.unipi.it

Abstract. High Performance Computing (HPC) has taken on an increasingly important role in recent years since it enables to model efficiently the complex behaviour of Systems, Components, and Structures (SSCs) and their interactions, and to perform simulations of the operational scenarios faster. Particularly in the nuclear field, HPC is currently used and expected to be utilized in nearly all areas, such as design optimization, safe and long-term safe operation, integrity evaluation, etc. The latter, as an example, requires an exhaustive SSCs assessment, which in turn requires higherfidelity models, both in terms of resolution and physics, increased knowledge of the physics of the phenomena or mechanisms affecting them, parametric studies, and better estimation of margins by reducing uncertainties and optimising operations.
In this paper some case studies are presented and described.
The first case consists in the modelling of creep behaviour of a fuel element. Finite Element Modelling (FEM) on hardware designed for parallel computation allows to simulate increasingly complex physical systems and reduce the computation time of the performed transient non-linear analyses. Results point out on the reliable prediction of creep behaviour obtained with a surrogate model. Understanding the thermo-mechanical properties of structural and nuclear materials and their changes provided essential for the plant safe operation as well as the Uncertainty Quantification (UQ).
The second case deals with a thermal-hydraulics application in support of Generation IV (GEN IV) Liquid Metal Fast Reactors (LMFRs). Results from RANS calculations will be compared with experimental data for a selected steady-state condition occurring in a rod-bundle geometry. UQ analyses are especially performed using deterministic sampling techniques to assess how much the results are affected by uncertainties in the input parameters.

1. Introduction

High-Performance Computing (HPC) is used practically in all areas of nuclear energy, as e.g. supporting the design and design improvement/retrofitting, enhancing material behaviour prediction, and understanding, safe operation, and life plant management of nuclear power plants whether fission or fusion. HPC is assuming an ever-increasingly important role in modelling as it enables us to perform simulations efficiently and faster of the behaviour of systems, structures, and components (SSCs), under operational scenarios, coupling various and different physics and scales [1]. Furthermore, the development of these new techniques combined with the availability of large computational resources has also allowed a significant development of new methods of machine learning and predictive analysis.

Figure 1 shows the three main areas of HPC (and scalable data science) [2] that can be applied in many scientific fields: algorithms, systems, and applications. The integration of all of them is essential to ensure HPC system operates more efficiently.

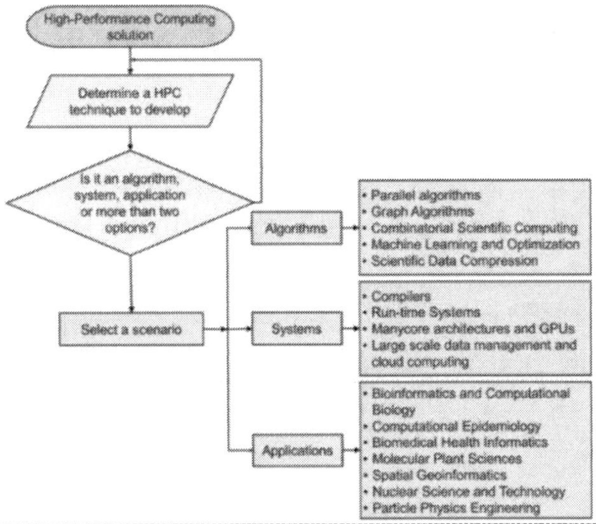

Figure 1. HPC data science organization (courtesy from [2]), which can be applied in many scientific fields

The adoption of HPC comes with its own set of challenges and limitations, such as e.g., the complex infrastructure and platform requirements, the scalability issues, the integration of multiple processors and accelerators, and the adoption of conventional/unconventional processor architectures required for non-intensive or heavy computations. Moreover, to perform multiphysics and multiscale simulations (heavy workloads), which can be characterized by the difficulty of precise prediction, with parallel programming, a robust and even more complex platform and cluster design is needed.

Another important aspect to consider is also the complexity of porting the existing codes, models and tools or rewriting them from scratch for the new architecture or fuel performance code (e.g., improving the reduced order models machine learning-based or surrogate models as did in the OPERA HPC project [3]) to cope with limitations and issues with programming and debugging tools for the accelerators.

By analysing the state of the art, it appears that HPC applications in the nuclear sector focus on the accuracy of simulation and reliability of data, used for validation purposes. Particularly, numerical modelling used for nuclear safety design and operation requires high-fidelity simulations. Advance the predictive capabilities of the numerical codes will improve significantly the understanding of component behaviour in a large variety of conditions. The knowledge acquired will be transferred into operational tools obtained an optimization of the plant operation.

The study of Shams et al. [5] focuses on the description of HPC facility at the Świerk Computing Centre (P) detailing the hardware configuration and software used for nuclear reactor design and safety and the computational requirements/capabilities.

Varè et al. [6] discusses the digitalization and virtual reality as key technologies to improve safety and performance of nuclear power. The study of Bakosi [7] describes the FE modelling by HPC of complex-geometry 3D flows and provides an overall fast numerical scheme based on an equivalent 1D finite volume scheme. This allows an effective use the computing resources even in the presence of hardware heterogeneities and dynamic parallel computational load. The study of Iannone [8] describes the infrastructure, providing technical details, and the performances of MARCONI, the new European HPC facility for Fusion, allowing modelling to exploit the latest available CPU technologies.

In this paper instead HPC is used to identify some key points for better estimation of safety margins by reducing simulation uncertainties and optimizing the plant operation. The paper consists of two main parts. The first one presents the improved HPC FE modelling of the creep mechanism of a fuel element, accounting for the thermo-mechanical loading during steady-state operation and transients, complex physical systems, and non-linear analyses material behaviour. The second part describes a thermal-hydraulics application (RANS calculations) in support of Generation IV (GEN IV) Liquid Metal Fast Reactors (LMFRs). This activity was carried out in the frame of the PNNR Flagship Project 4.2 "HPC tools for the simulation of nuclear systems".

2. FE creep modelling

Severe thermo-mechanical environments, such as relatively high temperatures, differential pressures, and corresponding hoop stress, influence largely the material behaviour of reactor components as microstructure is naturally dependent on the temperature and stress level it is facing during its lifetime.

As a results, important deformation mechanisms such as swelling, creep, pellet-clad interaction, fuel-overfragmentation and cracking can occur. Of all these, creep of fuel cladding is certainly the dominant mechanism of slow irreversible deformation process under the influence of stresses below the yield stress [9] [10].

From continuum point of view, a reliable prediction of creep requires the definition of the mechanical equilibrium equations, appropriate constitutive equations, not only based on Norton-Bailey model (e.g. for primary creep stage), and suitable initial and boundary conditions. Therefore, it is necessary to employ a new and flexible HPC modelling integrating possibly experimental, and in-line operational data in ways not possible with traditional experimentally based paradigm.

Figure 2 shows the overall dimensions of a PWR fuel rod and the FE model (only a section) that was developed starting from this geometry: the cladding and fuel gap is 0.1025 mm; the plenum's nominal height (61.0 mm) is filled with Helium at 2.31 MPa.

As first step, a mesh sensitivity study has been performed by varying the element types (i.e., shell, solid) and the mesh sizes (e.g., 960, 1920, 3840, and 4320 elements), and the core counts to determine the appropriate mesh density for accurate results. Given the axisymmetric geometry and considering the purpose of determining the performance of a reduced order model, it is believed that the model implemented with 4320 3D solid elements is the most appropriate for representing the thermo-mechanical behaviour of the fuel rod.

Analyses were executed on Workstation 1 (W1) - Intel(R) Core (TM) i9-10920X CPU @ 3.50GHz, 3504 MHz with 24 cores, and Workstation 2 (W2) - Intel(R) Core (TM) i9-9820, 3.30GHz-Turbo-Boost 4.20GHz.

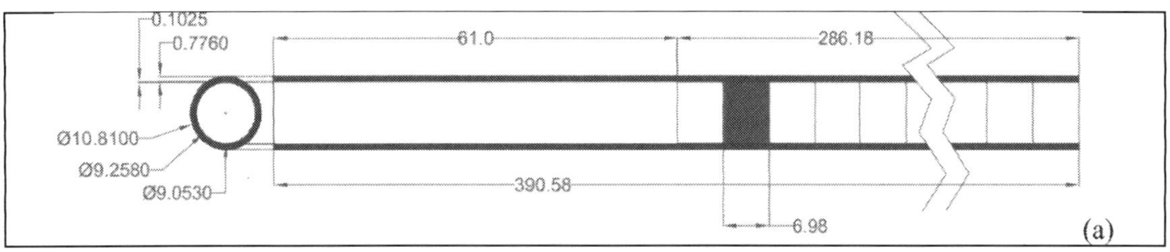

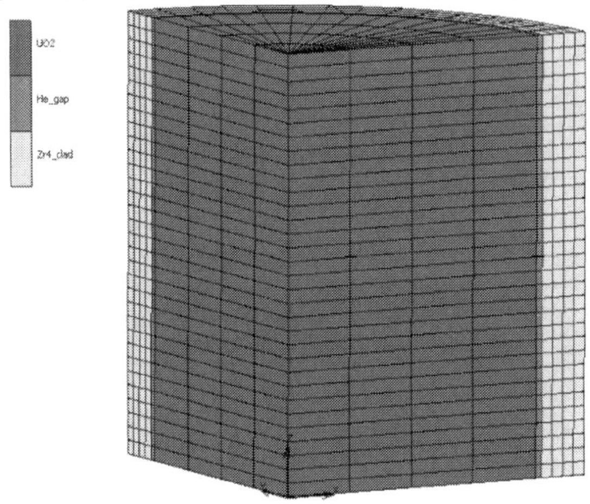

(b)

Figure 2. Fuel rod (UO2 pellet plus Zr-4 cladding) dimensions [mm] (a) and FE model (b). In the figure the red volume is the UO2 pellet, the light blue part represents the He gap, while the yellow annular volume represents the Zr-4 cladding.

Non-linear thermo-mechanical analyses were performed to simulate plasticity and creep mechanisms which are the most important responsible for accumulation of irreversible damage in the fuel rod. The secondary creep, which is characterized by a constant strain rate, is generally more important for temperatures in the range of 40 % and 70% of the melting temperature; at higher temperatures the creep mechanism is diffusion of atoms or Cobble or Nabarro-Herring creep, with a formula similar to the Norton law but with m=1 [12] [4]. To simulate creep and ensure unconditional stability regardless the time step, the fully implicit viscoplastic Maxwell model with creep strain rate ($\dot{\varepsilon}$) was implemented [12], [14]. The coefficients of the equation of the implicit power law model combined with regular plasticity are $A = 3.41\times10^{-33}$; $n = 3.3$; $m = 0.2$; $k = 0.8$. The UO2 behaviour was assumed isotropic with thermo-mechanical properties that automatically updated throughout the entire transient. This is supported by the recent studies on the fuel mechanical behaviour [15]. The boundary conditions reflect the typical normal operation conditions of the fuel rod [16], as e.g. coolant temperature and pressure equal to 287.7 °C and 15.52 MPa respectively, specific linear mean power of 42 kW/m, and the rod/pellets restraints [17]. The axial symmetry was also imposed. Staggered solution and the multifrontal direct solver were used for FE creep solution (0.01-time step, 200 increments) [18].

Using parallel processing, HPC-based FE model can distribute computational tasks across multiple processors, significantly reducing the analysis time.

Because of high thermal gradients within the pellet, internal thermal stresses are developing in the fuel. If one assumes that the average strength of a typical oxide fuel is about 150 MPa [19], one will find that the fuel pellet starts to fragment when the differential temperature across the surface of the pellet is higher than 120° C, i.e. under nominal conditions.

Results showed that the outward radial displacement of about $4 \cdot 10^{-5}$ m obstructing half gas gap. Moreover, the Von Mises stress overcome the allowable stress intensity [20] of 6 % and 30 % in the pellet and cladding respectively. The maximum value equivalent creep strain rate (of about $1.3\cdot10^{-4}$ s^{-1}) appeared in the innermost part of the pellet (see Figure 3), as expected, and aligns with reference [21]. In addition, from HPC standpoint, findings showed that increasing the number of cores high performance in terms of calculation time is obtained. However, no significant computational enhancement was observed beyond 8 cores.

Figure 4 shows the performance plots for the three discretized FE model; the vertical axis, termed "wall time", is the actual time taken from the start to the end of the simulation.

W1 resulted 15% to 20% faster than W2 to run complex nonlinear creep analyses. This difference in terms of performance is even more marked for models with a high number of elements.

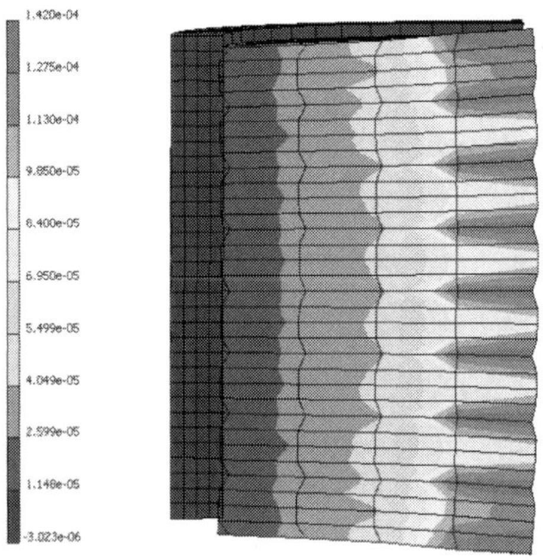

Figure 3. Equivalent creep strain (contour plot). The system unit is [hr^{-1}]. The external annular sector in blue represents the cladding deformation. The internal volume having a color ranging from red to blue represents the deformation of the pellet

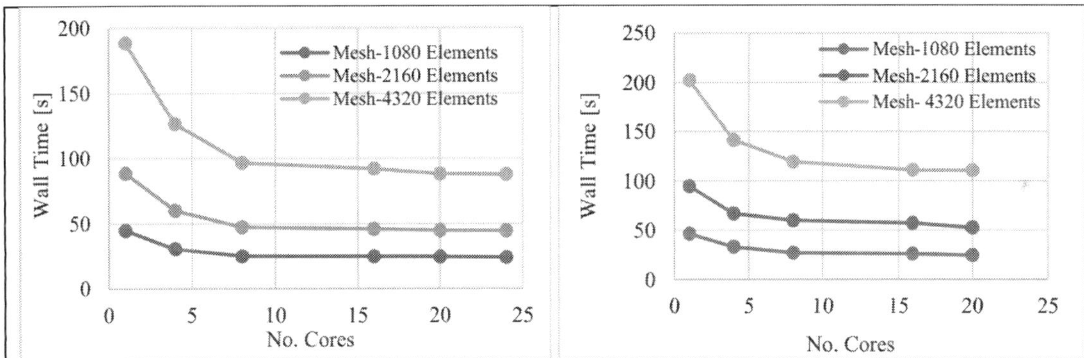

Figure 4. W1(left) and W2 (right) performance for the three discretized FE model

3. Thermal-hydraulics modelling and application

Concerning the thermal-hydraulics applications, the present paper focuses on CFD uncertainty quantification analyses in support of the development of GEN IV NPPs. In particular, a selected operating condition from the NACIE-UP Fuel Pin Simulator (FPS) [22], whose Fuel Pin Simulator (FPS) sketch design is reported in Figure 5 is considered. The FPS is a 19 rods wire wrapped electrically heated bundle in which LBE flows as the primary coolant. A simulation of this component allowing for a fine representation of each detail would require a very large computational cost, in the range of 50-100 million cells.

A nodalization for such a detailed model was developed, but its computational effort was later considered too large to be adopted for the copious number of calculations required for uncertainty quantification analysis. Therefore, a new model (see [23] for more reference), opting for a coarser representation of the wrapped wire geometry but allowing for the creation of a structured mesh, was

Journal of Physics: Conference Series **2940** (2025) 012025 doi:10.1088/1742-6596/2940/1/012025

developed to reduce the computational effort. The total cell count for the reduced cost model is in the range of 5 million.

Figure 6 shows a sample temperature distribution predicted by the code, the coarse shape of the wires, clearly not round, can be easily spotted in the referred figure.

As reported in [23], despite its reduced level of detail of the considered geometry, the model provides suitable capabilities in reproducing the thermal-hydraulic behaviour of the component at a definitively smaller computational cost. A validation of the approach was thus achieved.

Reducing the computational cost per run is very important for uncertainty quantification analyses owing to the large number of operating conditions to be considered in the excitation matrix. Indeed, excitation matrices, to prove their consistency with the addressed data, should fulfil several moments of the selected parameter distributions (average, variance, skewness…). This goal is typically achieved performing hundreds of calculations adopting a Montecarlo sampling, something that may lead to an incredibly large computational effort when dealing with CFD analyses. To prevent this, deterministic sampling methods were recently proposed in literature [24] providing tools to obtain uncertainty quantification analyses with a limited amount of selected operating conditions.

One of the key points of the deterministic sampling is the selection of a set of operating conditions fulfilling, by construction, the relevant moments of the addressed parameters distributions.

Figure 7 reports the excitation matrix considered for the present application obtained adopting the modified 4th order Hadamard approach, together with the considered variance for each investigated parameter.

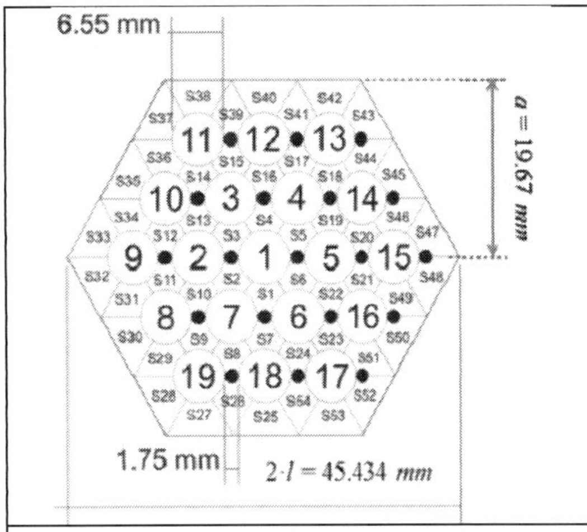

Figure 5. NACIE-UP facility sketch, taken from [23]

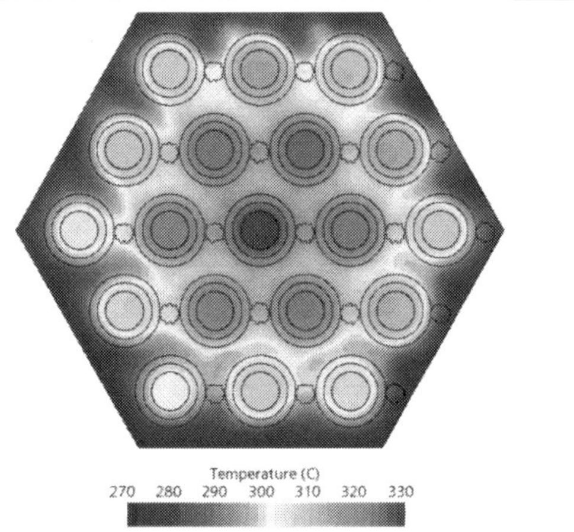

Figure 6. Temperature distribution for the FPS mid-section predicted by the reduced cost model.

$$\Sigma_{\mathrm{HAD}}^{T} = \begin{pmatrix} Sample & \dot{m}_{\mathrm{LBE}} & \mathbf{T}_{\mathrm{IN,FPS}} & \mathbf{Q}_{\mathrm{eff}} & \mathbf{Q}_{\mathrm{pre}} \\ HAD01 & \sigma_1 & \sigma_2 & \sigma_3 & \sigma_4 \\ HAD02 & -\sigma_1 & \sigma_2 & -\sigma_3 & \sigma_4 \\ HAD03 & \sigma_1 & -\sigma_2 & -\sigma_3 & \sigma_4 \\ HAD04 & -\sigma_1 & -\sigma_2 & \sigma_3 & \sigma_4 \\ HAD05 & \sigma_1 & \sigma_2 & \sigma_3 & -\sigma_4 \\ HAD06 & -\sigma_1 & \sigma_2 & -\sigma_3 & -\sigma_4 \\ HAD07 & \sigma_1 & -\sigma_2 & -\sigma_3 & -\sigma_4 \\ HAD08 & -\sigma_1 & -\sigma_2 & \sigma_3 & -\sigma_4 \end{pmatrix}$$

$\sigma_1 = \sigma_{mflow} = \pm 5\%\ of\ the\ reference\ flow$

$\sigma_2 = \sigma_{Tin} = \pm 1.5°C\ Inlet\ temperature\ uncertainty$

$\sigma_3 = \sigma_{power\_Active} = \pm 10\%\ of\ the\ reference\ Active\ region\ supplied\ power$

$\sigma_4 = \sigma_{power\_Upstream} = \pm 4\%\ of\ the\ reference\ Upstream\ region\ supplied\ power$

Figure 7. The 4th order modified Hadamard excitation matrix considered in the present work.

For each row, representing a distinct run, a perturbation of $\pm\sigma$ from the reference value is considered for each investigated parameter; e.g. for HAD01 the mass flow is $m_{flow1} = m_{flow,ref} + \sigma_{mflow}$ and for HAD02 the mass flow is $m_{flow2} = m_{flow,ref} - \sigma_{mflow}$.

The input parameters were selected after a sensitivity analysis process reporting that they are the most affecting the calculation outputs. Sigma values were instead provided by the experimentalists.

As it can be observed, this matrix allows investigating the effect of four input parameters while intrinsically assuring the fulfilment of mean and variance. This is confirmed by performing the average on each column, which turns to be zero, meaning that, on average, the reference mass flow rate is applied, i.e. thus fulfilling the considered distribution requirement on the zeroth moment (the average).

Figure 8 reports a sample of the results obtained for the calculations performed considering the first 4 rows of the excitation matrix. Here a comparison between the experimental measurements and the predicted values for temperature values inside the FPS is performed. On the x-axis the thermocouple number is reported, a label reporting its location inside the FPS is also included in the figure.

Results show that both overestimating and underestimating trends are predicted, thus reassuring about the approach capability in fulfilling the zeroth moment requirement (average). Figure 9 reports instead the result of the performed UQ approach obtained assembling all the predicted trends (HAD01 to HAD08) and calculating their average and root mean square. These values can be compared with experimental data showing that sufficiently reliable results were obtained, and the predicted uncertainty ranges size seems reasonable. A general increase of uncertainty is observed moving from left to right (i.e. from the inlet to the outlet) suggesting that a build-up of the uncertainty occurs along the axial direction. Almost all the experimental data lay in a $\pm\sigma$ range from the calculated average, thus suggesting that the adopted CFD model and approach are suitable for the addressed application. Further analyses are expected to be performed on other selected operating conditions and adopting different deterministic sampling approaches to further investigate and validate the considered UQ approach.

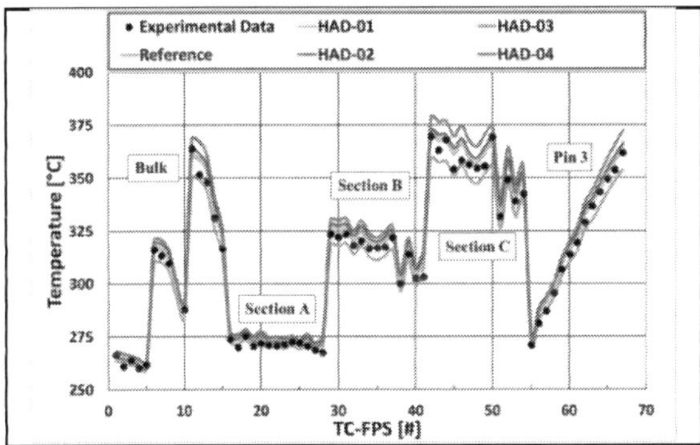

Figure 8. Experimental data vs. CFD predictions for the first 4 rows of the excitation matrix.

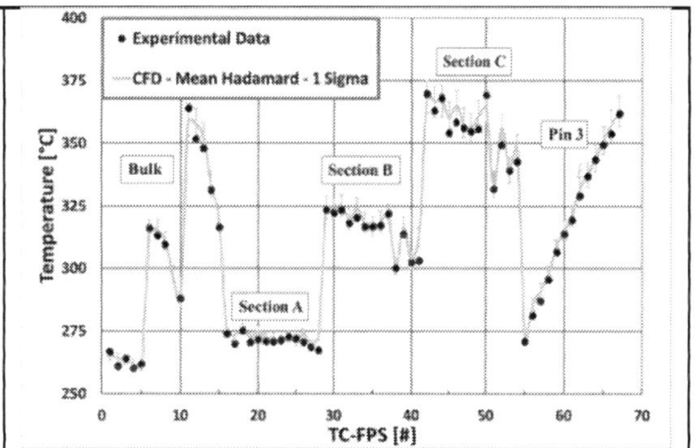

Figure 9. Results of the performed uncertainty quantification analysis.

4. Summary

Advantages and disadvantages of HPC have been explored with reference to two different application cases: thermo-mechanical analysis of creep for clad-pellet interaction and CFD uncertainty quantification analyses in support of the development of GEN IV NPPs. On the positive side, it emerged that HPC offers unparalleled computational power enabling to tackle complex problems and process massive amounts of data and complex operations with speed and efficiency. However, the adoption of HPC also comes with its own set of challenges and limitations, such as the complex infrastructure and platform requirements, energy consumption, scalability issues, integration of multiple processors and accelerators, and the adoption of conventional and/or unconventional processor architectures which are required for non-intensive and heavy computations, respectively.

Journal of Physics: Conference Series **2940** (2025) 012025 doi:10.1088/1742-6596/2940/1/012025

The findings showed that increasing the number of cores results in significant improvements in calculation time is obtained: W1 resulted 15% to 20% faster; performances increase with the model complexity. However, no significant computational enhancement was observed beyond 8 cores.

Results obtained from CFD uncertainty quantification analyses performed by considering the modified Hadamard excitation matrix show that both overestimating and underestimating trends are predicted (Figure 8), thus reassuring about its capability in fulfilling the first momentum requirement. As for UQ approach, sufficiently reliable results were obtained, and the predicted uncertainty ranges size seems reasonable. The main aim of the performed work is promoting the use of CFD in safety assessment analyses for nuclear power plants.

Further analyses are expected to be performed to investigate e.g. challenges posed by the development and integration of machine learning, AI, and deep learning techniques, and to quantify the uncertainty of predictions obtained by multi-scale and multi-physics approaches.

5. References

[1] Simonovski I, Clements J, Žefran B, Cimerman S. The role of High Performance Computing in the Nuclear Energy Sector. Nuclear Energy for New Europe 2015; Nuclear Society of Slovenia. Portorož (SL); 2015. p. 105. JRC97395

[2] Renteria del Toro, F. A, Hao, C (2024, march 10). High-Performance Computing (HPC) Framework and Deployment in Nuclear Energy Systems. International Youth Nuclear Congress 2022 (IYNC2022), Koriyama, Japan. https://doi.org/10.5281/zenodo.10801865

[3] EU project OperaHPC, OPEn HPC theRmomechanical tools for the development of eAtf fuels, https://www.operahpc.eu/

[4] Lo Frano, R, et al. Improved HPC Models For Fuel Performance Predictions, Nuclear Energy for New Europe 2024; Nuclear Society of Slovenia. 9-12 September Portorož (SL) 2024. ID 407

[5] Shams, A, et al. (2020). High-Performance Computing for Nuclear Reactor Design and Safety Applications. Nuclear Technology, 206 (2), 283–295. https://doi.org/10.1080/00295450.2019.1642683

[6] Varé, C, Morilhat, P (2020). Digital Twins, a New Step for Long Term Operation of Nuclear Power Plants. In: Liyanage, J., Amadi-Echendu, J., Mathew, J. (eds) Engineering Assets and Public Infrastructures in the Age of Digitalization. Lecture Notes in Mechanical Engineering. Springer, Cham. https://doi.org/10.1007/978-3-030-48021-9_11

[7] Bakosi, J, Open-source complex-geometry 3D fluid dynamics for applications with unpredictable heterogeneous dynamic high-performance-computing loads, Computer Methods in Applied Mechanics and Engineering, Volume 418, Part B, 2024, 116586, https://doi.org/10.1016/j.cma.2023.116586.

[8] Iannone, F, et al. MARCONI-FUSION: The new high performance computing facility for European nuclear fusion modelling, Fusion Engineering and Design, Volume 129, 2018, 354-358, https://doi.org/10.1016/j.fusengdes.2017.11.004.

[9] Ashby, M F, Frost, J (1982), Deformation-mechanism maps, the Plasticity and Creep of Metals and Ceramics, Oxford, Pergamon Press.

[10] Sattar, M, et al. Limitations on the computational analysis of creep failure models: A review, Engineering Failure Analysis, Volume 134, 2022, 105968, https://doi.org/10.1016/j.engfailanal.2021.105968.

[11] Miguel, MC, Vespignani, A, Zaiser, M, Zapperi, S Dislocation jamming and Andrade creep, Phys. Rev. Lett. 89, 165501, https://doi.org/10.1103/PhysRevLett.89.165501

[12] Masaharu, K,, Selected Topics on Material Strength and Thermally Activated Deformation Processes, Materials Transactions, 2015, Volume 56, Issue 2, 175-181, Released on J-STAGE January 25, 2015, https://doi.org/10.2320/matertrans.M2014386

[13] Lo Frano, R, et al. Preliminary Analysis of an Aged RPV Subjected to Station Blackout. Energies. 2021; 14(15):4394. https://doi.org/10.3390/en14154394

[14] Findley WN, Lai JS, Onaran, K. Creep and relaxation of nonlinear viscoelastic materials, North

Holland-Series in Applied Mathematics and Mechanics, Vol. 18, North-Holland Publishing Company, Amsterdam, 1976.

[15] Y. Zhang et al. (2012), Crack tip plasticity in single crystal UO2: Atomistic simulations, Journal of Nuclear Materials 430, 96–105, https://doi.org/10.1016/j.jnucmat.2012.06.044

[16] The Third Risø Fission Gas Project: Bump Test AN3 (CB8-2R). Technical Report Risø-FGP3-AN3, Risø, September 1990.

[17] Walker, CT, et al. Microbeam analysis of irradiated nuclear fuel, 2012, IOP Conf. Ser.: Mater. Sci. Eng. 32 012028 https://doi.org/10.1088/1757-899X/32/1/012028

[18] Turska, E, Wisniewski, K, Schrefler, B.A. Error propagation of staggered solution procedures for transient problems, Computer Methods in Applied Mechanics and Engineering, Volume 114, Issues 1–2, 1994, 177-188, https://doi.org/10.1016/0045-7825(94)90168-6.

[19] Nuclear Energy Agency, State-of-the-Art Report on Multi-scale Modelling of Nuclear Fuels, Nuclear Science NEA/NSC/R/(2015)5 October 2015, https://www.oecd-nea.org/upload/docs/application/pdf/2020-01/nsc-r2015-5.pdf

[20] Slagis, GC. ASME Section III Design-By-Analysis Criteria Concepts and Stress Limits. ASME. J. Pressure Vessel Technol. February 2006; 128(1): 25–32. https://doi.org/10.1115/1.2140797

[21] Bohaboy, PE, et al. Compressive Creep Characteristics Of Stoichiometric Uranium Dioxide. GEAP-10054, General Electric Co., Sunnyvale, Calif. Breeder Reactor Development Operation, 1 Jan. 1969. https://doi.org/10.2172/4716019.

[22] Di Piazza, I, Hassan, H, Benchmark specifications for NACIE-UP facility: non-uniform power distribution tests, rev. 2, IAEA CRP-I31038, Vienna, Austria (2023).

[23] Halim, O, et al. A comparative analysis of detailed and reduced CFD approaches to model wire-wrapped fuel bundles for LMFBRs applications. Annals of Nuclear Energy, Volume 211, February 2025, 110937. https://doi.org/10.1016/j.anucene.2024.110937

[24] Sahlberg, A. Ensemble for Deterministic Sampling with positive weights : Uncertainty quantification with deterministically chosen samples (Dissertation). (2016). Retrieved from https://urn.kb.se/resolve?urn=urn:nbn:se:uu:diva-297314

Acknowledgments

This work is supported by PNRR M4C2 - HPC, Big data and Quantum Computing (Simulazioni, calcolo e analisi dei dati e altre prestazioni - CN1) - CUP I53C22000690001 SPOKE 6 Multiscale modelling & Engineering applications.

Simulations of single and two-phase flows through helical tubes with different geometries and mass flow rates

M Giardina, P Buffa

Department of Engineering, University of Palermo, Viale delle Scienze, Ed. 6, Palermo, Italy

Corresponding author: mariarosa.giardina@unipa.it (Mariarosa Giardina)

Abstract. Predictions of pressure drops in helical pipes using the Relap5/mod3.3 code are explored, despite the code's limitations in modeling such geometries. To address this, the code was modified to incorporate helical tube geometries and correlations for frictional pressure drops under both single-phase and two-phase flow conditions. Comparisons with experimental data on pressure drops across test sections with varying curvature ratios, operational pressures, and mass flow rates were made. The study demonstrates that the modified Relap5 code significantly improves the accuracy of frictional pressure drop predictions for the examined experimental data.

Keywords: helical pipe, pressure drops, single flow, two-phase flow, Small Modular Reactor

1 Introduction

Various Generation III+ and Generation IV nuclear reactor projects are expected to use helically coiled pipes in Steam Generators (SG). In this context, some Small Modular Reactor (SMR) designs aim to incorporate once-through helically coiled SGs [1]. This integral layout enhances reactor compactness, improves heat transfer efficiency, and, by design, reduces the risks and consequences of various postulated accidents, such as Large Break LOCA events.

The curvature of helically coiled tubes has been shown to induce secondary flows, which enhance fluid mixing and improve heat transfer characteristics compared to straight tubes. However, this also leads to an increase in pressure drop, making its prediction more complex than for straight tubes.

Several researchers from the Engineering Department at the University of Palermo have conducted numerous studies focused on simulating single-phase and two-phase flows in vertical helical pipes (HPs) using the best-estimate Relap5 nuclear code, which initially lacked the ability to model components with helical geometries and the associated thermal-fluid dynamic phenomena.

The Relap5/mod3.3 release [2] has been modified as described in this paper to include a schematization for coiled tube geometries and integrate correlations for simulating frictional pressure drops under both single-phase and two-phase flow conditions.

The modified code has been validated against experimental data from the literature, as well as experiments conducted at various international and national research laboratories, with varying curvature, operational pressures, and mass flow rate conditions.

The results obtained demonstrate that the modified version of the code accurately predicts frictional pressure drops across a wide range of pressures and mass flow rates.

2 Correlations for predicting pressure drops in HPs

In previous studies, various authors have explored the issue of frictional pressure drops in both single-phase and two-phase flows in helical pipes (HP), using their experimental results to develop corresponding correlations.

Many of these correlations, typically suited for HPs with low pitch-to-diameter ratios and low surface roughness, establish a relationship between the friction factor, Reynolds number (Re), and the curvature of the helix, which is the ratio of pipe diameter to coil diameter, d/D.

Among the various approaches, Ito's correlations [3] for both laminar and turbulent single-phase flows are frequently considered some of the most dependable in the literature [4, 5]. For two-phase flow pressure drops, numerous correlations have been proposed.

Some researchers suggest using correlations originally developed for straight pipes, like the Lockhart-Martinelli (L-M) method, to estimate pressure drops in helical pipes by replacing the single-phase friction factor with adjusted correction factors for helical geometries [5].

Other researchers have developed correlations specifically for HPs, but these are often limited in their validity range and are based on dimensional parameters or models that do not maintain dimensional consistency. Additionally, when used outside their original development conditions, these correlations tend to produce significantly different results, demonstrating a lack of robustness to changes in flow parameters.

In this study, among the available correlations in the literature, the friction factor from Ito for single-phase flow is used, while for two-phase flow, the L-M correlations, as modified in [4], are adopted.

The following sections provide details about the correlations implemented in the Relap code for simulating HP, aiming to extend its applicability to various operating conditions in terms of working pressure and fluid flow rate.

2.1 Correlations for single phase flows

The single-phase experimental database includes approximately 2000 data points, covering Re from 40 to 145,000, with curvature ratios d/D ranging from 0.003 to 0.3.

Ito's correlations, applicable to both the laminar and turbulent regimes, provide the best fit to the experimental data, with errors of about 2% and 4% for the turbulent and laminar regimes, respectively.

Ito summarized the entire status of the fluid Fanning friction factor (f_f) for HPs as follows for laminar and turbulent flows, respectively:

$$f_f=\frac{344(d/D)^{0.5}}{\left\{1.56+\log_{10}\left[Re\left(\frac{d}{D}\right)^{0.5}\right]\right\}^{5.73}} \qquad 13.5\left(\frac{d}{D}\right)^{0.5}<Re<2000\left[1+13.2\left(\frac{d}{D}\right)^{0.6}\right] \tag{1}$$

$$f_f=0.076Re^{-0.25}+0.0075\left(\frac{d}{D}\right)^{0.5} \qquad Re>15000 \tag{2}$$

For Re number $13.5(d/D)^{0.5}$, friction factor for laminar flow in straight tubes can be considered valid:

$$f_f=16/Re \tag{3}$$

2.2 Correlations for two-phase flows

Early researchers primarily used the L-M method [6] to predict the frictional pressure drop in two-phase flow conditions.

Given that the L-M correlations were initially created for two-phase pressure drop in horizontal straight tubes, adjustments are necessary to align the results with experimental data for helical geometries [4, 7, 8].

In [4], by using the L-M method, a correlation factor derived from experimental data fitting for helically coiled tubes is proposed.

As is well known, the frictional pressure drop data can be correlated by the pressure drop multipliers ϕ_l and ϕ_g as a function of the L-M parameter χ, defined as follows:

$$\emptyset_l^2 = \frac{\left(\frac{dP}{dx}\right)_{TP,f}}{\left(\frac{dP}{dx}\right)_l} \qquad \emptyset_g^2 = \frac{\left(\frac{dP}{dx}\right)_{TP,f}}{\left(\frac{dP}{dx}\right)_g} \tag{4}$$

$$\chi^2 = \frac{\emptyset_g^2}{\emptyset_l^2} = \frac{\left(\frac{dP}{dx}\right)_l}{\left(\frac{dP}{dx}\right)_g} \tag{5}$$

$\emptyset_l$ and $\emptyset_g$ represent the two-phase friction multipliers for liquid-only and vapor-only flows, respectively. $(dP/dx)_{TP,f}$ denotes the frictional pressure gradient in two-phase flow, while $(dP/dx)_l$ and $(dP/dx)_g$ correspond to the pressure drops when only the liquid phase or gas phase flows through the channel, respectively.

It is important to specify that the pressure gradients (dP/dx)l and (dP/dx)g are evaluated assuming that each phase flows alone with its respective partial flow rate, i.e., G(1−x) for the liquid phase and Gx for the gas phase, with G being the total flow rate and x the quality. This approach differs from another common method in the literature, where the liquid and gas pressure gradients are calculated assuming that each phase flows alone but with the total flow rate G.

For straight pipes, the liquid and gas pressure drop multipliers are determined using the following correlations:

$$\emptyset_l^2 = 1 + \frac{c}{\chi} + \frac{c}{\chi^2} \tag{6}$$

$$\emptyset_g^2 = 1 + c\chi + c\chi^2 \tag{7}$$

The constant c in Eqs. (6) and (7) depends on the flow regimes of the liquid and vapor. To more accurately account for the effect of system pressure, as observed in the results reported in [4], the ratio (ρ_{mix}/ρ_l) is introduced as a correction factor in Eq. (6), consistent with approaches used in other studies [5].

The proposed correlation is as follows:

$$\emptyset_{l*}^2 = 0.032\ \emptyset_l^2\ De_l^{0.305} \left(\frac{\rho_{mix}}{\rho_l}\right)^{-0.51} \tag{8}$$

where $De_l = Re_l(d/D)^{0.5}$ is the Dean number of the liquid phase as it flows alone in the channel, used to account for the effect of the centrifugal force. The parameter $\emptyset_l^2$ is evaluated by Eq. (6) and the mixture density is determined by the relationship:

$$\rho_{mix} = \left[\frac{x}{\rho_g} + \frac{(1-x)}{\rho_l}\right]^{-1} \tag{9}$$

2.3 RELAP modification

Additional fortran subroutines were incorporated into Relap5/mod3.3 to facilitate the modeling of helical geometries for both single-phase and two-phase flow conditions. For single-phase flow, the subroutine responsible for calculating f_f was modified by integrating Eqs. (1) through (3).

In single-phase flow through straight pipes, a discontinuity in the friction factor is a well-documented physical phenomenon associated with the transition between laminar and turbulent flow at $Re \approx 2100$. In helical coils, however, this transition is typically smoother due to the effects of curvature and secondary flows, which tend to mitigate abrupt changes in the flow regime.

To address the discontinuity in the friction factor observed in Ito's correlations during the transition from laminar to turbulent flow, a new approach has been implemented, aiming to provide a more accurate representation of the gradual regime change in helical geometries.

This approach involves using a linear interpolation technique for f_f as a function of Re, applied between Re=2000[1+13.2(d/D)0.6] (Ito's upper limit for laminar flow) and Re=15,000 (Ito's lower threshold for turbulent flow).

For two-phase flow in straight pipes, the Relap5 friction model utilizes a two-phase multiplier approach. The multiplier is determined using the Heat Transfer and Fluid Flow Service (HTFS) correlation [9], which aligns with empirical data across a wide range of flow rates and flow regimes.

The HTFS correlation for the two-phase friction multiplier is defined by Eqs. (6) and (7), representing the liquid-alone and vapor-alone multipliers, respectively.

The constant c is evaluated as follows:

$$c = -2 + (28 - 0.3\,g^{1/2})\exp\left(-\frac{(\log_{10} k + 2.5)^2}{2.4 - g\,10^{-4}}\right) \tag{10}$$

where g is the total mass flow rate. The parameter κ is the Baroczy dimensionless index defined as:

$$k = \frac{\rho_g}{\rho_l}\left(\frac{\mu_l}{\mu_g}\right)^{0.2} \tag{11}$$

It is worth noting that the definition of the Baroczy dimensionless index in Eq. (10) closely resembles the square of the Lockhart-Martinelli parameter for turbulent-turbulent flow.

In the modified code, by combining Eqs. (4) through (8), the frictional pressure for two-phase flow configuration is computed by the following equation:

$$\left(\frac{dP}{dx}\right)_{TP,f} = \left[0.032\ De_l^{0.305}\left(\rho_{mix}/\rho_l\right)^{-0.51}\right]\left\{\left(\frac{dP}{dx}\right)_l + c\left[\left(\frac{dP}{dx}\right)_l\left(\frac{dP}{dx}\right)_g\right]^{0.5} + \left(\frac{dP}{dx}\right)_g\right\} \tag{12}$$

To calculate $(dP/dx)_l$ and $(dP/dx)_g$ of Eq. (12), Ito correlations for the friction factors, as previously described, were used. The constant c is evaluated by using Eq. (10).

In the following sections, the modified and unmodified versions of the Relap5 code will be referred to as *modified-R5* and *unmodified-R5*, respectively.

3 Results of validation for single phase flows

3.1 *Experiments by Ali (2001)*

The experiments presented in [10] were conducted using HPs with varying geometric dimensions, constructed from thick-walled polyethylene tubing. The test sections dimensions result in a d/D ranging from 0.0268 to 0.0519.

Ambient temperature water was used as the test fluid.

The nodalization for the Relap5 system code is based on an inclined pipe divided into subvolumes, which represents the HP test section. The length, tube diameter, and coil diameter are identical to those used in the experiments. A new flag in the Relap5 input file allows these geometric parameters to be specified for the coil pipe component.

As the experiments are carried out under steady-state conditions with a constant water mass flow rate, no pumps are included in the nodalization. The required experimental mass flow rates are maintained through a time-dependent junction that ensures the proper flow rate to the test section for each of the tests conducted.

Fig. 1 shows log-log plots of Euler number, $Eu = \Delta P/(2\rho v^2)$, against Re, based on experiments conducted with coil no. 1, along with results from both modified-R5 and unmodified R5. Fig. 2

illustrates the experimental Eu data versus Re for coils no. 5, 6, and 8, compared with predictions obtained using only modified-R5.

Analyzing all the results from the conducted studies, which are omitted here for brevity, it is evident that the modified-R5 accurately predicts the experimental data trends and values with an error margin not exceeding 10%.

In contrast, the unmodified-r5 produces results that significantly underestimate the experimental data.

3.2 Experiments by Coronel and Sandeep (2003)

The experiments were performed using a heat exchanger consisting of two concentric coils and covering a range of single-phase flow rates from $9.46\text{x}10^{-5}$ to $6.31\text{x}10^{-4}$ m^3/h [11]. Tap water at an initial temperature of 20°C was used as the working fluid and heated to specific endpoint temperatures within the heat exchanger's two coils. The endpoint temperatures were set to 20°C, 93.3°C, 121°C, and 149°C.

The coils geometries are reported in Table 1.

The Relap5 nodalization consists of two inclined pipes, each divided into 50 sub-volumes, connected to heat structure components that supply heat to the water flow. The water is pumped through the two coiled pipes, raising the fluid to the final temperature required for the experiment. The mass flow rates are controlled by a time-dependent junction that ensures the appropriate flow rate to the test section for each analyzed run.

Fig.s 3 and 4 present results showing the pressure drop per unit length as a function of flow rate, with varying endpoint temperatures in both the outer and inner coils.

Also in this case, upon examining the set of results, it can be highlighted that the modified-R5 allows for very accurate predictions of the experimental data, with an error that does not exceed 5%.

Table 1. Coil geometries in the heat exchanger [11].

Coil geometry	Outer Coil [m]	Inner Coil [m]
Outside diameter of tube	0.0127	0.0127
Inside diameter of tube	0.0102	0.0102
Length of coil	39.63	27.44
Diameter of coil	0.162	0.111

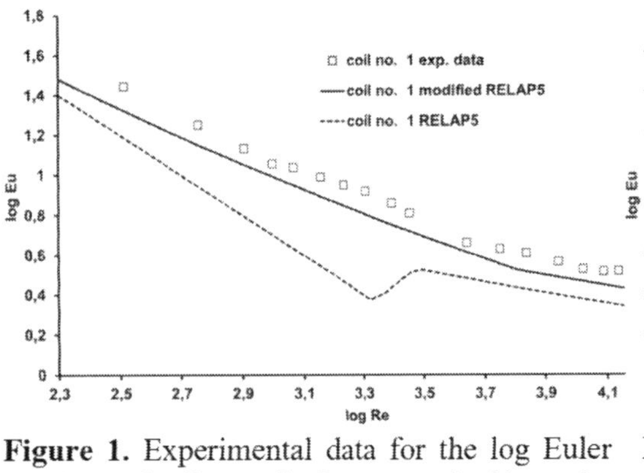

Figure 1. Experimental data for the log Euler numbers of coil no. 1 [11] compared with results from the modified-R5 and unmodified-R5.

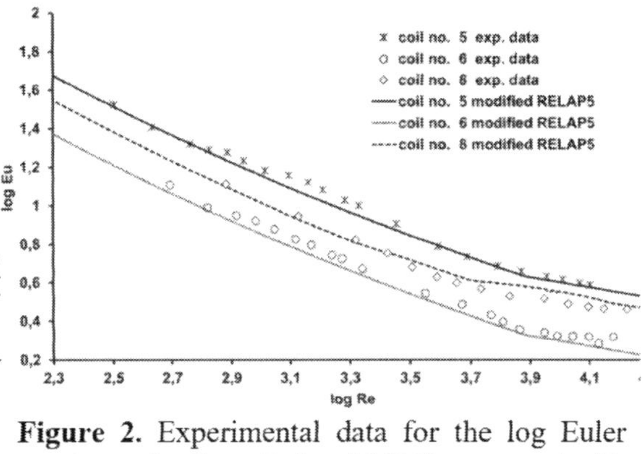

Figure 2. Experimental data for the log Euler numbers of coil no. 5, 6 and 8 [11] compared with results from the modified-R5.

4 Results of validation for two phase flows

4.1 HP with 1 m coil diameter under different test pressures

The experimental data examined in this section were conducted at SIET (Società Informazioni Esperienze Termoidrauliche) research laboratories at Piacenza [12].

The test section comprises a HP with 0.01253 m inner diameter, 1 m coil diameter, and 32 m length. Positioned before the test section is a pre-heater, which is an electrically heated coiled tube section designed to achieve the inlet target temperature. Along the coiled tube, nine pressure taps are distributed approximately every four meters, with eight differential pressure transducers connecting these taps.

The thermal-hydraulic conditions studied for the adiabatic two-phase flow pressure drops (without power supplied to the test section) include the following parameters: mass flux ranging from 192 to 772 [kg/m²s], pressures of 3, 4, and 6 MPa, and qualities from 0.17 to 0.84.

The Relap5 nodalization incorporates time-dependent volume components and junctions to define the inlet and outlet conditions of the helical coil tube loop. A proportional-integral-derivative (PID) controller is implemented through control variable cards to regulate the inlet pressure at the test section.

For brevity, only a subset of results is presented in Fig.s 5 and 6. These figures compare predictions from the modified-R5 and unmodified-R5 with experimental data of pressures measured along the coil at a channel inlet pressure of 3 MPa and various inlet two-phase mass flow rates. As depicted in Fig. 5, the data points closely align with the 45-degree line, demonstrating good agreement between the experimental measurements and the predictions obtained using the modified-R5.

Table 2 summarizes the root mean square error (RMSE) of pressure predictions along the test section obtained by using modified-R5 and unmodified-R5, for the examined experimental inlet channel pressures.

The modified-R5 improves predictions of experimental data, with an RMSE that is smaller than that obtained by the unmodified-R5.

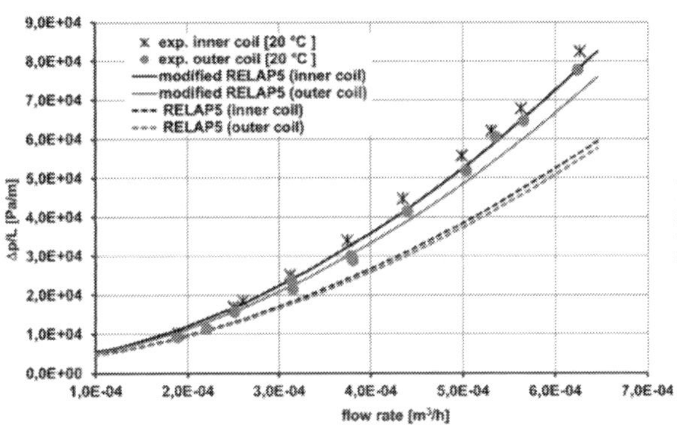

Figure 3. Experimental pressure drops in the outer and inner coils as a function of flow rate, at 20 °C endpoint temperature [12], and results by modified-R5 and unmodified-R5.

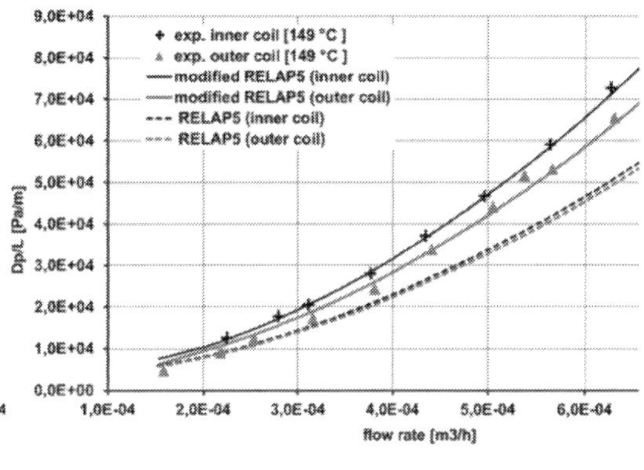

Figure 4. Experimental pressure drops in the outer and inner coils as a function of flow rate, at 149 °C endpoint temperature [12], and results by modified-R5 and unmodified-R5.

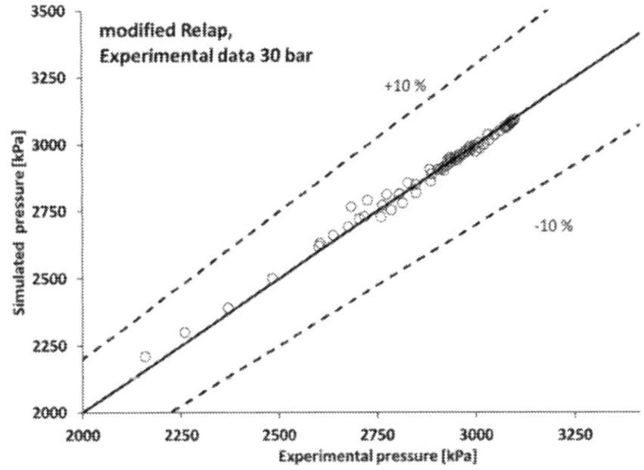

Figure 5. Pressures along the coil pipe predicted by the modified-R5 compared to experimental data [12].

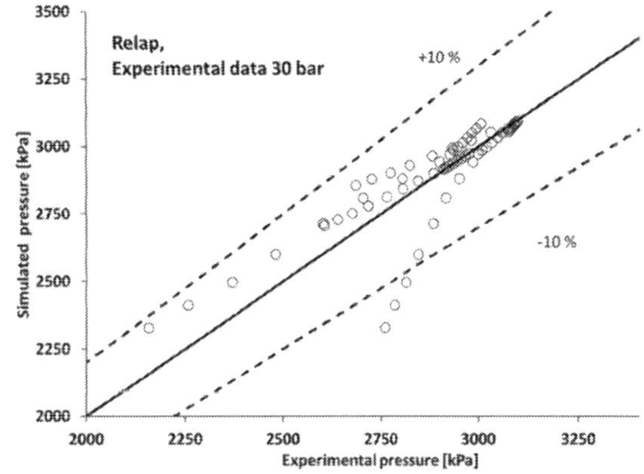

Figure 6. Pressures along the coil pipe predicted by the unmodified-R5 compared to experimental data [12].

Journal of Physics: Conference Series **2940** (2025) 012026 doi:10.1088/1742-6596/2940/1/012026

Table 2. RMSE of pressure drops calculated by modified-R5 and unmodified-R5.

Test Pressure [MPa]	RMSE, modified Relap P [kPa]	RMSE, unmodified Relap P [kPa]
3	22.52	112.73
4	28.66	102.83
6	17.8	69.38

4.2 HPs with 0.64 m and 1 m coil diameters under atmospheric pressure

The experimental facility at the Department of Energetics of Polytechnic of Torino consists of an air/water flow loop, to which various helical test sections were connected [13].

The helical test sections, constructed from plexiglass to allow observation of flow patterns, consist of two pipes with inner and outer diameters of 0.012 m and 0.018 m, respectively. The first pipe, referred to as "coil 1," has 10.25 m length, 0.64 m diameter, and 0.485 m pitch. The second pipe, "coil 2," features 10.07 m length, 1 m coil diameter, and 0.79 m pitch. The curvature ratios, d/D, for the two coils are 0.01875 and 0.012, respectively. Water was circulated at flow rates ranging from 200 to 800 l/h, while the air flow rate varied from a minimum of 0.043 g/s to a maximum of 0.26 g/s. Pressure drops were recorded using eight pressure transducers positioned along the length of the coiled tubes. The test data for each coil diameter were simulated by setting the water flow rate and adjusting the air flow rate. This process was repeated for all the different water flow rates tested.

Fig.s 7 and 8 present comparisons between experimental pressure drop data along the test section and predicted values for various two-phase mass flow rates. Notably, discrepancies are observed in the predictions from both the modified-R5 and unmodified-R5 codes when compared to many of the experimental measurements. These differences are evident for both coil diameters.

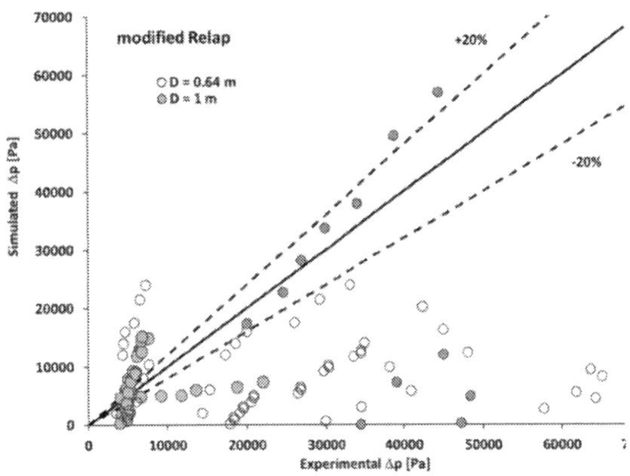

Figure 7. Predicted pressures along the coiled pipe using the modified-R5 compared to experimental data at atmospheric pressure [13].

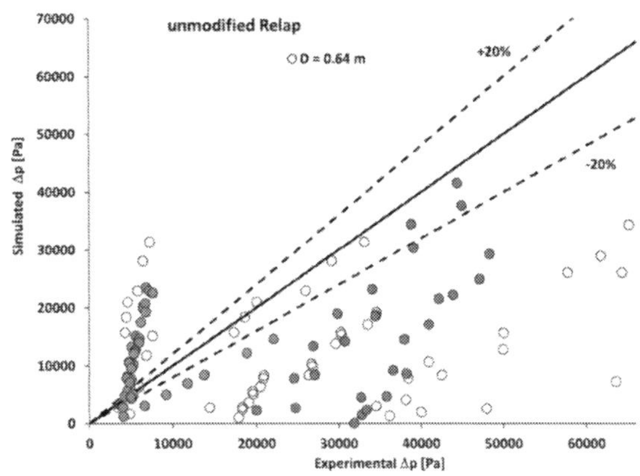

Figure 8. Predicted pressures along the coiled pipe using the unmodified-R5 compared to experimental data at atmospheric pressure [13].

5 Conclusions

The Relap5/Mod3.3 code was used to predict pressure drops in vertical HPs for both single-phase and two-phase flows. This code was enhanced with specific correlations designed to capture key aspects of the complex fluid behavior in these systems.

For single-phase flow, the friction factor correlations from Ito were applied, while for two-phase flow, a modified version of the L-M method, as presented in [5], was used. Experimental data from various international and national research laboratories were employed to validate the model.

The updated Relap5 code enhances the accuracy of pressure drop predictions in HPs for different single-phase flow configurations.

In the case of two-phase flows, good results are obtained for operating conditions characterized by high pressures, while the results are insufficient for conditions at atmospheric pressure. However, the latter result was to be expected because the model reported in [5] was obtained using experimental data characterized by operating pressures higher than 2 MPa.

This aspect will be the subject of further investigations aimed at understanding the applicability of the modified Relap5 code even at lower pressures.

References

[1] International Atomic Energy Agency, 2005, Innovative small and medium sized reactors: design features, safety approaches and R&D trends, IAEA-TECDOC-1451.
[2] Fletcher CD et al., RELAP5/MOD3 Code manual, NUREG/CR–5535, INEL–95/174, Idaho National Engineering Laboratory.
[3] Ito H, 1959, Friction factors for turbulent flow in curved pipes, ASME J. Basic Eng. Trans., 81.
[4] Colombo et al., 2015, A scheme of correlation for frictional pressure drop in steam–water two-phase flow in helicoidal tubes, Chem. Eng. Sci., 123, 460-473.
[5] Ferraris DL, Marcel CP, 2020, Two-phase flow frictional pressure drop prediction in helical coiled tubes. Int. J. of Heat and Mass Transfer, 162, 120372.
[6] Lockhart RW, Martinelli RC, 1949, Proposed correlation of data for isothermal two-phase two-component flow in pipes Chem. Eng. Prog., 45, 39-48.
[7] Guo L, Feng Z, Chen X, 2001, An experimental investigation of the frictional pressure drop of steam-water two-phase flow in helical coils, Int. J. Heat Mass Transfer, 44 (2001), 2601-2610.
[8] Castiglia F, Giardina M, Morana G, De Salve M, Panella B, 2012, Analyses of single- and two-phase flow pressure drops in helical pipes using a modified RELAP5 code, Nuclear Engineering and Design, 250, September 2012, pp 585-591
[9] Chaxton KT, Collier JG, Ward JA, 1972, H.T.F.S. Correlation for Two-Phase Pressure Drop and Void Fraction in Tubes, AERE-R7162, 1972.
[10] Ali S, 2001, Pressure drop correlations for flow through regular helical coil tubes, Fluid Dyn. Res. 28, 295-310.
[11] Coronel P, Sandeep KP, 2003, Pressure drop and friction factor in helical heat exchangers under nonisothermal and turbulent flow conditions, J. of Food Process Engineering, 26(3), 285-302.
[12] Santini L, 2008, Thermalhydraulic issues of IRIS nuclear reactor helically coiled Steam Generator and Emergency Heat Removal System, PhD Thesis, Dipartimento di Ingegneria Nucleare, Politecnico di Milano.
[13] Bertani C, De Salve M, Malandrone M, Orio M, Panella B, 2007, Studio sperimentale del deflusso bifase aria-acqua in un tubo elicoidale, XXV Congresso Nazionale UIT sulla Trasmissione del Calore, Trieste, 18-20 June 2007.

Acknowledgments

The authors wish to acknowledge Dipartimento Energia, Politecnico di Torino, and Dipartimento Energia, Politecnico di Milano, for providing the experimental data used in this article.
These data have significantly enriched the quality and robustness of our analyses, thereby contributing to the validity of the presented results.

Development of a numerical model of $CaCl_2$ hydration and dehydration reaction for thermochemical energy storage

E Baccega[1,*], Y Zhang[2,*], Y Su[2] and M Bottarelli[1]

[1] Department of Architecture, University of Ferrara, Via Quartieri 8, 44121 Ferrara, Italy
[2] Department of Architecture and Built Environment, University of Nottingham, University Park, Nottingham NG7 2RD, United Kingdom

[*] e-mail: eleonora.baccega@unife.it, Yong.Zhang@nottingham.ac.uk

Abstract. Periodical variations experienced by buildings' energy demand and supply might be tackled by means of Thermal Energy Storage (TES) systems. Among these, thermochemical heat storage (TCHS) by means of thermochemical material (TCM) is one of the most promising, allowing the greatest energy storage density per unit volume, negligible heat losses and long-term storage. As part of the Horizon Europe project "ECHO", a preliminary 2D model of the TCHS system was developed with the software COMSOL Multiphysics as a tool to support the real-scale prototype design, anticipating the behaviour of the system, as well as the implementation of the control algorithm. The model was implemented to work in both charge (dehydration) and discharge (hydration) without having to change the reaction kinetics equation from time to time, only based on the instant-by-instant boundary conditions. This ability to switch from hydration to dehydration and vice versa at any time makes it possible to estimate the performance of the TCM even in the event of not being completely hydrated or dehydrated, and also to establish some relationship between the TCM charging and discharging times.

1. Introduction

The greatest energy-consumer sector is the building and construction one [1], accounting for about 75% of primary energy supply still based on fossil fuels [2], more than 30% of the consumptions on a global scale and almost 40% of the total energy-related CO_2 emissions [3]. This energy demand, however, experiences periodical variations and the same applies to energy supply, especially in case of renewable sources, whose fluctuations occur from daily to seasonal scales. Demand and supply are hardly synchronized and this non-correspondence can prevent the full exploitation of energy from renewable sources.

Thermal Energy Storage (TES) is among the most promising technologies to increase the efficiency of renewable energy sources [4]. TES, in fact, allows thermal energy to be stored by heating or cooling a medium so that the stored energy is available at a delayed time, thus overcoming the mismatch between energy supply and demand. In general, TES systems involve three steps, namely thermal charging, thermal storing and thermal discharging [5], and can be done in three ways, namely sensible heat storage (SHS), latent heat storage (LHS) and thermo-chemical storage (TCHS). Among these, TCHS emerges due to the higher energy storage density [6], smaller volume required [7] and negligible heat losses during storage [8] due to the conversion between chemical potential and heat energy.

Journal of Physics: Conference Series **2940** (2025) 012027 doi:10.1088/1742-6596/2940/1/012027

Nomenclature list	
$\alpha(t)$	reaction advancement factor, -
$\frac{\partial \alpha}{\partial t}$	reaction kinetics, s^{-1}
C_0	initial vapor concentration inside TCM, mol/m^3
C_f	final vapor concentration inside TCM, mol/m^3
$C_{(t)}$	vapor concentration inside TCM at time t, mol/m^3
c_s	salt specific heat, J/(kg·K)
c_v	vapor specific heat, J/(kg·K)
D_g	water vapor diffusion coefficient, m^2/s
ε	salt bed porosity, -
ΔH	reaction enthalpy, J/mol
λ_s	salt thermal conductivity, W/(m·K)
M_s	salt molecular mass, g/mol
M_v	vapor molecular mass, g/mol
p_{eq}	equilibrium pressure, Pa
p_{ref}	reference pressure, Pa
p_v	vapor pressure, Pa
ρ_s	salt density, kg/m^3
ρ_v	vapor density, kg/m^3
$\dot{q}$	volumetric heat source, W/m^3
R	universal gas constant, J/(mol·K)
ΔS	reaction entropy, J/(mol·K)
S_w	mass source of water vapor, kg/(m^3·s)
z	stoichiometric number, -

In a thermo-chemical system, a reactant ($A{\cdot}xB$) absorbs heat and then decomposes into two products (A and B) in the so-called charging phase, during which the absorbed heat is converted in stable chemical potential energy, which is then stored. The reversed reaction, renamed discharging phase, allows the recombination of A and B and the conversion of the chemical potential energy back into thermal energy. This reversible reaction is described as:

$$A \cdot xB_{\text{solid}} + \Delta H \rightleftarrows A \cdot (x-y)B_{\text{solid}} + yB_{\text{gas}} \tag{1}$$

TCHS can be done by means of the so-called thermochemical materials (TCMs), such as salt hydrates like magnesium chloride ($MgCl_2{\cdot}6H_2O$) [9], calcium chloride ($CaCl_2{\cdot}6H_2O$) [10], strontium bromide ($SrBr_2{\cdot}6H_2O$) [11] or magnesium sulphate ($MgSO_4{\cdot}7H_2O$) [12] which are widely used due to their low cost, non-toxicity and regeneration temperature below 150°C.

Among these, $CaCl_2$ emerges as one of the most promising, due to its accessibility, cost-effectiveness, chemical stability, non-toxicity but most importantly its exceptional water adsorption capacity and energy storage potential. For these reasons, $CaCl_2$ is considered to be highly suitable in case of TES applications in buildings.

Depending on the boundary conditions in terms of temperature and vapor pressure, hydration and dehydration of CaCl2 occurs in multiple steps. Each of these steps can be expressed through the general reactions:

$$\begin{aligned} CaCl_2 \cdot 6H_2O &\rightleftarrows CaCl_2 \cdot 4H_2O + 2H_2O \\ &\rightleftarrows CaCl_2 \cdot 2H_2O + 4H_2O \\ &\rightleftarrows CaCl_2 \cdot H_2O + 5H_2O \rightleftarrows CaCl_2 + 6H_2O \end{aligned} \tag{2}$$

Journal of Physics: Conference Series **2940** (2025) 012027 doi:10.1088/1742-6596/2940/1/012027

However, pure salt hydrates are characterised by poor heat and mass transfer properties, which cause the relatively low thermal efficiency of the TCES system. Moreover, agglomerate and swelling of pure salt hydrates due to excessive absorption of water result in poor system cycle stability [13]. To solve these problems, researchers have developed the so-called SIM (*salt in matrix*), made of salt hydrates embedded in porous adsorbent materials, e.g., zeolites [14], expanded vermiculite [15] or expanded graphite [16]. The porous structure and the high thermal conductivity of the host matrix can effectively improve heat and mass transfer performance of TCMs, avoiding at the same time agglomeration and expansion, thus enhancing the system thermal efficiency and cycle stability [17]. On one side, the addition of inert materials can greatly extend the number of cycles of the salt, but on the other side, since the matrix is not involved in the adsorption and desorption process, the overall energy density of SIM is generally lower than that of pure salt.

The above-mentioned challenges can be addressed not just at material scale, but also at reactor scale. A reactor, in fact, should be carefully designed to facilitate heat and mass transfer and maximise the energy density while preventing the corrosion of the components and the agglomeration of the material. Moulakhnif et al. [18] classified TCM reactors into three main categories, namely packed bed reactors, modular reactors and moving bed reactors.

Packed bed reactors are the most used due to their simplicity, relatively low cost and easy maintenance. Researchers have focused on packed bed with the aim of minimizing heat losses and enhancing heat and mass transfer. For instance, in [19] a packed bed reactor was designed at lab scale using zeolite and cement composites along with strontium chloride, being able to enhance the power output and the temperature lift, reaching energy densities of 108-138 kWh/m^3. In [20] a packed bed reactor using magnesium chloride was designed, concluding that decreasing the reactor area ratio the charging time can be reduced but the pressure drop is increased. In [21] a metal mesh-packed reactor was used, obtaining an increase of the thermal efficiency and of the energy storage density. Modular reactors are studied because of their flexibility which allows the enhancement heat and mass transfer. In [22] a modular packed bed reactor with 8 layers was built using strontium bromide as salt, investigating heat and mass transfer changes based on different boundary conditions. Results showed the permeability of the reactive bed changing with the density changing. In [23] an open sorption pipe using calcium chloride embedded in vermiculite was designed, concluding that increasing the input absolute humidity could lead to an increase of the power output, but experiencing heat losses to the environment. In [24] a packed bed reactor was designed made of two reactors that could be used both in series as well as in parallel, obtaining interesting results in terms of power output and material energy density. In moving bed reactors, where the TCM is continuously moving, heat transfer inside the bed is enhanced and more uniform temperatures can be obtained. In [25] a moving bed reactor was designed using strontium bromide as storage. Despite reaching uniform temperatures of the bed, the material suffered from salt agglomeration. In [26] a circular moving bed reactor was realised with a vibrating bed, obtaining an increase of heat and mass transfer and, thanks to the vibration of the bed, preventing salt agglomeration.

As part of the Horizon Europe project "ECHO" (Efficient Compact Modular Thermal Energy Storage System) [27], a novel TES system based on a closed-loop thermochemical reactor will be developed. In order to support the design of this novel TES system, anticipating the behaviour of the TCM, preliminary numerical models were developed. Focus of this part of the research was the development of a model for both charge and discharge of the TCM that was able to automatically switch from one to the other, solely based on the boundary conditions. The first results obtained from the simulations are hereafter reported.

2. Numerical model

The first attempts to simulate the behaviour of the hydrated salt were done by only implementing the volume filled with salt, hereafter named TCM bed, using the software COMSOL Multiphysics. With the intention of proceeding in parallel with the ongoing experimental activities in the laboratory, the TCM bed in the numerical model has the same dimensions as the one in the experimental set up. The bed is modelled therefore as a cylinder, with diameter of 0.1 m and height of 0.05 m. To keep the

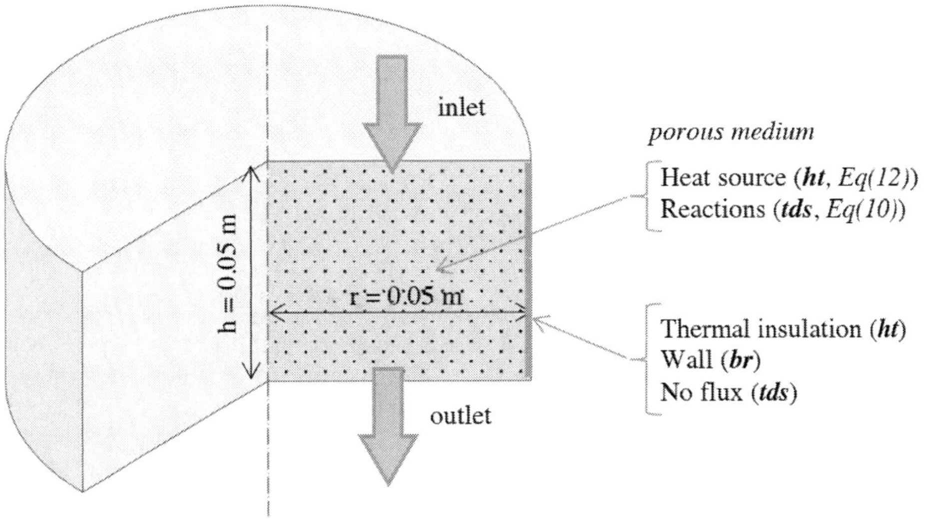

Figure 1. Geometry implemented in COMSOL Multiphysics along with the conditions imposed: ***ht*** stands for the "*Heat transfer in porous media*" module, ***br*** stands for the "*Brinkman Equations*" fluid flow module, while ***tds*** stands for the "*Transport of diluted species in porous media*" module.

numerical model as simple as possible, the geometry was modelled as 2D-axial symmetric. For the fluid flow across the porous TCM bed the "*Brinkman Equations*" module was used, for the heat storage and release the "*Heat Transfer in porous media*" module was used, while for the water vapor adsorption and desorption the "Transport of diluted species in porous media" module was used. In Figure 1 the geometry modelled and the conditions assigned are depicted.

Although the charging and discharging reactions of $CaCl_2$ occur in multiple steps as in equation (2), at first explorations were carried out by considering only one step, from a complete anhydrous state up to the complete hydration, at 6 molecules of water, and vice versa. The TCM bed is considered as immobile, with a spatially uniform bed porosity. Moreover, at this stage the thermo-physical properties, namely the thermal conductivity and the specific heat, are not function of temperature.

2.1. Mathematical formulation

2.1.1. Reaction kinetics. The reaction kinetics of the salt hydrate, in this case $CaCl_2$, depends on three variables, namely temperature T, a time-dependent reaction advancement factor α and pressure p as follows [28]:

$$\frac{\partial \alpha}{\partial t} = k(T) \cdot f(\alpha) \cdot h(p) \tag{3}$$

The reaction advancement factor α reflects the overall conversion degree and is defined between 0, corresponding to a complete dehydration state of the salt, and 1, when the salt is completely hydrated, given by:

$$\alpha(t) = \frac{C_{(t)} - C_0}{C_f - C_0} \tag{4}$$

where C_0 and C_f are the initial and final vapor concentration inside the TCM, respectively. $f(\alpha)$ is function of the advancement factor, that can be expressed as:

$$f(\alpha) = \alpha^c \cdot (1 - \alpha)^d \tag{5}$$

Journal of Physics: Conference Series **2940** (2025) 012027 doi:10.1088/1742-6596/2940/1/012027

where c and d are two constant parameters. More specifically, during charging c is equal to 1 and d is equal to 0, while during discharging c is equal to 0 and d is equal to 1.

As regards the pressure dependence, the term $h(p)$ can be expressed in case of hydration as:

$$h(p) = 1 - \frac{p_{\mathrm{eq}}}{p_{\mathrm{v}}} \tag{6}$$

while during dehydration as:

$$h(p) = \frac{p_{\mathrm{v}}}{p_{\mathrm{eq}}} - 1 \tag{7}$$

where p_v and p_{eq} are the partial and the equilibrium pressure of vapor, respectively. With the assumption that mass transfer and chemical reaction occur rapidly enough to keep the process at the prevailing temperature, the equilibrium pressure p_{eq} and temperature T can be correlated by the Clausius-Clapeyron relationship:

$$\ln(c) = -\frac{\Delta H}{R \cdot T_{\mathrm{ref}}} + \frac{\Delta S}{R} \tag{8}$$

where p_{ref} is the reference pressure [Pa], ΔH is the reaction enthalpy [J/mol], T_{ref} is the reference temperature [K], ΔS is the reaction entropy [J/(mol·K)].

2.1.2. Mass conservation. With the assumption that no mass is exchanged between the TCM bed and the surrounding environment, the density of the hydrated salt decreases throughout the charging process while the density of the dehydrated salts and water vapor increase. The mass transfer equation for the water vapor can be defined as:

$$\varepsilon \frac{\partial \rho_{\mathrm{v}}}{\partial t} = S_{\mathrm{w}} - \nabla(\rho_{\mathrm{v}} \vec{u}) + D_{\mathrm{g}} \Delta \rho_{\mathrm{v}} \tag{9}$$

where ε is the porosity of the salt bed, D_g is the water vapor diffusion coefficient in the porous bed while S_w is the mass source, which can be expressed as follows:

$$S_w = z \rho_{\mathrm{s}} \frac{\partial \alpha}{\partial t} \frac{M_{\mathrm{v}}}{M_{\mathrm{s}}} \tag{10}$$

where z is the stoichiometric number of the reaction and $\frac{M_{\mathrm{v}}}{M_{\mathrm{s}}}$ is the ratio between the molecular mass of vapor and that of the salt.

2.1.3. Energy conservation. The energy balance in the salt bed can be defined as:

$$(1-\varepsilon)\rho_{\mathrm{s}} c_{\mathrm{s}} \frac{\partial T}{\partial t} = \nabla(\lambda_{\mathrm{s}} \nabla \mathrm{T}) - \rho_{\mathrm{v}} c_{\mathrm{v}} \vec{u} \nabla \mathrm{T} + \dot{q} \tag{11}$$

where ρ_s is the density of the salt, c_s and c_v are the specific heat of the salt and of the water vapor, respectively, λ_s is the salt thermal conductivity and $\dot{q}$ is the energy source term that can be expressed as:

$$\dot{q} = \pm z \frac{\partial \alpha}{\partial t} \frac{\rho_{\mathrm{s}}}{M_{\mathrm{s}}} \Delta H \tag{12}$$

In case of hydration the sign is positive, hence the reaction is exothermic, while during dehydration, which is endothermic, the sign is negative.

2.2. Charging and discharging conditions

In order for the model to switch autonomously from charge to discharge of the TCM, it was necessary to define under which conditions the TCM was actually being charged rather than discharged. The factors affecting the reaction of the material are the temperature and the vapor content of the air passing through it.

The charging of the TCM, hence its dehydration, can be done by working on the air passing through it: (i) increasing the temperature, (ii) decreasing the vapor content, (iii) both of the above-mentioned. On the contrary, the discharging of the TCM can be done: (i) decreasing the temperature of the air passing through, (ii) increasing the vapor content, (iii) both decreasing the temperature and increasing the vapor content. In Figure 2 the light blue area indicates that the material at those conditions in terms of temperature and vapor pressure is hydrated, while the pink one indicates that the TCM is dehydrated. The darker areas graphically highlight the aforementioned conditions for charging (a) and discharging (b).

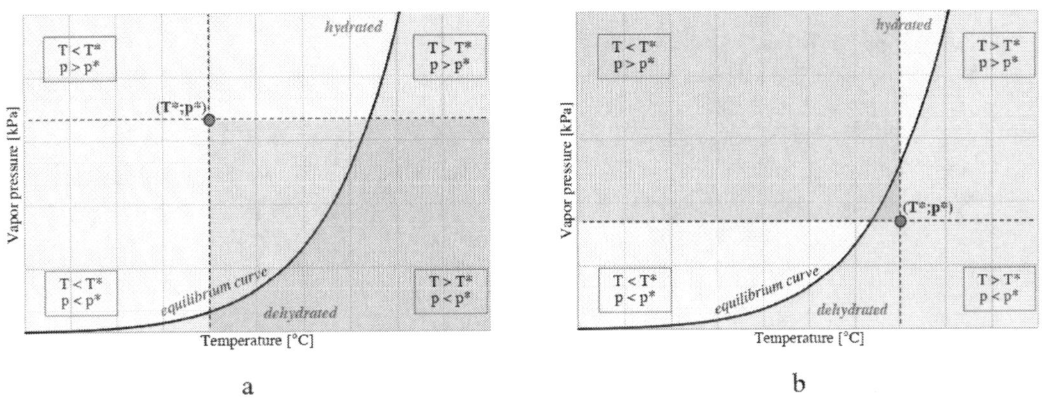

Figure 2. Darker rectangular areas depicting the TCM charging (a) and discharging (b) conditions.

The conditions above identified are summarised in Table 1, as assigned to the model.

Table 1. Temperature and vapor pressure conditions set in the numerical model

	$T_{inlet} > T^*$	$T_{inlet} < T^*$	$T_{inlet} = T^*$
$C_{inlet} > C^*$	-	**hydration**	**hydration**
$C_{inlet} < C^*$	**dehydration**	-	**dehydration**
$C_{inlet} = C^*$	**dehydration**	**hydration**	-

T_{inlet}: temperature of inlet air passing through the TCM
C_{inlet}: Vapor concentration of inlet air passing through the TCM
T*: instantaneous TCM temperature
C*: instantaneous vapor concentration

3. Results

At first simulations were carried out to verify whether the model was able to switch between charge and discharge solely based on the boundary conditions. However, to give more consistency to the implemented model, the results were compared to the data acquired during the experimental tests carried out in lab. As previously anticipated, the simplification of considering both hydration and dehydration as a single step was adopted, hence the comparison with experimental data is purely qualitative. It should also be pointed out that for this comparison the properties of the TCM have been slightly adjusted so that the multiple reactions could be merged into a single one. The key parameters assigned to the model are reported in Table 2.

For the comparison, experimental data shown in Figure 3 were used. The test consisted of an initial hydration of the TCM, followed by a partial dehydration and then a hydration phase again. During the hydration phases, a humidifier was used to increase the absolute humidity of the air passing through the TCM. Throughout the entire test ambient air was also constantly heated by means of an electrical heater before passing through the TCM bed. However, during the hydration phases a decrease in the inlet temperature is visible, which was due to evaporative cooling.

Table 2. Key parameters assigned to the model

Parameters	Description	Value	Unit
M_v	Molecular mass of water vapor	18.02	g/mol
M_s	Molecular mass of $CaCl_2$	110.99	g/mol
ρ_s	Density of $CaCl_2$	280*	kg/m^3
λ_s	Thermal conductivity of $CaCl_2$	0.6	W/(m·K)
c_s	Specific heat of $CaCl_2$	800	J/(kg·K)
D_g	Gas diffusion coefficient	$1.685 \cdot 10^{-6}$	m^2/s
ΔH	Reaction enthalpy	65	kJ/mol
ΔS	Reaction entropy	170	J/(mol·K)
R	Ideal gas constant	8.314	J/(mol·K)
u	inlet air velocity	0.3	m/s
ε	TCM bed porosity	0.8	-
z	stoichiometric number	6	-

* at this stage the vermiculite was not considered, hence the density was adjusted so that the amount of reacting salt was the same as the experimental tests

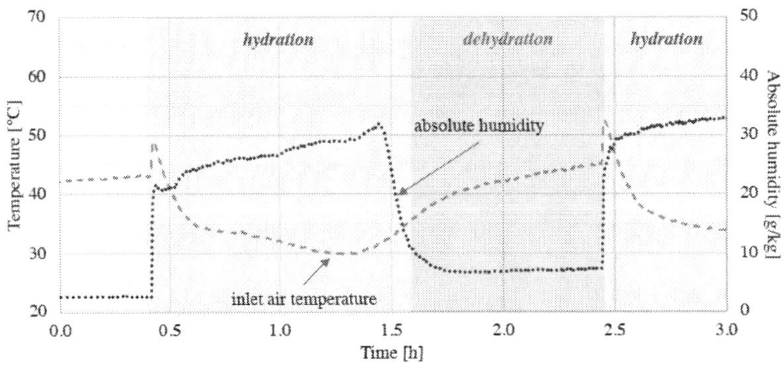

Figure 3. Boundary conditions in terms temperature and absolute humidity of the inlet air through the TCM.

The results obtained in terms of reaction advancement factor, α, and reaction rate, $\partial\alpha/\partial t$, are reported in Figure 4. Starting from a complete dehydration of the TCM, the first hydration phase between 0.5 h and 1.5 h brought the reaction at about 60%, which means that the material absorbed about 60% of the total amount of water that could be absorbed. As regards the reaction rate, the first peak can be ascribed to the drastic increase in the absolute humidity of the inlet air, which then remained more regular between 20 g_v/kg_{as} and 30 g_v/kg_{as}. During the second phase, the humidifier was turned off but the electrical heater was kept constant, hence the dehydration reaction is slightly visible due to the low inlet air temperature, at about 40°C-45°C. The reaction advancement factor barely decreased and also the reaction rate was close to 0 s^{-1}. During the third and last phase, the humidifier was turned on again and the TCM continued to hydrate. Starting from a reaction advancement factor close to 0.6, this last phase brought the reaction at almost 0.8, thus the reaction was not complete yet. As before, the peak that can be observed at the beginning of the second hydration phase can be ascribed to the drastic increase in the absolute humidity of the inlet air.

The comparison in terms of temperature was carried out at this stage by comparing the temperature monitored close to the middle of the TCM bed with the simulated one in the same position. It was decided not to compare the experimental and the simulated outlet temperatures because the sensor in the experimental set up was not placed close to the TCM bed. From this comparison, depicted in Figure 5a, it emerges that the experimental and simulated trends are similar, but there are consistent differences, even more than 5 K. It is worth reiterating that at this stage of the research it was decided to proceed by simplifying the various reaction steps, and adopting a single value for the reaction enthalpy and reaction entropy inevitably leads to deviations of the simulated data from the experimental ones.

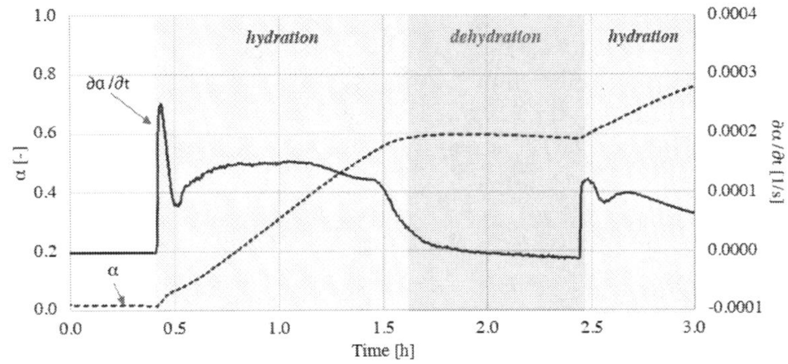

Figure 4. Reaction advancement factor (α) and reaction rate ($\partial\alpha/\partial t$) obtained from the simulation.

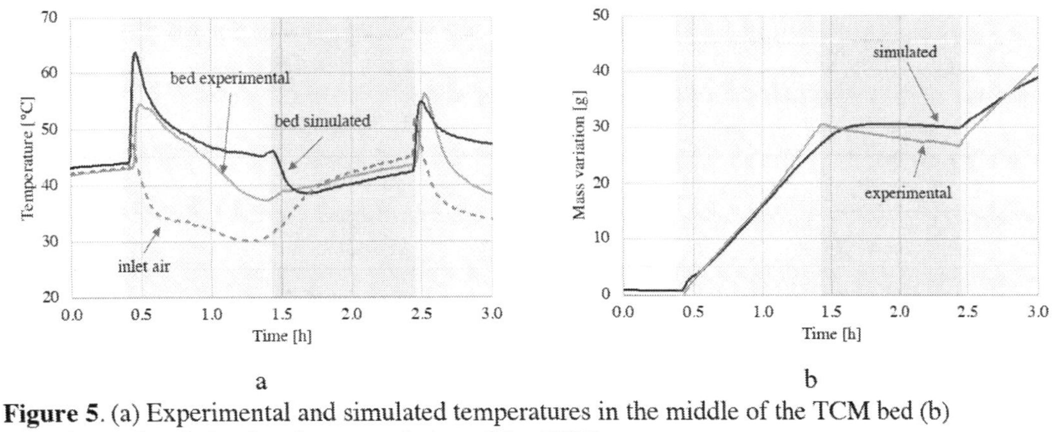

Figure 5. (a) Experimental and simulated temperatures in the middle of the TCM bed (b) Experimental and simulated mass variation of the TCM.

The comparison between experimental and simulated mass variation of the TCM is reported in Figure 5b. The trend of the two curves is quite similar, with the most visible difference during the dehydration phase, and at the end of three hours the experimental mass variation is of 41.2 g while the simulated one is 38.9 g, with a difference therefore of 2.3 g.

Based on these results, despite the simplification and the merging of the different reaction steps into a single one, it can be concluded that the conditions set allow the model to automatically switch from charge to discharge and vice versa. Further efforts will be focused on the implementation of the different reaction steps as reported in equation (2). Consequently, more extensive explorations will be carried out regarding the behaviour of the TCM and in order to establish relationships between charging and discharging times.

4. Conclusions

As part of the Horizon Europe project "ECHO", a preliminary numerical model was realized using COMSOL Multiphysics in order to support the design of a novel closed loop thermochemical reactor. At this stage the focus of the research was the implementation of a model for both charging (dehydration) and discharging (hydration) of the TCM. The first results obtained from the simulations confirmed the model was able to automatically switch from charging to discharging and vice versa, solely based on the boundary conditions. To give more consistency to the results, a comparison between experimental and simulated data was carried out.

Despite the simplifications adopted at this stage, the trends of the experimental and numerical curves are quite similar and, in particular, the simulated TCM mass variation deviates by only 2.3 g from the experimental one. Further efforts will be devoted on the implementation of the different hydration and dehydration steps, which will then be accurately validated against experimental data.

5. References

[1] Sinka M, Bajare D, Jakovics A, Ratnieks J, Gendelis S and Tihana J 2019 *Energy Build* **195** p 205
[2] Lizana J, Chacartegui R, Barrios-Padura A and Valverde J M 2017 *Applied Energy* **203** p 219
[3] Asadi E, da Silva M G, Antunes C H and Dias L 2012 *Energy Build* **44** p 81
[4] Navarro L, de Gracia A, Niall D, castell A, Browne M, McCormack S J, Griffiths P and Cabeza L F 2016 *Renew Energy* **88** p 526
[5] Akeiber H, Nejat P, Majid M Z A, Wahid M A, Jomehzadeh F, Famileh I Z, Calautit J K, Hughes B R and Zaki S A 2016 *Renew Sustain Energy Rev* **60** p 1470
[6] Michel B, Mazet N, Mauran S, Stitou D and Xu J 2012 *Sol Energy* **47** p 553
[7] Lin J, Zhao Q, Huang H, Mao H, Liu Y and Xiao Y 2021 *Sol Energy* **214** p 149
[8] Aydin D, Casey S P and Riffat S 2015 *Renew Sustain Energy Rev* **41** p 356
[9] Chen W, Li W and Zhang Y 2018 *Appl Therm Eng* **135** p 95
[10] Ait Ousaleh H, Sair S, Mansouri S, Abboud Y, Zahouily M, Faik A, and el Bouari A 2022 *J Energy Storage* **49**
[11] Ding B, Xu C, Liao Z and Ye F 2021 *J Energy Storage* **42**
[12] Miao Q, Zhang Y, Jia X, Tan L and Ding Y 2021 *Sol Energy* **220** p 432
[13] Shkatulov A I, Houben J, Fischer H and Huinink H P 2020 *Renew Energy* **150** p 990
[14] Wang Q, Xie Y, Ding B, Yu G, Ye F and Xu C 2019 *Sol Energy Mater Sol Cell* **200**
[15] Ding B, Xu C, Liao Z and Ye F 2021 *J Energy Storage* **42**
[16] Li W, Klemes J J, Wang Q and Zeng M 2021 *J Clean Prod* **285**
[17] Li W, Klemes J J, Wang Q and Zeng M 2020 *Renew Energy* **157**
[18] Moulakhnif K, Ousaleh H A, Sair S, Bouhaj Y, El Majd A, Ghazoui M, Faik A and El Bouari A 2024 *Energy Build* **318**
[19] Clark R-J and Farid M 2022 *Applied Energy* **316**
[20] Hawwash A A, Hassan H, and El Feky K 2020 *Appl Therm Eng* **168**
[21] Ji W, Zhang H, Liu S, Wang Z and Deng S 2022 *Renew Energy* **194**
[22] Michel B, Mazet N and Neveu P 2016 *Applied Energy* **180**
[23] Aydin D, Casey S P, Chen X and Riffat S 2016 *Energy Convers Manag* **121** p 321
[24] Johannes K, Kuznik F, Hubert J-L, Durier F and Obrecht C 2015 *Applied Energy* **159**
[25] Farcot L, Le Pierres N and Fourmigue J-F 2019 *J Energy Storage* **26**

[26] Wyttenbach J, Bougard J, Descy G, Skrylynk O, Courbon E, Frere M and Bruyat F 2018 *Applied Energy* **230**
[27] https://echo-euproject.eu
[28] Fopah Lele A, Kuznik F, Rammelberg H U, Schmidt T and Ruck W K L 2015 *Applied Energy* **154** p 447

Acknowledgments
This work was financially supported within the ECHO project (Efficient Compact Modular Thermal Energy Storage System) funded by the European Union's Horizon Europe research and innovation programme under Grant Agreement No. 101096368.

Journal of Physics: Conference Series **2940** (2025) 012028 doi:10.1088/1742-6596/2940/1/012028

Stability analysis of an anisotropic uniformly heated/cooled porous layer

A Barletta[1,*], **M Celli**[1], **P V Brandão**[1], **S Lazzari**[2], **E Ghedini**[1]

[1] Department of Industrial Engineering, Alma Mater Studiorum Università di Bologna, Viale Risorgimento 2, 40136 Bologna, Italy

[2] Department of Architecture and Design, Università di Genova, Stradone S. Agostino 37, 16123 Genova, Italy

* email: antonio.barletta@unibo.it

Abstract. The flow resulting from the combination of forced and natural convection within a horizontal porous layer filled with Newtonian fluid is analyzed. The layer boundaries are subjected to uniform and symmetrical heating/cooling. The porous layer is anisotropic such as the permeability is function of the coordinates, and a fully developed flow occurs in the layer, where the temperature is a function of both spanwise and streamwise directions. The latter steady basic state undergoes the onset of thermal instability for adequate values of the Darcy–Rayleigh number, with the neutral stability condition being heavily influenced by the Péclet number, which is linked to the mass flow rate of the basic state. A streamfunction approach is applied and the normal modes method is employed to investigate the linear stability of the basic state. The system of ordinary differential equations obtained is tackled numerically and the critical values of the governing parameters are studied as functions of the Péclet number and of the permeabilities fraction.

1. Introduction

The emergence of buoyancy driven convection in fluid saturated porous layers has been extensively studied. Within this field, the foundational works by Horton and Rogers [1], along with Lapwood [2], were the first to establish the key characteristics of this phenomenon. These researchers explored and solved what is now referred to as the Horton–Rogers–Lapwood (HRL) or Darcy–Bénard (DB) problem. They identified the threshold values of the governing parameters for the triggering of buoyancy driven instability in a porous medium heated from below, which is the porous equivalent of the Rayleigh–Bénard problem. The basic state studied by HRL is a still fluid characterized by a negative transverse, to the porous layer, temperature gradient. The Darcy–Rayleigh number is employed to define the stability threshold. The modified HRL problem where the basic state is characterized by a moving fluid had been studied by Prats [3]. He found that the moving fluid does not affect the threshold value of the Darcy–Rayleigh number.

When symmetric heat fluxes at the walls (either heating or cooling) are introduced in the HRL problem, a stationary rest state no longer exists. In this case, a non–zero mass flow rate is required to balance the energy under steady conditions. With a given horizontal pressure gradient and the given thermal boundary conditions, the temperature depends both on the streamwise and on the spanwise directions. Barletta [4] studied the stability of this configuration, focusing on an isotropic porous layer, where permeability is uniform in all directions. Anisotropic porous materials, however, are common in real–world applications, and the growing field of 3D printing yields an easier fabrication of engineered porous media displaying these features. The stability of anisotropic porous layers against the onset of thermal convection has been investigated for classical configurations in [5, 6].

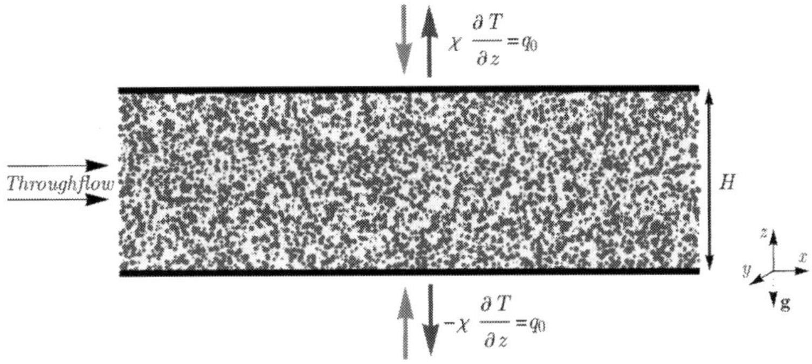

Figure 1. Qualitative drawing of the problem investigated

This study is focused on examining how non isotropic porous structures impact the triggering of buoyancy driven instabilities when the thermal boundary conditions are of symmetric heat flux. To this end, the value of streamwise direction permeability is considered to be different to the permeability of the spanwise direction. The stability analysis is carried out by integrating the ordinary differential equations obtained by assuming that the disturbances superposed to the basic state are in the form of Fourier modes. To reduce the complexity of the problem, the dependence on the streamwise coordinate y is parametrized by imposing an inclined basic flow. Given the two–dimensional nature of the stability problem, a streamfunction–temperature formulation is used. The neutral stability curves and critical values are presented for given values of the basic flow inclination angle, of the Péclet number and of the ratio of the two different permeabilities employed.

2. Mathematical model

Consider a porous layer saturated with fluid, having a height of H and extending infinitely in the horizontal directions x and y. The layer boundaries are both heated or both cooled (see Fig. 1) and a horizontal throughflow is assumed. To study the mass transport, Darcy's law is employed where we apply the Oberbeck-Boussinesq approximation [7]. The local energy balance equation is a conduction/convection equation for fluid saturated porous media while the local mass balance equation is the one for incompressible fluids, namely

$$
\begin{gathered}
\nabla \cdot \mathbf{u} = 0, \\
\frac{\mu}{K_h} u = -\frac{\partial p}{\partial x}, \quad \frac{\mu}{K_h} v = -\frac{\partial p}{\partial y}, \quad \frac{\mu}{K_z} w = -\frac{\partial p}{\partial z} + \rho_0 g \beta (T - T_0), \\
\rho_0 c \left(\sigma \frac{\partial T}{\partial t} + \mathbf{u} \cdot \nabla T \right) = \chi \nabla^2 T, \\
z = 0: \qquad w = 0, \quad -\chi \frac{\partial T}{\partial z} = q_0, \\
z = H: \qquad w = 0, \quad \chi \frac{\partial T}{\partial z} = q_0.
\end{gathered} \tag{1}
$$

The nomenclature follows the classical choice of (x, y, z) for the position vector, $\mathbf{u} = \{u, v, w\}$ for the Darcy velocity, p for the pressure, T and t for temperature and time, μ for the viscosity, K_h for the streamwise direction permeability, K_z for the spanwise direction permeability, ρ_0 for the density of the saturating fluid, T_0 for the Oberbeck-Boussinesq temperature, g for the modulus of $\mathbf{g}$, β for the thermal expansion coefficient, c for the specific heat capacity, σ for the heat capacity ratio, χ for the effective thermal conductivity, and $q_0 \gtrless 0$ for the heat flux.

Journal of Physics: Conference Series **2940** (2025) 012028 doi:10.1088/1742-6596/2940/1/012028

The scaling employed to render the governing equations dimensionless is:

$$\frac{(x,y,z)}{H} \to (x,y,z), \quad \frac{\chi}{\rho_0 c H^2 \sigma} t \to t, \quad \frac{\rho_0 c H}{\chi} \mathbf{u} \to \mathbf{u}, \quad \frac{\rho_0 c K_h}{\chi \mu} p \to p,$$
$$\frac{T - T_0}{\Delta T} \to T, \quad \text{for} \quad \Delta T = \frac{\chi \mu}{\rho_0^2 c \beta g K_h H}. \tag{2}$$

By applying Eq. (2) to Eq. (1) one obtains

$$\nabla \cdot \mathbf{u} = 0,$$
$$u = -\frac{\partial p}{\partial x}, \quad v = -\frac{\partial p}{\partial y}, \quad w = -a \frac{\partial p}{\partial z} + aT,$$
$$\frac{\partial T}{\partial t} + \mathbf{u} \cdot \nabla T = \nabla^2 T,$$
$$z = 0: \qquad w = 0, \quad \frac{\partial T}{\partial z} = -Ra,$$
$$z = 1: \qquad w = 0, \quad \frac{\partial T}{\partial z} = Ra, \tag{3}$$

with the Darcy–Rayleigh number and the fraction between the two permeabilities defined as

$$Ra = \frac{\rho_0^2 c \beta g K_h H^2 q_0}{\chi^2 \mu}, \quad a = \frac{K_z}{K_h}. \tag{4}$$

Since $q_0 \gtrless 0$ then also $Ra \gtrless 0$.

2.1. *Stationary flow*

The steady state which stability is investigated is solution of Eq. (3), namely

$$\mathbf{u}_b = \left[Pe - \frac{Ra(2z-1)}{Pe} \right] (\cos\varphi\, \mathbf{e}_x + \sin\varphi\, \mathbf{e}_y),$$
$$\nabla T_b = \frac{2Ra}{Pe} (\cos\varphi\, \mathbf{e}_x + \sin\varphi\, \mathbf{e}_y) + \left[Ra(2z-1) - \frac{2Ra^2(z^2 - z)}{Pe^2} \right] \mathbf{e}_z, \quad \nabla p_b = (-u_b, -v_b, T_b), \tag{5}$$

where the subscript "b" stands for "basic state" and $(\mathbf{e}_x, \mathbf{e}_y, \mathbf{e}_z)$ are the unit vectors in the Cartesian directions. The definition of the Péclet number is set to obtain the average velocity on the layer cross section:

$$Pe = \int_0^1 \mathbf{u}_b \cdot \mathbf{n}\, \mathrm{d}z, \quad \text{for} \quad \mathbf{n} = \cos\varphi\, \mathbf{e}_x + \sin\varphi\, \mathbf{e}_y. \tag{6}$$

The velocity and temperature profiles in Eq. (6) reach a singularity for vanishing values of Pe. This behavior is expected since throughflow is necessary to attain a steady state when simmetric Neumann conditions are imposed at the walls. It is worth noting that this fundamental state is identical to the one analyzed by Barletta [4] for the isotropic case $a = 1$.

In the context of a mixed convection steady state solution a inclination angle φ to the x–direction is employed to parametrize the perturbation dependence on the y–direction. This angle varies within the range $[0, \pi/2]$ where $\varphi = 0$ corresponds to transverse modes, while $\varphi = \pi/2$ results in longitudinal modes. Additionally, it is important to point out that only the gradient of temperature T_b is given: inside the local energy balance equation and the boundary conditions (3) just the derivatives of the temperature are present thus the temperature scale is not defined.

3. Stability analysis

The analysis is carried out by adding perturbations to the basic state in the form of normal modes. As previously stated, we assume to parametrize the dependence on y–direction by employing the angle φ. Since the problem is formally two dimensional, it is useful to define the disturbances by using the streamfunction Ψ, as follows:

$$u = u_b + \varepsilon \frac{\partial \Psi}{\partial z}, \quad v = v_b + \varepsilon V, \quad w = -\varepsilon \frac{\partial \Psi}{\partial x}, \quad T = T_b + \varepsilon \Theta. \tag{7}$$

The streamfunction formulation makes the local mass balance an identity while one may straightforward obtain, from Darcy's law in (3), that $V = 0$. By considering the y component of the curl of Darcy's law in (3), one can encompass the presence of the pressure. By considering just terms of order $O(\varepsilon)$, the linearized equations for the disturbances are derived from (3), (5), and (7), namely

$$\begin{gathered}
\frac{\partial^2 \Psi}{\partial x^2} + a\frac{\partial^2 \Psi}{\partial z^2} + a\frac{\partial \Theta}{\partial x} = 0, \\
\frac{\partial \Theta}{\partial t} + \frac{\partial T_b}{\partial x}\frac{\partial \Psi}{\partial z} - \frac{\partial T_b}{\partial z}\frac{\partial \Psi}{\partial x} + u_b \frac{\partial \Theta}{\partial x} = \frac{\partial^2 \Theta}{\partial x^2} + \frac{\partial^2 \Theta}{\partial z^2}, \\
z = 0, 1: \qquad \Psi = 0, \quad \frac{\partial \Theta}{\partial z} = 0.
\end{gathered} \tag{8}$$

Streamfunction and temperature are thus defined in terms of normal modes as

$$\Psi(z,t) = i\, f(z)\, e^{i(kx-\omega t)}\, e^{\eta t}, \qquad \Theta(z,t) = h(z)\, e^{i(kx-\omega t)}\, e^{\eta t}, \tag{9}$$

where k is the wavenumber of the single Fourier mode, η the growth rate of the mode and ω the angular frequency of the mode. By employing (9), the (8) can be rewritten, namely

$$\begin{gathered}
a f'' - k^2 f + a k h = 0, \\
h'' - \left(k^2 + i k u_b + \eta - i\omega\right) h - i\frac{\partial T_b}{\partial x} f' - k\frac{\partial T_b}{\partial z} f = 0, \\
z = 0, 1: \qquad f = 0, \quad h' = 0.
\end{gathered} \tag{10}$$

Barletta [4] proved that longitudinal modes ($\varphi = \pi/2$) are the most unstable modes. We thus assume $\varphi = \pi/2$, leading to $u_b = 0$ and $\partial T_b/\partial x = 0$, as can be inferred from (5). With this assumption (10) reduces to:

$$\begin{gathered}
a f'' - k^2 f + a k h = 0, \\
h'' - \left(k^2 + \eta - i\omega\right) h - k\left[Ra(2z-1) - \frac{2Ra^2(z^2-z)}{Pe^2}\right] f = 0, \\
z = 0, 1: \qquad f = 0, \quad h' = 0.
\end{gathered} \tag{11}$$

It is worth noting that the system (11) does not change when the transformation

$$Ra \to -Ra, \quad z \to 1 - z \tag{12}$$

is applied. The invariance of the system (11) on the Darcy–Rayleigh number sign simplifies the analysis since lets one treat just the case $Ra > 0$.

The system of ordinary differential equations (11) is treated numerically by using the shooting method [8, 9]. The objective is to identify the critical values of the angular frequency ω and of the Darcy–Rayleigh number Ra for given values of the wavenumber k, of the Péclet number Pe, and of the permeabilities fraction a. The shooting method is performed by employing the software *Mathematica 14.0* [10]. To determine how the Darcy–Rayleigh Ra number depends on the wavenumber k for null temporal growth rates $\eta = 0$, an additional condition to eliminate the scale invariance of (f, h) is imposed, $h(0) = 1$. Finally, the boundary condition $h'(1) = 0$ is employed to obtain the pair (Ra, ω) through a root–finding process.

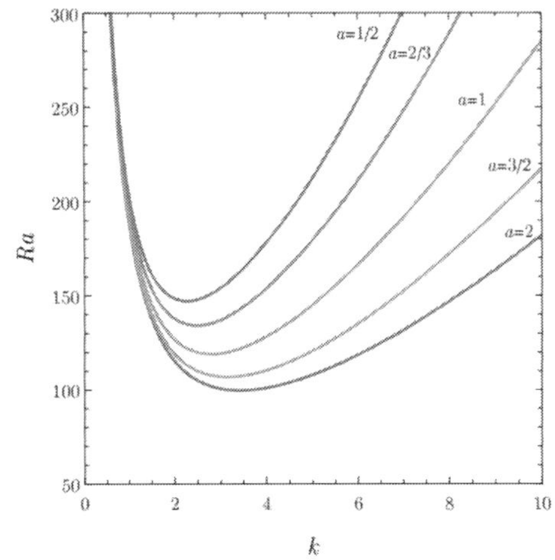

Figure 2. Darcy–Rayleigh number as a function of the wavenumber k for given values of a when $Pe \gg 1$

	$a = 1/2$		$a = 1$		$a = 2$	
Pe	k_c	Ra_c	k_c	Ra_c	k_c	Ra_c
25	2.95157	232.193	3.31888	163.326	3.85083	127.081
50	2.40635	159.844	2.90868	126.769	3.51080	104.888
100	2.30341	150.041	2.82692	120.866	3.44169	100.958
1000	2.27058	147.139	2.80097	119.081	3.41980	99.7531
∞	2.27025	147.110	2.80071	119.064	3.41958	99.7411

Table 1. Ra_c and k_c as functions of Pe

4. Discussion of the results

There is an intriguing asymptotic scenario that focuses on basic states characterized by $Pe \gg 1$. This situation becomes evident in equations (11) by assuming that $Ra^2/Pe^2 \to 0$. For this particular case, the curves $Ra(k)$ for null growth rate (neutrally stable condition) are presented in Fig. 2 for various permeabilities fraction a. Compared with the case ($a = 1$), we can clearly observe destabilization for $a > 1$ and a stabilization for $a < 1$. The configurations prescribing linear instability are, in fact, those values of k and Ra which identify those points, in Fig. 2, lying above the curves: the threshold values of the governing parameters for the onset of thermal convection are obtained by finding the minima of the curves. These minima let one define the critical conditions for the triggering of convection (k_c, Ra_c). Therefore, linear instability is only possible when $Ra > Ra_c$ and, from Fig. 2, it is clear that Ra_c decreases with a. This phenomenon can be physically interpreted recalling that larger values of a correspond to higher vertical permeability relative to horizontal permeability. Higher values of vertical permeability thus enhance the formation of buoyancy–driven convection cells, thereby promote instability.

Figure 3 illustrates, by fixing the permeabilities fraction $a = 1/2, 1$ and 2, the destabilizing impact of Pe: as Pe increases, the curves move downward. Figure 3 displays that the case $Pe \to \infty$ can be reasonably estimated by employing $Pe = 1000$ and even $Pe = 100$.

It is worth noting that, for the solutions reported, the angular frequency value is always null, $\omega = 0$. One can thus conclude that just non–oscillatory modes may occur, or in other words, that the principle of exchange of stabilities applies.

Numerical values of k_c and Ra_c are listed in Table 1: both a and Pe have a destabilising effect since Ra_c decreases with increasing values of the two parameters.

Journal of Physics: Conference Series **2940** (2025) 012028 doi:10.1088/1742-6596/2940/1/012028

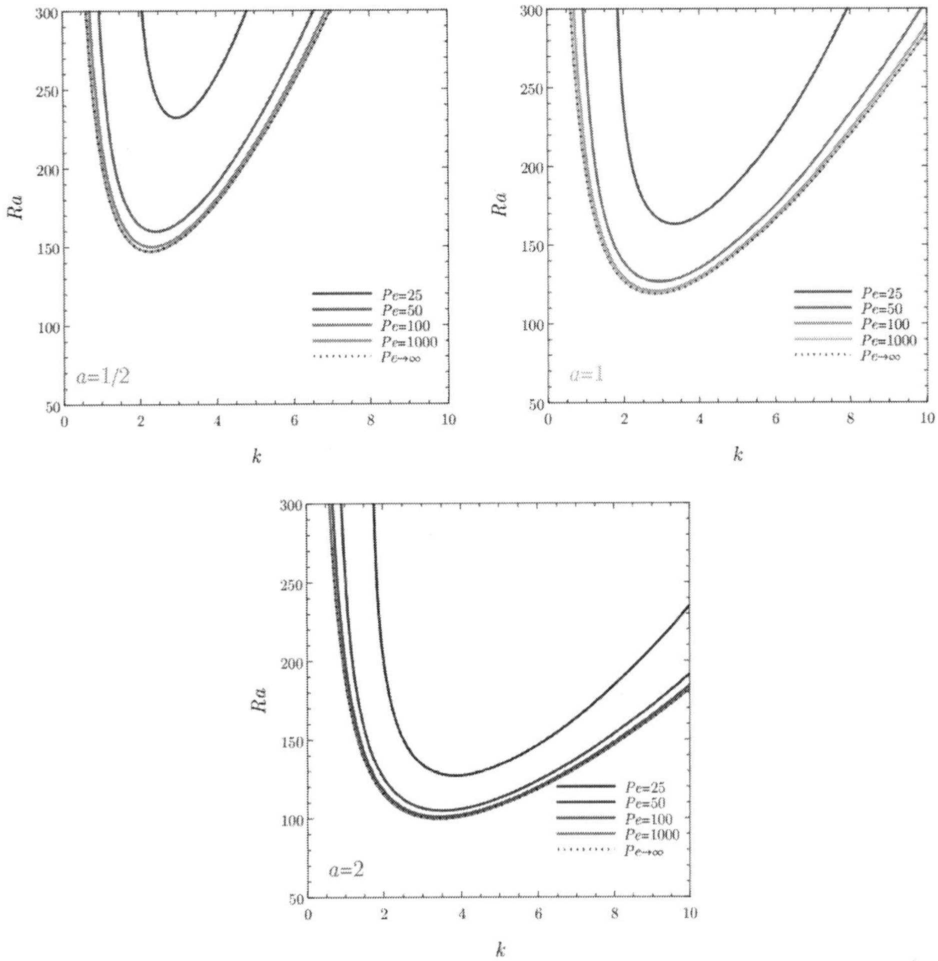

Figure 3. Darcy–Rayleigh number as a function of the wavenumber k for $a = 1/2, 1, 2$ and fixed values of Pe

5. Conclusions

The analysis of linear instability in mixed convection steady flows within a porous layer where the boundaries are subjected to uniform and symmetrical heating/cooling is performed. This analysis employed Darcy's law where the Oberbeck–Boussinesq approximation is applied. The porous layer displays different permeabilities in the spanwise and streamwise directions. A permeabilities fraction a was introduced to represent the relationship between the spanwise direction and streamwise direction permeabilities. The Darcy–Rayleigh number Ra is related to the heat flux applied to the layer walls.

The steady state which stability has been investigate is driven by both the buoyancy force and by a pressure gradient: the intensity of the flow is driven by the Péclet number Pe. The stability analysis is performed by using normal mode method. It was found that both the permeabilities fraction a and the Péclet number Pe contribute to destabilization. Specifically, a higher valued of the spanwise direction permeability compared with the streamwise direction permeability leads to unstable state for lower values of the Darcy–Rayleigh number Ra. Similarly, the destabilizing role of the Péclet number implies that increasing the net mass flow rate induces the triggering of instability for lower values of Ra. Additionally, the solutions obtained for the limiting case $Pe \gg 1$ comes out to be a valuable and accurate approximation for those case characterized by $Pe > 100$.

References

[1] Horton C W and Rogers Jr F T 1945 *Journal of Applied Physics* **16** 367–370
[2] Lapwood E R 1948 *Mathematical Proceedings of the Cambridge Philosophical Society* **44** 508–521
[3] Prats M 1966 *Journal of Geophysical Research* **71** 4835–4838
[4] Barletta A 2012 *Transport in Porous Media* **92** 419–437
[5] Trew M and McKibbin R 1994 *Transport in porous media* **17** 271–283
[6] Storesletten L and Rees D A S 2019 *Fluids* **4** 75
[7] Barletta A 2022 *Mechanics Research Communications* **124** 103939
[8] Straughan B 2013 *The Energy Method, Stability, and Nonlinear Convection* (Springer)
[9] Barletta A 2019 *Routes to Absolute Instability in Porous Media* (Springer)
[10] Wolfram Research, Inc 2024 Mathematica, Version 14.0 URL `https://www.wolfram.com/mathematica`

doi:10.1088/1742-6596/2940/1/012029

Topology optimization using variable-porosity metal foams: design of conduction paths for cooling a heat generating disc and comparison with tree-shaped architectures

N Bianco[1], **A Fragnito**[2*], **M Iasiello**[1] **and G M Mauro**[2]

[1]Dipartimento di Ingegneria Industriale, Università degli Studi di Napoli Federico II, Napoli, Italy
[2]Dipartimento di Ingegneria, Università degli Studi del Sannio, Benevento, Italy

* Corresponding author: afragnito@unisannio.it

Abstract. Topology optimization (TO) is a design algorithm providing the optimal material layout within a design domain to minimize/maximize an objective function. In thermal science, for instance, it can be used to optimize the design of heat sinks to minimize thermal compliance, entropy generation, average temperature, etc. Recently, classical TO frameworks have been enhanced in multi-material TO in order to include more materials, thereby enhancing the degrees of freedom of the system, and thus ensuring better thermal performance. This work implements both TO and multi-material TO (MMTO) to address a benchmark heat conduction problem, *i.e.,* the cooling of a circular heat generating volume through heat conduction paths. The heat generation is uniform in the disc, the rim is adiabatic, while the centre is set at a fixed temperature – Dirichlet boundary condition – and serves as heat sink. In TO the choice is between void and high-conductivity material, while in MMTO variable-porosity metal foams are integrated. The interpolation of the materials' thermal conductivity is conducted via an ordered solid isotropic material penalization (SIMP) algorithm. The distinction between materials is attained by setting different thresholds in the interpolation and projection functions. The dimensionless global thermal resistance and domain average temperature are alternately addressed as objective functions to be minimized at equal weight of the system. The findings unveil that MMTO outperforms TO, which outperforms constructal tree networks, considering in the latter case different configurations of different complexity.

1. Introduction

Devices that dissipate heat through conduction form the backbone of various engineering applications. Optimizing their design is a current priority, and numerous techniques are employed for this purpose: simulated annealing [1], genetic algorithm [2], constructal law principles [3], topology optimization [4], generative design [5], and others. These methodologies have been applied to various benchmark cases, *i.e.,* i) internal heat source plus 1st type boundary condition; ii) heat flux boundary and 1st type boundary; iii) two 1st type boundaries; iv) arbitrary heat source and boundaries of any types [6]. However, employing an optimization algorithm presupposes the identification of an objective function, which is closely tied to the problem at hand. Various objectives have been considered: average temperature [7],[8] maximum temperature [9], variance [10], thermal compliance [11]. Nevertheless, the choice remains a subject of ongoing research because it heavily depends on boundary conditions and problem formulation. Furthermore, if we consider applications where temperature is not the sole predominant factor, it becomes evident that the technique to be used can vary. For instance, if weight is an important parameter, one might opt for multi-material topological optimization to further enhance performance [12]. One notable technique discussed in the literature is the ordered SIMP interpolation [13], which has been formulated to effectively address the complexities of multi-material topology optimization. It consists of a systematic arrangement of candidate materials based on their normalized mass densities, followed by the interpolation of material properties such as modulus of elasticity and cost [13].This approach allows for the integration of materials as porous media, foams, and periodic structures. By exploiting different porosities, designs with more degrees of freedom can be realized while still adhering to the weight constraint of known solutions. In this work, we apply this methodology to a known case, a disk with uniform generation [3], to quantify the improvement associated with the introduction of this methodology. The aim is to promote importance of porous structures in additive manufacturing, their lightweight nature, and their ability to improve damage tolerance for several applications, *e.g.*, electronic, aerospace.

2. Methodology

This work shows the potential of ordered multi-material strategy [13] for optimizing the topology of variable-porosity structures with a focus on addressing thermal conduction challenges of the uniformly heated disk problem. The assessment of MMTO benefits is performed though a comparison with tree-shaped architecture and classical TO design.

2.1. Uniformly heated disk

The primary objective is to define a two-dimensional network of conductive material able to increase heat dissipation within the disk design domain. Therefore, the steady-state heat equation with a uniform heat source (Q_{source}) of 10^6 W/m^3 is here solved. The problem refers to a disk geometry [4], identified by an internal radius (r_{int} = 8 mm) and an external one (r_{ext} = 30 mm). Two homogeneous boundary conditions complete the differential problem, *i.e.*, 1st type boundary condition applied on the internal boundary (Γ_1), and adiabatic condition on the others (Γ_2). The dimensionless temperature $T^* = (T - T_{ref})/(Q_{source}\, r_{ext}^2\, /k_{max})$ and the dimensionless radius $r^* = r/\, r_{ext}$, make the heat conduction equation as follows:

$$\frac{1}{r^*}+\frac{\partial}{\partial r^*}\left(r^*\frac{\partial T^*}{\partial r^*}\right)+\frac{1}{r^{*2}}\frac{\partial}{\partial \phi}\left(\frac{\partial T^*}{\partial \phi}\right)+1=0 \quad in\,\Omega \tag{1}$$

$$T^*=0 \quad on\,\Gamma_1 \tag{2}$$

$$\partial T^*/\partial \mathbf{n}=0 \quad on\,\Gamma_2 \tag{3}$$

Geometry and all details about numerical formulation are shown in Fig. 1. It is worth noting that the domain third dimension is supposed equal to the external radius of the disk (r_{ext}).

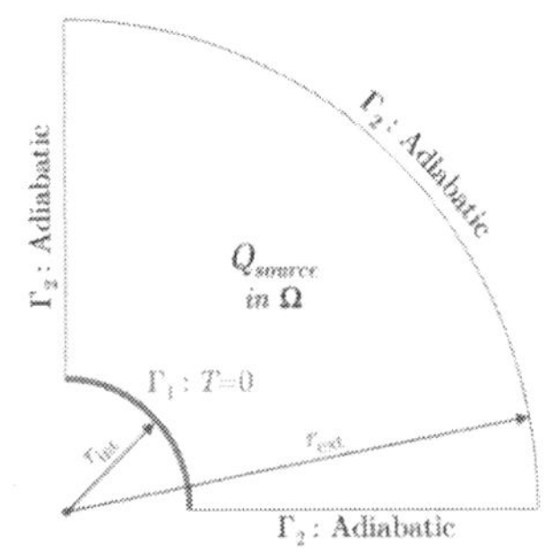

Figure 1. Design domain of a uniformly heated disk.

2.2. *Tree-shaped network*

The tree-shaped network recalls the constructal theory [3], and particularly the case with one bifurcation. The design has been generated according to a parametric optimization [4], that considers a similar problem (differs only on the value of boundary conditions). This design covers the disk domain with a solid material volume fraction equal to 30%, *i.e.*, ratio between solid area and total one. The expected progression of solid material from the internal boundary to the whole domain should resemble natural processes. According to the constructal law, optimizing global performance necessitates the equilibrium of various resistances. Consequently, the system evolves to facilitate accessibility to the driving force imposed upon it. Therefore, the distribution of solid material will confirm the appearance of veins' network observed in a leaf. According to the parametric optimization results, the tree-shaped configuration does not change with the considered objective function, *i.e.*, dimensionless thermal resistance or average temperature.

2.3. *Topology optimization*

Topology optimization (TO) – in its density-based formulation [14] – requires the definition of a design variable (γ), which ranges from 0 to 1, where 0 is the void and 1 is the solid. During the iterative process of topology optimization, the design variable's value in each cell evolves autonomously based on the governing equations and constraint conditions. Consequently, adjacent cells may exhibit notable disparities in the design variable values, known as the checkerboard effect. To address this issue, a density filter radius is established to refine the results of topology optimization [14]. This refinement process involves using a Helmholtz-type partial differential equation, expressed as:

$$-r^2\nabla^2\tilde{\gamma}+\tilde{\gamma}=\gamma \tag{4}$$

where, γ is the design variable, $\tilde{\gamma}$ is the smoothed scalar function, r is the diffusion coefficient of the filter radius. The diffusion coefficient has been set in order to make the filter radius equal to the length scale of the tree-shaped design, *i.e.*, 0.6 mm. By doing this, the TO design will have solid feature comparable in size with tree-shaped branches thickness. The additional projection step [14] is introduced to reduce the grayscale, using the hyperbolic tangent function.

$$\bar{\tilde{\gamma}}=\frac{tanh(\beta\eta)+tanh(\beta(\tilde{\gamma}-\eta))}{tanh(\beta\eta)+tanh(\beta(1-\eta))} \tag{5}$$

Here, β is the projection steepness, while η is the projection point. The function interpolation technique is employed to establish a connection between thermal conductivity and the design variable. Specifically, solid interpolation of material penalization (SIMP) is utilized to generate a seamless function for thermal conductivity, outlined as follows:

$$k(\gamma)=k_{\min}+(k_{\max}-k_{\min})\gamma^n \tag{6}$$

The thermal conductivities are k_{min} = 0.1 W/m K and k_{max} = 47 W/m K, respectively. The penalty factor (n) determines the steepness of this function. In this work we deal with two benchmark optimization problems, *i.e.*, minimize dimensionless thermal resistance [3] and domain average temperature, constraining the total weight (w):

$$\min_{\gamma}\ \frac{T_{\max}-T_{\min}}{Q_{source}r_{\text{ext}}^{2}/k_{\max}}=\frac{\left(\int_{\Omega}T^{p}d\Omega\right)^{1/p}}{Q_{source}r_{\text{ext}}^{2}/k_{\max}} \tag{7}$$
$$s.t.\quad w(\gamma)\le w_{ref}$$

$$\min_{\gamma} \int_{\Omega} T^{*}(\gamma) d\Omega \quad s.t. \quad w(\gamma) \leq w_{ref} \tag{8}$$

where w_{ref} refers to the mass of the 30% of volume that can be occupied by the high conductive material in order to perform a fair comparison with others designs. For thermal resistance formulation, the p-norm is chosen to evaluate the maximum temperature, as the latter is not differentiable and thus not compatible with gradient based optimization [10]. An exponent equal to 15 is chosen to approximate max operator, without falling into instability related to excessive values. $T_{\min}$ is omitted as its value is equal to the one of the internal boundary, *i.e.*, $T^{*} = 0$.

2.4. Multi-material topology optimization

The aim of the ordered SIMP interpolation method [13] is to increase the flexibility of previous approaches while addressing issues with convergence, such as ineffective penalization of intermediate densities and complex post-processing for topology. Accurate assignment of material properties is critical in multi-material topology optimization with ordered SIMP interpolation. This study emphasizes thermal conductivity, which is the only thermal property required by Eq. (1), with the SIMP method applied solely to this parameter. Four materials are included: void, two metal foams with different porosities (ε), and solid, identified as M_1, M_2, M_3, and M_4, respectively. Table 1, based on data from [15], presents the thermal conductivity values for the foams. It also outlines the process for normalizing these values (k_N) by comparing them to the maximum conductivity of 47 W/(m K), and incorporating the effect of porosity.

Table 1. Thermal conductivity of the investigated materials

Material	$\varepsilon = 1 - \gamma$	k (W/(m K))	$k_N = k/k_{max}$
Void (M_1)	1.00	0.10	1/470 [13]
1st Metal foam (M_2)	0.98	2.90 [15]	0.062
2nd Metal foam (M_3)	0.85	11.0 [15]	0.23
Solid (M_4)	0.00	47.0	1.00

According to Tab. 1, the thermal conductivity of the M_{n+1} material is higher than the one of a material M_n. Similarly to Eq. (6), the equation for interpolated normalized modulus of thermal conductivity ($k_N = k/k_{max}$), for $\gamma_N \in [\gamma_m, \gamma_{m+1}]$, is as [13]:

$$k_N(\gamma_N) = k_{N,m} + (k_{N,m+1} - k_{N,m}) \left[\frac{(\gamma_N - \gamma_m)}{\gamma_{m+1} - \gamma_m} \right]^p \tag{9}$$

The projected density – equal to the normalized density – is written as:

$$\bar{\tilde{\gamma}}_N = \gamma_m + (\gamma_{m+1} - \gamma_m) \frac{tanh(\beta\eta) + tanh\left[\beta\left(\frac{\tilde{\gamma}_N - \gamma_m}{\gamma_{m+1} - \gamma_m}\right) - \eta\right]}{tanh(\beta\eta) + tanh[\beta(1-\eta)]} \tag{10}$$

Fig. 2 shows a graphical version of the presented functions:

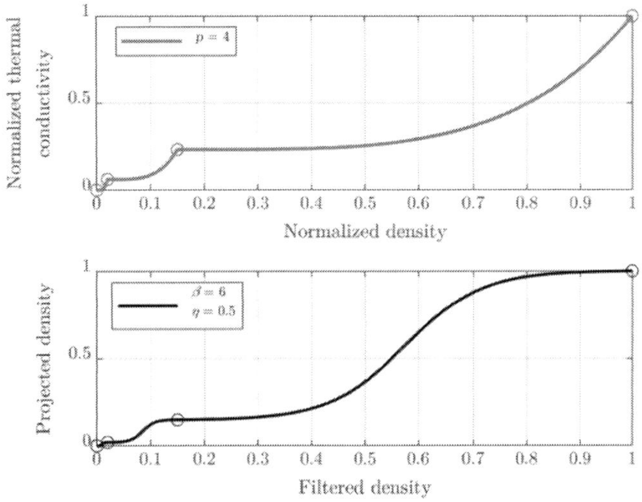

Figure 2. Normalized density (projection) and thermal conductivity (interpolation).

3. Results and discussion

This section highlights the differences in results obtained using tree-shaped, TO, and MMTO approaches applied to a uniformly heated disk. The goal is to illustrate how the objective function values – dimensionless thermal resistance and average temperature – vary with the number of materials incorporated into the model. This variation leads to distinct designs, with the MMTO approach achieving superior outcomes.

3.1. Optimized design

As widely proved in [10], minimizing average temperature and thermal compliance leads to similar and almost identical results (for this problem formulation). Conversely, optimizing dimensionless thermal resistance, defined with respect to the maximum temperature, generates design with different features and therefore different performance. Fig. 3 unveils resulting designs for objectives stated in Eqs. (7-8), for tree-shaped, TO and MMTO cases. As specified before, the tree-shaped network consists of a structure repeated 10 times along the angle, defined by a first branch that then divides into two secondary branches. This allows to uniformly cover the whole design domain. As aforementioned, we consider the same design in both cases. The topology optimization outcome changes depending on the considered objective. In the first case, *i.e.,* thermal resistance, one can spot two main central branches and other two side branches. The tendency is to reduce the peak temperature, and therefore all secondary branches tend to cover the design domain as possible. On the other hand, in the second case, *i.e.,* average temperature, one can spot 5 main branches – with an average thickness similar to each other and less than the branches of the previous design – that are evenly distributed within the domain. So, the first main difference is that priority in trying to cover the domain is given to the main branches, whereas in the previous case it was from the secondary branches. A separate matter comes from MMTO results. As shown in Fig. 3, among the four materials that can be used, the MMTO mainly takes advantage of the solid and porous ones. This is a consequence of the difference in thermal conductivity values. Since there is no constraint on device price, there is no reason to comprise void as a possibility. Therefore, in the first case, *i.e.,* thermal resistance, only the M_4 (solid) and M_3 materials are used. The solid one stratifies near the boundary to promote heat transfer where it is most needed. On the other hand, in the second case, *i.e.,* average temperature, one can clearly spot the three materials (M_1, M_2, and M_3), sequentially ordered, from the inside to the outside, respectively.

Journal of Physics: Conference Series **2940** (2025) 012029 doi:10.1088/1742-6596/2940/1/012029

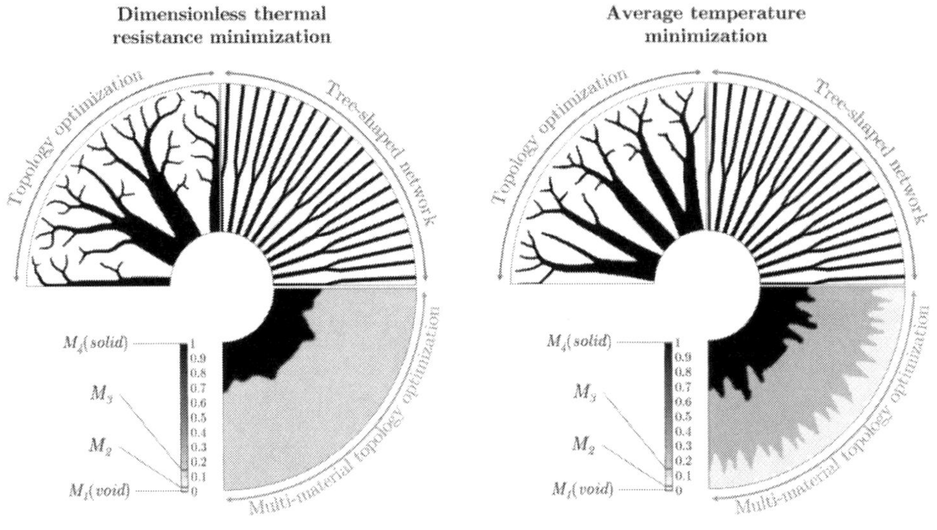

Figure 3. Design results for tree-shaped, topology optimization and multi-material topology optimization approaches, when minimizing dimensionless thermal resistance.

3.2. *Optimized temperature fields*

Temperature distribution (dimensionless) for the considered cases and for both objectives are shown in Fig. 4. On the left side, the temperature fields are referred to minimization of dimensionless thermal resistance, while on the right size to average temperature reduction. All results are presented with the same dimensionless temperature scale, from 0 to 35. Based on the mathematical formulation, minimizing the dimensionless thermal resistance leads to lower peak temperature, which is confirmed by the presented results.

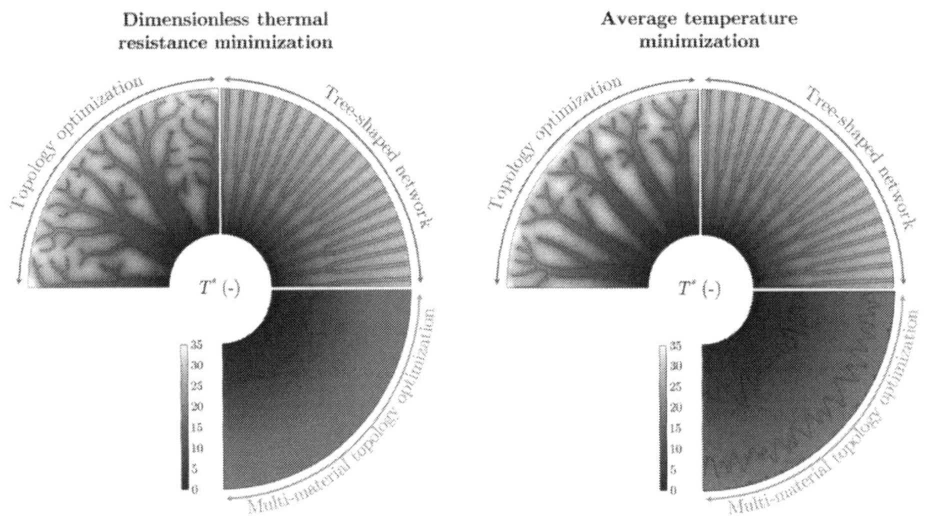

Figure 4. Temperature fields for tree-shaped, topology optimization and multi-material topology optimization approaches, when minimizing dimensionless thermal resistance.

Comparing TO results for both objectives, one can find a higher number of hotspots in the design that minimize average temperature. Similarly, for the MMTO configuration, the single porous material ensures a lower maximum temperature. Thus, what can be seen is that for both TO and MMTO, minimizing the mean temperature causes the algorithm to prefer more material near the inner boundary. Doing so – however – reduces the availability of material in the outer zones, leading to higher maximum temperatures. Despite this, the benefit of the former compensates for the disadvantage of the latter, establishing better performance. To quantify TO and MMTO benefits with respect to tree-shaped design, Tab. 2 lists both R_{th} and T^*_{avg} data for all considered designs. Reducing the average temperature leads to better results, as proved by the percentage reduction provided in Tab. 2. The dimensionless average temperature of 20.91 of the tree-shaped is reduced by 7.51% by TO, and 37.6% by MMTO, respectively, when minimizing R_{th}. The improvement is even higher when minimizing the average temperature: by 8.99% by TO, and 43.8% by MMTO, respectively.

Table 2. Objective function values for various objective functions.

Configuration	*Dimensionless thermal resistance,* R_{th} (-)	*Average temperature,* T^*_{avg} (-)
Minimize (R_{th})		
Tree-shaped	0.0562	20.91
Topology optimization	0.0559 (-0.53%)	19.34 (-7.51%)
Multi-material topology optimization	0.0545 (-3.02%)	13.05 (-37.6%)
Minimize (T^*_{avg})		
Tree-shaped	0.0562	20.91
Topology optimization	0.0561 (-0.18%)	19.03 (-8.99%)
Multi-material topology optimization	0.0548 (-2.49%)	11.75 (-43.8%)

Finally, an example of the qualitative aspect of the designs obtained - only for the case where T^*_{avg} is minimized - is shown in Fig. 5

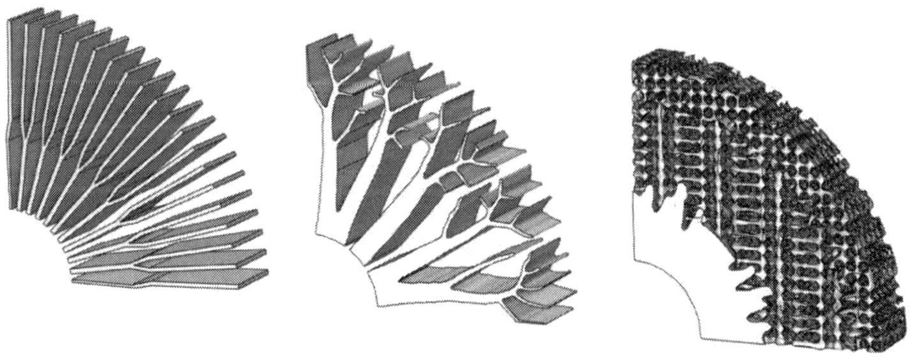

Figure 5. Qualitative appearance of the optimized designs obtained with the described methodologies.

Conclusions

The work presented here underscores the significant advancements in the field of topology optimization (TO) and multi-material topology optimization (MMTO) within the context of thermal management, particularly for the optimization of heat conduction paths in systems requiring efficient thermal dissipation. By focusing on a benchmark heat conduction problem – namely, the cooling of a circular heat-generating volume – the study contrasts the performance of classical TO against the more advanced MMTO approach, integrating different-porosity metal foams to enhance the system's thermal performance. Key to this investigation is the application of an ordered solid isotropic material penalization (SIMP) algorithm for the interpolation of materials' thermal conductivity, ensuring a clear distinction between materials through thresholds in the interpolation and projection functions. This approach aims at the minimization of dimensionless global thermal resistance and the domain average temperature. The findings of this study confirm MMTO designs – with its augmented material flexibility and design freedom – as best solutions, reducing average temperature compared to classical TO configurations, *i.e.,* about 7.5% and 9%, and even more if compared to tree-shaped design, *i.e.,* about 37% and 44%. This benefit is attributed to the more strategic distribution of materials enabled by MMTO, which directly contributes to an enhanced thermal performance by minimizing critical metrics such as dimensionless thermal resistance and average temperature.

Acknowledgments
This work was supported by the Italian Government MUR Grant No. 2022TZ3S4M – PRIN 2022 "TOpology & Material Optimization using additive manufactuRing to maximize stRuctural and thermal perfOrmances of building Walls – TOMORROW" as concerns the development of a comprehensive methodological framework for CFD simulation and optimization of heat transfer systems/components. In addition, the work was supported by the Italian Government MUR Grant No. P2022NYPHL – PRIN 2022 PNRR "VISIONS: eVolutIonary deSign of InnOvative heat traNsfer deviceS" – Funded by European Union – Next Generation EU, as concerns the energy analysis and optimization of the phase change materials, metal structures, and their combination.

References
[1] Chen K, Xing J, Wang S and Song M 2017 Heat source layout optimization in two-dimensional heat conduction using simulated annealing method *Int. J. Heat Mass Transf.* **108** 210–9
[2] Boichot R and Fan Y 2016 A genetic algorithm for topology optimization of area-to-point heat conduction problem *Int. J. Therm. Sci.* **108** 209–17
[3] Rocha L A O, Lorente S and Bejan A 2006 Conduction tree networks with loops for cooling a heat generating volume *Int. J. Heat Mass Transf.* **49** 2626–35
[4] Lange F, Hein C, Li G and Emmelmann C 2018 Numerical optimization of active heat sinks considering restrictions of selective laser melting *COMSOL Conference 2018 Lausanne* pp 0–7
[5] Lohan D J, Dede E M and Allison J T 2017 Topology optimization for heat conduction using generative design algorithms *Struct. Multidiscip. Optim.* **55** 1063–77
[6] Su C-J, Zhao T and Guo Z-Y 2023 Generalized entransy dissipation and its application in heat conduction optimizations with arbitrary boundary conditions *Int. J. Heat Mass Transf.* **216** 124534
[7] Gersborg-Hansen A, Bendsøe M P and Sigmund O 2006 Topology optimization of heat conduction problems using the finite volume method *Struct. Multidiscip. Optim.* **31** 251–9
[8] Bianco N, Fragnito A, Iasiello M and Mauro G M 2023 Design of PCM-based heat sinks through topology optimization *J. Phys. Conf. Ser.* **2509** 012001
[9] Wu S, Zhang Y and Liu S 2019 Topology optimization for minimizing the maximum temperature of transient heat conduction structure *Struct. Multidiscip. Optim.* **60** 69–82
[10] Lohan D J, Dede E M and Allison J T 2020 A study on practical objectives and constraints for heat conduction topology optimization *Struct. Multidiscip. Optim.* **61** 475–89
[11] Yan S, Wang F and Sigmund O 2018 On the non-optimality of tree structures for heat conduction *Int. J. Heat Mass Transf.* **122** 660–80
[12] Bianco N, Cherella N, Fragnito A, Iasiello M and Mauro G M 2024 Multi-material topology optimization of innovative microchannel heat sinks equipped with metal foams *Int. J. Heat Mass Transf.* **222** 125201
[13] da Silveira O A A and Palma L F 2022 Some considerations on multi-material topology optimization using ordered SIMP *Struct. Multidiscip. Optim.* **65** 261
[14] Bendsøe M P and Sigmund O 2004 *Topology Optimization* (Berlin, Heidelberg: Springer Berlin Heidelberg)
[15] Iasiello M, Bianco N, Chiu W K S and Naso V 2019 Thermal conduction in open-cell metal foams: Anisotropy and Representative Volume Element *Int. J. Therm. Sci.* **137** 399–409

Estimation of flow boiling heat transfer coefficient in enhanced tubes. Benchmark correlations and ANN approach

A F Passarelli[1], R Mastrullo[1], R Revellin[2], L Viscito[1] and A W Mauro[1,*]

[1] Department of Industrial Engineering, Federico II University of Naples, P.le Tecchio 80, 80125 Naples, Italy
[2] INSA Lyon, CNRS, CETHIL, UMR5008, 69621 Villeurbanne, France

* Corresponding author: wmauro@unina.it

Abstract. This paper addresses the critical gap in predicting the heat transfer coefficient during flow boiling in enhanced tubes, where the use of conventional correlations and predictive methods developed for smooth surfaces do not usually provide satisfactory results. For such purpose, a comprehensive database was collected from existing literature, including a wide range of operating conditions and enhanced tube geometries from several independent sources. The dataset includes mass flow rates spanning from 50 to 1000 kg/m^{2}s, vapor qualities from the onset of boiling (x=0.0) to the dry-out occurrence and beyond (x=0.99), reduced pressures from 0.05 to 0.80, and tube diameters (measured up to the fin tip) from 0.7 to 11.9 mm, for a total amount of approximately 3000 data points.
Existing flow boiling heat transfer coefficient predictive methods for enhanced tubes were implemented and tested with the present dataset, proving a limited accuracy for most of them mainly in case of testing beyond the specific parameter ranges they were developed for. Extrapolation frequently resulted in statistically poor or even non-physical outcomes. Several artificial neural network models were then developed, according to sensitivity analysis approach to look for potential input parameters and network structures. Specifically, two approaches were employed: a standard neural network model and a correlated informed neural network (CINN), integrating physical correlations into the network's architecture, thus informing the model with physical principles that govern the heat transfer process. Despite a lower overall accuracy, the correlated informed neural network demonstrated superior reliability than standard one, resulting in an instrument to improve the accuracy of existing correlations.

Keywords: *flow boiling, enhanced tubes, heat transfer coefficients, correlations, ANN methods*

1. Introduction

Nowadays, internal micro-fin tubes are widely employed in the refrigeration and air-conditioning industry because of their great potential about two-phase heat transfer enhancement, resulting in components downsizing and refrigerant charge reduction, with the latter being a crucial topic because of the new environmental policies that will lead to a wide diffusion of low-GWP flammable refrigerants. Indeed, this type of enhanced tube allows a larger heat transfer surface than a smooth tube with the same internal diameter. In particular, taking into account flow boiling, the thermal performance is enhanced thanks to a promoted flow turbulence with a delayed dry-out occurrence, in particular for helicoidal micro-fins [1]. Moreover, micro-fins can help nucleate boiling, since the micro-grooves can behave as nucleation site, mitigating flow boiling suppression [2].

Therefore, several experimental studies have been conducted through years to investigate the effect of different micro-fin geometries on flow boiling physics with the development of some predictive methods that are structured to be general but are actually validated with small datasets [3]- [4] or they are case-based on particular geometries, operative conditions and fluids [5] - [6]. In both cases, the poor general accuracy drives the research towards the development of new predictive methods [7].
Moreover, given the complexity of the phenomenon and the recent great attention to artificial intelligence, some studies have been conducted to investigate the accuracy of machine learning tools for flow boiling in tubes and heat exchangers [8] [9] [10].
Having accurate and general predictive models means to have reliable energy systems simulations tools, consequently it is possible to design efficient and sustainable plants. In this regard, starting from a large experimental database collected from literature, the aim of the work is to present a comparison between benchmark flow boiling correlations and Artificial Neural Networks, highlighting strength and drawbacks of both, evaluating also the perspective of integrating physical correlations into the network's architecture.

2. Database description

The database described in this section has been compiled from approximately 30 papers in the literature and is summarized in Table 1. It includes a wide range of operating conditions, with mass flow rates spanning from 50 to 1000 kg m^{-2}s, vapor qualities from the onset of boiling (x=0.0) to the dry-out occurrence and beyond (x=0.99), reduced pressures from 0.05 to 0.80, and tube diameters (measured up to the fin tip) from 0.7 to 11.9 mm, for a total amount of approximately 3000 data points.

Table 1: Database summary.

Authors	Fluids	Saturation temperature $[°C]$	Fin tip diameter, D_{ft} [mm]	Heat flux $[kWm^{-2}]$	Mass Flux $[kgm^{-2}s^{-1}]$	Number of points
[5]	R513A	20	2.4	12-60	300-800	72
[11]	R1233zd	30	4.3	15-60	100-300	67
[12]	R1234ze	30	2.4	10-50	375-940	95
[13]	R134a	30	3.4	10	190-755	54
[14]	R410A	10-15	6.36	5-15	200-400	30
[15]	R32	15	1.98-2.49-2.96	2.5-40	100-400	468
[16]	R744	0-5	4.1-8.68	6-20	318	180
[17]	R407C	5-30	8.08	75	17-254	199
[18]	R32-R410A	5-20	4.2	12-51	200-600	116
[19]	R134a	15-22	8.62	10	190-381	47
[20]	R1234yf	30	3.4	10-50	190-940	95
[21]	R1234ze	30	3.4	10-50	190-940	99
[22]	R32	15	3.25	5-20	50-400	203
[23]	R1233zd-R245fa	30	4.2	15-90	100-400	164
[24]	R134a	5	8.22-8.46	3.2-17.8	90-340	39
[25]	R600A	-10 / -20	6.36	9-15	20-40	78
[26]	R410A	10	6.08	10-20	100-300	56
[27]	R22–R407C	2-6	8.92	6-14	100-300	55
[28]	R134a	5	8.52	5	100-500	167
[29]	R717	-20	9.87	40	50-150	14
[30]	R407C	10	6.18-11.48	10-20	200-300	134
[31]	R744	29.2-30.1	6.9	10-29	150-250	81
[32]	R744	5	8.12	4-60	75-500	30
[33]	R744	15	1.77	9-18	360-720	185
[34]	R744	1-15	1.38	1.67-4.17	100-600	195
[35]	R744	10	2.82	10-30	190-770	71
[36]	R290	-24 - -4	6.26	10-73	70-190	107
[37]	R290	5-10	4.26-4.32	15-34	250-500	80

The distribution of fluids shown in Figure 1(a) highlights a large number of points for R32 and carbon dioxide, while the histograms referred to the main operating parameters shown in Figure 1(b) present a quite uniform distribution for the vapor quality, with a predominance of positive temperature and mass fluxes lower than 500 kg m^{-2}s.

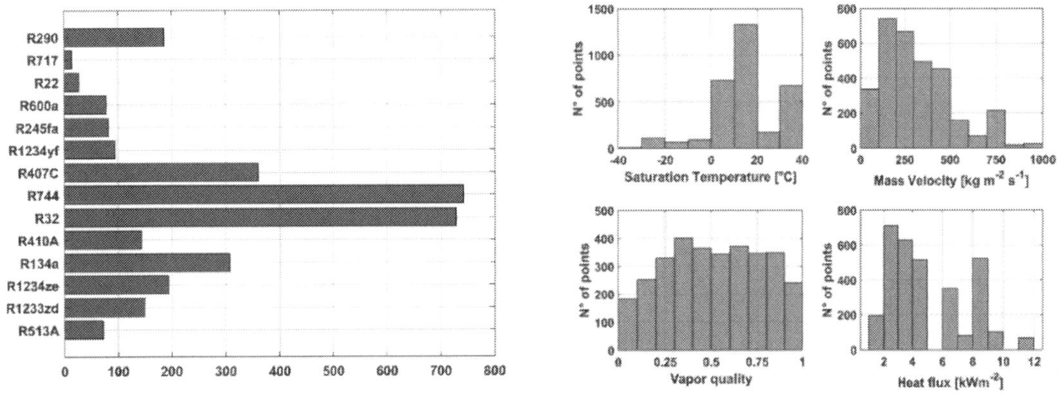

(a) (b)

Figure 1. Fluids distribution (a) and experimental conditions histograms (b).

The helical micro-fin geometry is used in all the references and its main geometrical features are shown in Figure 2(a) while in Figure 2(b) the related histograms are presented, resulting in peak distribution for the helix angle and the fin pitch.

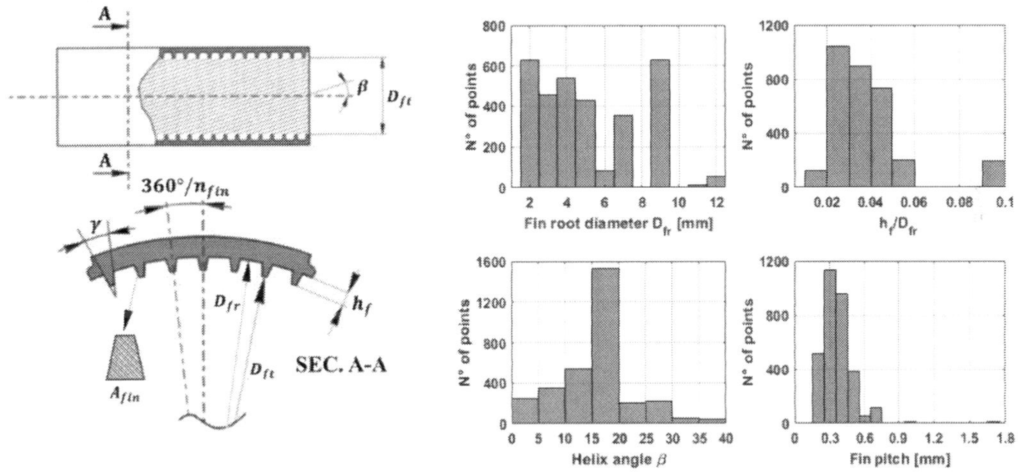

(a) (b)

Figure 2. Geometrical features of a micro-fin tube (a) and related database distribution (b).

Since the micro-fins presence leads to the definition of fin and root diameters, it is possible to define a reference equivalent diameter as proposed by several authors [38]:

$$D_{eq} = \left[4\left(\frac{\pi D_{fr}^2}{4} - n_{fin}A_{fin}\right)\frac{1}{\pi}\right]^{\frac{1}{2}} \quad (1)$$

3. Predictive methods, state of the art

One of the first correlations for micro-fin tubes was developed by Thome et al. [3]. This method has been created to be applied to any micro-fin geometry, but it was developed starting from a quite small database. Starting from the recurrent form of the heat transfer coefficient as a sum of convective (h_{cb}) and nucleate contributions (h_{nb}) [39], this correlation presents a couple of parameters that consider the micro-fin features:

$$h = \left[\left[(h_{cb} \cdot E_{RB})^3 + h_{nb}^3\right] \cdot E_{mf}\right]^{1/3} \tag{2}$$

with E_{RB} involving all the main geometrical features of the micro-fin geometry with the liquid film Reynolds number (Re_D) and the Prandtl number of liquid phase (Pr_L), while E_{mf} takes into account the improvement of the nucleate boiling with a reference mass flux G_{ref} of 500 kg m^{-2}s:

$$E_{RB} = \left\{1 + \left[2.64(Re_D)^{0.036}\left(\frac{h_f}{D_f}\right)^{0.212}\left(\frac{p_f}{D_f}\right)^{-0.21}\left(\frac{\beta}{90°}\right)^{0.29}(Pr_L)^{-0.024}\right]^7\right\}^{\frac{1}{7}} \tag{3}$$

$$E_{mf} = 1.89\left(\frac{G}{G_{ref}}\right)^2 - 3.7\left(\frac{G}{G_{ref}}\right) + 3.02 \tag{4}$$

The micro-fin structure can lead to capillarity effect at low mass fluxes that can improve the heat transfer coefficient. Indeed, Cavallini et al. [4] proposed the following correlation for pure fluids:

$$h = h_{nb} + h_{cb} + h_{cap} \tag{5}$$

where the convective contribution presents a parameter Rx that is an enhancement factor of the Nusselt number for single-phase flow in smooth tubes ($Nu_{cv,st}$) multiplied by some recurrent parameters in flow boiling such as the Bond (Bo) and the vapor-phase Froude (Fr_V) numbers:

$$h_{cb} = \left(\frac{k_L}{D}\right)Nu_{cv,st}Rx^{2.14}(Bo \cdot Fr_V)^t\left(\frac{0.01}{D_{ft}}\right)^{0.59}\left(\frac{100}{G}\right)^Z \tag{6}$$

and it that takes into account the micro-fin geometry features such as micro-fins number n_f, apex angle γ, helix angle β, height h_f and fin tip diameter of the tube D_{ft}:

$$Rx = \left\{\frac{2h \cdot n\left[1 - sin\left(\frac{\gamma}{2}\right)\right]}{\pi D_{ft} cos\left(\frac{\gamma}{2}\right)} + 1\right\} \cdot \frac{1}{cos\beta} \tag{7}$$

Instead, the capillarity contribution h_{cap} is greater than zero only if the mass flux is less than 100 kg m^{-2}s^{-1}. Starting from this method, some updated versions have been developed through last years to adapt the correlation also to mixtures [40] and particular fluids as done by Diani et al. [21]. Despite the latter was developed for a particular fluid, it is one of the most quoted methods in literature since it has a good accuracy also when applied to other fluids.

The same peculiarity is true for the Rollmann and Spindler correlation [6], developed for R407C and R410A with a statistical approach based on the most influencing parameters involved (liquid-only Reynolds number Re_{LO}, boiling number Bo and Pr_L) without taking into account the micro-fin geometry:

$$Nu = 1.2\left(-\frac{3.7}{\mathrm{Pr}_L^2} + 0.71\right)Re_{LO}^{\frac{2}{3}} \cdot \left[\ln(Bo) + 12.17\right] \cdot x^{\left(-\frac{3.7}{\mathrm{Pr}_L^2}+0.71\right)} \tag{8}$$

However, this correlation can lead to unphysical results for low liquid Prandtl numbers, thus some modifications to the coefficient are needed to make this correlation suitable for all refrigerants and operational conditions as it is clarified by the authors.

Hence, Thome and Cavallini models represent the theorical benchmark for the development of new correlations. Indeed, other quoted methods are those developed by Li et al. [41] - [7] that present the same basic features shown in Equations (1)-(5).

4. Artificial Neural Networks as a predictive method

Given some input data, a machine learning tool has the objective of providing the corresponding output. Through internal mathematical functions and without any physical assumptions, these tools can overcome the complex non-linear relationship between variables, resulting in the potential of giving accurate results only through "learning". Among this methods, artificial neural networks link input and output with a more or less complex network of neurons which are mathematical functions as shown in Figure 3 and whose coefficient should be determined through a training process.

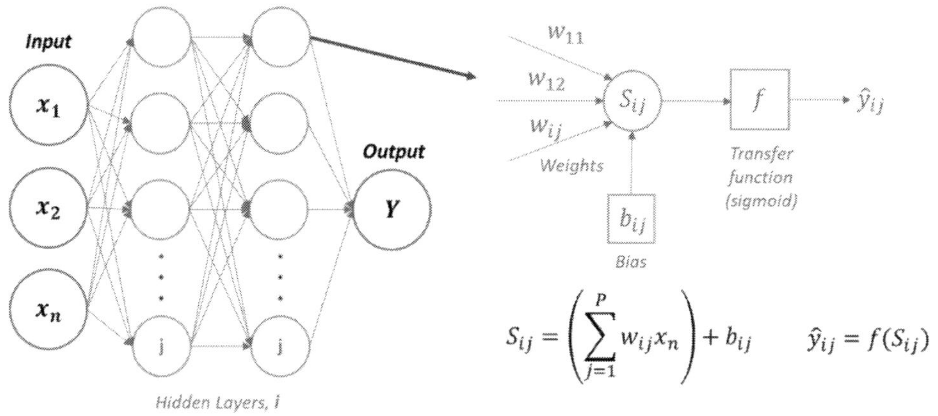

Figure 3. Artificial neural network's architecture and neuron function.

For the majority of physical problems, where it is possible to collect experimental data, the training process is called "supervised", thus through giving the sets of input-outputs and fixed the resolution algorithm, the network will be defined "trained" when the mean error between given outputs and predicted outputs is minimized. In particular, figure 3 shows a feed-forward network with fully connected layers that is generally recommended in case of regression problems [42] such as the flow boiling heat transfer coefficient prediction.

4.1. Best neural network

To find the most accurate neural network, several sets of input parameters and network architectures have been simulated. In particular, sets of input parameters could be made by a group of dimensional (such as liquid and vapor densities ρ, viscosities μ, thermal conductivity k or surface tension σ), not dimensional parameters (such as reduced pressure P_{red}, Martinelli parameter X_{tt} or the previous shown in section 3) or both of them and also the output could be dimensional or not (h in $[kWm^{-2}K^{-1}]$ or Nusselt number Nu). Networks structures involved in this analysis are made of a number of layers varying from 1 to 4 and a number of neurons from 10 to 80, with layers made by the same number of neurons or having a tree structure, reducing the number of neurons from the first to the last layer with a factor varying from 1 to 2. The recursive feature elimination (RFE) technique is used to find the best input parameters set instead of using sensibility analysis based on Pearsons or Kendall parameters with arbitrary thresholds [43]. Therefore, starting from all the influencing parameters of flow boiling, the best networks have been built using the groups of inputs shown in Table 2.

Table 2: Selected input parameters for different input types.

Input type	Input Parameters	Output
Dimensional	$G, T_{sat}, \dot{q}, x, \beta, \rho_v, \mu_v, k_l$	h
Non-dimensional	$P_{red}, Re_{LO}, Re_L, Xtt, Rx, Bo$	Nu
Mixed	$G, T_{sat}, \dot{q}, x, \beta, f_p, \rho_v, \rho_l, \mu_l, \mu_v, \sigma, k_l, D_{eq}, Fr_v, P_{red},$ $Re_{LO}, Re_l, Xtt, Rx, \rho_l\rho_v^{-1}, \mu_l\mu_v^{-1}, h_f D_{fr}^{-1}, Bo$	h

4.2. Correlated informed neural networks

Despite giving different input parameters, using a correlation result as an input could be seen as a way to join different inputs in an ordered form, providing more "physical" information to the network, avoiding the risk of unphysical results as highlighted in previous research [44]. The concept of feeding an ANN with physical information is known as "Transfer Learning" that should be done in different ways [45]- [46]. Therefore, starting from the predictive methods described in the second part, several neural networks, using input sets made up by one correlation result at a time and some other parameters taken from table 2, have been trained following the approach mentioned in the previous section in order to understand the potential of this method.

5. Results: ANNs vs Correlations

The assessment of micro-fin tubes methods and both ANNs developed in the previous paragraph is shown in Table 3 in terms of mean absolute percentage error (MAPE), mean relative percentage error (MRPE) and number of points within the $\pm 30\%$ error band.

$$MAPE = \frac{1}{N}\sum_{i=1}^{N}\left|\frac{(h_{predicted} - h_{experimental})}{h_{experimental}}\right| \cdot 100 \quad (9)$$

$$MRPE = \frac{1}{N}\sum_{i=1}^{N}\frac{h_{predicted} - h_{experimental}}{h_{experimental}} \cdot 100 \quad (10)$$

Table 3: Assessment of the most quoted predictive methods and the best artificial neural networks.

	N. points	MAPE [%]	MRPE [%]	$\delta_{30\%}$ [%]
Cavallini et al. [4]	3146 (2947)	162 (114)	157 (109)	28 (30)
Diani et al. [21]	3146 (2947)	94 (51)	71 (25)	51 (54)
Rollmann et al. [6]	2408	54	-6	40
Tang et al. [7]	3146 (2947)	114 (43)	87 (15)	41 (44)
Thome et al. [3]	3146 (2947)	92 (50)	46 (-0.1)	41 (44)
Wu et al. [41]	3146 (2947)	195 (49)	155 (7)	54 (57)
ANN (80-64-51) *	3146	8.6	0.7	96
CINN (50-25) **	3146	30	16	2453 (78.4)
* 3 layers with 80, 64 and 51 neurons each; ** 2 layers with 50 and 25 neurons each				

The best correlation in terms of MAPE seems to be the Rollman et al. [6] one, but the number of points is reduced since this method is not suitable for some conditions and fluids. Therefore, the best correlation in terms of trade-off between MAPE and $\delta_{30\%}$ is the one by Diani et al. [21]. As a proof of how some particular operative conditions can lead to huge errors for correlations, Table 3 show a second set of statistical results between brackets that refer to the total amount of points except particular conditions of very low mass fluxes and high heat flux [17]. Hence, avoiding these common outliers, Diani et al. [21], Wu et al. [41] and Tang et al. [7] correlations show quite better results. However, the best artificial neural network (mixed inputs with 3 layers) has much better statistical results than correlations, while the best CINN network (Cavallini et al. [4] and Rx as inputs with 2 layers) leads to a strong accuracy improvement of the starting predictive method. Figure 4 shows the comparison between experimental and predicted heat transfer coefficients for the best correlation and ANN/CINN.

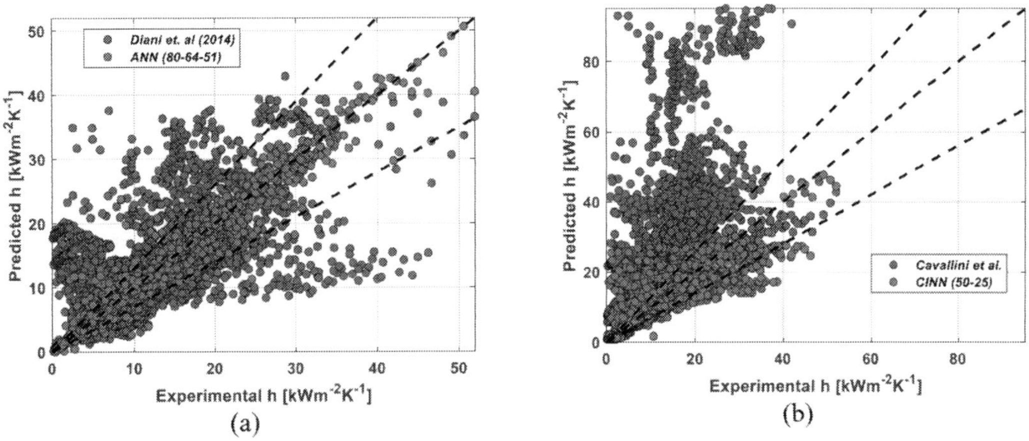

Figure 4. Experimental vs predicted heat transfer coefficient for Diani et al. and the best ANN (a) and Diani et al. and the best CINN compared with the input correlation (b).

Despite the great accuracy that artificial neural networks can reach, the lack of a physics base has to be accounted as one of the main risks when using an ANN for extrapolation outside the database used in the training process. At this purpose, Figure 5 shows a local analysis comparison of the best predictive methods, ANN and CINN first for one experimental test included in the database (a) and then for a slightly different set of operative conditions not included in the database (b).

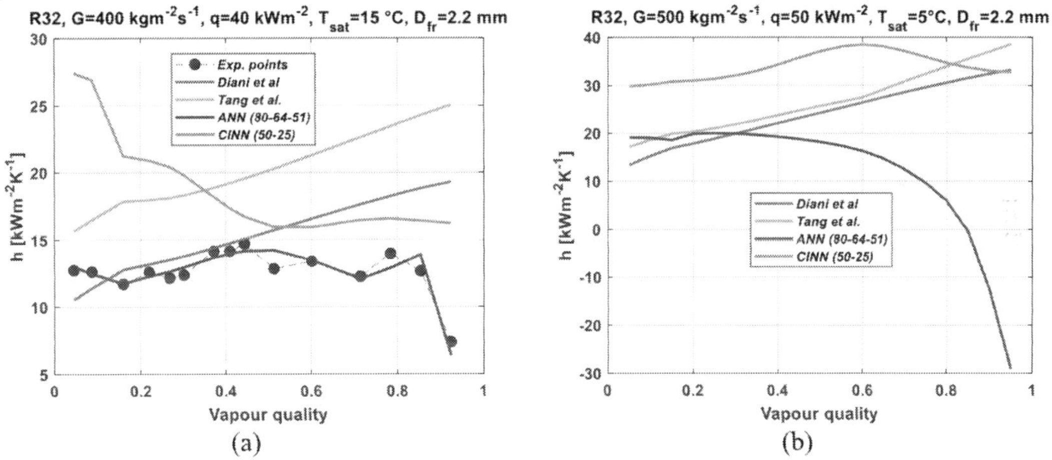

Figure 5. Local heat transfer coefficient analysis for one of the experimental conditions within the database (a) and for similar conditions not involved in the database (b).

It is clear that the ANN brings unphysical results while the CINN network is more connected to the physics due to the predictive method given in input, resulting in a higher reliability for extrapolations.

6. Conclusions

In this work, a comprehensive database regarding experimental heat transfer coefficients for flow boiling in internal micro-fin tubes has been collected in order to investigate and compare the accuracy and reliability of the most common predictive methods in literature and artificial neural networks.

The assessment presented show that traditional predictive methods have a poor generality, with the best statistical results for the Tang et al. [7] correlation with a MAPE of 43% (excluding some common outliers for all the correlations) while the Wu et al. [41] correlation presents the higher number of points

(54%) in the ±30% error band. Instead, the best ANN reach very high accuracy with a MAPE of 8.6% and 96% of points in the ±30% error band and it is made by both dimensional and non-dimensional inputs with 3 layers of 80-64-51 neurons. Since an ANN is based only on mathematical functions without any connection with the physics of the phenomenon, some attempts have been done to understand the potential of giving some physical information to the network through using a correlation result as input. The best result for a CINN network shows an intermediate accuracy (MAPE of 30%) between the best ANN and the best correlation, but the reliability in extrapolation is much higher than the best ANN, avoiding unphysical results.
Therefore, the current analysis has shown that ANNs have a great predictive potential but they should be used only in case of sets of input very similar to the ones used to train the network, while having a training process with a large and comprehensive database is helpful to enhance accuracy and enlarge the field of application. Instead, if extrapolations are needed and the starting database is not general, using a correlated informed technique could be useful to have more accurate results instead of using only a physics-based predictive method.

References

[1] Yang C-M, Hrnjak P 2019 Int. J. Heat Mass Transfer 144 118654.
[2] Thome J R 1996 Int. J. Refrig. 19 435-457.
[3] Thome J R 1997 Convective flow and pool boiling Conf..
[4] Cavallini A, Del Col D, Rossetto L 2006 ECI Int Conf.
[5] Diani A, Rossetto L 2020 Int J. Heat and Mass Transfer 162 120402.
[6] Rollmann P, Spindler K 2016 Int J. Thermal Sciences 103 57-66.
[7] Tang W, Li W 2018 Int J. Heat and Mass Transfer 126 1067-1078.
[8] Zhu G, Wen T, Zhang D 2021 Int J. Heat and Mass Transfer 166 120783.
[9] Qiu Y, Vo T, Garg D, Lee H, Kharangate C R 2023 Int J. Heat and Mass Transfer 202 123728.
[10] Longo G A, Mancin S, Righetti G et al. 2020 Int J. Heat and Mass Transfer 160 120204.
[11] Righetti G, Longo G A, Zilio C, Akasaka R, Mancin S 2018 Int J. Refrig. 91 69-79.
[12] Diani A, Mancin S, Cavallini A, Rossetto L 2016 Int J. Refrig. 69 272-284.
[13] Mancin S, Diani A, Rossetto L 2014 45 En. Procedia 608-615.
[14] Lee E J, Kim N H 2016 Int. J. Heat and Mass Transfer 92 283-297.
[15] Jige D., Inoue N. 2019 Int. J. Heat and Mass Transfer 134 566-573.
[16] Cho J M, Kim M S 2007 Int J. Refrig. 30 986-994.
[17] Deb S, Mahesh K P,Das M, Das D C, Pal S 2023 Int J. Th. Sciences 188 108239.
[18] Longo G A, Mancin S, Righetti G, Zilio C 2021 Int J. Refrig 515-525.
[19] Celen A, Cebi A, Dalkilic A S 2018 Int Comm. Heat and Mass Transfer 93 21-33.
[20] Diani A, Mancin S, Rossetto L 2015 Exp. Thermal Fluid Science 66 127-136.
[21] Diani A, Mancin S, Rossetto L 2014 Int J. Refrig. 47 105-119.
[22] Jige D, Sagawa K, Iizuka S., Inoue N 2018 Int J. Refrig. 73-82.
[23] Righetti G, Longo G A, Zilio C, Mancin S 2019 Int J. Refrig. 104 344-355.
[24] Colombo L P M, Lucchini A, Muzzio A 2012 Int J. Refrig. 2150-2165.
[25] Moon S H, Lee D, Kim M, Kim Y 2022 Int J. Heat and Mass Transfer 190 122769.
[26] Yang C-M, Hrnjak P 2018 Int J. Heat and Mass Transfer 117 924-939.
[27] Kuo C-S, Wang C-C 1996 App. Th. Engineering 16 719-731.

Journal of Physics: Conference Series **2940** (2025) 012030 doi:10.1088/1742-6596/2940/1/012030

[28] Filho E P B, Barbieri P E L 2011 Exp Thermal and Fluid Science 35 832-840.
[29] Kabelac S, de Buhr H-J 2001 Int J. Refrig. 24 41-50.
[30] Passos J C, Kuser V F, Haberschill P, Lallemand M 2003 Exp. Th. and Fluid Sc. 27 705-713.
[31] Zhao X, Bansal P 2012 Int J. Thermal Sciences 59 38-44.
[32] Schael A-E, Kind M 2005 Int J. Refrig. 28 1186-1195.
[33] Dang C, Haraguchi N, Hihara E, 2010 Int J. Refrig. 33 655-663.
[34] Wu X, Zhu Y, Tang Y 2015 Int J. Refrig. 59 281-294.
[35] Gao L, Honda T 2006 Int Refrig. Air Cond. Conference 770.
[36] Zhao C, Guo H, Xue H, Nie F, Gong M, Yang Z 2023 Int J. Refrig. 155 195-206.
[37] Allymer E, Pardinas A A, Eikevik T M, Hafner A 2020 Applied Th Eng. 181 115880.
[38] Mainil A K, Sakamoto N, Ubidiyah H, Kariya K, Miyara A, 2022 Int. J Refrig. 144 342-353.
[39] Chen J C 1966 Ind. Eng. Chem. Proc. Des. Dev. 5 322-329.
[40] Chamra L M, Mago P J 2007 Journal Mech. Eng. Science 221 443-454.
[41] Wu Z, Wu Y, Sunden B, Li W 2013 Exp. Thermal and Fluid Science 44 398-408.
[42] Aggarwal C C, Neural Networks and Deep Learning: A Textbook, 2018 Springer.
[43] Bard A, Qiu Y, Kharangate C R, French R 2022 Applied Th. Engineering 210 118305.
[44] Mauro A W, Revellin R, Viscito L 2024 Int J. Heat and Mass Transfer 221 125106.
[45] Zhu G., Liu S, Zhang D, Chen W, Li J, Wen T 2024 Int J. Heat and Mass Transfer 220 125020.
[46] Montanez J A, Barroso J M et al. 2022 Int J. Heat and Mass Transfer 194 123017.

Journal of Physics: Conference Series **2940** (2025) 012031 doi:10.1088/1742-6596/2940/1/012031

Multiphase boiling models of propane in a horizontal pipe: application of CFD techniques

Lorenzo Giunti[1], Federico Gigliotti[1], Francesco Giacomelli[2], Lorenzo Talluri[1] and Adriano Milazzo[1*]

[1] Department of Industrial Engineering, University of Florence, Italy
[2] Danfoss S.r.l., via Natale Salieri, 33-35, 40024 Castel San Pietro Terme, Bologna, Italy

*E-mail: adriano.milazzo@unifi.it

Abstract. This paper investigates the capabilities of CFD models, available in the commercial software Ansys, used in the study of the propane evaporation heat transfer coefficient trend. In particular, the flow in a horizontal pipe, at various average vapor qualities, has been considered. The scheme adopted in the definition of the problem is based on the Eulerian multiphase approach and RPI-CHF wall heat partitioning. A preliminary benchmark assessment, essential for validating the mathematical model, has been conducted with reference to boiling water in vertical pipes. The prediction is in consistent agreement with experimental results if a modification in some of the fundamental parameters is made. The validation has to be regarded as a propaedeutic study for future research related to more complex geometries, such as channels between corrugated plates of compact heat exchangers.

1. Introduction

The legislative requirements for environmental protection and concern about global climate change have prompted some countries to adopt measures towards a limitation in the use of Hydro-Fluoro-Carbon (HFC) refrigerants and alternative solutions which are more environmentally friendly.

Hydro-Carbons (HC) have been considered as real substitutes for HFCs as they show good thermodynamic and transport properties; in particular, propane (R290) has gained strong attention due to its physical characteristics similar to R22, with better heat transfer coefficient during the evaporation phase [1], ODP equal to 0 and GWP smaller than 3. Other advantages include compatibility with the majority of the HFCs-based equipment materials and miscibility with common compressor lubricants. Therefore, R290 represents a suitable choice for a charge switch.

The literature studies mainly give attention to the experimental test of performances of R290 in simple geometries. Wang et al. [2] carried out boiling heat transfer and frictional pressure gradient measurements in a smooth horizontal pipe at various saturation temperatures, heat and mass fluxes conditions. Choi et al. [3] and Maqbool et al. [4] reported the convective boiling pressure drops and local heat transfer coefficient HTC in horizontal and vertical mini-channels, respectively. Concerning more complex systems, the HTC during the vaporisation of R290 has been investigated by Longo [5] in a brazed plate heat exchanger with chevron corrugations, demonstrating substantial dependence on heat flux and outlet conditions.

Nevertheless, for a better physical understanding of the process and specific industrial applications, numerical simulations should be performed, developing new methods for multi-phase flows and computational techniques.

Chen et al. [6] provided an in-depth review of numerical simulations of boiling heat transfer, covering a wide range of scales, from the smallest to the largest relevant to the phenomenon, and emphasized the potential accuracy improvements achievable through a multi-scale approach. Specifically, Computational Fluid Dynamics (CFD) is a powerful tool capable of handling a broad spectrum of conditions, in terms of characteristic lengths and boiling heat transfer scenarios. Setoodeh et al. [7] developed a bubble dynamics model incorporating inertia- and thermal diffusion-controlled regimes to describe bubble growth within the Euler-Euler framework and validated it using experimental data from boiling flows of R12 refrigerant in a vertical pipe and R134a in an annular geometry. Colombo & Fairweather [8], adopting a similar approach, highlighted the need for improvements in modelling bubble break-up and coalescence phenomena, along with capturing the average velocities in the near-wall zone of annular channels, to obtain more reliable predictions. Končar & Krepper [9] conducted CFD simulations of R-113 flow in a vertical annular channel with an internally heated tube, showing that a two-phase wall function approach ensures a reasonable representation of the flow profiles at the wall. In the same context, few studies are devoted to propane boiling analysis. Qi et al. [10] employed a mathematical approach based on the k-ε turbulence model for the two-phase heat transfer of subcooled flow of R290 at low saturation temperature, in a 1.56 m long pipe. Dai et al. [11] used the Volume Of Fluid (VOF) methodology for exploring propane vapor-liquid flow behaviour in a horizontal smooth copper pipe.

In this paper, a novel numerical technique, implemented within the commercial software ANSYS Fluent 2022 R2, for predicting propane thermodynamics during the boiling process is presented.

The study of the phenomenon is divided into two cases, based on the initial temperature of the liquid: subcooled boiling, if it is below the saturation temperature and saturated boiling when the values are equal. In the latter case, the evolution of the flow regime inside the channel has also been taken into account, as it affects the heat transfer at the wall, approaching the single-phase convection with the onset of the dry-out condition. As depicted in Figure 1, the CFD model that is able to cover all the different situations is Critical Heat Flux, namely CHF-RPI, derived from the study of Kurul & Podowski [12] and implemented in the Eulerian multi-phase approach, which

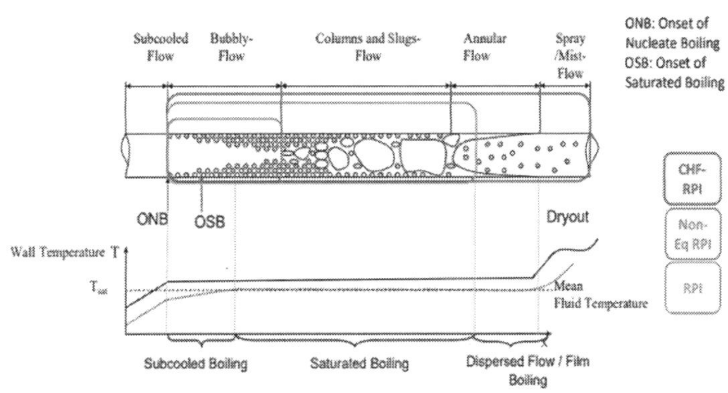

Figure 1. Boiling flow regimes and models applicability [13].

solves the governing equations for each phase, requiring correlations for the empirical closure of coefficients of the interfacial forces in the momentum balance.

2. Numerical methods and approaches

In this chapter, the mathematical and physical modelling used in this work is presented. The main specifications of the boiling model described hereafter are derived from the work done by Bianchini et al. [14] regarding the assessment of numerical procedures for multiphase heat transfer in different regimes.

2.1 Interfacial momentum equation terms

Four interfacial forces have been introduced during the analysis.

The Drag, as hydrodynamic friction between the liquid phase q and the dispersed phase p, is:

$$\vec{F}_D = K_{pq}|\vec{V}_p - \vec{V}_q| \tag{1}$$

where K_{pq} is proportional to a C_d coefficient, which was expressed by Ishi [15], selecting the lower value between the viscous regime and the distorted one.

The Lift, that pushes the small bubbles towards the wall, leading to void fraction peaking close but away from it, is:

$$\vec{F}_L = -C_L\alpha_p\rho_q(\vec{V}_q - \vec{V}_p) \times (\nabla \times \vec{V}_q) \tag{2}$$

where C_L is estimated with the Tomiyama correlation, modified by Frank et al. [16] and depends on the Eötvös number.

The Wall lubrication, instead, tends to prevent the bubbles from touching the solid surface, acting in the opposite direction to F_L and it is:

$$\vec{F}_W = C_{wl}\alpha_p\rho_q|\vec{V}_q - \vec{V}_p|_{\backslash\backslash}^2\vec{n}_w \tag{3}$$

with the subscript meaning that the velocity components are taken tangentially to the wall, whose unit normal is $\vec{n}_w$ and C_{wl} was determined by Antal [17].

Turbulent dispersion represents the impact of eddies in the continuous phase on the bubbles, and it promotes the homogenization of their distribution. The model proposed by Lopez de Bertodano [18] was applied:

$$\vec{F}_{td} = C_{td}\rho_q k_q \nabla\alpha_p \tag{4}$$

where k_q represents the turbulent kinetic energy in the continuous phase and C_{td} is a costant defaulting to a value of 1.

Finally, turbulence interaction between the induced one by particles and shear is formulated according to Troshko & Hassan [19] by adding the following source terms to the respective equations:

$$\Pi_k = C_{ke} \sum K_{pq} \left|\vec{V}_p - \vec{V}_q\right|^2 \tag{5}$$

$$\Pi_\varepsilon = C_\varepsilon \frac{1}{\tau_p} \Pi_k \tag{6}$$

with $C_{ke} = 0.75$ and $C_\varepsilon = 0.45$, τ_p is the characteristic time of the induced turbulence.

2.2 Modelling of wall boiling

While the temperature of the vapor with the basic RPI model is not a result of the calculation but it is fixed at its saturation value, CHF-RPI treats it as an unknown variable, included in the solution process.

The entire heat flux transferred from the solid wall to the fluid is divided into four contributions:

$$\dot{q}_w = (\dot{q}_l + \dot{q}_q + \dot{q}_e) f(\alpha_l) + \dot{q}_v (1 - f(\alpha_l)) \tag{7}$$

The first term on the right-hand side is due to liquid-phase convection and it is:

$$\dot{q}_l = h_l (T_w - T_l)(1 - A_b) \tag{8}$$

followed by the quenching factor:

$$\dot{q}_q = \frac{2\lambda_l}{\sqrt{\pi \beta_l \tau}} (T_w - T_l) A_b \tag{9}$$

where λ_l is the conductivity, τ the bubble departure time and β_l is the liquid diffusivity.

Then, $\dot{q}_e$ is the component associated with evaporation:

$$\dot{q}_e = V_d N_w \rho_v h_{fv} f \tag{10}$$

V_d is the volume of the bubble which depends on the bubble departure diameter D_w, N_w is the active nucleate site density, h_{fv} is the latent heat of evaporation and $f = \tau^{-1}$.

The vapor phase convection $\dot{q}_v$ is analogous to $\dot{q}_l$. With A_b, the area fraction occupied by nucleating bubbles has been specified as, as:

$$A_b = \frac{K N_w \pi D_w^2}{4} \tag{11}$$

the constant K is calculated using Del Valle & Kenning's findings [20]. The release frequency f is based on an empirical correlation [21] as well as D_w [22].

The most relevant parameters in the analysis, as shown by the results in the next sections, are represented by N_w and the function $f(\alpha_l)$ which is based on the local liquid volume fraction α_l. The first one is usually expressed as follows:

$$N_w = C^n (T_w - T_{sat})^n \tag{12}$$

with $C = 210$ and $n = 1.805$ as proposed by Lemmert & Chawla [23].

Journal of Physics: Conference Series **2940** (2025) 012031 doi:10.1088/1742-6596/2940/1/012031

Tentner et al. [24] suggested the following definition for $f(\alpha_l)$:

$$1 - f(\alpha_l) = f(\alpha_v) = \begin{cases} 0 & \alpha_v < \alpha_{v1} \\ 0.5\left[1 - \cos\left(\pi \dfrac{\alpha_v - \alpha_{v1}}{\alpha_{v2} - \alpha_{v1}}\right)\right] & \alpha_{v1} \le \alpha_v < \alpha_{v2} \\ 1 & \alpha_v \ge \alpha_{v2} \end{cases} \tag{13}$$

remembering that $\alpha_v = 1 - \alpha_l$.

Equation (13) shows that the heat flux terms in (7) are weighted by the relative local amount of vapor; the breakpoints are set to $\alpha_{v1} = 0.9$ and $\alpha_{v2} = 0.95$ by default, so $\dot{q}_w = \dot{q}_v$ for $\alpha_v \ge \alpha_{v2}$. In the standard RPI, $f(\alpha_l)$ is always equal to 1 and $\dot{q}_w = \dot{q}_l + \dot{q}_q + \dot{q}_e$.

3. Results and discussion

3.1 Validation of the model

To assess the adequacy of the simulation and acquire a deep understanding of how the model operates, RPI-CHF was preliminarily applied to two literature cases: the experiment of Bartolomej et al. [25] and Hoyer [26].

The former, which is illustrated in Figure 2a) together with boundary values, describes the sub-cooled boiling of water in a pipe whose internal diameter D_i is 15.4 mm and length $L = 2$ m; in the latter, measurements on externally heated pipes were carried out varying the diameter, mass flux and thermal power, the geometry selected is the one with $D_i = 10$ mm and $L = 7$ m and the data are shown in Figure 2b). In the second test section, a higher vapor volume fraction (VF) is achieved at the outlet implying the reaching of the critical flux region.

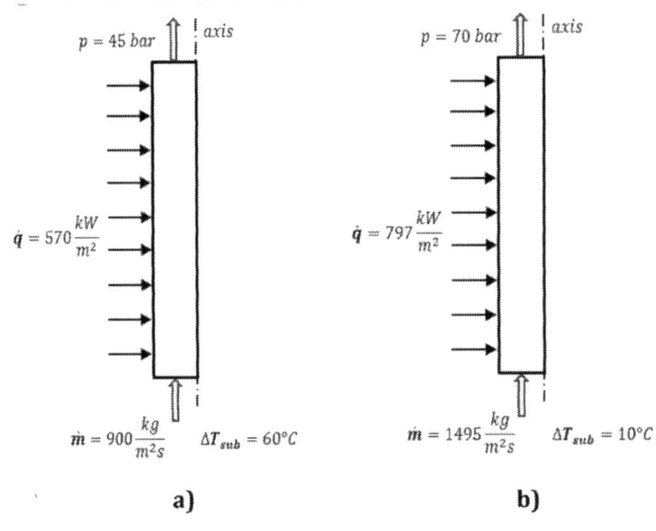

Figure 2. Experimental test sections schemes.

For both of them, a 2D axisymmetric computational domain has been created, meshed with hexahedral cells; a steady-state analysis has been conducted and liquid thermophysical properties have been varied with temperature, keeping saturation conditions constant along the pipe as the pressure level does not vary significantly. The mixture k-ω SST turbulence model was implemented. Negligible effects on the results were observed using either first or second-order discretization for each governing equation.

The profiles of the calculated VF at the centreline and wall temperature T_w, in simulating Bartolomej's case, are reported in Figure 3 for different models. Standard RPI reproduces quite well the real trend, even though it tends to slightly overestimate VF, while RPI-CHF fails in the prediction of T_w, even though it should be expected to have the same behaviour. A change could be brought by acting on N_w parameters, according to equation (12), and on an additional correction within the code that enhances the bubbles nucleation, getting better results with a redefined RPI-CHF modified.

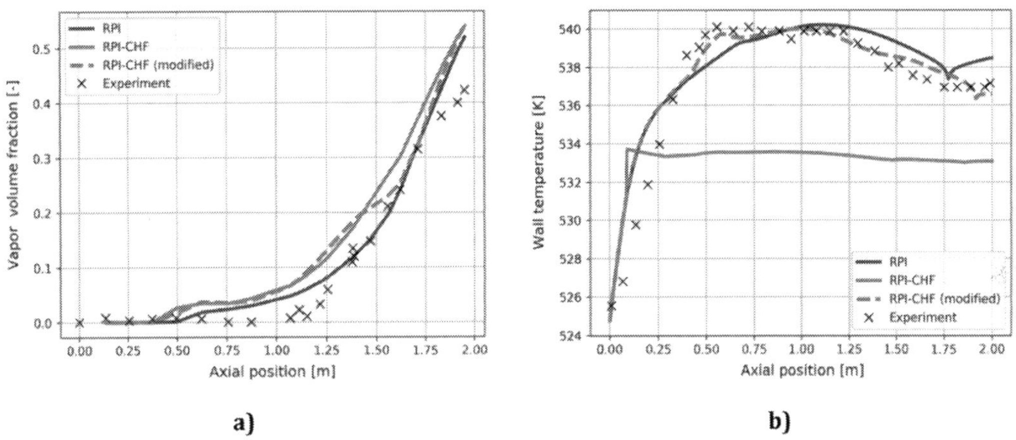

Figure 3. Bartolomej et al. – Comparison of void fraction and wall temperature trends.

In Hoyer's test, the modified model seems to not match the experimental curve of wall temperature at higher VF, while RPI-CHF provides a consistent solution, properly identifying the position of dry-out, as depicted in Figure 4. However, in light of the foregoing and considering the whole range of vapor qualities which can be experienced in an evaporator, RPI-CHF is the most appropriate approach to adequately represent the boiling process.

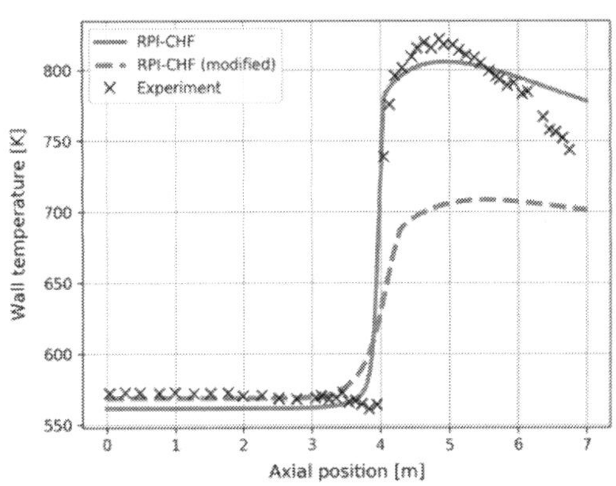

Figure 4. Hoyer – Comparison of wall temperature trends.

Journal of Physics: Conference Series **2940** (2025) 012031 doi:10.1088/1742-6596/2940/1/012031

To provide a clearer understanding of the key differences between the models described above, their main characteristics are listed in Table 1. This summary highlights the distinct features of each model, enabling a comparison of their functionality.

Table 1. Characteristics of RPI-based models.

	Wall treatment	Description
Standard	$\dot{q}_w = \dot{q}_l + \dot{q}_q + \dot{q}_e$	Vapor temperature set to its saturation value
CHF	$\dot{q}_w = (\dot{q}_l + \dot{q}_q + \dot{q}_e) f(\alpha_l) + \dot{q}_v (1 - f(\alpha_l))$	Convective heat flux of the vapor phase included
CHF (modified)	$\dot{q}_w = (\dot{q}_l + \dot{q}_q + \dot{q}_e) f(\alpha_l) + \dot{q}_v (1 - f(\alpha_l))$	Bubble nucleation enhancement deactivated

3.2 Boiling propane analysis

The test rig studied by Allymehr [27] has been selected: he measured the heat transfer coefficient (HTC) of R290 at various average vapor qualities x, setting the inlet saturation temperature, mass flow rate and heat flux in a system with a smooth copper pipe electrically heated, located at Thermal lab of the Department of Energy and Process Engineering of Norwegian University of Science and Technology.

The computational domain consists of the solid geometry and fluid volume, meshed with polyhedral elements, adding a proper number of inflations layers near the wall to ensure getting dimensionless distance y^+ greater than 30 as a general recommendation for wall functions in boiling cases [13], with the mixture k-ω SST turbulence model implemented.

The pipe, with $D_i = 4.2$ mm and $L = 0.5$ m, has been lengthened upstream of the inlet section to allow flow stratification, as illustrated in Figure 5 for one of the eight simulated test points, and backward heat diffusion. Due to gravity, the liquid settles down at the bottom where the VF is lower than at the top, the difference between these two values becomes smaller approaching the outlet and increasing x too, with a minimum experienced by the last point having everywhere $VF > 0.99$.

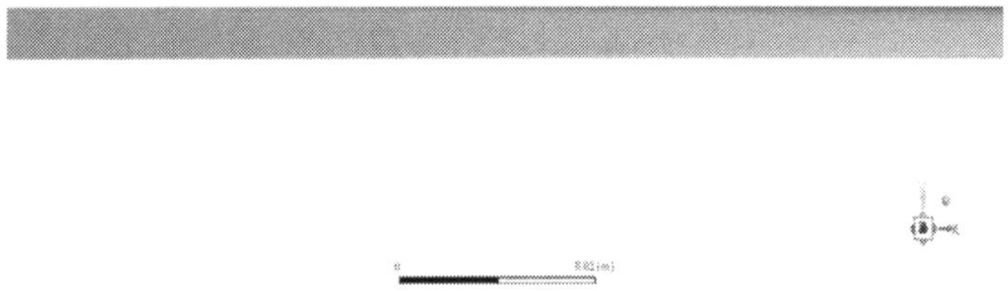

Figure 5. Vapor volume fraction contour in the adiabatic entrance.

Conjugate heat transfer calculations in steady state were executed with an imposed heat flux at the external surface and boundary conditions as listed in Table 2; vapor quality x_{in} at the inlet has been determined by the relationship between x_{in}, $\dot{q}_w$, h_{fv} and G knowing x, with $x = x_{in} + 0.5\Delta x$; pressure at the outlet has been derived from pressure drops.

Table 2. Boundary conditions for the investigated scenario.

	Variable	Unit of measure	Value
Inlet temperature	T_{sat}	°C	5
Mass flux	G	kg/(m²s)	379.41
Internal heat flux	$\dot{q}_w$	kW/m²	31.88

With the default settings for the blending function $f(\alpha_l)$, breakpoints and nucleate site density, the model is not able to correctly match the HTC trend for $0.1 < x < 0.7$, indeed dry-out occurs at a quality less than expected and T_w rises along the pipe, reaching a high value near the outlet section.

Therefore, a modification to the blending function has been made to delay the situation of single-phase convection with vapor only and N_w parameters have been adjusted to avoid wall temperature underestimation. This is also justified by the fact that almost all the implemented boiling definitions are valid for water in vertical channels. Figure 6 presents the CFD results for the HTC trend and error, compared with the experimental data.

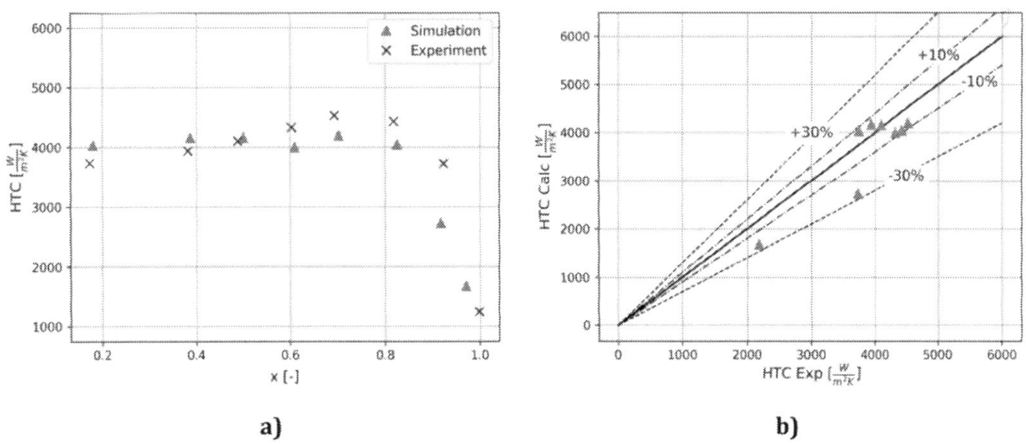

Figure 6. Comparison between calculated and experimental HTC.

As shown in Figure 6a), the location of HTC maximum and its following drop for $x > 0.7$, is well reproduced. The most significant component in Equation (7) is $\dot{q}_e$ being beyond 60% of $\dot{q}_w$ since $x < 0.8$; it grows for $0.1 < x < 0.7$ and starts to decrease for $x > 0.8$, with the process

Journal of Physics: Conference Series **2940** (2025) 012031 doi:10.1088/1742-6596/2940/1/012031

that becomes driven by single-phase convection with vapor. The lower wall temperature reached at HTC maximum is the reason for liquid heat transfer reduction in the above-mentioned range, even if the velocity is higher as the flow rate is fixed. The quenching factor is the second most significant contribution at the lowest x when $\dot{q}_v$ is nearly zero.

HTC errors are less than 10% regarding the first six test points and the greatest one is about 26% at $x = 0.92$.

4. Conclusions

In the present work, an investigation of current available CFD multiphase models for testing boiling propane in a horizontal pipe has been proposed, with the specific goal of assessing the most relevant characteristics affecting the solution.

A good qualitative agreement was found with the experiments even though a proper change of default settings has to be applied, taking into account the difference in both the arrangement and the type of fluid considered to calibrate the fundamental correlations and values on which the scheme is based.

An accurate analysis of the wall heat flux partitioning and the weight of individual terms has been performed, to comprehend the basics of the transfer phenomenon and identify how it operates.

Future research is necessary to extend the calculation to other wall heat and mass flux conditions and to evaluate more suitable sub-model equations, like bubble departure diameter and release frequency, for R290.

References

[1] Granryd E., 2001, *Hydrocarbons as refrigerants*, International Journal of Refrigeration, Vol. 4, pp. 15-24

[2] Wang S., Gong M. Q., Chen G. F., Sun Z. H., 2014, *Two-phase heat transfer and pressure drop of propane during saturated flow boiling inside a horizontal tube*, International Journal of Refrigeration, Vol. 41, pp. 200-209

[3] Choi K., Pamitran A. S., Oh J., Saito K., 2009, *Pressure drop and heat transfer during two-phase flow vaporization of propane in horizontal smooth minichannels*, International Journal of Refrigeration, Vol. 32, pp. 837-845

[4] Maqbool M. H., Palm B., Khodabandeh R., 2013, *Investigation of two phase heat transfer and pressure drop of propane in a vertical circular minichannel*, Experimental Thermal and Fluid Science, Vol. 46, pp. 120-130

[5] Longo G. A., 2010, *HC-290 (Propane) Vaporisation Inside a Brazed Plate Heat Exchanger*, International Refrigeration and Air Conditioning Conference, paper 1009

[6] Chen Y., Yu B., Lu W., Wang B., Sun D., Jiao K., Zhang W., Tao W., 2024, *Review on numerical simulation of boiling heat transfer from atomistic to mesoscopic and macroscopic scales*, International Journal of Heat and Mass Transfer, Vol. 225, paper 125396

[7] Setoodeh H., Ding W., Lucas D., Hampel U., 2021, *Modelling and simulation of flow boiling with an Eulerian-Eulerian approach and integrated models for bubble dynamics and temperature-dependent heat partitioning*, International Journal of Thermal Sciences, Vol. 161, paper 106709

[8] Colombo M., Fairweather M., 2016, *Accuracy of Eulerian–Eulerian, two-fluid CFD boiling models of subcooled boiling flows*, International Journal of Heat and Mass Transfer, Vol. 103, pp. 28-44

[9] Končar B., Krepper E., 2008, *CFD simulation of convective flow boiling of refrigerant in a vertical annulus*, Nuclear Engineering and Design, Vol. 238, pp. 693-706

[10] Qi D., Zhang J., Gao W., 2020, *Numerical Study on Two-Phase Heat Transfer and Flow of Subcooled Flow Boiling of R290 in A Vertical Tube*, World Scientific Research Journal, Vol. 6, Issue 7

[11] Dai Y., Zhang N., Tian S., Du H., 2020, *Numerical study of flow boiling heat transfer of propane in a horizontal smooth copper tube*, Heat Transfer Research, Vol. 51, pp. 1653-1667

[12] Kurul N., Podowski M. Z., 1991, *On the Modeling of Multidimensional Effects in Boiling Channels*, In Proceedings of the 27th National Heat Transfer Conference, Minneapolis, USA

[13] ANSYS Fluent Learning Hub, ANSYS, Inc., Canonsburg, PA

[14] Bianchini C., Da Soghe R., Mazzei L., Caggiano G., Angelucci M., 2021, *Assessment of CFD Models for Multiphase Heat Transfer in Different Boiling Regimes*, Proceedings of ASME Turbo Expo 2021, Turbomachinery Technical Conference and Exposition, GT2021

[15] Ishii M., 1979, *Two-fluid model for two-phase flow*, 2nd International Workshop on Two-phase Flow Fundamentals, RPI, Troy, NY

[16] Frank T., Shi J. M., Burns A. D., 2004, *Validation of Eulerian Multiphase Flow Models for Nuclear Safety Applications*, 3rd International Symposium on Two-Phase Flow Modeling and Experimentation, Pisa, Italy

[17] Antal S. P., Lahey R. T., Flaherty J. E., 1991, *Analysis of phase distribution in fully developed laminar bubbly two-phase flow*, International Journal of Multiphase Flow, Vol. 17, pp. 635–652

[18] Lopez de Bertodano M., 1991, *Turbulent Bubbly Flow in a Triangular Duct*, Ph.D. Thesis, RPI, Troy, New York

[19] Troshko A., Hassan Y. A., 2001, *A Two-Equation Turbulence Model of Turbulent Bubbly Flow*, International Journal of Multiphase Flow, Vol. 27, pp. 1965–2000

[20] Del Valle V. H., Kenning D. B. R., 1985, *Subcooled flow boiling at high heat flux*, International Journal of Heat and Mass Transfer, Vol. 28, pp. 1907–1920

[21] Cole R., 1960, *A Photographic Study of Pool Boiling in the Region of the Critical Heat Flux*, AIChE, Vol. 6, pp. 533-538

[22] Tolubinski V. I., Kostanchuk D. M., 1970, *Vapor bubbles growth rate and heat transfer intensity at subcooled water boiling*, 4th International Heat Transfer Conference, Paris, France

[23] Lemmert M., Chawla L. M., 1977, *Influence of flow velocity on surface boiling heat transfer coefficient in Heat Transfer in Boiling*, E. Hahne and U. Grigull (Eds.), Academic Press and Hemisphere, New York, USA

[24] Tentner A., Lo S., Kozlov V., 2006, *Advances in Computational Fluid Dynamics Modeling of Two Phase Flow in a Boiling Water Reactor Fuel Assembly*, International Conference on Nuclear Engineering, Miami, Florida

[25] Bartolomei G. G., Brantov V. G., Molochnikov Yu. S., Kharitonov Yu. V., Solodkii V. A., Batashova G. N., Mikjailov V. N., 1982, *An experimental investigation of true volumetric vapor content with subcooled boiling in tubes*, Thermal Engineering, Vol. 29, pp. 132–135

[26] N. Hoyer, 1998, *Calculation of dryout and post-dryout heat transfer for tube geometry*, International Journal of Multiphase Flow, Vol. 24, pp. 319-334

[27] Allymehr E., 2017, *Experimental Investigation of evaporation of propane (R290) in small pipes*, Master Thesis, School of Industrial and Information Engineering, Milano

Journal of Physics: Conference Series **2940** (2025) 012032 doi:10.1088/1742-6596/2940/1/012032

Two-Phase Pressure Drops Analysis in a Chevron-Type Corrugated Channel Using Image Segmentation for Void Fraction Estimation

S Passoni[1,*], A Ferrario[1], R Mereu[1] and F Inzoli[1]

[1] Politecnico di Milano, Department of Energy, via Lambruschini 4A, 20156, Milano, Italy

* Corresponding author: stefano.passoni@polimi.it

Abstract. This study examines the pressure drops and flow patterns in a Chevron-type Compact Plate Heat Exchanger (CPHE) with air-water two-phase mixture under adiabatic conditions. A transparent replica channel, made from epoxy resin, was used to mimic the heat exchanger's fluid dynamics and allow for optical access. The channel featured a 63° corrugation angle, 2.5 mm depth, and 8.9 mm pitch, yielding a global enlargement factor of 1.17. A particular inlet section was designed to ensure even phase mixing and distribution. Operating conditions varied from 6 to 365 kg/m²s for water and 0.02 to 5 kg/m²s for air, expanding the range of previous studies. High-framerate back-illuminated visualizations provided detailed observations of the flow patterns. Moreover, frame-by-frame void fraction was estimated via HSV video analysis adopting a convolutional neural network (U-Net) to produce binary masks and estimate gas bubble volumes and via a volume-reconstruction algorithm exploiting the pseudo-2D shape of the channel. A novel correlation for Darcy's friction factor based on Reynolds number was developed for single-phase flow, complemented by an analysis of two-phase pressure drops using a two-phase multiplier correlated with the Lockhart-Martinelli parameter. The incorporation of void fraction data significantly improved their modelling accuracy.

1. Introduction

Plate Heat Exchangers (PHEs) have gained widespread recognition in the industrial sector for their efficiency and compact design. These devices consist of a pack of gasketed or brazed corrugated plates that enhance heat transfer thanks to the increase turbulence promoted by the corrugated profile. Additionally, a correlation exists between the enhancement of heat transfer performance and the incidence of pressure drops. Specifically, larger corrugation angles are known to improve heat transfer efficiency at the cost of increased pressure drops. Conversely, smaller corrugation angles tend to minimize pressure drops but may compromise heat transfer effectiveness. Nevertheless, the primary advantage of PHEs lies in their performance-to-footprint ratio, demonstrating greater efficiency than other industrial heat exchangers in just a fraction of the volume. Compact plate heat exchangers are currently gathering significant attention for diverse applications, including their potential role as steam generators in Small Modular Nuclear Reactors (SMRs) [1] where spaces are very limited and volume is what drives the choice of a specific configuration. Consequently, because of this and other potential applications, the industrial and scientific focus has gained more and more interest towards the study of two-phase flows within these components. The Chevron-Type stands as the most prevalent corrugation pattern in PHEs. It features a sinusoidal corrugation profile that creates a herringbone pattern. The plates align with opposing patterns to form a closed channel with contact points determined by the corrugation angle. Key geometrical parameters are the plate length and the plate width that characterize the dimension of the plates whereas the corrugation profile is defined by the Chevron angle, corrugation depth wavelength, which are directly linked to the heat exchanger's performance as highlighted previously.

Numerous studies have focused their attention on the complex dynamics of two-phase flow within Plate Heat Exchangers aiming at developing valuable correlations for accurately predicting pressure drops [2 - 7]. All these works tend to investigate similar operating conditions, mainly overlapping in the region defined by gas and liquid superficial velocities in the range of 0.2-6 m/s and 0.03–0.4 m/s respectively. A graphical representation of this condition is given in Figure 1, where the experimental range of all the cited studies is compared with the database produced in this work. Moreover, the region of very low gas and liquid superficial velocities is not covered by any of the presented works despite being of major interest. Therefore, this work aims at analyzing an extensive range of operating conditions to create a uniform and coherent database that will allow a more comprehensive analysis of two-phase pressure drops and broaden the range of validity of the proposed correlations. Furthermore, the majority of research documented in the literature assumes homogeneous flow, thereby neglecting the slip between phases. In an effort to move beyond this significant hypothesis, the present study employs image analysis to extract details on void fraction, which will then be utilized to refine pressure data processing.

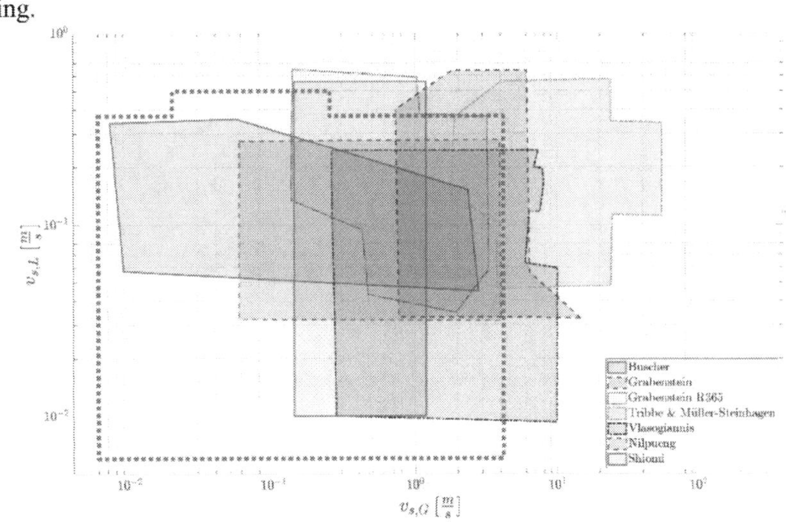

Figure 1. Experimental ranges of the previous studies compared with the one of the present work (highlighted with red bold dashed line).

2. Materials and Methods

2.1. Plant Layout

The adiabatic loop depicted in Figure 2 was used for both flow visualization and pressure drop experiments. The system consists in an open-loop air circuit and a closed-loop water one. Water flow was driven by a centrifugal pump and measured by rotameters, while compressed air flow was regulated to desired pressures and measured by rotameters. The setup included a bypass line to ease start-up and shut-down operations and a mixing section for fluid mixing before entering the visualization area. The flow was distributed and mixed via T-junctions before passing through a plenum and the corrugated channel, returning then to the feeding tank. Experiments were carried out with varying water (G_w=6 to 365 kg/m²s) and air (G_{air}=0.02 to 5 kg/m²s) mass fluxes under controlled temperature conditions to mitigate pump-induced heating. Pressure drops were recorded using differential pressure transmitters and post-processed with MATLAB®. Flow visualization was performed in a back-illuminated setup featuring a flicker-free LED panel and the two-phase flow was captured using a high-speed camera at 1900 fps, allowing detailed observation of the dynamic flow patterns. The tests were conducted in an up-flow configuration and the tested plate had a corrugation angle of 63° (from flow direction), a corrugation depth of 2.5 mm and a wavelength equal to 9 mm. The total corrugation width and length were 182 mm and 320 mm respectively. Table 1 and 2 summarize the characteristics of the instrumentation used.

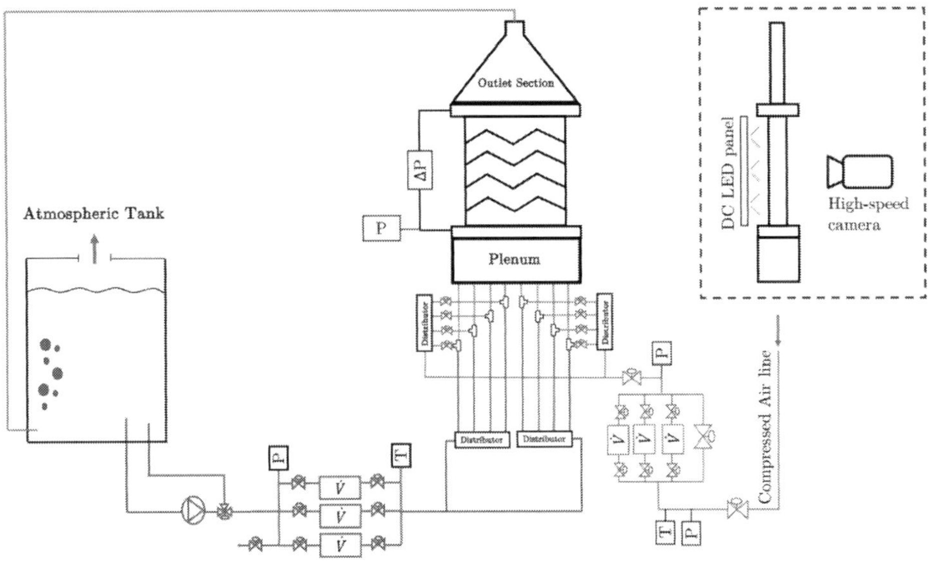

Figure 2. A detailed schematic of the loop employed during experiments

Table 1. Characteristics of the differential pressure transmitters.

Model	Klay DP-4000	Klay DP-4000	Klay DP-4000	SETRA 267
Fluid	Air/Water	Air/Water	Air/Water	Air
Range	0 – 60 mbar	0 – 400 mbar	0 – 2000 mbar	0 – 10 mbar
Accuracy	0.075% F.S.	0.075% F.S.	0.075% F.S.	1% F.S.

Table 2. Characteristics of the flowmeters.

Asametro	P13-2500	P13-2600	P13-2900	E5-2500	E5-2500	E5-2500
Fluid	Water	Water	Water	Air	Air	Air
Range	10 – 100 L/h	40 – 400 L/h	250 – 2500 L/h	4 – 190 nL/h	100 – 800 nL/h	400 – 4000 nL/h
Uncertainty	± 3% F.S.	± 3% F.S.	± 3% F.S.	± 2% F.S.	± 2% F.S.	± 2% F.S.

2.2. Image analysis

As already introduced, image analysis was employed to extract void fraction information from the experimental videos. To this aim, we employed an encoder-decoder style neural network (U-net) to tackle the task of semantic image segmentation. This architecture was initially engineered and trained for image segmentation in the biomedical field [8] but, thanks to its transfer-learning capabilities, it can be re-trained with a low number of manually annotated images. In particular, the training for this work was performed by fine-tuning the model's weights initially trained on front-lit images of subcooled flow boiling, with a total of 13 manually annotated frames extracted from the experimental videos. Prior to annotation, background subtraction and contrast enhancing procedures were performed. Model's performance was then evaluated by assessing standard pixel-level metrics like accuracy, precision, recall and F1-score, with each one surpassing the threshold of 90% quality. The obtained binary masks were further processed in MATLAB® to perform blob analysis. Figure 3 reports an example of image processing. A novel volume reconstruction algorithm was designed and implemented to estimate the void fraction by calculating the three-dimensional volume of individual gas bubbles from their corresponding two-dimensional footprints, allowing for accurate determination of the void fraction through volume ratio analysis. This methodology is based on the classification of gas bubbles into small and large clusters based on their equivalent diameter and average channel depth across their footprint and adapting the reconstruction method accordingly. The volume of small bubbles was approximated by using a revolving ellipsoid to reconstruct their shape while large gas clusters' volume was computed by coherently summing the depth map of the channel previously computed in positive and negative directions across bubble footprint. Moreover, to provide a more accurate representation of the bubbles' morphology, the borders were smoothed with an exponentially decaying smoothing function. This algorithm exploits the pseudo-2D shape of the corrugated channel and it is superior to an area-based estimation of the void fraction as this latter methodology does not account for how bubbles distribute in the channel's cross-section. Figure 4 illustrates an exemplary application of the volume reconstruction algorithm on an single frame. Frame-specific void fraction value was then computed as by summing the volume of all the bubbles reconstructed and the by dividing it by the total volume of the channel. Then, by averaging statistically enough frames, the average void fraction ($\bar{\alpha}$) of the specific experimental test is computed.

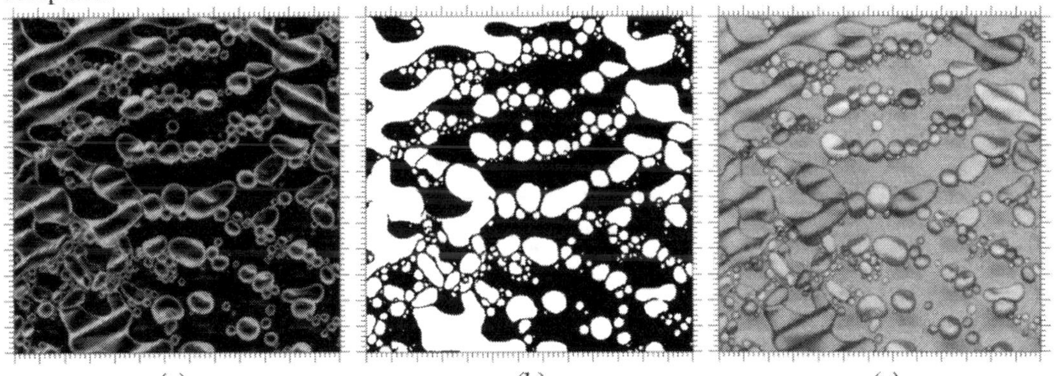

Figure 3. Image post-processing example. Data for liquid superficial velocity 0.006 m/s and gas superficial velocity of 0.158 m/s. Original image with background subtracted and contrast enhanced (a), Binary mask from U-Net (b) and connected component labelling (c). Images cropped to squared aspect ratio. Minor axes ticks in *mm*.

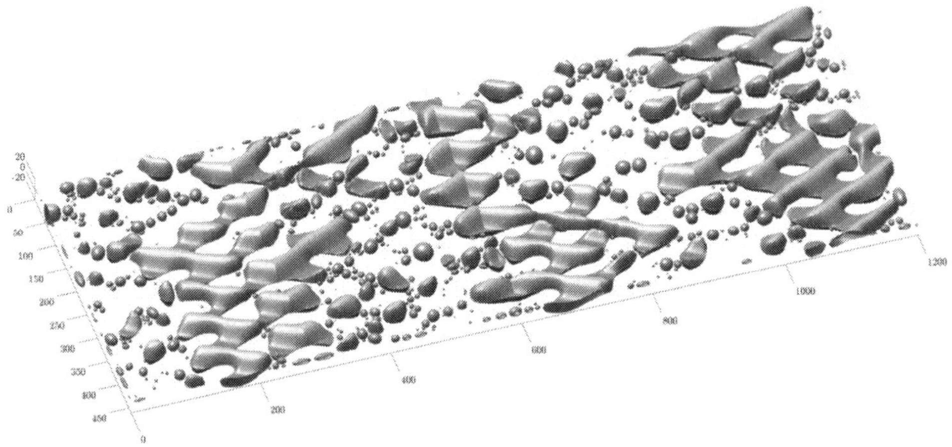

Figure 4. Result of the volume reconstruction algorithm applied to an entire frame.

2.3. Data reduction

Pressure drop measurements were conducted under both single and two-phase flow conditions. For the first case, the results were elaborated in the form of Darcy's friction factor plotted against the Reynolds number. Test performed both single gas and liquid phase were performed and, when using gas as the operating fluid, the mass flow rate was carefully corrected to account for deviation from calibration conditions. The friction factor was computed according to equation (1), where ΔP represents the measured differential pressure drop, D_h the hydraulic diameter, A_f the frontal area, Q the volumetric flow rate and $L_p the\ plate\ length.$

$$f_d = \frac{2\Delta P\ D_h A_f^2}{\rho\ Q^2\ L_p} \tag{1}$$

Two-phase pressure drops instead required the isolation of the frictional component of the pressure drop by removing the accelerative and the gravitational components from the measurements. The first can be simply neglect as it has little to no impact for adiabatic flows while the second was computed as reported in equation (2), calculating the mixture density with the void fraction obtained from image analysis.

$$\Delta P_{grav} = \rho_m g\, L_p \qquad \text{with} \qquad \rho_m = (1-\bar{\alpha})\rho_L + \bar{\alpha}\rho_G \tag{2}$$

The Lockhart-Martinelli approach was employed to present frictional pressure drop data in a dimensionless manner. This method was preferred over the definition of a two-phase friction factor as it does not depend on an additional model for defining two-phase viscosity for the two-phase Reynolds number. The chosen method treats the fluids as if flowing independently and relates single-phase to two-phase pressure drops through a specific multiplier factor. It introduces two dimensionless parameters: the Lockhart-Martinelli parameter (X) and the Two-phase Multiplier (Φ_{TP}), both defined in equation (3). In these equations the subscripts L and G stand for single-phase operation in either liquid or gas phase respectively while TP stands for two-phase tests. This approach is useful since it allows the direct computation of the two-phase frictional pressure drops given a correlation and the single-phase pressure loss.

$$X = \sqrt{\frac{\left(\frac{\Delta P}{L_p}\right)_L}{\left(\frac{\Delta P}{L_p}\right)_G}} \qquad \Phi_{TP} = \sqrt{\frac{\left(\frac{\Delta P}{L_p}\right)_{TP}}{\left(\frac{\Delta P}{L_p}\right)_L}} \tag{3}$$

Once the parameters used to process the measurement are defined, the calculation of the expanded measurement uncertainty is needed. In accordance with the guidelines outlined in the JCGM 100:2008 standard [9], this study adhered to best practices for experimental uncertainty analysis and reporting. For each data point acquired, expanded uncertainties were calculated and visually represented as error bars on the respective plots. Table 3 reports the minimum, average and maximum uncertainty for the main four parameters used to describe the results.

Table 3. Minimum, average, and maximum values of the main variable's expanded uncertainty

	Expanded Uncertainty		
	Min	Avg	Max
Re	2.49%	5.16%	15.00%
f_d	4.99%	12.60%	38.78%
X	0.08%	6.33%	47.78%
f_{TP}	0.06%	0.95%	17.62%

3. Results and Discussion

After segmenting all the images using the methodology introduced previously, the void fraction estimation algorithm was applied to post-process the binary images. The results are presented as void fraction contour map plotted against the superficial velocity of both phases in Figure 5. This map represents the void fraction distribution and is generated by fitting the discrete image analysis data using a locally-weighted smoothing linear regression (LOWESS) technique. At low liquid superficial velocities, the contour lines are nearly vertical, gradually decreasing in slope as the liquid flow rate increases. Hence, if one moves vertically at a constant gas velocity (i.e., gas flow rate), the void fraction decreases as expected.

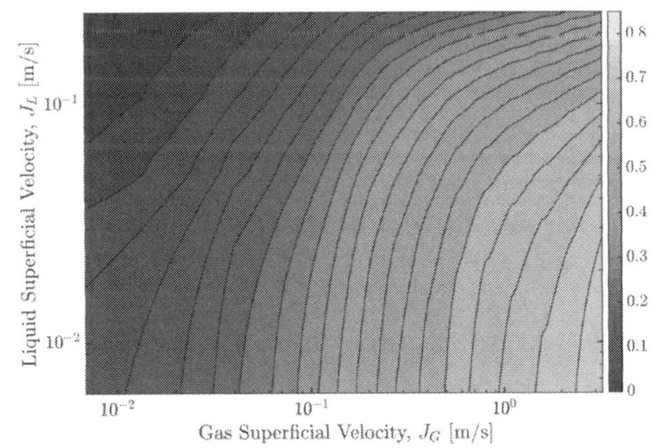

Figure 5. Contour map of void fraction obtained from image analysis.

Regarding single-phase pressure drops measurements, the friction factor data are reported in figure 6(a). This image displays the data obtained with single-phase liquid and gas flow as well as the equation of the best fit curve: $f_d = 217.94Re^{-0.74} + 3.99$. As expected, the results from the two fluids overlaps very well and the proposed correlation, besides accurately interpolating the experimental points, presents a MAPE (mean absolute percentage error) of 5.23%. Figure 6(b) shows instead the relationship between the two-phase multiplier and the Lockhart-Martinelli parameter. Data from the void fraction map presented above were used to eliminate the gravitational component of the pressure drop without relying on the homogeneous flow assumption and the results showed a far less scattered database with the suggested approach with a significant increase in interpolation accuracy passing from R^2=0.773 to R^2=0.954, proving the homogeneous flow assumption to be a very strong one. The best-fitting correlation for the proposed data adopts the format suggested by Chisholm [10], correlating the Lockhart-Martinelli parameter with the two-phase multiplier as shown in the equation 4

$$\Phi_{TP}^2 = 1 + \frac{C}{X} + \frac{1}{X^2} \tag{4}$$

where the coefficient C is equal 9.92 after fitting the data points. The observed value falls within the range delineated by Chisholm for viscous-viscous and turbulent-viscous flow regimes in circular conduits, residing between the lower and upper limits of 5 and 10, respectively [10].

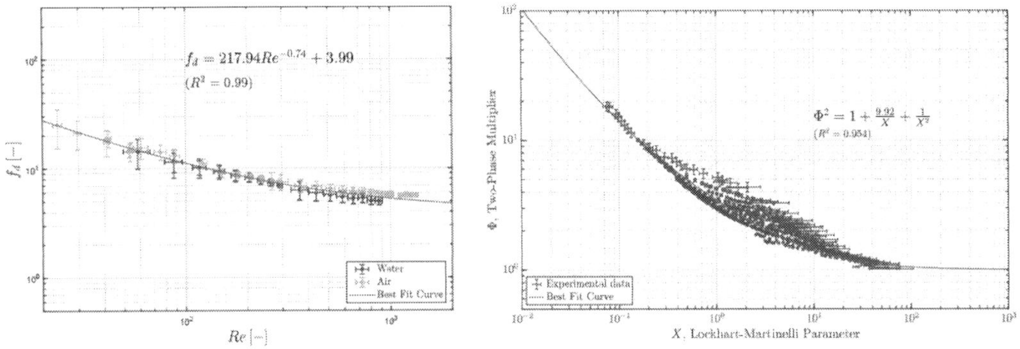

Figure 6. Single-phase friction factor plotted against Reynolds number (a) and two-phase multiplier against Lockhart-Martinelli parameter (b).

4. Conclusion

This work proposed an experimental study of pressure drops in a chevron-type corrugated channels operated with an adiabatic mixture of air and water. The plant layout used for the experiments was described and the data reduction methodology as well. During two-phase operations, besides measuring the differential pressure drop and monitoring all the needed variables, high-framerate videos were captured to visualize the flow pattern in detail. These data were processed according to an innovative methodology that blends the use of machine learning for image segmentation and a novel volume reconstruction algorithm to estimate the void fraction. This approach is more accurate than a standard area-based one and it allowed a more precise modelling of the two-phase frictional pressure drop yielding correlations that more closely align with the experimental data, thereby enhancing the accuracy of the predictions.

5. References

[1] Kang, J., Bak, J. Y., Lee, B. J., Chung, C. K., & Yun, B. (2022). Numerical investigation of a plate-type steam generator for a small modular nuclear reactor. Nuclear Engineering and Technology, 54(8), 3140-3153.

[2] Buscher, S. (2021). Two-phase pressure drop and void fraction in a cross-corrugated plate heat exchanger channel: Impact of flow direction and gas-liquid distribution. Experimental Thermal and Fluid Science, 126, 110380.

[3] V. Grabenstein, A.-E. Polzin, and S. Kabelac, "Experimental investigation of the flow pattern, pressure drop and void fraction of two-phase flow in the corrugated gap of a plate heat exchanger," International Journal of Multiphase Flow, vol. 91, pp. 155–169, 2017.

[4] Tribbe, C., & Müller-Steinhagen, H. M. (2001). Gas/liquid flow in plate-and-frame heat

exchangers-Part I: Pressure drop measurements. Heat transfer engineering, 22(1), 5-11.

[5] Vlasogiannis, P., Karagiannis, G., Argyropoulos, P., & Bontozoglou, V. (2002). Air–water two-phase flow and heat transfer in a plate heat exchanger. International Journal of Multiphase Flow, 28(5), 757-772.

[6] Nilpueng, K., & Wongwises, S. (2010). Two-phase gas–liquid flow characteristics inside a plate heat exchanger. Experimental thermal and fluid science, 34(8), 1217-1229.

[7] Shiomi, Y., Nakanishi, S., & Uehara, T. (2004). Characteristics of two-phase flow in a channel formed by chevron type plates. Experimental Thermal and Fluid Science, 28(2-3), 231-235.

[8] Ronneberger, O., Fischer, P., & Brox, T. (2015). U-net: Convolutional networks for biomedical image segmentation. In Medical image computing and computer-assisted intervention–MICCAI 2015: 18th international conference, Munich, Germany, October 5-9, 2015, proceedings, part III 18 (pp. 234-241). Springer International Publishing.

[9] Jcgm, J. C. G. M. (2008). Evaluation of measurement data—Guide to the expression of uncertainty in measurement. Int. Organ. Stand. Geneva ISBN, 50, 134.

[10] Chisholm, D. (1967). A theoretical basis for the Lockhart-Martinelli correlation for two-phase flow. International Journal of Heat and Mass Transfer, 10(12), 1767-1778.

Acknowledgments
This work has received funding from the Euratom research and training programme 2014-2018 under 525 Grant Agreement No. 847553.

Journal of Physics: Conference Series **2940** (2025) 012033 doi:10.1088/1742-6596/2940/1/012033

Assessing RELAP5 performances for the simulation of Containment Wall Condenser

G. Grippo[1,*], M. Polidori[2]

[1] Laboratory of Montecuccolino-DIN, Alma Mater Studiorum University of Bologna, via dei Colli 16, 40136 Bologna, Italy.
[2] NUC-ENER-SIC division, ENEA Research Center, Via dei Mille 21, 40129, Bologna, Italy.
[*] e-mail: gianmarco.grippo2@unibo.it

Abstract. The new designs of Nuclear Power Plants are increasingly emphasizing the implementation of passive systems to fulfill several safety functions. Such systems operate with low driving forces, and often employ components that require a "change state" signal for actuation. These features may lead to complete or partial actuation failures, altering the driving force conditions and consequently reducing the operational ranges and reliability of the systems themselves. However, even though the operational modes differ from those of active systems, the safety standards remain equivalent. To study these possible situations, the computer system codes must be able to simulate the phenomena related to the specific system, and they should be assessed using data both from separate and integral experimental facilities.
Within the framework of the EU2020 PASTELS (Passive Systems: Simulating the Thermal hydraulics with Experimental Studies) project, a series of natural circulation experiments have been conducted in the PASI facility, an open loop Containment Wall Condenser (CWC), varying initial and boundary conditions to support the validation of system codes. As project partner, ENEA developed the PASI input deck for RELAP5/mod3.3 to assess the code's capability in simulating complex loop components, such as the sparger, and related thermohydraulic phenomena like water flashing and geysering within the open loop. This paper presents the post-test calculations conducted with RELAP5 for test 2 (quasi-static test) and test 6 (replicating the drain phase of the water storage tank).

Keywords: *Containment Wall Condenser, Thermal hydraulic, Passive systems, RELAP5, Natural circulation*

1. Introduction

The Nuclear Power Plant (NPP) containments are provided with safety systems in order to maintain the integrity of the structure during accidental scenario. The driving phenomena of this type of systems is the natural circulation which allow to remove the heat power inside the containment. However, this type of phenomenology is unstable and the understanding of the operational working ranges of this type of systems is a key element to increase their reliability [1]. Therefore, experimental data performed with different working conditions are needed to understand these ranges and to have some insights on the computational code, as RELAP5 [2], to understand the simulations gap and the required updates to improve their capability to predict the natural circulation driving phenomena. In the framework of the

HORIZON 2020 PASTELS project [3], the work done was directed to identify the physical mechanisms that could reduce the heat transfer efficiency of a CWC and, parallelly, assess and improve the simulation tools using new experimental data to support the code validation. In particular, a series of tests have been conducted on the PASI facility; it has been developed and built at LUT university and represent a low-pressure open loop CWC [4] [5] [6]. Therefore, a tests matrix has been defined addressing the common concerns related to passive systems and considering several operating conditions on the containment side. Within this framework, ENEA was carried out to perform the post-test simulations of all the tests matrix; the input deck has been developed with RELAP5/mod3.3 code and, after the calibration of the model using one of the 10 tests, the same input deck has been used to perform the simulations of the whole tests' matrix. The aim of this work is to give a first insights on the physical behaviours driving a CWC and understand the state of the art of system code, as RELAP5, to simulate different operational working conditions [7]. Only two tests will be analysed: the test 2, a quasi-static test performed to understand the behaviour of the systems during different power conditions on the containment side, and test 6, a drainage test to investigate the influence of the pool gravity head on the loop performance.

2. PASI experimental facility

The PASI facility, developed by LUT University, is a low-pressure open loop designed to test a CWC [6]. This setup aims to analyse the physical mechanisms that might impact the CWC's safety function and aid in validating system codes.

The PASI test facility, illustrated in **Figure 1**, takes as reference system the passive containment heat removal system of an AES-2006 type pressurized water reactor. The height scale of the PASI facility is 1:2 relative to the reference system and the main components include the Pressure Vessel (PV), which simulates containment conditions, Heat Exchanger (HX), Water Pool (WP), and interconnecting riser and downcomer pipelines. The vessel is not a scaled-down version of the prototypical system; therefore, the mixture of steam, gasses and condensate cannot be representative of a real system; however, it has been designed to maintain and simulate realistic heat transfer conditions. In addition, it features three large ducts for instrumentation, Non-Condensable Gases (NCGs) injection, or steam injection; in the setup used during the experimental campaign related to PASTELS project [3], the lower duct has been selected as steam input. The bottom of the vessel houses the condensate water drain system, consisting of two flowmeters, a selector valve, a condensate tank with two level switches, and a purge valve. A safety valve is installed in the vessel's upper head to prevent pressure from exceeding the maximum set-point of 5 bar.

Inside the pressurized vessel, the HX tube section and hot collector are installed and directly connected to the riser pipeline located outside the vessel. The HX section includes a hot/cold collector and a bundle of 15 heat exchange tubes, linking it to the hot/cold legs of the open loop (riser/downcomer pipeline). The tubes bundle is not entirely installed inside the PV as the lower part is outside at the PV bottom and do not participate to the heat transfer. The water pool is located at the higher loop elevation and serves as a reservoir within the natural circulation loop system. On one side of the pool there is a sparger, shown in **Figure 1**, which is a shroud that surrounds and extends over the top of the riser pipeline. It has been designed to prevent fast condensation and water hammer phenomena. It consists of two concentric shroud and the inner one continues from the top of the riser, while the outer shroud starts at higher elevation. Additional information about the experimental facility, the instrumentation employed for the measurements and the related uncertainties can be found in [4].

3. PASI model for RELAP5/mod3.3

The PASI input deck, depicted in **Figure 1**, has been nodalized using the RELAP5/mod3.3 [2] thermohydraulic code, which has been developed through a collaboration between the United States Nuclear Regulatory Commission (USNRC) and a consortium of international and domestic organizations involved in the International Code Assessment and Applications Program (ICAP) and its successor, the Code Application and Maintenance Program (CAMP).

RELAP5/mod3.3 employs a finite difference numerical scheme to solve coupled fluid dynamics and heat transfer problems. The code implements a one-dimensional, transient, two-fluid model capable of simulating two-phase steam-water mixtures with non-condensable components in the steam phase. The fundamental equations comprise continuity, momentum, and energy equations for both phases, solved in terms of time and volume-averaged dependent variables. Moreover, the code features a comprehensive library of component models including pipes, valves, pumps, heat structures, and control systems, making it particularly suitable for simulating complex thermal-hydraulic systems. It also incorporates special process models for phenomena such as form losses, choked flow, and non-condensable gas transport, which are essential for accurately representing the physical behavior of natural circulation systems like the PASI facility.

Specifically, in the model, the pressure vessel is divided into four zones, represented by two branch components (upper and lower plenum) and two 11-cell pipe components for the middle section. The vessel's depressurization line is simulated with a trip valve and a time-dependent volume connected to the upper plenum. This trip valve opens when the pressure vessel reaches the maximum set-point of 5 bar, and the time-dependent volume represents the sump under environmental conditions. The vessel's lower plenum connects to two distinct lines: one for steam injection, allowing steam entry to achieve various test conditions of pressure, temperature, and mass flow; the other is a drain system, including a drain line, reclaim tank, purge line, and equalization tube. The reclaim tank collects condensate formed during tests and expels it from the pressure vessel. The purge line is active when the condensate reaches a target level of 0.4 m, while the equalization tube balances pressure between the reclaim tank and the pressure vessel. The pressure vessel is thermally coupled with the HX section (**Figure 1**), which includes the HX in vessel and the hot collector. The 15 HX tubes are represented by a single pipe. The riser pipeline connects the HX section with the pool sparger, divided into three hydraulic components: two pipe components for the central and annular channels, and a branch component linking the central channel, annular channel, pool dome, and middle pool. A single junction simulates the sparger pressure distribution hole, directly connecting the central and annular channels. The pool is divided into five zones for slice nodalization and thermal coupling with the sparger, using three branch components and two pipes connected horizontally by eight single junctions. The pool discharge line adjusts the pool level at the Start of Transient (SoT). The pool dome connects to an external boundary condition with air under environmental conditions, while the pool bottom connects to the sparger's annular channel, simulating the water flow path through the sparger shroud, and to the downcomer pipeline section. The downcomer pipeline consists of two pipe components connected by a ball valve. The valve is modelled as a servo valve with forward and reverse Cv coefficients, calibrated using experimental data (internal communications) to determine the k-loss based on the stem position [7].

To set the correct pressure drop through the loop, several loss coefficients were added to the input deck. A sensitivity analysis of the pre-transient phase of Test 1 (with the valve fully open) was conducted to determine the appropriate loss coefficients for the pressure distribution hole, enlargements, and narrowing of various sections in the loop (collector entry/exit, pool entry/exit, sparger central channel exit, and annular channel entry/exit). All heat structures in the model include the corresponding insulation material (if present). Given that previous work indicated RELAP5 tends to underestimate heat exchange in the presence of NCGs, several sensitivity analyses were conducted on fouling factors on both the HX tube and shell sides, as well as the correlation used ("Vertical bundle without crossflow") for calculating the heat transfer coefficient, to enhance the calibration of heat exchange between the vessel environment and the HX pipe. The best results were obtained using the default fouling factor value of 1.0 (no fouling) [8].

The model was calibrated with Test 1, and the final input deck version was used to simulate all subsequent tests, ensuring consistency throughout the PASI test matrix.

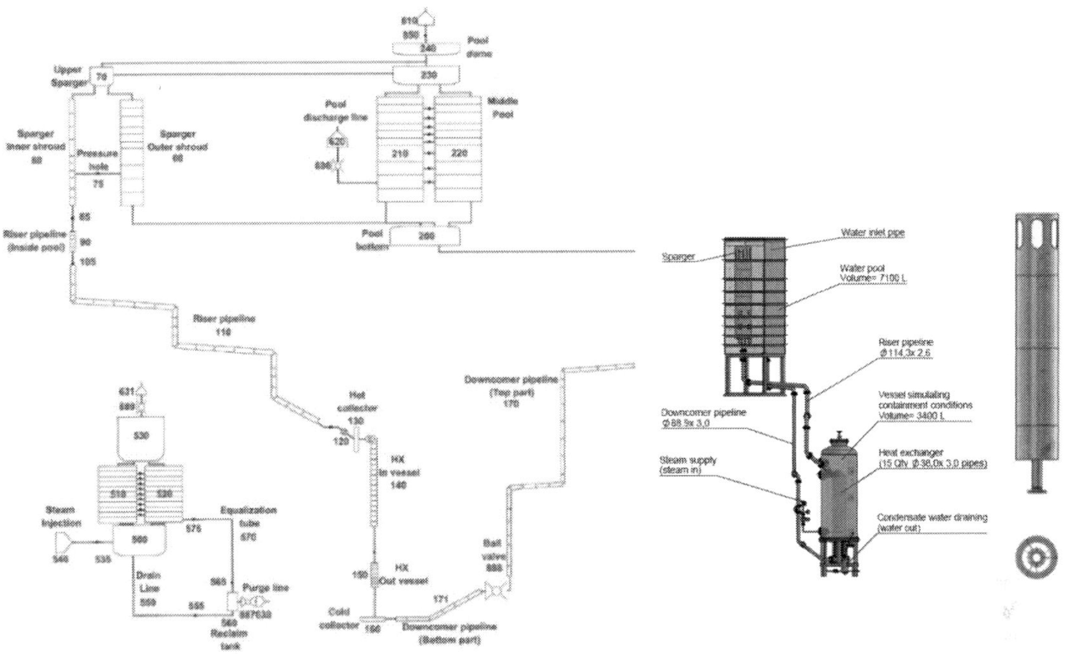

Figure 1. PASI R5 input deck and schemes *of* PASI facility layout and sparger

4. Test simulations

The tests performed with PASI facility cover different transient scenarios, aiming to demonstrate the relevant physical phenomena of such system. The test matrix is composed by 10 tests, divided into different representative scenario in order to investigate and analyse several physical behaviour: characterization of cooling system behaviour with different injected steam power, understanding the importance of the pool gravity head on the loop behaviour, study the behaviour of the loop with different sparger geometries, analyse the startup phase and the effect on the transient of the cold water refill in the pool.
In the present paper only two tests will be considered and analysed; quasi-static test 2 at fixed power in the pressure vessel, and test 6, which concerns the drainage of the pool water.

4.1. Test 2 – Description

During the pre-transient phase, the steam supply line injects a heating power of 200 kW into the PV, with subsequent reduction to 100 kW reaching the test 2 nominal power set-point. Then, the transient starts with different phases characterized by different ball valve opening ratio, varying the loop hydraulic resistance. When the water exiting from the top of the sparger central channel reaches saturation conditions, an oscillating two-phase flow begins, driven by a combination of geysering and flashing phenomena. Flashing occurs when there is a rapid vaporization of fluid due to a decrease in static pressure, which brings subcooled liquid from the cold collector into superheated vapor. Geysering phenomena results from high local pressures which led to fluid expulsion, potentially escalating into a boiling eruption at higher heating powers, characterized by the intense release of steam bubbles. During these phenomena, may be observed vapor eruption, subsequent vapor condensation, and reverse flow. The oscillations caused by vapor eruption interrupt the flashing, preventing it from becoming widespread. While flashing can enhance circulation due to the generated driving head, it also suppresses geysering by reducing the likelihood of steam slug formation. Additionally, higher flow rates induced by flashing further inhibit geysering. Examining a single oscillation period, as shown in **Figure** *2*, it reveals two peaks. The higher peak is preceded by a slight increase in flow rate due to local boiling at the HX outlet (geysering). The bubbles produced rise along the riser pipeline and condense quickly.

Journal of Physics: Conference Series **2940** (2025) 012033 doi:10.1088/1742-6596/2940/1/012033

When the riser water temperature reaches saturation, flashing occurs. The flashing front moves downward, increasing the density gradient, and consequently, the mass flow rate in the open loop increases, peaking at the highest point of the period. The cycle then restarts due to the lack of driving force caused by cold water from the downcomer pipeline flooding the HX pipes. In **Figure *3*** is reported the simulated flowrate in good qualitative agreement, while the discrepancies will be discussed in the following.

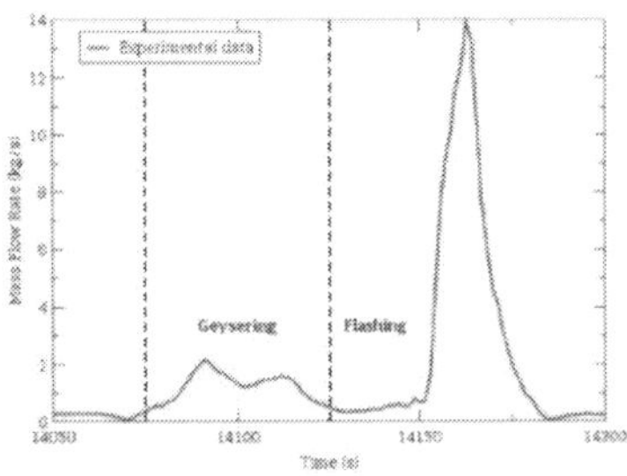

Figure 2. Experimental flowrate

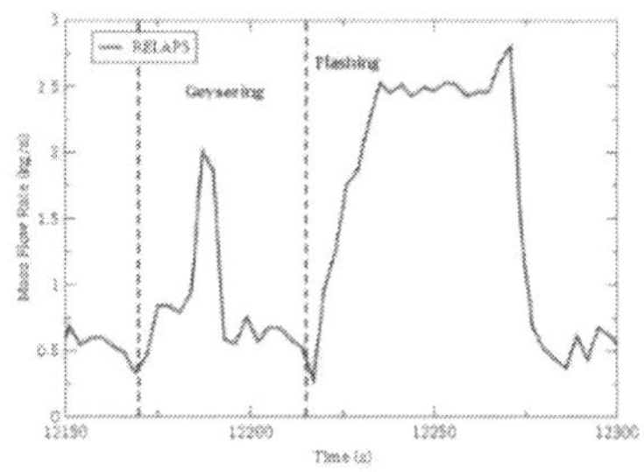

Figure 3. R5 flowrate

4.2. Test 2 – Results

In order to characterize the heat exchange between the PV and the open loop, the nodalization has been improved to optimize the thermal hydraulic conditions and to allow the natural convection inside the PV. The pressure and temperature in the upper head of the vessel, in ***Figure 4*** and ***Figure 5***, show the code capability to predict experimental data both qualitatively.

During the initial 8500 seconds, the pressure closely follows the experimental curve, albeit with a slight overestimation of 0.5 bar in the first 5000s, then the discrepancy reduced. Subsequently, shifting the power injection from 200 kW to 100 kW, the calculated pressure underestimated experimental data while maintaining sufficiently accurate predictions. This discrepancy arises from an inherent limitation of RELAP5 to predict precise temperature stratification in pool volumes, leading to an overestimation of temperature at the HX inlet/outlet, and also a non-specified decreasing in the experimental input steam power. Therefore, in R5 input deck has been implemented a linear decreasing shown in the pressure behaviour. Consequently, the reaching of saturation conditions at the end of the sparger central channel, as shown in ***Figure 6***, is anticipated against the experimental behaviour, as well as the start of the mass flowrate oscillations, shown in ***Figure*** *7*. This behaviour changes the driving force and subsequently the heat exchange in the HX section; therefore, when the injected power in the PV decrease, the loop already achieves the saturation conditions at the sparger, increasing the heat exchange and slightly decreasing the PV pressure.

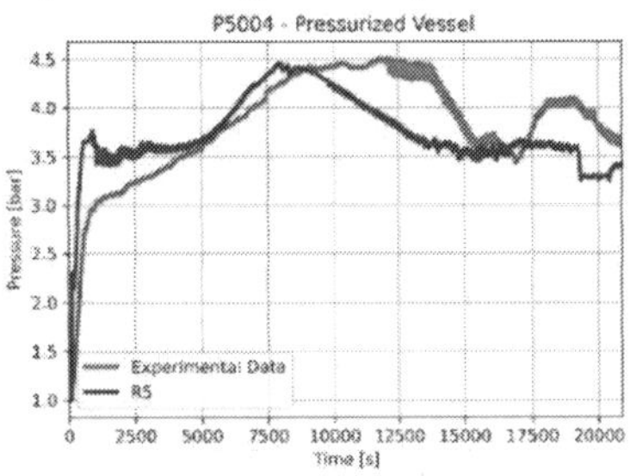

Figure 4. Pressure behaviour in PV

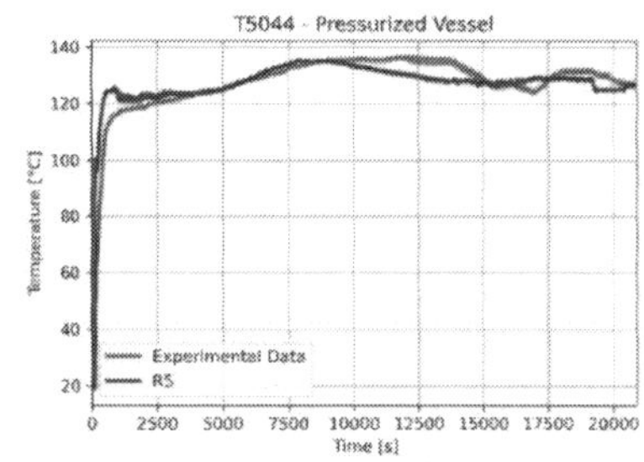

Figure 5. Temperature behaviour in PV

The magnitude of the mass flowrate oscillations is underestimated probably due to input limitation regarding the mixing phenomena between the hot and cold collector and an underestimation of the density gradient in the HX section. This behaviour is related to the excess of mixing occurring in the pool, as previously described. The flow coming from the water storage tank through the downcomer pipeline presents a higher temperature compared with the experimental data, causing a reduction of the driving forces. However, from a qualitative point of view, the code predicts the mass flow behaviour during the transient phase; in fact, the amplitude of the oscillations decreases when the loop flow resistance increases as shown in **Figure** *7*.

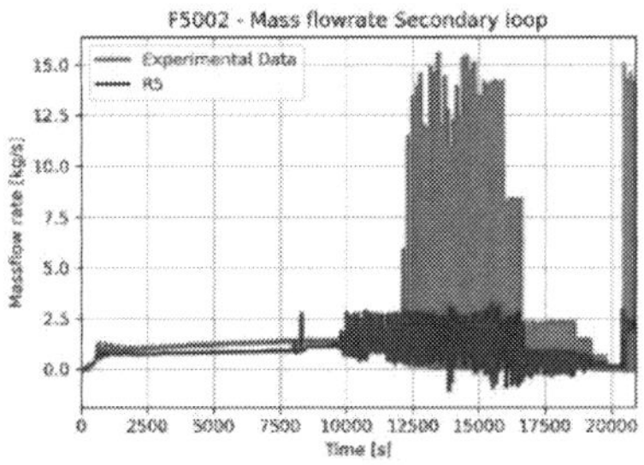

Figure 6. Massflow rate in the open loop

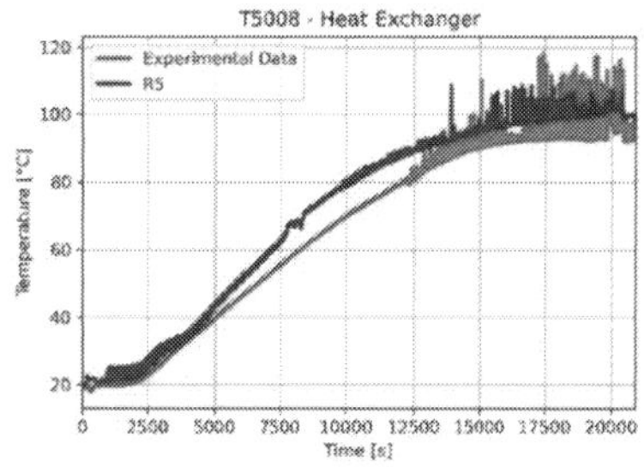

Figure 7. Inlet temperature in HX section

4.3. Test 6 – Description

Test 6 investigates the behavior of the facility during specific operational conditions, particularly focusing on the influence of the water pool level on system behavior. The test begins with an initial setup phase where the steam supply line injects a heating power of 200 kW into the pressure vessel for 400 seconds to achieve the desired conditions. After the initial heat-up phase, where the temperature in the pool is stabilized, the power injection is reduced to 100 kW for the duration of the test.

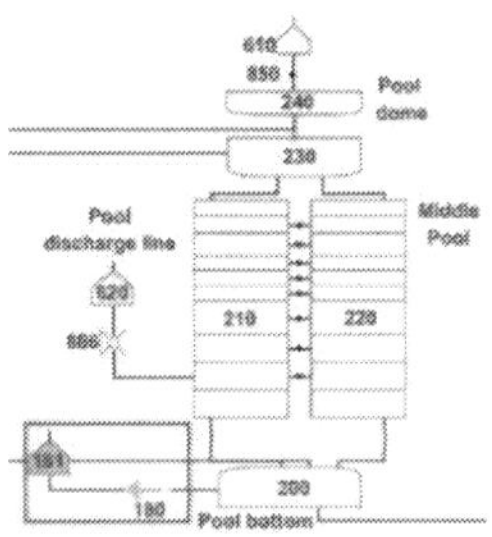

Figure 8. Drainage line in pool section

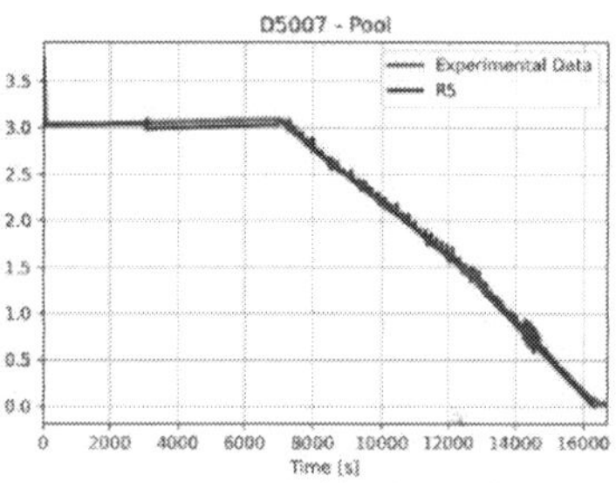

Figure 9. Level in the pool

The containment vessel is filled with air at atmospheric pressure, and the test aims to observe the dynamic behaviour of the system under these controlled conditions. To accurately simulate the test 6, the time-dependent junction allowing the steam injection in the PV has been stabilized to impose the correct mass flow rate and ensure proper power injection. Regarding the pool level, junction (886) is utilized during the initial 100 seconds of the simulation to achieve the correct pool set-point water level; while a time-dependent junction (180) is inserted to recreate the conditions of the test accurately (**Figure** *8*). The boundary conditions that regulate the test phase are derived from available experimental data. Consequently, the RELAP5 simulation curve for mass flow rate has been discretized to match the experimental conditions accurately (**Figure** *9*).

4.4. Test 6 – Results

After the pre-transient phase of 6200 s, the drain phase lasts up to 16700 s.
The pressure behaviour in the PV is represented in **Figure** *10*. During the pre-transient phase, the pressure stabilizes slightly below 4 bar, aligning closely with the experimental data. Around the 6000s, there is a notable pressure decrease observed both in the experimental data and the RELAP5 simulation due to the transition from the startup phase to the transient phase. Then it maintains a steady state with minor fluctuations observed in the experimental data; however, the RELAP5 model shows a consistent underestimation of the results after the 8000s, when the drainage of the pool section starts.

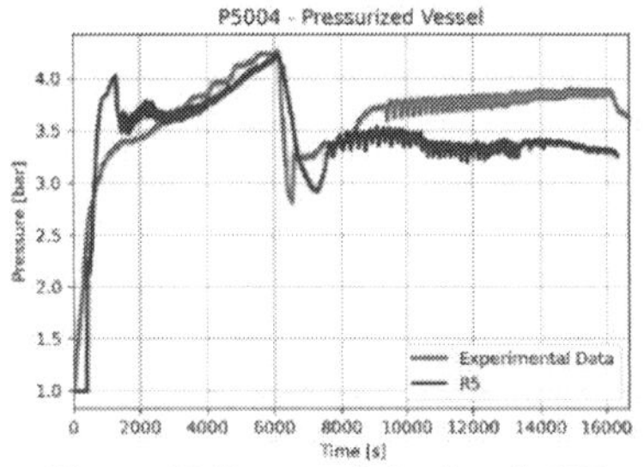

Figure 10. Pressure behaviour in PV

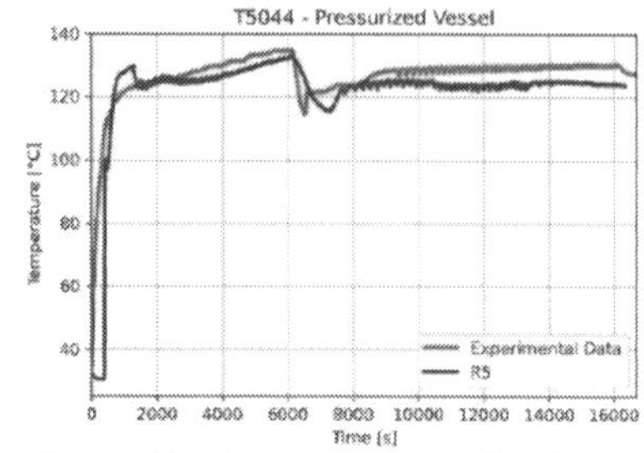

Figure 11. Temperature behaviour in PV

The temperatures at the HX outlet/hot and inlet/cold collectors are shown in ***Figure 12*** and ***Figure 13***. Regarding the outlet temperature, RELAP5 results begin to diverge, particularly beyond 10000s, where the drainage phase is started yet, reducing the HX temperature is higher than the experimental data. The RELAP5 simulation tends to overestimate these oscillations compared to the experimental data. This discrepancy may be due to the code's simplified representation of the complex heat transfer dynamics and internal recirculation within the heat exchanger. In the first part of the pre-transient, the temperature of the cold collector, which is placed outside the PV, increases due to an internal circulation between the heat exchange tubes, which mixed the hot and cold water. In the RELAP5 input deck, the HX tube bundle is simulated as only one thermal hydraulic pipe and, for this reason, the model does not consider the internal recirculation phenomenology and may explain the overestimation of the outlet temperatures in the HX section. During the transient phase, as previously described for the test 2, the saturation conditions are reached early starting the oscillating flow driven by the flashing and geysering early respect the experimental data.

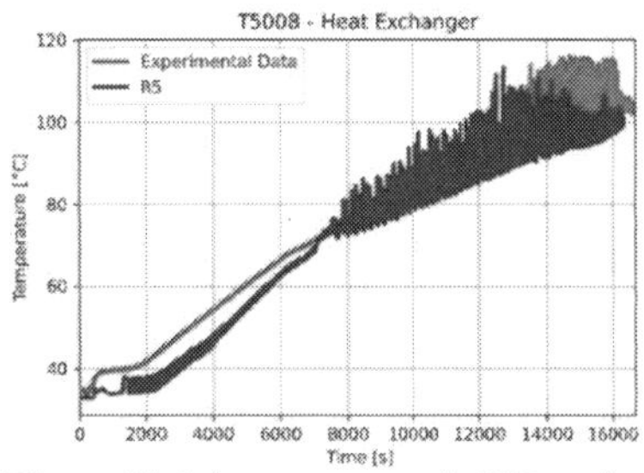

Figure 12. Inlet temperature in HX section

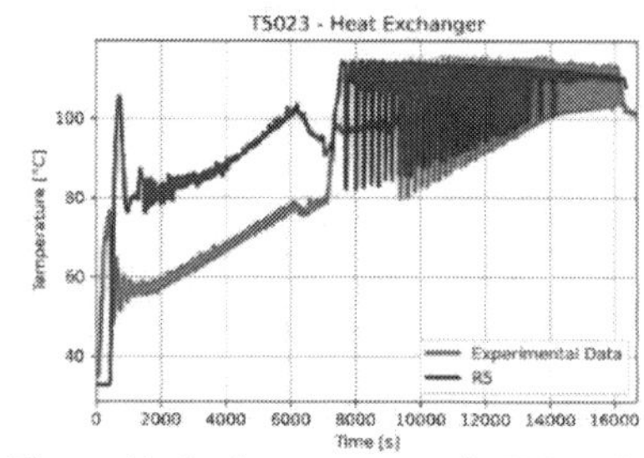

Figure 13. Outlet temperature in HX section

During the pre-transient phase, an increase of the HX outlet temperature is observed due to the start of the steam injection into the PV.

5. Conclusions

In this study, the performance of a low-pressure natural circulation system was investigated using the thermal-hydraulic code RELAP5/mod3.3. The analysis involved a series of tests conducted at the PASI facility developed by LUT, aimed at assessing the system behaviour under several natural circulation conditions. The experimental data obtained were used to verify the capability of RELAP5 to reproduce the thermal-hydraulic phenomenology characteristic of natural circulation systems.

The analysis demonstrates that while RELAP5 could effectively capture the dynamic behavior of the system during the pre-transient and transient phases, some discrepancies in predicting the pressurized vessel pressure were noted, attributed to the inherent limitations of the code in accurately predicting thermal stratification in the open loop pool. This phenomenon affects the transient, overestimating the internal loop temperature, leading to earlier saturation conditions, and altering the geysering and flashing processes, as well as heat exchange. The current pool nodalization scheme, dividing the pool into five distinct volumes to enhance natural circulation modeling and hydraulic coupling with the sparger section, appears to introduce excessive thermal mixing, suggesting that future investigations should focus on optimizing this aspect of the model.. These findings contribute to the broader goals of the

PASTELS project, supporting European stakeholders in the nuclear industry by providing insights into the physical behavior of passive systems and validating numerical tools crucial for design and safety demonstrations. The integration of experimental studies with numerical simulations using RELAP5/mod3.3 contributes to the validation and understanding of system performance under several transient conditions.

Future activities will focus on conducting comprehensive sensitivity analyses on nodalization, with particular attention to changing the pool and vessel section discretization, attempting to capture the characteristic phenomena, such as pool thermal stratification and oscillating mass flow rate in the HX, assessing the impact of nodalization choices on code performance through systematic comparison with experimental data. However, the results obtained from the tests at the PASI facility enhance the knowledge base for designing and optimizing passive safety systems in the view of possible innovative NPP design licensing needs which uses passive mitigation strategies in case of accidental events.

Acronyms: PASTELS (*Passive Systems: Simulating the Thermal hydraulics with Experimental Studies*), CWC (*Containment Wall Condenser*), NPP (*Nuclear Power Plant*), PV (*Pressure Vessel*), HX (*Heat Exchanger*), WP (*Water Pool*), NCG (Non-Condensable Gas), USNRC (*United States Nuclear Regulatory Commission*), ICAP (*International Code Assessment and Applications Program*), CAMP (*Code Application and Maintenance Program*), SoT (*Start of Transient*).

Acknowledgments

The PASTELS project has received funding from the Euratom research and training programme 2019-2020 under grant agreement No 945275. This text reflects only the author's view and the Commission is not responsible for any use that may be made of the information contained therein.

ENEA carried out the research activity with RELAP5/mod3.3 code and SNAP, obtained in the framework of the ENEA-ISIN agreement signed on March 23rd 2020 as part of the General Arrangement between the United States Nuclear Regulatory Commission (US-NRC) and the Italian National Inspectorate for Nuclear Safety and Radiation Protection (ISIN).

References

[1] INTERNATIONAL ATOMIC ENERGY AGENCY, 'Passive Safety Systems and Natural Circulation in Water Cooled Nuclear Power Plants'. Vienna 2009.
[2] Information Systems Laboratories, Inc., *RELAP5/MOD3.3 CODE MANUAL VOLUME I: CODE STRUCTURE, SYSTEM MODELS, AND SOLUTION METHODS*. 2001.
[3] 'PAssive Systems: Simulating the Thermal-hydraulics with ExperimentaL Studies - Home'. Accessed: Feb. 01, 2024.
[4] V. Riikonen, J. Telkkä, A. Räsänen, E. Kotro, and K. Tielinen, 'Research Report LUT University Nuclear Engineering'.
[5] V. Kouhia *et al.*, 'General description of the PASI test facility, second edition'.
[6] V. Kouhia, V. Riikonen, O.-P. Kauppinen, J. Telkkä, and J. Hyvärinen, 'PASI – A test facility for research on passive heat removal', *Nuclear Engineering and Design*, vol. 383, p. 111417, Nov. 2021, doi: 10.1016/j.nucengdes.2021.111417.
[7] J. Telkkä, V. Riikonen, F. David, and M. Puustinen, 'Deliverable 4.4: PASI-CWC test results'.
[8] G. Grippo and M. Polidori, 'RELAP5 Simulations of the PASI Experiment on Containment Wall Condensation', in *2024 International Congress on Advances in Nuclear Power Plants (ICAPP)*, Las Vegas, NV: American Nuclear Society, 2024, pp. 736–745. doi: 10.13182/T130-44635.

Journal of Physics: Conference Series **2940** (2025) 012034 doi:10.1088/1742-6596/2940/1/012034

Numerical investigation on mixed convection with nanofluids in vertical channels with opposing moving plate

B Buonomo[1,*], **O Manca**[1] , **S Nardini**[1] **and G Sarli**[1]

[1] Dipartimento di Ingegneria, Università degli Studi della Campania "Luigi Vanvitelli", Via Roma 29, 81031, Aversa (CE), Italy

* Corresponding author: oronzio.manca@unicampania.it

Abstract. A numerical investigation on 2-dimensional steady-state mixed convection in a vertical channel asymmetrically heated with water-alumina (Al_2O_3) nanofluids is performed to evaluate the thermal and fluid dynamic behavior. The interest in this problem is related to several technological applications such as nuclear reactors and electronic components. Nanoparticles have a diameter of 30 nm and four different values of concentration (from 0 to 6%). The computational domain is made up of the channel and two lateral reservoirs, which simulate the external ambient. The channel is bounded by two vertical parallel plates with a constant aspect ratio. One of the plates is heated with a constant heat flux, while the other is considered adiabatic and is moving in the opposite direction of the buoyancy force. The governing equations are numerically solved with finite volume method by means of the ANSYS-Fluent code. Four different values of wall heat flux (with $Ra=7\times10^2$ to 10^6) and six different velocities of the unheated plate (with $Re=1$ to 10^2) are considered. Results are presented in terms of velocity, temperature, pressure and stream function. Finally, correlations for the prediction of the average Nusselt number along the heated plate are presented.

1. Introduction

Nowadays, a relatively new class of materials is being used to increase heat transfer coefficient. They are called nanofluids because they consist of nanoparticles (such as copper, aluminum oxide, zinc oxide (ZnO), titanium oxide, and their mixtures) suspended in a base fluid (with water being the most common choice). These particles enhance the surface to volume ratio and, through their movement, improve performance in heat transfer [1]. Numerous studies have been conducted, both experimentally and theoretically, on the thermophysical properties of different nanofluids to develop their applications in the heat transfer industries. The first study was performed by Masuda et al. [2], in which the heat transfer coefficient and viscosity of water with very small particles of Al_2O_3, SiO_2 and TiO_2 were evaluated. Choi [3], in 1995, was the first to use the term "nanofluids". Since the beginning of new millennium, nanotechnology techniques have been improved, and the behavior of nanofluids has been widely investigated, confirming their efficiency in both natural [4] and forced [5] convection. As a result, mixed convection, which is a combination of both, has emerged as another application of nanofluids. Despite the significant research efforts, there is no general law to predict this enhancement since every case is different. A vertical channel with a moving plate is a typical geometry in mixed convection [6,7]. However, unlike other systems like porous structures [8], vertical cones [9], or rectangular enclosures [10], the vertical configuration seems to be relatively unexplored with nanofluids. In this paper, mixed convection in a vertical channel with an opposing moving plate, using water/Al_2O_3 nanofluid as the working fluid, is numerically investigated. Different values of wall heat flux, plate velocity, and aspect ratio have been considered, and thermo-fluid dynamic parameters have been evaluated in each case. Temperature, pressure, and stream function fields have been found. determined. Additionally, temperature, velocity, and wall shear stress profiles have been calculated. Finally, correlations for predicting the average Nusselt number along the heated plate are proposed.

2. Model description

The investigated system comprises two parallel plates with an assigned distance between them. The heated plate, at constant imposed wall heat flux $\dot{q}_w$, is stationary while the indefinite parallel plate has a constant velocity u_p in the opposite direction with respect to the gravity.

2.1. *Physical domain and assumptions*

Figure 1a shows the physical domain with the two plates. L is the length of the channel, while b is the distance between the two plates. The aspect ratio of the channel is constant: L/b=5. The moving indefinite plate is assumed adiabatic. The gravity effect has been considered in the direction of the channel since it is vertical.

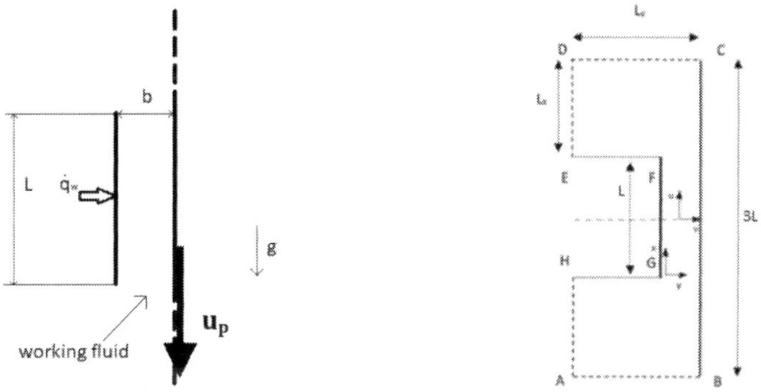

(a) (b)

Figure 1. Physical (a) and computational (b) domain on the system.

2.2. *Computational domain and governing equations*

To represent the system in the code, the computational domain shown in Figure 1b is chosen. It differs from the physical domain due to the presence of two reservoirs, each with dimensions of L_x in the direction of the channel and L_y in the orthogonal direction. These reservoirs simulate the external ambient conditions, with the 'pressure inlet' at the bottom and the 'pressure outlet' at the top. Since L_x equals L, the total length of the domain in the vertical direction is $3L$, with L_y being half of L_x. The thermophysical properties of the working fluid are considered constant with temperature, and the Boussinesq approximation has been employed. Viscous dissipation and radiation effects are neglected. The Single-Phase Model has been utilized for the nanofluid, as it is valid within the concentration range considered in the simulations, from 0% to 6% [11]. Therefore, the nanofluid is treated as a homogeneous mixture in local thermal equilibrium with no slip motion between the fluid and particles resulting in only one velocity field [11]. Under these conditions, the governing equations consist of the typical equations for a steady-state 2-dimensional laminar flow with mixed convection [12], namely the continuity (Equation 1a), the momentum in the x-direction and y-direction (Equations 1b-c), and the energy (Equation 1d).

$$\frac{\partial u}{\partial x}+\frac{\partial v}{\partial y}=0 \tag{1a}$$

$$u\frac{\partial u}{\partial x}+v\frac{\partial u}{\partial y}=-\frac{1}{\rho_{nf}}\frac{\partial p}{\partial x}+\frac{\mu_{nf}}{\rho_{nf}}\left(\frac{\partial^2 u}{\partial x^2}+\frac{\partial^2 u}{\partial y^2}\right)+g\beta_{nf}(\Delta T) \tag{1b}$$

$$u\frac{\partial v}{\partial x}+v\frac{\partial v}{\partial y}=-\frac{1}{\rho_{nf}}\frac{\partial p}{\partial y}+\frac{\mu_{nf}}{\rho_{nf}}\left(\frac{\partial^2 v}{\partial x^2}+\frac{\partial^2 v}{\partial y^2}\right) \tag{1c}$$

$$u\frac{\partial T}{\partial x}+v\frac{\partial T}{\partial y}=\frac{k_{nf}}{\rho_{nf}c_{p,nf}}\left(\frac{\partial^2 T}{\partial x^2}+\frac{\partial^2 T}{\partial y^2}\right) \tag{1d}$$

Boundary conditions reported in Table 1 are needed to solve the Equations (1-4).

Table 1. Boundary conditions of the problem.

	AB	BC	CD	DE	EF	FG	GH	HA
u	$p=p_t$	$u=u_p$	$p=p_s$	$u=0$	$u=0$	$u=0$	$u=0$	$u=0$
v	$v=0$	$v=0$	$v=0$	$p=p_s$	$v=0$	$v=0$	$v=0$	$p=p_t$
T	$T=T_0$	$\frac{\partial T}{\partial y}=0$	$T=T_0$	$T=T_0$	$\frac{\partial T}{\partial x}=0$	$-k\frac{\partial T}{\partial y}=q_w$	$\frac{\partial T}{\partial x}=0$	$T=T_0$

2.3. Properties of nanofluids

The coefficient of the governing equations written above are the thermophysical properties of the working fluid, which is a nanofluid made of water (H_2O) with nanoparticles of alumina (Al_2O_3). All the nanoparticles have the same diameter (30 nm), but their volume fraction can have four different values, ranging from 0% (pure water) to 6%. First, the properties of these two separate materials are needed. Density ρ, specific heat at constant pressure c_p, dynamic viscosity μ, thermal conductivity k and volumetric thermal expansion coefficient β of alumina and water are given in Table 2.

Table 2. Thermophysical properties at the reference temperature of 293.16 K [13].

	ρ [kg/m^3]	c_p [J/(Kg*K)	μ [Pa*s]	k [W/(m*K)]	β [K^{-1}]
Nanoparticles (Al_2O_3)	3880	773	-	36	8.5*10^{-6}
Base fluid (H2O)	998.2	4182	9.93*10^{-4}	0.597	2.1*10^{-4}

Equations for the prediction of the thermophysical properties of the nanofluid are found in literature [14,15]. The volume fraction, or concentration, of the nanofluid is written as:

$$\varphi=\frac{V_{np}}{V_{bf}+V_{np}} \tag{2}$$

For the evaluation of the thermodynamic properties of the working fluid, the Boussinesq approximation [16] is used. As a result, the density varies only with temperature. The initial density ρ_0 is the weighted average of those of the base fluid (water) and the nanoparticles (alumina) and β_{nf} is the volumetric thermal expansion coefficient of the nanofluid and is a weighted average as well [17]:

$$\rho_{nf}=\rho_0-\beta_{nf}\rho_0(T-T_0)\quad \rho_0=\rho_{bf}(1-\varphi)+\rho_{np}\varphi\quad \beta_{nf}=\frac{(1-\varphi)(\rho\beta)_{bf}+\varphi(\rho\beta)_{np}}{\rho_0} \tag{3}$$

Moreover, also the specific heat at constant pressure $c_{p,nf}$ is a weighted average [18]. On the other hand, as for the dynamic viscosity μ_{nf}, the relation by Corcione [19] is used, where d_{np} is the diameter of the particles and d_{bf} is another empirical correlation, with M being the molecular weight of the water, N being the Avogadro number and $\rho_{bf,0}$ being the mass density of the base fluid calculated at T=293 K:

$$c_{p,nf} = \frac{(1-\varphi)\left(\rho c_p\right)_{bf} + \varphi\left(\rho c_p\right)_{np}}{\rho_0} \qquad \mu_{nf} = \frac{\mu_{bf}}{1-34.87(d_{np}/d_{bf})^{-0.3}\varphi^{1.03}} \qquad d_{bf} = 0.1\left(\frac{6M}{N\pi\rho_{bf,0}}\right)^{1/3} \tag{4}$$

Finally, the correlation by Vajjha and Das [20] is used for the thermal conductivity k_{nf}:

$$k_{nf} = k_{bf}\frac{k_{np}+2k_{bf}-2(k_{bf}-k_{np})\varphi}{k_{np}+2k_{bf}+(k_{bf}-k_{np})\varphi} + 5\times10^4\gamma\varphi\rho_f c_{p,bf}\sqrt{\frac{k_B T}{\rho_p d_{np}}}f(T,\varphi) \tag{5}$$

The Boltzmann constant k_B is equal to 1.38×10^{-23} J/K. The quantities γ and $f(T,\varphi)$ are the following:

$$\gamma = 8.4407(100\varphi)^{-1.07304} \qquad f(T,\varphi) = (2.8217\times10^{-2}\varphi + 3.917\times10^{-3})\frac{T}{T_{fr}} - (3.0669\times10^{-2} + 3.923\times10^{-3}) \tag{6}$$

In eq. (6) T_{fr} is the melting point of ice, equal to 273.15 K. In Table 3 The estimated values by means of eqs. (3)-(6) are reported.

Table 3. Thermophysical properties of the nanofluid at the reference temperature of 293.16K

	ρ [kg/m^3]	c_p [J/(Kg*K)]	μ [Pa*s]	k [W/(m*K)]	β [K^{-1}]	*Pr*
φ=0%	998.2	4182	9.93*10^{-4}	0.597	2.10*10^{-4}	6.96
φ=2%	1055.8	3931	1.19*10^{-5}	0.660	1.95*10^{-4}	7.11
φ=4%	1113.5	3707	1.51*10^{-5}	0.694	1.82*10^{-4}	8.07
φ=6%	1171.1	3504	2.07*10^{-5}	0.730	1.70*10^{-4}	9.94

The dimensional parameters have been converted into the following dimensionless parameters:

$$X = \frac{x}{b} \quad Y = \frac{y}{b} \quad U = \frac{u}{u_p} \quad \Psi = \frac{\psi\rho_{nf}}{\mu_{nf}} \quad \theta = \frac{(T-T_0)k_{nf}}{q_w b} \quad p^* = \frac{p}{\frac{1}{2}\rho_{nf}u_p^2} \tag{7}$$

Dimensionless groups have also been calculated:

$$Gr = \frac{g\rho_{nf}^2\beta_{nf}q_w b^4}{k_{nf}\mu_{nf}^2} \quad \mathrm{Re} = \frac{\rho_{nf}u_p b}{\mu_{nf}} \quad \mathrm{Pr}_{nf} = \frac{\mu_{nf}}{c_{p,nf}k_{nf}} \quad Pe = \mathrm{Re}\,\mathrm{Pr}_{nf} \quad Ra = Gr\,\mathrm{Pr}_{nf} \quad Ri = \frac{Gr}{\mathrm{Re}^2} \quad Nu = \frac{hb}{k_{nf}} = \frac{1}{\theta} \tag{8}$$

2.4. Data reduction

To obtain the required Grashof number, $\dot{q}_w$ was modified for each case. Two different values of b were chosen (0.005 for $Gr\leq10^4$; 0.01 for $Gr=10^5$) to prevent $\dot{q}_w$ from being too small or too high, and each time the value of L was adjusted accordingly. Similarly, u_p was modified for each case to obtain the required Reynolds number from 1 to 10^3.

2.5. Numerical procedure

The governing equations were numerically solved by means of the finite volume method using the ANSYS Fluent (2020 R2 version) [21] code in 2D mode with the "Double Precision" option, employing the Semi Implicit Method for Pressure-Linked Equation (SIMPLE) scheme [22] to couple pressure and velocity components, in steady-state condition with a laminar and incompressible flow. The residuals related to the equations should be less than 10^{-4} for continuity and velocity components and 10^{-8} for energy to satisfy the assigned converging criteria. Three different simulations were performed using three different uniform meshes, built with rectangular cells, with φ=6%, Gr=10^4 and Re=10^2, and evaluating each time the average Nusselt number along the heated plate to find the best compromise between the error and the computational time. The meshes were obtained varying the "Number of divisions" parameter in L and b, reported respectively as n_x and n_y in Table 4a. Since the meshes are uniform, the divisions in L_x are equal to n_x, while in L_y are five times n_x. A mesh with 120x60 elements in the channel (and, thus, with 300 divisions in L_y) was chosen since its error with respect to the finest mesh is less than 1%.

Table 4. Grid dependence analysis (a) and validation of the numerical model (b).

	Nodes	Nu_{avg}	Error (%)
60 x 30	23911	6.1719	2.19
120 x 60	94621	6.0843	0.74
240 x 120	376441	6.0394	-

(a)

$Ra(b/L)$	Nu_{avg}	Nu_{avg} [23]	Error (%)
10^3	3.1897	3.2247	1.09
10^4	6.2785	6.3096	0.49

(b)

In order to validate the employed model, a comparison with the results given in [23] using air (Pr=0.71) with the same geometry, initial, and boundary conditions has been provided. Results in two different cases with u_p=0 and relative errors have been reported in Table 4b.

3. Results and discussion

First, dimensionless temperature fields are depicted in Fig. 2. The impact of the opposing moving plate is advantageous only when Ri=0.1, indicating that the convection can be considered forced. However, when $Ri\approx1$, this effect counteracts buoyancy, making the moving plate less favourable as the results in terms of heat transfer are superior with natural convection (Ri=10). Nevertheless, the utilization of a nanofluid leads to a reduction in the dimensionless temperature. As clarified in the last of the Equations 8, this parameter represents the reciprocal of the local Nusselt number. Consequently, convective heat transfer is enhanced with the nanofluid.

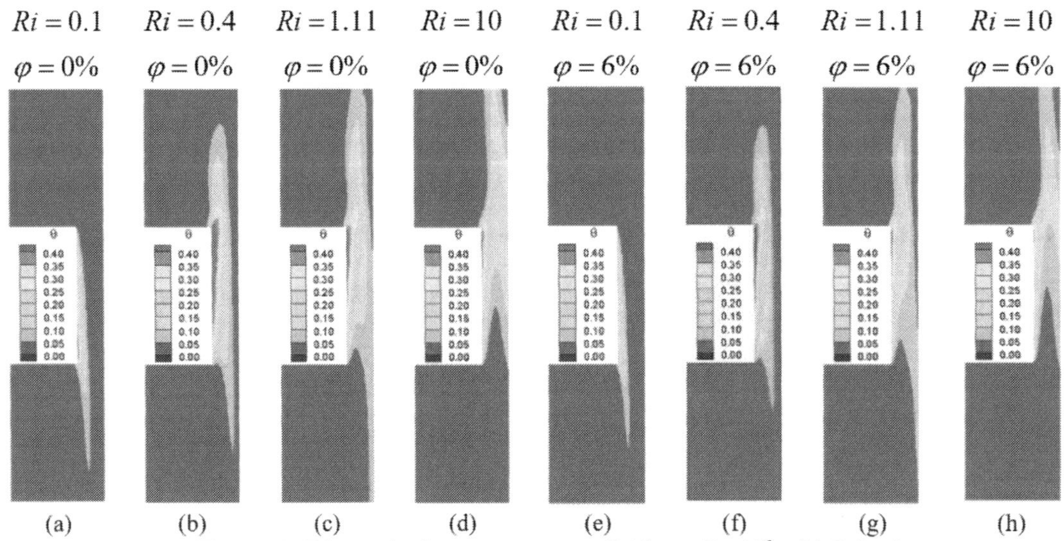

Figure 2. Dimensionless temperature fields at Gr=10^3 with L/b=5.

Dimensionless pressure fields are shown in Figure 3, along with stream function lines. The highest absolute values of this parameter are obtained with Ri=10, with a strong depression in the channel. Absolute values at constant Richardson number are slightly lower with the nanofluid. Dimensionless stream function lines reveal the presence of vortexes in the channel for Ri>0.1. This is because of the buoyancy and the moving plate contrasting each other. This effect explains the worsening of the heat

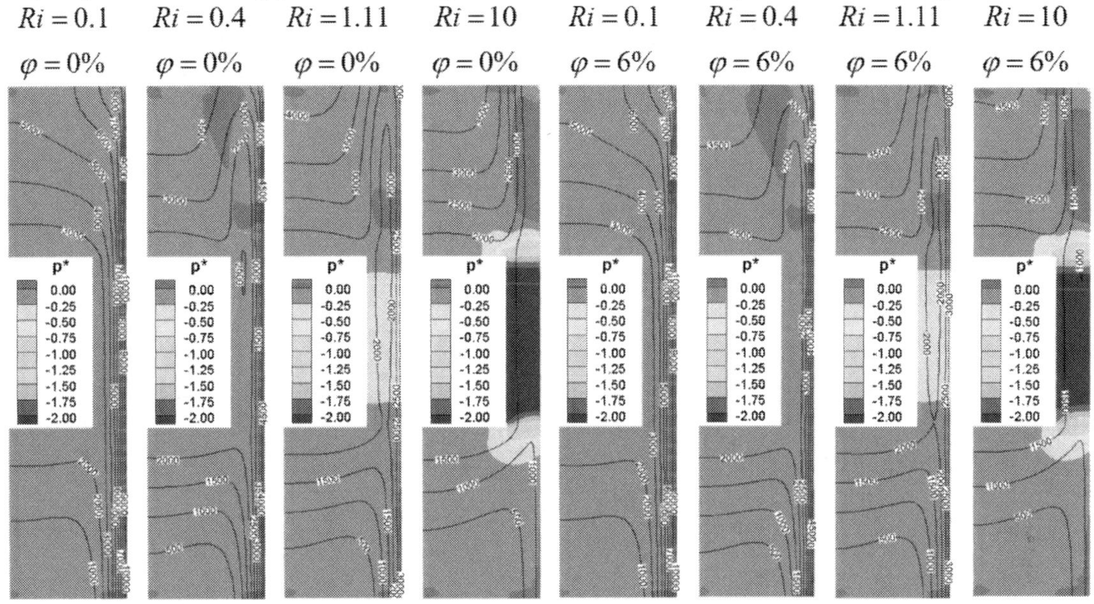

(a) (b) (c) (d) (e) (f) (g) (h)

Figure 3. Dimensionless pressure fields and stream function lines at Gr=10^3.

transfer at the medium values of the Richardson number. With Ri=10, the effect of the moving plate is weak, and the convection can be considered natural with a good approximation. Only small vortexes are created near the outlet of the channel, one of them being visible in Figure 3h.

Dimensionless temperature profiles on the heated wall (Figure 4a) validate the effectiveness of nanofluids, as this parameter decreases with increasing concentration. Two types of profiles can be observed, depending on the Richardson number. For Ri=10, the maximum temperature is reached near the outlet of the channel (X=4.6), whereas for Ri=0.1, the trend is reversed, and the maximum temperature is near the inlet of the channel (X=0.4). However, the maximum value is approximately 5% lower for Ri=0.1, confirming the beneficial impact of the high-velocity plate. Dimensionless x-velocity profiles at the inlet (Y=0) and exit (Y=5) cross sections, as shown in Figure 4b, elucidate the reasons for the different temperature fields obtained. With Ri=0.1, the fluid closely follows the moving plate, as this

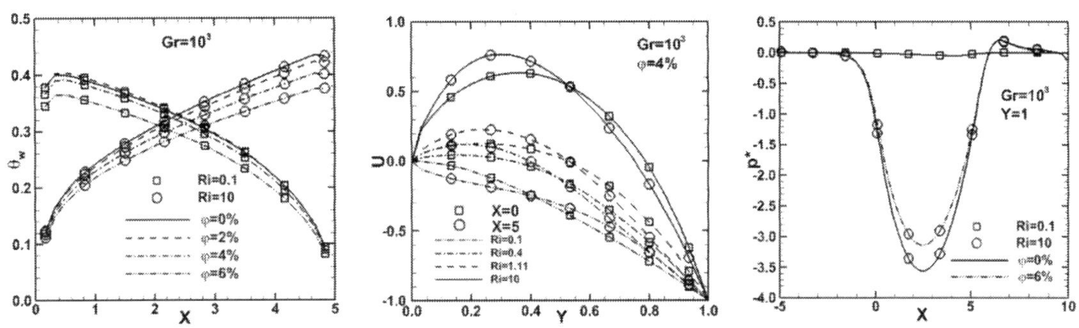

(a) (b) (c)

Figure 4. Dimensionless temperature, velocity, and pressure profiles.

motion overcomes the buoyancy force, resulting in a monotonically decreasing profile. Conversely, for Ri>0.1, a parabolic profile is observed, with the maximum at varying abscissas depending on the

Richardson number. As the *Ri* increases, the maximum moves closer to the heated wall. In Figure 5c, dimensionless pressure profiles on the moving plate are depicted. These profiles confirm the depression in the channel at Ri=10, which is more pronounced when φ=0% in dimensionless terms.

Average Nusselt number along the heated plate has been evaluated for different Grashof and Reynolds numbers, concentrations, and aspect ratios to find correlations for its prediction.

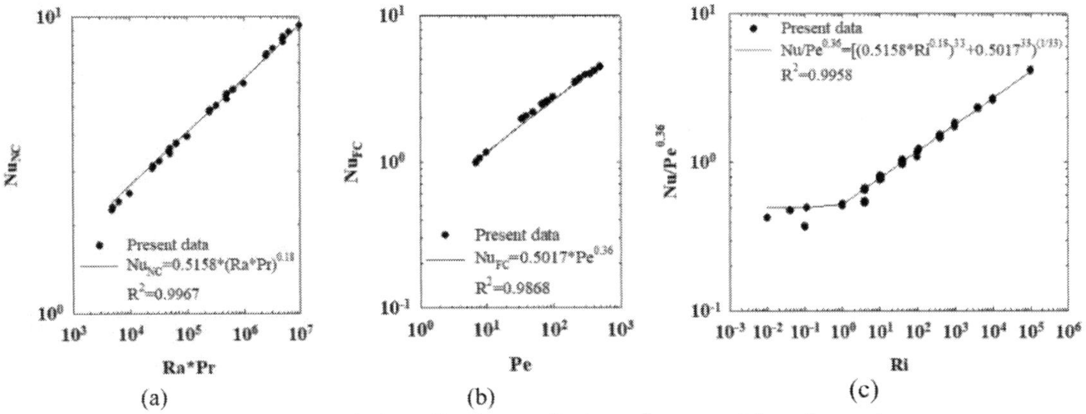

Figure 5. Correlations for the prediction of average Nusselt number.

Simulations with a fixed plate were conducted to obtain results in natural convection, leading to the correlation shown in Fig. 5a. The average Nusselt number along the heated plate varies with a power of *Ra*Pr*. Conversely, simulations without gravity were performed to obtain results in forced convection, resulting in the correlation depicted in Fig. 5b, based on the Peclet number, which is a function of the Prandtl number (as clarified in one of Equations 8), itself varying with the concentration of the nanofluid. The results in mixed convection with opposing moving plates can be summarized in a composite correlation, following the model proposed by Andreozzi et al. [23]:

$$\frac{Nu}{Pe^{0.36}} = \left[\left(\frac{Nu_{NC}}{Pe^{0.36}} \right)^p + \left(\frac{Nu_{FC}}{Pe^{0.36}} \right)^p \right]^{(1/p)} \tag{9}$$

An iterative method found p=33, which results in the correlation

$$\frac{Nu}{Pe^{0.36}} = \left[(0.5158 Ri^{0.18})^{33} + 0.5017^{33} \right]^{(1/33)} \tag{10}$$

presented also in Fig. 5c. However, results with $Ri \approx 1$ are valid for $Gr \geq 10^4$ since only for those values the vortexes does not occur.

4. Conclusion

An analysis of mixed convection in a vertical channel with opposing moving plates has been conducted. All the results indicate an improvement in the system's performance compared to natural convection with a fixed plate, especially when the plate velocity is high enough to achieve a low Richardson number (less than 0.1). However, this issue does not seem to arise when the inlet wall flux is high ($Gr \geq 10^4$). The efficiency of the nanofluid is supported by the presence of the Prandtl number in all the correlations identified. The practical implementation of such a system is contingent upon conducting an economic cost/benefit analysis.

5. References

[1] Hwang K S, Jang S P and Choi S U S 2009 Flow and convective heat transfer characteristics of water-based Al_2O_3 nanofluids in fully developed laminar flow regime *Int J Heat Mass Transf* **52** 193–9

[2] Masuda H, Ebata A, Teramae H and Hishimura N 1993 Alteration of thermal conductivity and viscosity of liquid by dispersing ultra-fine particles, *Netsu Bussei (Japan)* **4** 227–33

[3] Choi S U S 1995 Enhancing thermal conductivity of fluids with nanoparticles *ASME Fluids Engineering Division* 99–103

[4] Siddique I, Sadiq K, Jarad F, Al Mesfer M K, Danish M and Yaqoob S 2022 Analysis of Natural Convection in Nanofluid Flow through a Channel with Source/Sink Effect *J Nanomater* **2022**

[5] Ahmed H E, Salman B H and Kerbeet A S 2019 Heat transfer enhancement of turbulent forced nanofluid flow in a duct using triangular rib *Int J Heat Mass Transf* **134** 30–40

[6] Kang B H and Jaluria Y 1994 Heat transfer from continuously moving material in channel flow for thermal processing *J Thermophys Heat Trans* **8** 546–54

[7] Chamkha A J, Takhar H S and Nath G 2001 Effect of buoyancy forces on the flow and heat transfer over a continuous moving vertical or inclined surface *International Journal of Thermal Sciences* **40** 825–33

[8] Hindebu B, Makinde O D and Guta L 2022 Unsteady mixed convection flow of variable viscosity nanofluid in a micro-channel filled with a porous medium *Indian Journal of Physics* **96** 1749–66

[9] Yahaya R I, Mustafa M S, Arifin N M, Pop I, Wahid N S, Ali F M and Mohamed Isa S S P 2024 Mixed convection hybrid nanofluid flow over a stationary permeable vertical cone with thermal radiation and convective boundary condition *ZAMM Zeitschrift fur Angewandte Mathematik und Mechanik*

[10] El hadoui B and Kaddiri M 2024 Enhancing the convective heat transfer in vertical and horizontal rectangular enclosures using nanofluids: The crucial role of aspect ratio *Physics of Fluids* **36**

[11] Kefayati G and Bassom A P 2021 A lattice Boltzmann method for single- and two-phase models of nanofluids: Newtonian and non-Newtonian nanofluids *Physics of Fluids* **33** 102008

[12] Mansour M A and Bakier M A Y 2013 Free convection heat transfer in complex-wavy-wall enclosed cavity filled with nanofluid *International Communications in Heat and Mass Transfer* **44** 108–15

[13] Rohsenow W M, Hartnett J P (James P) and Cho Y I 1998 *Handbook of heat transfer* (McGraw-Hill)

[14] Khanafer K, Vafai K and Lightstone M 2003 Buoyancy-driven heat transfer enhancement in a two-dimensional enclosure utilizing nanofluids *Int J Heat Mass Transf* **46** 3639–53

[15] Chon C H, Kihm K D, Lee S P and Choi S U S 2005 Empirical Correlation Finding the Role of Temperature and Particle Size for Nanofluid (Al2O3) Thermal Conductivity Enhancement *Appl Phys Lett* **87** 1–3

[16] Boussinesq J V 1897 *Théorie de l'écoulement tourbillonnant et tumultueux des liquides dans les lits rectilignes a grande section* vol 1 (Paris (France): Gauthier-Villars)

[17] Sidik N A C, Yazid M N A W M, Samion S, Musa M N and Mamat R 2016 Latest development on computational approaches for nanofluid flow modeling: Navier-Stokes based multiphase models *International Communications in Heat and Mass Transfer* **74** 114–24

[18] Hentschke R 2016 On the specific heat capacity enhancement in nanofluids *Nanoscale Res Lett* **11** 1–11

[19] Corcione M 2011 Empirical correlating equations for predicting the effective thermal conductivity and dynamic viscosity of nanofluids *Energy Conversion and Management* vol 52 (Elsevier Ltd) pp 789–93

[20] Vajjha R S and Das D K 2009 Experimental determination of thermal conductivity of three nanofluids and development of new correlations *Int J Heat Mass Transf* **52** 4675–82

[21] Anon 2020 *ANSYS Fluent Tutorial Guide (2020 R2 Release)* (USA)

[22] Patankar S V. 1980 *Numerical Heat Transfer and Fluid Flow* (New York, NY, USA: Hemisphere Publishing Corporation)

[23] Andreozzi A, Bianco N, Manca O and Naso V 2008 Effect of a moving plate on heat transfer in a uniform heat flux vertical channel *Int J Heat Mass Transf* **51** 3906–12

Acknowledgments

This work was supported by the Italian Government MUR Grant No. P2022NYPHL – PRIN 2022 PNRR "VISIONS: eVolutIonary deSign of InnOvative heat traNsfer deviceS" – Funded by European Union – Next Generation EU.

AUTHOR INDEX